DR. A. F. McDONAGH
1120 HSW
Department of Medicine
University of California
San Francisco, Calif. 94122

Colour and Life

Light gave to aimless matter Life
And light to Life is All-in-all,
Source of all sweet to feed the strong,
Transducible in many forms,
Filtered by chromes of varied kind
Which take the appropriate wavebands out
To energise Life's cyclic paths,
Anabolic, catabolic,
Transmitting clear residual hues.

So spectral hues inhere to Life
And vision, first in black and white,
Came to discern the rainbow's range,
Each having valence for some eye
Of food, of danger, shelter, mate.

In consequence came camouflage,
Advertisement, aesthetic tints,
And colour, first mere residue
Of light robbed of some useful band,
Became the 'many splendoured thing'
The visual cortex renders it.

Artist and lover ask not why
But chemist and biologist
Pursue the rainbow legend still
And trust to find the crock of gold.

Arthur E. Needham

The Significance of Zoochromes

With 54 Figures

Springer-Verlag
New York · Heidelberg · Berlin 1974

Arthur E. NEEDHAM
Department of Zoology
South Parks Road
Oxford/Great Britain

The cover illustrations show: Pseudaposematic (Batesian) mimicry by palatable insects (cockroaches) of unpalatable Chrysomelidae (leaf beetles) and Coccinellidae (lady-bird beetles) which have conspicuous aposematic (warning) colour patterns. On close examination the roaches are detectable by such features as the long, tapering antennae and relatively thin elytra. (From *Mimicry in Plants and Animals.* By W. WICKLER. London: Weidenfeld and Nicolson 1968)

ISBN 0-387-06331-5 Springer-Verlag New York · Heidelberg · Berlin
ISBN 3-540-06331-5 Springer-Verlag Berlin · Heidelberg · New York

This work is subject to copyright. All rights are reserved, whether the whole or part of the material is concerned, specifically those of translation, reprinting, re-use of illustrations, broadcasting, reproduction by photocopying machine or similar means, and storage in data banks.
Under § 54 of the German Copyright Law where copies are made for other than private use, a fee is payable to the publisher, the amount of the fee to be determined by agreement with the publisher.
© by Springer-Verlag Berlin · Heidelberg 1974. Library of Congress Catalog Card Number 73-81287. Printed in Germany.
The use of registered names, trademarks, etc. in this publication does not imply, even in the absence of a specific statement, that such names are exempt from the relevant protective laws and regulations and therefore free for general use.
Typesetting, printing and bookbinding: Druckerei Georg Appl, Wemding.

To
All those whose interest, industry and insight made this work possible
and to
All who may be encouraged to build on their foundations

Preface

As the title indicates, the theme of this book is the functions of biochromes in animals. Recent works on zoochromes, such as those of D. L. Fox (1953), H. M. Fox and Vevers (1960) and Vuillaume (1969), have been concerned primarily with the chemical nature and the taxonomic distribution of these materials, and although function has been considered where relevant this has not been the centre of interest and certainly not the basis for the arrangement of the subject matter. Functional significance is a profitable focus of interest, since it is the one theme which can make biochromatology a discrete and integral subject, and because it is the main interest in all biological fields.

At present chromatology seems to be a particularly schizoid subject since it is clear that in metabolic functions biochromes are acting in a chemical capacity whereas integumental pigments function mainly biophysically, in neurological and behavioural contexts. It is profitable to attempt an integration by studying the functions of as many chromes as possible, from all aspects.

Much is known about those concerned in the various redox aspects of metabolism, including oxygen-transport, and the adaptive value of these is unquestioned. By contrast, early studies on the integumental chromes led Poulton (1890) to the conclusion that many of them serve no useful purpose – this is a common opinion also concerning the chromes of internal organs – and there has been a tendency to perpetuate the view by those who regard 'purpose' or 'teleonomy' (Pittendrigh, 1958) as tantamount to teleology (Verne, 1926; Vuillaume, 1969). Biochromes therefore have become a test case for the general validity of Natural Selection, and for the principle that it induces 'purpose' into biological activities; and the aim here is to assess the situation as fully as possible.

As a preliminary to the assessment it is necessary to understand the basis of colour in molecules and then to survey the chemical structure and properties of the different classes of zoochrome, followed by their distribution among the taxa of animals and within the body. In the subsequent chapters their known or probable physiological, biochemical, reproductive and developmental functions are considered in turn. Some pathological conditions are considered as being also informative. Further evidence on their significance is then sought from their evolution as preserved in chromogenesis and in its genetic basis. As the codification of their past history, this leads on to the consideration of evolutionary aspects in general and a final assessment.

In Chapter 2 it becomes evident that the inital study of biochromes should be chemical and a quotation from VERNE (1926, p. 544) therefore is particularly significant since, like his predecessors in the field, he was still mainly concerned with camouflage and other 'visual-effect' functions of biochromes: "Nous arrivons ainsi à la conclusion que tous les problèmes pigmentaires, même les plus complexes, tels que ceux ayant trait à l'ethologie et à la génétique, sont dominés par des questions chimiques et physicochimiques . . . . c'est là que l'avenir nous réserve la solution de la plupart d'entre eux."

The visual effect functions therefore evolved as an epiphenomenon, quite indirectly related to the basic chemical functions of biochromes. They have evolved into a second major group of functions, which must be given proportionate consideration. There is a real connection between the two groups and this is shown to be the function of light-perceptor organs.

Oxford, 1974   A. E. NEEDHAM

Contents

I. General

Chapter 1
Introduction 3

Chapter 2
Basis and Significance of Colour in Molecules 5
  2.1 Physical Basis of Colour 5
  2.2 The Electronic Basis of Colour 9
  2.3 Other Optical Properties of Biochromes 17
    2.3.1 Fluorescence and Phosphorescence 17
    2.3.2 Dichromatism or Dichroism 19
    2.3.3 Other Relevant Properties 20
  2.4 Conclusion 21

II. The Nature and Distribution of Zoochromes

Chapter 3
The Chemical Classes of Zoochrome 25
  3.1 Chromoproteins and Other Conjugates 29
  3.2 Carotenoids 33
    3.2.1 Properties 35
  3.3 Fuscins 35
  3.4 Chroman Biochromes 37
    3.4.1 Flavonoids 38
  3.5 Ternary Quinones 40
    3.5.1 Properties 44
  3.6 Metalloproteins 47
    3.6.1 Copper Proteins 49
    3.6.2 Iron Proteins 50
  3.7 Pyrrolic Zoochromes 52
    3.7.1 Properties 55
  3.8 Indolic Zoochromes 59
    3.8.1 Indigotins 59
    3.8.2 Other Indolic Oligomeric Chromes 60
    3.8.3 Indolic Melanins 60
    3.8.4 Other Zoochromes of the Indolic Melanin Pathway 62
  3.9 Ommochromes (Phenoxazones) and Related Biochromes 63
    3.9.1 Properties 64
    3.9.2 Other Zoochromes of the Kynurenine Pathway . 65
  3.10 Other N-heterocyclic Zoochromes: Purines, Pterins and Flavins (Isoalloxazines) 66

IX

  3.10.1 Purines 67
  3.10.2 Pterins 68
  3.10.3 Flavins (Isoalloxazines) 71
 3.11 Other Zoochromes 73
 3.12 Zoochromes in Relation to pH 77
 3.13 Conclusion 80

Chapter 4
The Taxonomic Distribution of Zoochromes 81
 4.1 Distribution between Animals and Plants 81
 4.2 Distribution between the Phyla and Classes of Animals 82
  4.2.1 Conclusion 90
 4.3 Distribution in the More Obsure Phyla and Classes . 91
  4.3.1 Conclusion 92
 4.4 Distribution between Lower Taxonomic Units 92
 4.5 Conclusion 93

Chapter 5
Distribution and State of Zoochromes in the Body 95
 5.1 Distribution of Zoochromes between Organs, Tissues and Fluids of the Body 95
  5.1.1–5.1.16 Individual Organs, etc. 97
  5.1.17 Conclusion 109
 5.2 The Location of Chromes within Organs and Tissues . 111
  5.2.1 Conclusion 113
 5.3 The Location and State of Zoochromes within Cells 113
  5.3.1 The Chromasome 118
  5.3.2 Protozoan Chromes 123
  5.3.3 State of Zoochromes in Storage and Scavenging Cells 123
  5.3.4 Conclusion 125
 5.4 Biochromes and the Solid State 125
 5.5 The Stability of Zoochromes 126
 5.6 General Conclusion 127

III. Physiological Functions of Zoochromes

Chapter 6
The Functions of Integumental Zoochromes 131
 6.1 Integumental Colour-Change 136
  6.1.1 Chromogenic Change 136
  6.1.2 Chromomotor Colour-Change 138
  6.1.3 The Mechanism of Chromomotor and Chromogenic Responses 141
   6.1.3.1 Primary Responses 141
   6.1.3.2 Secondary Responses 142

    6.1.3.3 The Response-Mechanism as a Whole . . 154
 6.2 Other Functions of Integumental Chromes 156
  6.2.1 Protection against Radiation 156
  6.2.2 Mechanical Protection 158
  6.2.3 Chemical Defence 158
  6.2.4 Thermal Regulation 160
 6.3 Conclusion 161

Chapter 7
Zoochromes and Sensory Perception 163
 7.1 Light Perception 163
 7.2 Colour Vision 168
 7.3 Other Zoochromes of the Retina 170
 7.4 Simple Photoperceptors 173
 7.5 Zoochromes of Other Perceptor Organs 176
 7.6 Conclusion 177

IV. Biochemical Functions of Zoochromes

Chapter 8
Oxidation-Reduction Functions of Zoochromes 181
 8.1 The Biochromes of the Standard Respiratory Pathways 184
  8.1.1 The Molecular Basis of the Action of Respiratory Chromoproteins 188
 8.2 Other Redox Reactions Catalysed by Chromoprotein Enzymes 192
 8.3 Redox Functions of Other Zoochromes 197
 8.4 Photochemical Redox Catalysis by Zoochromes . . . 206
 8.5 Biochromes and Bioluminescence 207
 8.6 Redox Zoochromes that are Vitamins 209
 8.7 Conclusion 210

Chapter 9
Zoochromes that Transport Oxygen 211
 9.1 Physiological Properties of Oxygen-transport-Zoochromes 213
  9.1.1 Haemoglobins 213
  9.1.2 Haemerythrins 224
  9.1.3 Haemocyanins 224
 9.2 The Molecular Basis of O_2-Transport 228
  9.2.1 Haemoglobins 229
  9.2.2 Haemerythrins 233
  9.2.3 Haemocyanins 235
  9.2.4 Comparison between the Classes of Oxygen-carriers 237

    9.3 Other Metalloproteins and Oxygen-transport 238
    9.4 Conclusion 240

Chapter 10
Zoochromes in Other Aspects of Metabolism 241
    10.1 Zoochromes and Digestive Fluids 241
    10.2 Zoochromes as End-products of Metabolism 242
    10.3 Conclusion 246

Chapter 11
Pathological States Involving Zoochromes 247
    11.1 The Destruction of Zoochromes by Light 247
    11.2 Photodynamic Actions of Biochromes 247
    11.3 Qualitative Chromopathies 249
    11.4 Deficiency and Excess of Dietary Chromes 250
    11.5 Deficiency and Excess of Endogenous Chromes . . . 252
    11.6 Parasitism and Zoochromes 255
    11.7 Conclusion 259

V. The Significance of Zoochromes for Reproduction and Development

Chapter 12
Zoochromes in Reproduction and Development 263
    12.1 Reproduction 263
        12.1.1 Chromes of the Reproductive Organs 263
        12.1.2 Integumental Chromes 263
        12.1.3 Pubertal and Seasonal Changes 264
        12.1.4 Reproductive Metabolism 264
        12.1.5 Control of Reproduction 268
        12.1.6 Conclusion 269
    12.2 Zoochromes in the Ontogenesis of Animals 269
        12.2.1 Effects of the Classes of Biochrome on Embryogenesis 270
        12.2.2 Pigment Formation by the Embryo 274
        12.2.3 Chromogenic Changes during Metamorphosis 275
        12.2.4 Conclusion 276

VI. Evidence from Chromogenesis in the Individual

Chapter 13
The Evidence from Chromogenic Pathways 281
    13.1 Known Chromogenic Pathways 281
        13.1.1 Carotenoids 281
        13.1.2 Chromones and Flavonoids 282
        13.1.3 Ternary Quinones 284
        13.1.4 Porphyrins 286

13.1.5 Indolic Melanins 288
13.1.6 Ommochromes 289
13.1.7 Purines, Pterins and Isoalloxazines 290
13.2 Adaptive Features of Chromogenic Pathways . . . 293
13.2.1 The Primality of Chromogenic Pathways . . . 293
13.2.2 Sophistication of Chromogenic Pathways . . . 294
13.2.3 Economy in Chromogenic Pathways 295
13.2.4 Chromogenic Pathways in the Different Taxa 295
13.3 Interrelationships between Chromogenic Pathways 296
13.3.1 General Similarities between Chromogenic Pathways 296
13.3.2 Paths Beginning in a Common Precursor . . . 296
13.3.3 Connections between Pathways 297
13.3.4 Convergence between Pathways 297
13.3.5 Interaction between Members of Different Chromogenic Paths 297
13.3.6 Biochromes Acting as Coenzymes in Chromogenic Paths 298
13.4 Conclusion 299

Chapter 14
Evidence from the Control of the Supply of Biochromes . . . 301
14.1 Uptake and Handling of Exogenous Chromes 301
14.1.1 Carotenoids 301
14.1.2 Other Exogenous Zoochromes 304
14.1.3 Chromes Acquired Both Exogenously and Endogenously 306
14.1.4 Conclusion 307
14.2 The Control of Chromogenesis 307
14.2.1 Feedback Control of Chromogenesis 307
14.2.2 The Systemic Control of Haem-Synthesis . . . 309
14.2.3 The Control of Phase-coloration in Insects . . . 311
14.2.4 The Control of Exogenous Integumental Chromes 312
14.2.5 The Indirectness of Chromogenic Controls . . 312
14.2.6 The Chain of Intermediation in Chromogenic Controls 313
14.2.7 Conclusion 314

Chapter 15
The Development of Integumental Colour-Patterns 315

Chapter 16
The Genetic Basis of Chromogenesis 323
16.1 Genetic Colour-Polymorphism 324
16.2 General Properties of Chromogenic Mutants 325
16.3 The Alleles at Individual Loci 326

XIII

  16.4 Alleles at Chromoprotein Loci 327
  16.5 The Number and Types of Chromogenic Loci 328
  16.6 Interactions between Chromogenic Loci 329
  16.7 Chromogenes and Taxonomy 329
  16.8 Conclusion 331

VII. Evolutionary Evidence and General Assessment

Chapter 17
Zoochromes and Evolution 335
  17.1 Fossil Zoochromes 336
  17.2 Prebiological Evolution of Organic Chromes 337
  17.3 Chromes and the First Discrete Organisms 338
  17.4 The Evolution of Chromes in Animals 341
  17.5 Conclusion 344

Chapter 18
General Assessment 345
  18.1 Outstanding Adaptations of Zoochromes 346
  18.2 Chromatological Problems 347
    18.2.1 Problems Concerning Specific Classes of Integumental Chrome 349
    18.2.2 Problems Concerning Chromes Other than those of the Integument 351
  18.3 Unidentified Zoochromes 354
  18.4 Conclusion 354

Bibliography 355
Subject and Systematic Index 405

Acknowledgements

I am deeply indebted to the Editors of the Series, and particularly to Professor W. S. HOAR who encouraged me to undertake the project, devoted much time to the M. S. at intermediate stages and gave very valuable advice on the first draft. I wish to thank also a number of colleagues for help and encouragement, through discussions, supply of references and materials and other generosity. In particular I thank Professor W. H. BANNISTER and Dr. J. V. BANNISTER, Dr. J. BARRETT, Dr. EILEEN H. BEAUMONT, Dr. P. C. J. BRUNET, Dr. M. J. COE, Professor G. Y. KENNEDY, Dr. P. L. MILLER, Dr. J. D. MURRAY, Dr. A. C. NEVILLE, Dr. R. N. PAU, Dr. V. R. SOUTHGATE and Dr. E. J. WOOD. For photography I am indebted to Mr. J. S. HAYWOOD and Mr. R. TANNER and for photomicrography to Mrs. SUE PHIPPS. To the staff of the Radcliffe Science Library I owe a great, and also a longstanding, debt for unstinting help and patience in the pursuit of obscure literature and difficult references.

For a number of figures and tables in the book I acknowledge with gratitude permission of authors and publishers to use their originals, as follows:

Fig. 5.1: McGraw-Hill Book Co., Inc., 330 West 42nd. St., New York, N. Y. 10036 for permission to use Fig. 80, A, B, C on p. 294 of The Invertebrata, Vol. I, by the late L. H. HYMAN.

Fig. 5.2: Dr. J. P. GREEN, School of Biological Sciences, University of Malaya, Kuala Lumpur, Malaysia, Dr. M. R. NEFF, Summit High School, Summit, N. J. 07901 and the Longman Group, Ltd., Journals Division, for the use of fig. 1 on p. 141 of the paper by J. P. GREEN and M. R. NEFF in Tissue and Cell **4**, 137–171 (1972).

Fig. 5.4: The Secretary, Company of Biologists Ltd., Zoology Department, Downing St., Cambridge for the use of Fig. 1, Plate 19 and Fig. 5, Plate 20 of the paper by L. J. PICTON in the Quarterly Journal of Microscopical Science **41**, 263–302 (1899).

Fig. 5.5: Professor D. L. FOX, Scripps Institution of Oceanography University of California, San Diego, La Jolla, California 92037 and the Cambridge University Press for the use of Fig. 8, p. 28 of the book Animal Biochromes and Structural Colours by D. L. FOX (1953).

Fig. 5.6: Messrs. DOIN, Editeurs, 8, Place de l'Odéon, 75006, Paris, for the use of the figure on p. 297 of the book Les Pigments dans l'Organisme Animale by J. VERNE (1926).

Fig. 5.7: Mme. J. BRETON-GORIUS, Unité de Recherche sur

les Anémies, I. N. S. E. R. M. U-91, Hopital Henri-Mondor, 94-Créteil, France and Masson et Cie, 120 Boulevard St. Germain, Paris, VIe for the use of Plates IX and XI of the paper by J. Breton-Gorius in Annales des Sciences Naturelles, Zoologie et Biologie Animale (12), **5**, 211–272 (1963).

Fig. 5.8 A and Fig. 5.9: Professor A. S. BREATHNACH, Dept. of Anatomy, St. Mary's Hospital Medical School, London W2 and Academic Press, Inc. 111, Fifth Avenue, New York, 10003 for the use of Figs. 12, 13, 14, 15 on pp. 372, 374 of the article by A. S. BREATHNACH in the book Pigments in Pathology, pp. 353–394, edited by M. WOLMAN (1969).

Fig. 5.10: Dr. W. S. HARTROFT and Dr. E. A. PORTA and Academic Press for the use of Fig. 3, p. 215 of the article by E. A. PORTA and W. S. HARTROFT in the same book, pp. 191–235.

Fig. 9.4: Dr. M. F. PERUTZ, F. R. S. and Macmillan (Journals) Ltd., Little Essex St., London W. C. 2 for the use of Parts 1 and 6 of Fig. 4, p. 732 of the paper by M. F. PERUTZ in Nature (London) **228**, 726–739 (1970).

Fig. 13.4: Academic Press (London), Inc., for the use of Fig. 4, p. 243 of my article in Comparative Biochemistry of Nitrogen Metabolism, Vol. I, pp. 207–297, edited by J. W. CAMPBELL (1970).

Fig. 15.1: The Company of Biologists Ltd., Zoology Dept., Downing St., Cambridge for the use of Text-fig. 7 A p. 65 of my paper in the Quarterly Journal of Microscopical Science **84**, 49–72 (1942).

Fig. 15.3: Professor C. H. WADDINGTON, F. R. S., Buchanan Professor of Animal Genetics, Institute of Animal Genetics, West Mains Road, Edinburgh EH9 3JN and for the use of Fig. 55, p. 202 of the book New Patterns in Genetics and Development by C. H. WADDINGTON (1962).

Table 3.8: Dr. E. J. WOOD, Department of Biochemistry, University of Leeds, 9, Hyde Terrace, Leeds LS2 9LS and the Editors of Chest Piece, Royal University of Malta, Msida, Malta for the use of Table I of the article by E. J. WOOD in Chest-Piece, March 1968, pp. 5–13.

Table 3.10: Professor T. W. GOODWIN, F. R. S., Department of Biochemistry, University of Liverpool and Academic Press, (London) Inc., for the use of Table 2, p. 172 of the book Biosynthesis of Vitamins and Related Compounds by T. W. GOODWIN (1963).

Table 9.3: Dr. J. D. JONES, Department of Zoology, University of Sheffield and Pergamon Press Ltd., Headington Hill Hall, Oxford OX3 OBW for the use of the table on p. 33 of the article by J. D. JONES in Problems in Biology, Vol. I, pp. 9–90, edited by G. A. KERKUT (1963).

Table 9.4: Professor C. L. PROSSER, Department of Physiology and Biophysics, University of Illinois, 524 Burrill Hall, Urbana,

Ill. 61801 and W. B. SAUNDERS Co., West Washington Sq., Philadelphia, Pa. 19105 for the use of the table on p. 206 of Comparative Animal Physiology by C. L. PROSSER and F. A. BROWN, Jr. (1961).

Abbreviations

ACh	Acetylcholine
ACTH	Adrenocorticotropic hormone of pituitary
AHP	2-amino-4-hydroxy pteridine ("Pterin")
AMP	Adenosine monophosphate (adenylic acid)
AQ	Anthraquinone
ATP	Adenosine triphosphate (adenyl pyrophosphate)
BQ	Benzoquinone
Br	Bilirubin
Bv	Biliverdin
C	Percentage saturation of oxygen-carrier with oxygen
$C_1, C_2, \ldots$	Method of indicating number of carbon atoms in molecule or part of molecule
cal	small calorie (4.1868 joules)
CD	Circular dichroism
ChlHb	Chlorohaemoglobin (Chlorocruorin)
Cy	Cytochrome
d	Atomic orbital: orbital of third sub-shall of each shell of electrons orbiting the atom; energy level higher than that of p-orbitals
2d, 4d	Ecto- and meso- somatoblasts in embryos of animals with spiral cleavage
δ-ALA	Delta-amino laevulinic acid
Δ^5-steroids:	Steroids with the one double bond of the sterane nucleus situated between carbon 5 and carbon 10
DNA	Deoxyribonucleic acid
DOPA	Dihydroxyphenylalanine
DPG	Diphosphoglyceric acid
E	Energy of one quantum
E	Photochemical effect
e	Electron
E'_0	Standard electrode potential, at pH 7.0, of aqueous solution of redox system containing equal activities of oxidised and reduced species of the system
EM	Electron microscope
EPR	Electron paramagnetic resonance (= ESR)
ESR	Electron spin resonance (= EPR)
eV	Electron-volt (2.18×10^{-18} joules): energy of electron in potential gradient of one volt
FAD	Ribitylflavin-adenine dinucleotide
Fl	Ribitylflavin

FMN	Ribitylflavin mononucleotide
FP	Flavoprotein
γ-rays	High energy non-particulate radiation, electromagnetic
GMP	Guanosine monophosphate (guanylic acid)
G-7'-RP } G-9'-RP	Guanosine monophosphate with the ribose phosphate at positions 7' and 9' respectively of the guanine
GTP	Guanosine triphosphate (guanyl pyrophosphate) Planck's constant
Hb	Haemoglobin
Hcy	Haemocyanin
Hery	Haemerythrin
HOK	Hydroxy kynurenine
i. r.	Infra red
K	Kynurenine
K-cal	Kilocalorie: 1000 calories
λ	Wavelength of radiations
$\mu = \mu m =$	$m \times 10^{-6}$
$\mu g =$	$g \times 10^{-6}$
M-shell	Third shell of orbiting electrons of atom, numbered from innermost (K) shell
m	Metre
m	Meta-disubstituted aromatic ring
Mb	Myoglobin
Me	Methyl group (H_3C-)
MetHb	Haemoglobin in which the iron is oxidised to ferric, destroying physiological activity. Similarly MetHcy and MetHery
m-RNA	Messenger ribonucleic acid
MSH	Melanin-stimulating hormone, causing melanosome dispersal and proliferation in lower vertebrates
mV	Millivolt (10^{-3} volts)
N-shell	Fourth shell of orbiting electrons of atom
n-orbital:	Non-bonding molecular orbital
n	Index of interaction between units of molecule of oxygen-carrier
NAD	Nicotinamide-adenine dinucleotide
NADP	Nicotinamide-adenine dinucleotide phosphate
nm	nanometres ($m \times 10^{-9}$)
NQ	Naphthoquinone
o	Ortho-disubstituted aromatic ring
ORD	Optical rotatory dispersion
π-orbital:	Molecular orbital originating from two or more p-orbitals of constituent atoms of molecule
π^*-orbital:	Anti-bonding π-orbital of molecule, containing an excited electron
p-orbital:	Orbital of second sub-shall of each shell of electrons

XIX

	orbiting the atom; energy level higher than that of s-orbitals
p	Para-disubstituted aromatic ring
P_{50}	Oxygen pressure at which a carrier is 50% saturated with oxygen
PGA	Pteroylglutamic acid (folic acid)
pH	Negative logarithm of hydrogen ion concentration: gram-ions per litre
PQ	Polycyclic quinone
PrHb	Protohaemoglobin
Q	Quinone
Qol	Quinol
R	Molecular refraction
R_1, R_2	Retinals 1 and 2 (aldehydes of vitamins A_1 and A_2)
RF	Ribitylflavin (6,7-dimethyl-9-ribityl isoalloxazine)
RNA	Ribonucleic acid
s-orbital:	Subshell of lowest energy value in each shell of electron-orbitals of the atom
σ-orbital:	Molecular orbital originating from overlap of s- and p-orbitals of constituent atoms
σ^*-orbital:	Anti-bonding σ-orbital of molecule, containing excited electron
u	Speed of light
UQ	Ubiquinone (Coenzyme Q)
u. v.	Ultra-violet
u. v. f.	Ultraviolet fluorescence
v	Wave-frequency of radiations (reciprocal of wavelength)
∞	Infinity

I. General

Chapter 1

Introduction

Few properties of living organisms evoke popular admiration and curiosity more than their colours. They fire the interest of the biologist at least as powerfully, and often for much the same reasons: – their conspicuousness, or the converse, and the many provocative questions they raise. A high proportion of zoologists have been drawn to spend at least some of their time investigating biochromes, whether from these or from other aspects.

Collectively biochromes constitute a very significant chemical component of the living organism and, since they act as their own 'indicator' or 'label', they are relatively easy to study. There is considerable justification for the now famous opinion of Sir F. GOWLAND HOPKINS (FOX, 1953, p. 266) that any student in whom the brilliant and surprisingly water-soluble pigment of the turaco's feathers (p. 88) evoked no excitement was unlikely to become a dedicated biochemist. It may seem paradoxical that HOPKINS' students, and most biochemists since, have devoted their attention mainly to colourless materials; these have smaller, simpler molecules than biochromes and constitute the bulk of the substrates of metabolism. They were logically the correct group for initial studies and so HOPKINS had a colourful vision of the future, when the structure and functions of large and relatively complex molecules synthesised by living organisms would be the centre of interest. Biochromes are the end-products of rather long biosynthetic pathways and they are coenzymes rather then substrates.

Biochromes are produced in a number of distinct biosynthetic pathways and do not constitute a single natural chemical group. For the organic chemist therefore they are an heterogenous, illogical assemblage and so very few chemists have been interested in biochromatology as an integral subject. A number of leading organic chemists have attempted, as much as is possible, the study of particular classes of biochrome, and have usually found that one class alone constituted a life's work. The debt of biology to such basic research is enormous.

In their biological functions and distribution biochromes do constitute a somewhat more integral group of materials but even here there is enough multiplicity to dissuade many biologists from recognising a discrete science of biochromatology to which, for instance, a specific journal should be devoted, or on which an integral teaching course could be based. On the one hand there are the tissue haemoproteins, flavins, etc., and the blood pigments, all concerned with some aspect of respiratory catabolism. In plants there is also the group of chromes concerned in photosynthesis, i.e., the fundamental anabolic process which initiates the other phase of the metabolic cycle of living organisms. On the other hand there are the integumental pigments concerned in camouflage and advertisement (sematic) functions (POULTON, 1890; COTT, 1940), what may be termed generically the *visual-effect* functions. Many of these integumental chromes belong to chemical classes which function also in metabolism and are synthesised in common for both purposes. Others are peculiar to the integument, so that we have here a further type of heterogeneity. The integumental chromes interest the neuroeffector physiologist, ethologist, morphologist, and evolutionist more than the biochemist. The retinal perceptor, retinene, is a specialist in this group but, significantly enough, a cycle of molecular changes is involved in its neurosensory function. The perceptor chromes thus provide one direct link between the two main groups of zoochrome.

In addition there are a number of zoochromes the functional significance of which is to varying degrees obscure. This is true particularly of many of the chromes of the deeper organs and tissues, and of some in the body fluids. These should be the centre of interest since the probability is that they will prove to serve functions not previously recognised and this might strengthen the validity of a discrete science of biochromatology. In any case they have crucial importance. It is largely on account of these that a number of biologists still think many biochromes are casual and useless accumulations, due merely to faulty metabolic and excretory mechanisms, or to poor control of what is absorbed from the food. It may be noted that the dual origin of zoochromes, exogenous (dietary) and endogenous, is another type of heterogeneity, which must be considered.

In most contexts of this book the terms *biochrome* (Fox, 1944) and *zoochrome* are preferable to *pigment*, which unfortunately is rather deeply rooted in tradition. Biochrome is a more scientific and precisely definitive term, while pigment has a prior usage in art and industry, where it usually means a particulate and insoluble coloured material or even a mixture of such materials, used as a paint for external application (Allen, 1971, p. 13). Some zoochromes indeed are in solid or semi-solid form, within or outside cells, and often there are two or more different chromes superimposed, if not actually mixed, in the same cell. It might be reasonable to use 'pigment' in these situations. On the other hand the blood zoochromes are mostly in solution, often in the plasma, and for them in particular, pigment seems an unsatisfactory term. Many carotenoids, and some other chromes, exist *in vivo* in a third form, dissolved in lipid, in intracellular vacuoles, and again 'pigment' is inappropriate.

Most biochromes are quite readily soluble in some solvents and they are rendered insoluble *in situ* only for good biological reasons. 'Pigment' seems most usefully applied only to materials of solid or indefinite physical states, to mixtures of chromes and to unanalysed and unknown coloured materials. A biochrome then is defined as a specific chemical substance with a coloured molecule, synthesised by living organisms, and a zoochrome is any such chrome found in the bodies of animals.

This excludes structural colours, due to a variety of optical properties emerging at the supramolecular level of magnitude (Fox, 1953). These are relevant to the present purpose since they affect the retina of a beholder in the same way as true chromes and so have come to function alongside them in camouflage and sematic functions. They will not be considered except where, as in the case of the purines, they are properties of a specific chemical class of material which also behaves in other respects as true chromes and shares common control mechanisms with these. For an account of the principles which determine structural colours, reference may be made to Fox (1953), who also suggested that the term biochrome should be restricted to materials in which colour is an intrinsic property of the molecule.

The leading question, how a molecule produces colour, must be answered as fully as possible since the significance of biochromes is probably always dependent on this, even where colour itself seems irrelevant. It is common to all chromes, irrespective of chemical class and it brings us nearest to a discrete science of chromatology. It is the theme of the next chapter.

Chapter 2

The Basis and Significance of Colour in Molecules

> "... chromophoric groupings may impart both colour and increased reactivity or chemical instability to the same molecule. ... Colour and biochemical activity are, in such instances, two interlocked effects of the same fundamental molecular phenomenon."
>
> D. L. Fox, Science 100, 470 (1944)

2.1 Physical Basis of Colour

Intrinsic colour is due to a molecule differentially absorbing light of one or more particular wavelengths, or more usually wavebands, in the visible range, i.e., between 400 and 750 nm. The residual light transmitted to an observer's eye is the 'colour' normally attributed to that molecule, although in fact it is the complement of that which the molecule absorbs

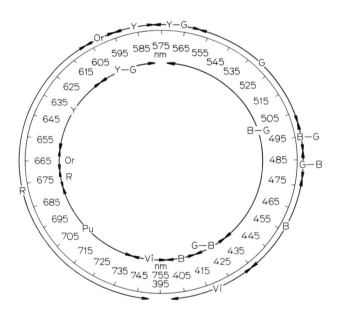

Fig. 2.1. The visible spectrum represented as the subjective colour-circle. Around the outside of the wavelength scale are represented the corresponding subjective hues for the average person. Around the inside of the scale is shown the spectrum derived from the usually accepted view of the hues subjectively complementary to those around the outside. The two are not identical, perhaps indicating minor errors in the view but equally possibly that the subjective colour ring is not a precise circle with a precisely linear scale of wavelengths. The colours are indicated by initial letters

and utilises (Fig. 2.1). Some explanation is required of the fact that this transmitted, residual spectrum should so often appear to be a single, simple hue; this is due to the essential subjectivity of 'colour'.

The subjective sensation of 'white light' is caused by a simultaneous perception of the full range of visible light waves in the same relative intensities as in the solar radiation reaching the Earth's surface. When such a beam of white light is sorted out, according to wavelength, in a *spectral display,* the average human eye distinguishes six distinct hues, or colours, as six broad, contiguuous wave-bands. Starting from the long wavelength end of the spectrum these are red, orange, yellow, green, blue and violet. They appear to grade into their neighbours so that a sensitive eye and a good central mechanism can recognise and memorise a number of intermediate hues, though subjectively these seem to be less discrete and 'real' than the main colours. Indeed, orange already looks like a mixture of red and yellow, green of yellow and blue, and violet of blue with a tinge of red, so that there is an *a priori* hint of three primary colours and of the three retinal wavelength-discriminators considered in chapter 7. However, the latter are maximally sensitive to red, green and blue wavebands, respectively, and not to red, yellow and blue.

With this relatively gross analyser any substance which absorbs reasonably strongly somewhere in the yellow band will appear to transmit blue light and *vice versa.* If absorption is in the middle of the spectrum, in the green, then the transmission appears purple, which is a mixture of red and blue and is a subjective addition to the spectral six. Subjective colour-vision therefore conveniently is represented by the now familiar colour circle (Fig. 2.1) in which red is juxtaposed to violet at the opposite end of the perceived range. It is noteworthy that the shortest rays themselves give a slight component of red in the sensation so that the circle is inbuilt to the neurosensory system and is not simply a textbook convenience. When any particular hue on the circle is absorbed by a material the residual transmission appears to approximate to the complementary colour, on the opposite radius of the circle. This relationship is not precise if the wavelength scale is uniformly represented round the circle (Fig. 2.1), and no doubt the retinal discriminators lack precise 'symmetry' in some aspects of their discrimination (p. 168). In addition there is considerable individual variation in the subjective impression, ranging down to frank colour-blindness. Fig. 2.1 gives, on the outer circle, a generally acceptable compromise concerning the band-widths of the subjective hues, and on the inner circle the complementary colours as recognised in a residual transmission spectrum. These are:

Band absorbed	Violet	Blue	Gr.-Bl.	Bl.-Gr.	Gr.	Yel.-Gr.	Yel.	Orange	Red
Colour transmitted	Yel.-Gr.	Yel.	Orange	Red	Pur.	Violet	Blue	Gr.-Bl.	B.-Gr.

It will be noticed that no single absorption band will give a pure green residual transmission, yet green transmitting biochromes and artificial chromes are very common; they absorb in two bands, near the two ends of the spectrum. Absorption in two or more regions elsewhere in the spectrum likewise gives transmission approximating to a single hue, subjectively.

In all cases there will be darkening, i.e., addition of 'black', in proportion to the number and extent of the bands absorbed. With only three primary discriminators any two absorption bands are likely to affect two of these more than the third, which determines the subjective hue transmitted. If all three are significantly affected the subjective effect is of a reduction in transmission throughout the spectrum; shades or dark tones (BATTERSBY, 1964) result from this addition of black to the transmission. Thus in fact the primitive, intensity-perception simply of black-and-white interacts with colour-vision. Red darkens to brown, green to olive, etc. The final stage for all is black but the average human eye can recognise and discriminate a number of shades of darkening for each hue. If absorption is not very intense anywhere in the spectrum, then the subjective impression is that the transmitted hue is mixed or diluted with white: red becomes pink and Oxford blue becomes Cambridge blue. Again a number of these tints or light tones of each hue can be discriminated, from full saturation to almost white. Altogether, therefore, human colour vision is very rich and efficient in spite of its apparently crude perceptor basis (p. 168). For technical purposes discrimination has been greatly extended and improved by spectroscopic and other instruments, so that today any personal shortcomings can be offset.

Many biological materials are colourless, for instance most of the carbohydrates and their derivatives, the hydroxy-acids, glycerides and most other lipids, the simple proteins and the nucleic acids. Moreover in bulk and in their functions these include the most important of biological materials. The molecules of most, though not all, minerals of the body-fluids and skeletons also are colourless so that biochromes constitute a special category just because they absorb within the visible range. Colourless materials do absorb elsewhere in the electromagnetic spectrum: all materials absorb strongly in one or more regions (JENKINS and WHITE, 1957) and more commonly in the u.v. and i.r. than in the visible range. Most biochromes absorb in all three ranges. The pattern and intensity of absorption are related to the frequency of the radiation (JENKINS and WHITE, l. c. pp. 442, 613) so that most materials absorb strongly in the u.v.

On earth water is an important molecule which absorbs very strongly in both the ultraviolet (u. v.) and infrared (i. r.) ranges (STEVEN, 1963) so that its 'window' over the visible range contrasts with its almost total cut-off outside this range (WEISSKOPF, 1968). There is a little absorption also in the long-wave visible so that thick layers of water are greenish and a layer 30 metres thick will absorb all the red light. Water has a small window also in one region of the u.v. but on earth today this is made irrelevant by the absorption spectrum of ozone in the upper atmosphere; virtually no radiation outside the visible range penetrates far into the hydrosphere where life originated. This emphasises the basic importance of materials which absorb in the visible range, i.e., of biochromes. Atmospheric oxygen and nitrogen also contribute to i.r. and u.v. absorptions by the environment, and they again absorb virtually nothing in the visible range, at ordinary temperatures (WEISSKOPF, 1968). By contrast, organically bound O and N play a most important part in molecular absorptions in this range.

Light waves or particles (photons) are characterised by discrete quanta of energy, the magnitude of which is inversely related to wave-length, λ, by the relation $E = h/\lambda$, where h is Planck's constant, and directly related to frequency, v, since all light travels at the same speed, $u: \lambda v = u \approx 186,000$ miles per second (3×10^8 m/sec). A photon will promote any molecular activation and chemical change the energy for which matches fairly closely its quantal value, the proviso being that quanta must be absorbed in all-or-none fashion.

The low-energy i.r., or heat, radiation of the sun is abundant and has considerable value to living organisms, by promoting molecular vibrations and rotations and so speeding all their activities; moreover various groupings in a molecule can each absorb heat and pool the resulting activation (MASON, 1961). Indeed, in molecules which are particularly easily excited, such as some ribitylflavin complexes (p. 206), thermal energy may even be adequate to induce electron-transitions and overt chemical reaction (KOSOWER, 1966). This actual induction of a reaction by heat is to be distinguished from the mere acceleration of reactions actually induced by other agencies. Ribitylflavin is a very important biochrome (p. 193) and some others, in particular melanin (p. 60), also are very efficient at absorbing heat-energy. A dramatic example of thermal activation is the so-called thermochromatic response of substances such as bixanthylene and bixanthone, which become reversibly coloured on heating (JAFFE and ORCHIN, 1962). Bixanthone is a dimer of anthraquinone (p. 40) and in fact the natural anthraquinones (p. 42) also show this response: kermesic acid is yellow in cold, and red in hot aqueous solution, and the alkali salts of the anthraquinols are pale pink when cold but deep red in hot water (FEIGL, 1966.)

These molecular activations by heat are exceptional, however; for most reactions, energy of higher quantal value is necessary to lift an electron from its ground state to an orbital of higher energy. Quanta of 750nm red light are adequate to drive a reaction demanding 38.1 kcal of activation-energy per mole while those of 400nm violet light are equivalent to 71.4 kcal per mole. This range is adequate for many of the reactions which take place in living systems, so that the potential value of biochromes is very evident. Actually many of the important bonds in biomolecules require rather higher quantal energy (ROSENBERG, 1955):

Bond	C=O	O-H	N-H	C-H	C-O	S-H	C-C	C-N	C-S	S-S
Energy required) to rupture bond) (kcal/mole)	145	118	104	100	90	87	82	98–77	72	64

A quantal value of 145 kcal/mole corresponds to u.v. light of 200nm. However, in most large resonant molecules these bond-rupture energies are considerably reduced and activation at the bond is possible at least by long u.v. waves and in most cases by visible light.

Actually visible light is not known to trigger many reactions except those of the large resonant molecules in living organisms (OSTER, 1968) and so its seems likely that modern biological systems have evolved in adaptation particularly to this range. Quite apart from the action of the constituents of the atmosphere and hydrosphere in trimming down the incoming solar radiation to the visible and i.r. ranges, the solar spectrum when it leaves the sun is already richest in light of the visible range; biologically it would be logical to exploit mainly this range. Photosynthesis and light-perception by animals (chapter 7), the two main biological photo-activations, make use of wavebands distributed over much of the visible range.

Light of wavelengths shorter than 350 nm has increasingly an ionising action, causing the complete ejection of electrons from atoms and molecules. This leads to molecular changes and chemical processes not easily controllable by living systems, which necessarily must be self-regulating; for that purpose electron-transitions and changes in covalent bonding are ideal. Ionising rays destroy living systems in the following order of increasing quantal value and effect: u.v., X-Rays, γ-rays and cosmic rays. Extant organisms and their ancestors for at least some 10^9 years have been so little exposed to heavy dosage in these ranges that there can have been no natural selection for resistance to them, even if in principle this is feasible. Some biochromes, such as the porphyrins and the polycyclic quinones, as well as various laboratory-synthesised dyes, can make even visible light damaging to living tissues (p. 247) by sensitising molecules not normally activated by quanta in this range. The phenomenon is an indication of ways in which biochromes can extend the range of reactions inducible in living systems; some of these extensions are controllable and not damaging.

It is known that nucleic acids are subject to mutational changes under the action of u.v. of the waveband (250–260 nm) absorbed by their bases, as well as by the action of activating chemicals. These changes mostly are damaging, but can include some which are useful for biological evolution. Less is known about possible useful results of absorption in the u.v. range by proteins (mainly at 280–300 nm) and other colourless biological materials; in the letter this may not be relevant to biochromes but in spirit it is (PULLMANN, 1972) since the electromagnetic spectrum is continuous and the visible range an abstraction, albeit a biologically significant one. The significant range may have been different in the past. It is possible that there was little oxygen and therefore presumably also little ozone, in the atmosphere and hydrosphere before modern plants evolved (OPARIN, 1957; 1960); at that time u.v. light may have penetrated the seas to a depth of 10 m, and any organisms living on the land will have been exposed to higher dosage, except in sheltered sites. At that time there may have been selection for living systems which made use of, and were not damaged by, this short wave radiation. The synthesis of vitamin D from precursors in the human skin is an example of the persistence of such use.

The value of coloured materials is not simply that they are activated by visible light but also that in turn they are able to activate other molecules with matched activation-energy levels. This essentially chemical transfer can continue through many steps, as for instance in the synthesis and subsequent catabolism of the materials resulting from the initial photoactivation of the chlorophyll system. Equally, of course, biochromes themselves can be activated by chemical as well as by photic activation; this can occur in darkness and explains the familiar fact that haemoglobin (p. 211), one of the most brilliant of biochromes, usually operates in the absence of any but an incidental, or a diffuse, exposure to light. Some biological (p. 18) and other molecules activated in this way actually re-emit the energy as light and in this sense chemiluminescence (p. 206) is the converse of photochemical activation. Colourless biological materials are mostly activated by chemical means.

The light energy absorbed by a biochrome may be re-emitted as light; for thermodynamic reasons this is always of lower energy, i.e., of longer wavelength. The emission is either as fluorescence or as phosphorescence (p. 17), according to its duration and the precise nature of the active state responsible (p. 12). This 'radiative transfer' is termed 'trivial' by chemists (LAMOLA, 1969), and for biologists, also, radiationless or chemical transfer is the useful form. Direct, radiationless transfer in fact is most common and in this sense, again, radiative transfer is the more trivial. *In vitro* many free biochromes are fluorescent but *in vivo* the emission is usually 'quenched' by conjugants such as proteins, metals, etc. Energy is thereby directed to chemical use.

2.2 The Electronic Basis of Colour

To understand more fully the properties which make some molecules absorb sufficiently in the visible range to appear coloured, and how this confers special reactivity on the molecule it is necessary to consider the electronic basis of the phenomena. Relevant orbital electrons of the molecule oscillate in resonance with those photons having quanta that match the energy required to raise one of them to a higher orbital. The electron makes the actual transition to this orbital if it absorbs the quantum, which is said to have raised it from its ground state to a first excited state with the necessary energy increment (Figs. 2.2, 2.4). The energy is put to further use when the electron returns to its ground state. Although the energy associated with molecular and atomic vibrations and rotations, and other innate movements, are small compared with electronic oscillations (Fig. 2.2), they serve as a useful 'fine adjustment', and can make a whole wave-band of photons sufficiently matched to the requirements of a particular electron; further they may help in matching the excited state of the latter to the requirements of an acceptor molecule. This emphasises one advantage of relatively large molecules, such as those of all biochromes, and it is one way in which the external energy required for bond-rupture is lowered (p. 8).

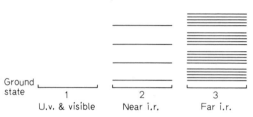

Fig. 2.2. Jablonski diagram of energy levels relative to ground-state of (1) the first excited state of an orbital electron (2) atomic and molecular vibrations and (3) the same plus molecular rotational movements. Below each column is given the range of electromagnetic radiation having energy quanta appropriate to excite the three types of movement

Only the labile, bonding and non-bonding, electrons in the outer shells of the constituent atoms are in a position to move out into an unoccupied orbital, and since most of the biological elements are members of the first two periods of the chemist's periodic table, the relevant electrons are mainly those of the 1s, 2s and 2p atomic orbitals or the 1δ, 2δ and 2π of the derived molecular orbitals. The transitional metals of the next period also play an important part in biochrome structure and function (p. 47) and so for them

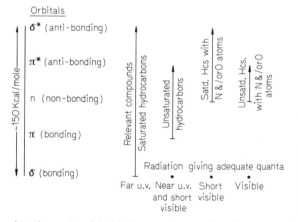

Fig. 2.3. *Left:* Relative energy-levels of five types of electronic orbital relevant to organic molecules.
Right: The easiest orbital transitions for electrons in four classes of organic molecule. The range of e.m. radiation having energy-quanta capable of exciting these transitions is given at the foot of each

the 3d atomic orbitals also are relevant. For the present, however, it is adequate to consider only δ, π and the lone-pair electrons of non-bonding orbitals, n. The biologically important n-orbitals are those of O and N atoms. On excitation electrons from these three groundstate orbitals can be raised into δ or π non-bonding orbitals, usually designated δ* and π* (Fig. 2.3). Biologically these are the only common transitions: their relative frequency is inversely related to the energy required.

The most energy-demanding transitions are from δ to δ*. These are the only orbitals and transitions possible in the simple hydrocarbons, which therefore absorb only in the far u.v., are never significantly coloured, are rarely excited in nature and play little part in living systems. Methane, CH_4, absorbs at 122 nm and ethane, H_3C-CH_3, at 135 nm, corresponding to quantal values in the region of 150 kcal/mole. However, there is a progressive bathochromic shift with increasing chain-length and also with any branching of the molecular chain. A bathochromic shift has the meaning that the absorption-bands move to a progressively longer wavelength; i.e., that all electron transitions demand correspondingly smaller energy quanta; the various accessory motions and atomic interactions are largely responsible. With N or O atoms substituted in the molecule the $n \to \delta^*$ transitions by their lone-pair electrons become possible, but even so virtually no aliphatic hydrocarbon derivatives are visibly coloured.

The smallest energy-increment is for $n \to \pi^*$ transitions (Fig. 2.3) and it is these which are most widely exploited in biological systems; this alone is enough to establish the outstanding importance of nitrogen and oxygen in biological compounds. The initial fixation of solar energy by photosynthesis involves this kind of transition (KASHA, 1961) and is adequa-

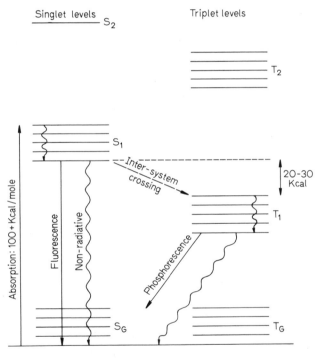

Fig. 2.4. Jablonski diagram of energy-levels relative to ground state of first and second excited singlet and excited triplet states of orbital electrons and of the appropriate molecular vibrations. The modes of transition between these various levels are indicated. G = ground state; S = singlet state; T = triplet state. Sinuous arrows denote non-radiative transitions (After LEERMAKERS, 1969)

tely powered even by the visible rays of longer wavelengths. As in commercial dyes, the most important *chromophor* and *auxochrome* groups of biochrome molecules are radicals containing N or O (Fig. 2.3). Next in facility is the $\pi \to \pi^*$ transition; π-orbitals are characteristic of unsaturated organic compounds and these are conspicuously common among biochromes.

The biologically important $n \to \pi^*$ and $\pi \to \pi^*$ transitions commonly show a further useful feature, namely an excited state of intermediate energy level (Fig. 2.4) and of correspondingly greater stability than the usual first excited state; consequently it lasts long enough to have a high probability of exciting other molecules (PORTER, 1961). It is called the triplet excited state and differs from the normal singlet excited state in having two electrons with parallel spins in the same orbital, and so is paramagnetic. This can be detected by the electron spin resonance (ESR) or *electron paramagnetic resonance* (EPR) signal produced upon excitation. Many biochromes even including melanin (COMMONER, 1954; SWAN, 1963) show this phenomenon. The magnitude of the EPR signal given by living systems is proportional to their metabolic activity (SZENT GYORGYI, 1960), and this shows what a dominant part biochromes play in metabolism.

The energy levels of the singlet and triplet excited states differ by 20–30 kcal/mole (WHELAND, 1955). Usually the singlet excited state is the first transition and the electron then returns to the ground state via the long-lived triplet state (Fig. 2.4). Direct transitions from ground to excited triplet state are 'forbidden' (GLASS, 1961, p. 819) because parallel spins contravene the Pauli principle. Approach must be from the first singlet excited state by 'intersystem crossing' (Fig. 2.4). The $n \to \pi^*$ transitions are generally characterised by brightly coloured, free radical excited states with an internal quenching of fluorescence, i.e., a useful absorption of the excitation within the molecule itself. Ribitylflavin (p. 188) shows all these properties particularly well (BERENDS et al., 1966) and so do the porphyranoprotein enzymes (p. 194).

The halogens have so many lone pairs of electrons, in n orbitals, that even their simple elemental diatomic molecules are brightly coloured (MASON, 1961). The ease of $n \to \pi^*$ transitions is increased by overlap between the orbitals of the two constituent atoms and this makes a major contribution to the colour. However, they are only monovalent so that thyroxine and other biological compounds with single halogen atoms as substituents are not coloured; dibromoindigotin (p. 59) is coloured even in the absence of its halogen. Halogens therefore are rare in biochromes and indeed in all biological compounds (NEEDHAM, 1965b.)

The quanta of short-wavelength visible light are adequate to promote the more demanding $\pi \to \pi^*$ transitions but only under favourable conditions (MASON, 1961). These conditions are very adequately provided by biochromes, with many double-bonded carbon atoms, $C = C$, in the molecule, since the activation energy required decreases markedly with every double bond added in a regular, *conjugated* fashion, i.e., regularly alternating with single bonds. This permits delocalisation of the π-electrons over the whole conjugated system. The groundstate energy increases progressively with the size of the system, whereas that of the first excited state increases by smaller steps, or even decreases progressively. The simplest unsaturated compound, ethylene, $H_2C = CH_2$, is colourless, its longest wave absorption-peak being at 162 nm, but a C_{10} chain absorbs at 343 nm and the C_{20} homologue at 476 nm, well into the blue region of the visible range. As in the alkane series mere increase in molecular size also contributes to this bathochromic shift.

The term bathochromic was originally applied to the colour transmitted which, once absorption has moved into the visible range, becomes progressively 'deeper' with each addition of a conjugated double bond. It is at first pale yellow (complementary to the violet absorbed) and then, in order, round the circle, yellow, orange, red, purple, violet, blue and green. Not everyone would agree that in this sequence the transmitted colour becomes progressively 'deeper' but no doubt the term was prompted also by the progressive darkening due to the fact that other absorption bands move into the visible range following in the wake of the first; indeed black is usually regarded as the final stage of the bathochromic series. The term has now come to mean any shift in absorption to a range of longer wavelengths, anywhere in the electromagnetic spectrum. The antonym is hypsochromic. The terms are not inappropriate, after all, since there is a fall in frequency and energy-value associated with the bathochromic shift.

A triple bond, as in the acetylene series of hydrocarbons, causes more electron-delocalisation and bathochromicity than a double bond so that the absorption band of longest wavelength by acetylene is at 210–250 nm compared with 162 and 135 nm respectively for ethylene and ethane (MASON, 1961). However, acetylenic bonds are rare in biological material since a conjugated system of triple and single bonds would be effectively a filament of elemental carbon (with the attendant electronic properties no doubt). Occasional triple bonds, in the carotenoids (p. 33), stress their rarity.

In the alkene, as in the alkane, series most substituents on the carbon chain also have a bathochromic effect and facilitate electron-transitions. When N or O is involved in a double bond both $n \to \pi^*$ and $\pi \to \pi^*$ transitions are possible; the most powerful chromophor substituent groups are: $-N = O$ (nitro), $-N = O$ (nitroso), $-N = N \to O$ (azoxy), $-N = N-$ (azo), $C = O$ (carbonyl), and $C = S$ (thiocarbonyl). However, the $n \to \pi^*$ transition suffers a hypsochromic shift in the alkene series and this tends to counteract the opposite trend for $\pi \to \pi^*$. The effect applies to both increase in size of conjugated system and addition of substituents. The polar solvents typical of biological systems have the same effect (KASHA, 1961), all solvents having some effect on the energy of solute molecules. The total result, in biological materials collectively, is that the energy-ranges for the two types of transition overlap, or 'mix', and this must be an important contribution to the degree of spontaneity of vital processes (NEEDHAM, 1965b; 1968c).

Owing to the delocalisation of π-electrons the molecule becomes a photoconductor (KASHA, 1959; ELEY and WILLIS, 1967) of the excitation energy or 'exciton', that induces an electron-transition in it. There results a charge-transfer, of variable potential up to the value (2 eV) associated with the complete transfer of an electron (SZENT GYORGYI, 1960). The charge transfer may produce a stable *free radical* (p. 188), which induces overt chemical reaction in a second type of molecule. Charge-transfer is restricted to the excited state because in the ground state there is negligible overlap and interaction between orbitals (KASHA, 1959).

The phenomenon is to be distinguished from *semiconduction*, which involves lower potentials, of the order of 1.0 eV, but on the other hand requires no light-excitation. Semiconduction probably does occur in the large crystalline arrays of solid proteins and other biological polymers and it will be seen (p. 62) that some zoochromes are in such a form *in vivo*. Photobleaching of chromes is most marked in the solid state (PATTERSON and PILLING, 1964; ELEY and WILLIS, 1967).

Typical electrical conduction as in metals and elemental carbon requires much smaller potentials and certainly does not need photo-excitation. These materials absorb or 'damp' light uniformly throughout the visible spectrum and so are grey or black (JENKINS and WHITE, 1957, pp. 430, 449), but they differ fundamentally from black biochromes such as melanin in their high reflectance. Matt black materials such as melanin have very low reflectance and high absorptance at the surface. This surface-absorptance is to be distinguished from the damping out by their free electrons of that light which does enter metals; biochromes by contrast are usually dielectric materials, with strong electrical insulating properties. They could make good condensers and the sub-microstructure of the outer segments of the retinal rods and cones (p. 163) may be relevant here.

Delocalisation of electrons is even more extensive in the conjugated double bond systems of aromatic penta- and hexacyclic ring-compounds, where the electrons can move freely round the rings; benzene has a more bathochromic absorption than the corresponding open chain olefine, hexatriene. However, ring-closure reduces the degree of polarity, or electrical asymmetry, of the molecule, and so also the dipole moment (see below), and a polycondensation of aromatic rings is necessary to offset this. Most biochromes are indeed polycyclic aromatic compounds, usually with side-chain substituents which further improve the polarisation. Ring-condensation increases molecular size and often the extent of the conjugated double bond system, and there is a progressive bathochromic shift with the degree of 'annelation' (Table 2.1), particularly if the annelation is linear so as to give maximal polarity. Pure benzene is colourless whereas pentacene is a deep violet-blue colour.

Table 2.1. Effect of increasing annelation on the absorption-spectrum of aromatic compounds. (From MASON, 1961)

Compounds	No. of rings	Wavelength of absorption-bands (nm)			
		β	p	α	t
Benzene	1	183	207	264	340
Naphthalene	2	221	289	315	470
Anthracene	3	255	379	–	670
Tetracene	4	274	471	–	975
Pentacene	5	310	576	428	1300

As usual, side-chain substituents containing nitrogen or oxygen have a strong bathochromic effect, particularly if they are so situated as to increase the *dipole moment* of the molecule. Since this is the product of the electrical potential difference between two points on the molecule and the distance separating these points, sidechains such as $-NH_2$, $-OH$ and $=O$, the familiar auxochrome, or colour-deepening groups (MACCOLL, 1947), give particularly high values.

Both N and O, but particularly N, can contribute bathochromicity as conjugately bonded members of the aromatic rings themselves and the rings of biochromes are mostly heterocyclic. Oxygen produces the necessary minimum of three bonds only if it becomes the oxonium cation, $=O^+-$, as in the anthocyanins (p. 38). Ring $-$ N atoms likewise are most commonly in the corresponding ammonium, $\equiv N^\pm$, form and so the molecule as a whole may be made strongly polar. Oxygen is the more valuable in side-chains since the amine/imine change of N does not provide the powerful redox property of the quinol/quinone change (p. 45). Obviously the dipole moment of a molecule is greatest when there is a maximal difference in electrical charge between opposite ends of the longest axis; highly polar molecules with a high axial ratio therefore are most bathochromic. Some alkanes have a very high axial ratio but no electrical polarity and therefore a very small dipole moment.

Like most commercial dyes, many biochromes have sufficiently strong polar substituents to form salts with appropiate ions. Most of these affect the dipole moment and the colour; the response to pH changes (Table 3.15) is one example. Another is the familiar effect of various heavy metal cations in producing blue hydrangeas; other flowers show similar responses provided they have an anthocyanin (p. 39) with two $-$ OH groups substituted *ortho* to each other on the ring system (HARBORNE, 1964). Also as among commercial dyes, anionic biochromes are more common than the cationic type; as substituents therefore the H^+ of acids and other cations are more important than $-OH^-$ and other anions.

Aromatic ring systems that are not fully condensed but simply linked by chains which do not break the conjugation can be even more powerfully bathochromic. This is exemplified particularly by the tetrapyrrolic porphyrins (p. 52) and bilatrienes (p. 56), the most potent of all biochromes. In porphyrins the methene bridges between the pyrrole rings (Fig. 3.6) make the whole super-ring a conjugated double bond system with complete delocalisation of electrons. Moreover the super-ring is ideally planar (p. 52) allowing maximal mobility for the delocalised electrons, because overlapping orbitals then are maximally confluent. On the other hand the porphyrin ring has a remarkable degree of symmetry which, as in the simpler benzene ring, reduces the dipole moment; the open-chain bilatrienes therefore are more bathochromic than the porphyrins, though even they have a degree of symmetry in the long axis which prevents the full exploitation of their high axial ratio. The whole of this tetrapyrrole group of molecules thus appears to be 'driving with the brakes on' but there may be a good biological reason for this (see below).

Biological molecules having fully or partially aromatic rings linked by saturated alkane chains are rather common. If short this chain may act as though partially unsaturated and sharing in the conjugation of the system. This is known as *hyperconjugation* (PULLMAN and PULLMAN, 1963) and is seen among luciferins (p. 207) and other metabolites with sub-chromatic properties.

A considerable number of inorganic compounds are coloured (FEITKNECHT, 1967). This depends largely on the formation of hydrated ions, or other complexes, but even so most of them are still much smaller than those organic molecules that are coloured. This is because there is a smaller percentage of σ and a higher proportion of π and n orbitals than in the hydrocarbon skeleton of the latter; the effect of the many n orbitals of the halogens has already been mentioned (p. 12). In addition, the coloured inorganic complexes are mostly those of the transitional metals, which have various unfilled orbitals available for electron-transitions. The atoms of these transitional elements have the further useful property that some of the orbitals show 'splitting', i.e., a varying energy level, so that the necessary energy-increment for electron-transitions can be very small. These properties are most strongly developed in metals of the first transitional series and are operative with organic and other ligands. These metals, particularly iron and copper, are important components of certain classes of biochrome (pp. 47, 53). The property is shared in a minor degree by metals of the Mg, Ca series of Group II.

The lowering of the energy increment for e-transitions in molecules with the various properties described is measured by the so-called *resonance energy* of the molecule (PULLMAN, 1972). This is as much as 150–200 kcal/mole for porphyrins. Paradoxically it confers thermodynamic stability (WHELAND, 1955) as well as the reactivity due to the small increment necessary for excitation. The molecule behaves as though it were more stable than the most stable identified structure, with shorter interatomic bonds, etc. Through charge-transfer (p. 13) it exists as a number of charged resonant species (MASON, 1961; KOSSOWER, 1966), which are the essential contributors to the bathochromicity of the molecule (FINAR, 1959, p. 745).

Resonance is essentially a polarisation phenomenon in fact (FINAR, 1959, p. 97; PULLMAN and PULLMAN, 1962) and helps to explain the significance of the dipole moment of the molecule. Again paradoxically, resonance is most evident in unsaturated, covalently bonded organic molecules and this emphasises the importance of the summation of many small energy contributions. Excited states have more charged forms than the corresponding ground states so that resonance lowers the energy of the former more than it does that of the latter; in this way (p. 10) the energy increment is lowered. Resonance is greatest in planar molecules, and this shows that it is related to the general delocalisation and mobility of the π-electrons.

The excited state has also a higher dipole moment than the ground state and another measure of the probability of an e-transition is the difference between these, the 'transition-dipole' or 'dipole moment of transition.' The intensity of light-absorption by a particular substance also is proportional to the transition dipole (FINAR, 1959, p. 744) and is another measure of the probability of e-transition in the individual molecule. Again both are related to the planarity of the molecule, which determines the ease with which charge-separation occurs. For this reason the all-*trans* isomers of olefinic substances are more activating than the *cis*-configurations (WHELAND, 1955).

The net dipole moment is the *vector* sum of the moments of all constituent interatomic bonds and so will be strongly favoured by coplanarity. Vector properties play an important part in e-transitions; the operative e-vibrations must be parallel to the electric vector of the exciting light (MAHLER and CORDES, 1966, p. 89). The relevance of this to biochromatic structures *in vivo* has scarcely been envisaged up to now. Ordered solids are known not to absorb light when the transition-moment of their molecules is perpendicular to the electric vector of the radiation. These vector properties no doubt are particularly important in connection with structural colours and other optical phenomena.

One further feature of importance is the breadth of the light absorption-bands of molecules, as compared with those of single atoms. The breadth again is directly proportional to molecular size and complexity. It is partly because of the width of the bands that the transmitted light often approximates to a single hue (p. 5). The position of the peak-absorption in the band is usually attributed to the operative e-transition, while the spread-over to neighbouring frequencies depends largely on atomic and molecular vibrations and molecular rotations, as well as on interactions between molecules, in which solvent molecules also play a part (WHELAND, 1955). These components also contribute to the total intensity of the absorption. In the visible and u.v. ranges the absorption-bands are thus properties of the whole molecule and the critical e-transition is inducible by a rather broad spread of quantal values both above and below the critical value.

The breadth of the absorption bands of large molecules is greatest and the intensity least for saturated aliphatic compounds, which absorb diffusely throughout the spectrum. The absorption by aromatic compounds is much more selective and intense (MASON, 1961) and this is seen in the spectra of most of the biochromes. The type of molecular structure which gives a high probability of e-transition in response to radiation of a particular quantal value must *ipso facto* show selective, or band, absorption. When a porphyrin comes to ligate a metal its pyrrole N-atoms all become equivalent, and a single absorption peak replaces the two peaks of a free porphyrin (JONES, 1967). Biochromes thus combine the selectivity of smaller molecules with a bathochromicity possible only in larger organic molecules; this could be regarded as the essence of their unique biological significance.

Conjugation of biochromes with other components, particularly protein (p. 29), enhances most of the essential chromatic properties discussed above but conjugants have so many other functions that they are best considered in those contexts.

The biological importance of all molecules and complexes with these properties has been emphasised by PULLMAN and PULLMAN (1962): "All essential molecules related to, or performing, fundamental functions of living matter are constituted of completely or partially conjugated resonant systems rich in delocalised π-electrons." It is significant, since in fact the interest of these authors was centred on biochemical function in general, and not on any particular chemical class, how closely this description fits the biochromes. The following chromes were among the metabolites cited as examples; pterins, porphyrins, quinones, carotenoids and melanins. Flavin and other co-enzymes also were discussed so that few important classes of biochrome were not included. Among colourless components they emphasised only nucleic acids, proteins and phosphate compounds.

2.3 Other Optical Properties of Biochromes

At present we know relatively little about other optical properties which are both generally characteristic of biochromes and also very relevant to their biological functions. Fluorescence, phosphorescence and dichromatism seem to merit most attention at this point.

2.3.1 Fluorescence and Phosphorescence

The origin of these forms of visible light-emission has been indicated (p. 9) and it is not surprising that most biochromes in the free state show one or both of these. Fluorescence may continue for 10^{-8} seconds after the cessation of incident irradiation, but phosphorescence for as long as 10^{-4} seconds (GLASS, 1961, p. 820) because it involves the relatively long-lived triplet state (p. 12). Fluorescence is particularly characteristic of biochromes (Table 2.2) but relatively few are at all strongly emitting *in vivo* so that useful quenching is the rule. From a study of their fluorescence much can be learned about the normal use of the energy absorbed by these materials (WEBER, 1960).

Both forms of emission have longer wave-lengths than the light absorbed (Stokes' Law), in accordance with the laws of entropy and with the Franck-Condon principle governing interatomic distances and molecular shape (ROSENBERG, 1955). One result of this is that substances absorbing in the u.v. range may fluoresce in the visible; this can be a valuable aid to the identification of many colourless materials. Most of the biochromes are excited to fluoresce by the longer u.v. waves in the region of 355 nm, so that the simple method of examination in such a beam, in a darkened room, is invaluable in their identification also. This method is necessary because fluorescence is rarely strong enough to be detected in daylight: the aphins (p. 44) are exceptional in this respect. Table 2.2 gives only the fluorescence when excited by this long-wave u.v. In response to u.v.-irradiation ribitylflavin fluoresces at a wavelength near its visible colour and this is true also for a chrome in the tentacles of the sea-anemone, *Anemonia* (FOX and VEVERS, 1960, p. 192). As in certain commercial products the coincidence may increase the normal intensity of the visible colour significantly and serve a useful purpose *in vivo*.

The harnessing *in vivo* of the energy of excited biochromes, so that their fluorescence is quenched, often depends on various components with which they have become conjugated. Proteins, rather generally, and the iron atom of haem, are outstanding examples of such quenchers. Most paramagnetic atoms, radicals and molecules have this power (GLASS, 1961) since electrons with unpaired spin absorb light radiation very strongly. Oxygen and in some states the transitional metals are outstandingly paramagnetic (p. 228). The oxygen and copper of oxyhaemocyanin absorb over precisely the fluorescence-emission range of tryptophan and this is now shown to be functionally an important amino-acid in Hcy (BANNISTER and WOOD, 1971): its fluorescence is internally quenched in fact.

Phosphorescence by a particular molecule typically occurs at a longer wavelength than its fluorescence and is much weaker. Both differences result from the lower energy level of the triplet, than of the singlet, excited state. Potentially it is more important biologically than fluorescence (ROSENBERG, 1955) because of its longer life and because free radicals are usually involved. The smaller energy quanta involved also can be advantageous. Phosphorescence-emission is virtually restricted to viscous solutions and rigid glasses (GLASS, 1961, p. 829), so that *in vivo* useful quenching is the rule.

Table 2.2. Fluorescence of Zoochromes

Zoochrome	Colour and wavelength of fluorescence	Reference
Carotenes	Orange-red	FOX and VEVERS, 1960, p. 192
Xanthophylls	Yellow	FOX and VEVERS, 1960, p. 192
Carotenoid acids	Blue	FOX, 1953, p. 69
Fuscins	Green, yellow, orange	PORTA and HARTROFT, 1969
Ceroid	490–510 nm	PORTA and HARTROFT, 1969
Cerumen chrome	Yellow (cold)	
	Green (hot)	FOX and VEVERS, 1960, p. 165
Neuron: Yellow	Violet-blue, green	FOX and VEVERS, 1960, p. 163
chrome	440–60, 530–60 nm	HYDÉN and LINDSTRÖM, 1950
Flavone of *Helix*	Lemon yellow	KUBIŠTA, 1950
Xanthoaphin	Green-yellow (acid) (+ V)	
	Pink (alk)	
Chrysoaphin	Green (acid) (+ V)	DUEWELL et al., 1948
	Crimson (alk)	
Erythroaphin	Orange red (+ V)	
Strobinin (*Adelges*)	Yellow	FOX and VEVERS, 1960, p. 144
Lanigerin (*Eriosoma*)	Green (acid)	FOX and VEVERS, 1960, p. 144
Stentorin & Stentorol	Red	FOX and VEVERS, 1960, p. 144
Zoopurpurin	Red	EMERSON, 1929
Fabrein	Red (+ V)	FONTAINE, 1934
Haemerythrin	Blue	MARRIAN, 1926
Porphyrins (free) and Mg-porphyrans	Crimson	LEMBERG and LEGGE, 1949
Haems	Quenched	
Bilins	None (untreated)	
Phaeomelanin	Yellow (weak)	FOX and VEVERS, 1960, p. 31
Erythromelanin	Yellow (weak)	FOX and VEVERS, 1960, p. 32
Eumelanin	None	
Kynurenine	Yellow	REBELL et al., 1957
Papiliochrome (?)	Yellow	COCKAYNE, 1924
Pterins	Red, yellow, green or blue	FOX and VEVERS, 1960, p. 152
Leucopterin	Violet	ZIEGLER-GUNDER, 1956
Ichthyopterin	Blue (450–70 nm)	JACOBSON and SIMPSON, 1946
Xanthopterin	Blue-green, yellow (acid)	FOX, 1953, p. 295
Chrysopterin	Green-yellow (on alumina)	FOX, 1953
	Blue-violet	JACOBSON and SIMPSON, 1946
Erythropterin	Violet	JACOBSON and SIMPSON, 1946
Ribitylflavin	Lemon yellow (515–530 nm)	NEEDHAM, 1966a
Lumichrome	Blue	WAGNER-JAUREGG, 1954
Lumiflavin	Yellow (alk)	WAGNER-JAUREGG, 1954
NADH	440–480 nm	CHANCE, 1964

Table 2.2 (continued)

Zoochrome	Colour and wavelength of fluorescence	Reference
Lampyrine	Rose pink (acid)	METCALF, 1943
	Vermillion (alk)	METCALF, 1943
Arenicochrome	Green	DALES, 1963
Uranidine (*Holothuria*)	Green (+ V)	MACMUNN, 1890
Violet of *Janthina*	Red (+ V)	MOSELEY, 1877
Green of *Anemonia*	Green	FOX and VEVERS, 1960, p. 169
Yellow of feathers of Psittaci	Green to yellow	VÖLKER, 1937

(+ V) indicates fluorescence excited also by visible light.

2.3.2 Dichromatism or Dichroism

A number of biochromes show this phenomenon in varying degree: in solution they appear to show different colours, according to the angle of viewing and to certain other optical conditions, such as the length of the light-path through the solution. From certain angles both colours are evident since reflected, scattered and transmitted rays all may be perceived. Chlorophyll shows the effect very clearly (FOX and VEVERS, 1960, p. 116) being red by reflected light and, if the solution is concentrated, also by transmitted light; this colour therefore depends on the number of molecules the beam has encountered rather than on path-length in itself. Dilute solutions and thin layers transmit green. Chlorohaemoglobin (p. 211) shows almost identical dichroism. Haemoglobin is more weakly dichroic, again red through concentrated solution but yellow through dilute solutions and thin layers, as for instance through a single erythrocyte, under the microscope. Free porphyrins show the phenomenon to varying degree.

Other dichroic zoochromes include the polycyclic quinone, *stentorin* (p. 86), blue-green by transmitted and reddish by reflected light (TARTAR, 1961). Carminic acid, a simpler polycyclic quinone (p. 43) is orange-red by transmitted, and a deep garnet-red by reflected, light (FOX, 1953). Astacene and some other carotenoids have a degree of dichroism (FOX, 1953, p. 97) and so has the violet chrome of the centipede; the phenomenon is rather widespread, therefore.

It is due to the differential absorption of short-wave relative to long-wave light increasing with the number of molecules encountered, thin layers and dilute solutions therefore transmitting the more hypsochromic hue. The hue changes discontinuously with concentration because the absorption by chromes is restricted to discrete wave-bands. The phenomenon is clearest, therefore, in biochromes with two well-spaced, strong absorption bands in the visible, and presumably depends on differential changes in the intensities of absorption in the two bands. They are said to have different absorption coefficients (JENKINS and WHITE, 1957, p. 463). Each absorption-band may have functional significance and dichroism therefore may be an index of functional bivalence. This index has limited value since a number of biochromes have more than two strong absorption bands whereas it is not easy to recognise the corresponding higher grades of polychromatism.

2.3.3 Other Relevant Properties

There are a number of properties which, like the bathochromic shift in light-absorption (p. 12), show a regular trend with the size of conjugated double bond system. One is 'optical exaltation' or *molecular refraction* (R), defined as $\frac{n^2 - 1}{n^2 - 2} \cdot \frac{M}{d}$ where n = refractive index, M = molar weight and d = density of the substance in question. This increases with the linear extent of the conjugated system and with the extent of linear annelation in polycondensed rings.

Another property is *resonance Raman scattering*, i.e., a particularly intense Raman scattering by the solution of a chrome in the region of an absorption band. This has permitted the detection of, and observations on, carotenoids *in situ*, among concentrations of other materials which preclude other optical techniques (GILL et al., 1970). Since Raman scattering causes a change in wavelength of the light waves concerned (THOMAS, 1971), it is in any case a phenomenon with theoretical chromatic significance and possibly with practical biological relevance. The wavelength-change may obey Stokes' Law, as in fluorescence but sometimes there is the opposite, anti-Stokes transition.

DOUZOU (1964) has discussed at length the value of a study of the phenomena of photochromatism and thermochromatism already mentioned (p. 8), the production of transient chromes by light-, and heat-, energy. The coloured intermediary compounds and radicals of active ribitylflavin (p. 188), and other chromes, can be produced in this way. The reversibility of photo- and thermochromatic responses makes them eminently suitable for biological exploitation: A photochromatic response could be a valuable camouflage-mechanism, since the amount of colour is related to the intensity of illumination, and it would be worth searching for possible natural instances. The coloured product tends to be ionic and water-soluble, in cases where the parent substance is not, so that the phenomenon could be very important in terminal respiration, as well as in other metabolic contexts.

Optical asymmetry is not peculiar to biochromes but these mostly have molecules large and complex enough to have at least one asymmetrical carbon atom. As in all such compounds in living organisms, there is a great preponderance of one of the two enantiomorphs due to any particular asymmetric carbon atom. In fact biochromes themselves rarely are optically active because their π-electron system has planar symmetry but they are usually conjugated with optically active moieties so that the system as a whole does rotate a plane-polarised beam. Optical asymmetry has a small but technically very useful effect on the properties of a coloured molecule (BUSH, 1971): just as optical rotation itself is due to differential rotation of dextro- relative to laevo-circularly polarised light so a coloured optically active molecule absorbs differentially the two oppositely polarised beams of light for wavelengths in the region of its absorption band (FINAR, 1964, p. 85). This *circular dichroism* (CD) has been studied with profit in haemocyanin (TAKASADA and HAMAGUCHI, 1968), the flavoprotein D-aminoacid oxidase and other biochromes. Like most optical properties the phenomenon is not restricted to the visible range.

Optical rotation, or circular birefringence varies with wavelength, a phenomena known as optical rotatory dispersion (ORD). In the molecule of a chrome having an asymmetric carbon atom near a carbonyl C = O, group the ORD curve changes sign in the middle of the absorption band of the chrome and this is known as the *Cotton Effect* or anomaly. The details of the ORD curve can give information about the carbonyl group and its environment in the molecule (FINAR, 1964, p. 10).

The non-optical properties of biochromes are more peculiar to the individual chemical classes and therefore will be considered in that relation, in the next chapter and in other contexts.

2.4 Conclusion

The colour usually attributed to a chrome is the transmitted complement, on the visual colour-circle, of the *hue* that it maximally absorbs. Dark *shades* are transmitted if there is some absorption throughout much of the visible spectrum, and pale *tints* if absorption is not total over any part of the range. Colourless materials absorbing in the u.v. can be regarded as 'u.v. chromes' and both u.v. and i.r. radiation do have biological importance; but the limits of our visible spectrum correspond rather closely with those of the 'window' of solar emission that reaches the Earth and so are biologically significant. Biological reactions particularly are geared to being triggered by quanta in the visible light range. Biochromes are excited by these (though also by chemical energy) and have been selected for the faculty of transducing the energy into metabolic reaction.

A molecule is excited when one of its outer orbital electrons is raised to a higher orbital by quanta of the appropriate value. Absorption in the visible range by most chromes is strongly focussed on one or more wavebands, i.e., on specific quantal values. The easiest electron-transitions, requiring the smallest *energy-increment* are $n \to \pi^*$, and next $\pi \to \pi^*$, both passing through long-lived *triplet* excited states of intermediate energy level after the initial *singlet excited state*. These transitions are characteristic of most biological materials but are particularly facilitated in biochromes. Here their necessary energy-increment is minimised by a number of *bathochromic* factors, mostly interrelated: (1) large molecular size, (2) conjugated double bonding, (3) polar structure and high dipole-moment, (4) planar molecular structure favoured by aromatic condensed ring-systems and permitting maximal delocalisation of π-electrons; this gives high resonance-energy, thermodynamic stability and facility for rapid charge-transfer, (5) oxygen and nitrogen atoms in side-chains and in aromatic rings, (6) atoms of transitional metals forming ligand fields in the molecule, (7) protein and other conjugants.

Characteristic of the excited states of biochromes is the formation of highly coloured, paramagnetic free radicals, particularly apt to pass on excitation to other molecules. Biochromes in general have dielectric, insulator and condenser, rather than conductor properties. The trivial loss of excitation-energy from biochromes as fluorescence or as phosphorescence is usually prevented *in vivo* by conjugants or by internal quenching. The energy increment for exciting these quenchers closely matches the quantal value of the emission.

Other optical properties have diagnostic and other value and are being used increasingly in the study of biochromes. Some may have functional significance *in vivo*.

II. The Nature and Distribution of Zoochromes

Chapter 3

The Chemical Classes of Zoochrome

The structural formulae of representative members of each of the fully characterised chemical classes of zoochrome are given in Fig. 3.1. The number of known classes is small and conforms to the general law of biochemical austerity (LWOFF, 1944; FLORKIN, 1960; NEED-

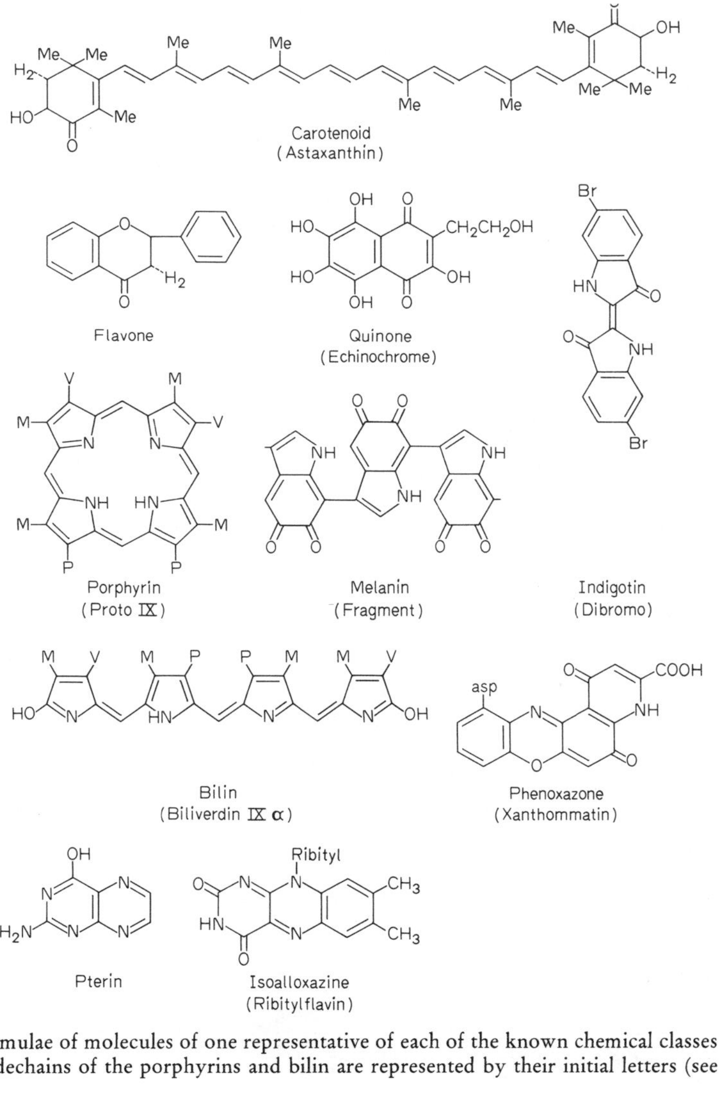

Fig. 3.1. Structural formulae of molecules of one representative of each of the known chemical classes of zoochrome. The sidechains of the porphyrins and bilin are represented by their initial letters (see Table 3.9)

HAM, 1965b) which governs the exploitation of materials by living organisms. The melanins, fuscins and simple metalloproteins are not represented since the structure of their chromo-

Table 3.1. Solubility of known classes of zoochrome

Class of zoochrome	Non-polar organic (petrol, benzene)	Neutral polar organic (acetone, alcohol)	Acid polar organic	Alkaline polar organic	Strong acid, aqueous	Dilute acid, aqueous	Neutral aqueous	Dilute alkaline aqueous	Strong alkaline aqueous
Carotenoids	S. (esp carotenes and esters).	S.: polar C.s. even in aqueous alcohol.	S. (esp. polar C. s.).	S.	Destroyed	Slowly destroyed	Carotenoproteins and carotenoglycosides S.	Free C.s. I.	?
Fuscins	Some S.	?	?	?	I.	I.	I.	I.	I.
Cerumen-chrome	I.	I.	S. in acetic acid.	Slightly S.	S.	S.	S.	S.	S.
Benzoquinones	S.	S.	S. (glacial acetic)	S. (unstable)	Unstable	Fairly S.	Slightly S.	Fairly S. unstable	Unstable
Naphthoquinones	Variably S.; polar NQs least.	Most are S.	S.	?	S. in c H_2SO_4	S.	I. but salts and protein-conjugates are S.	S.	?
Anthraquinones	I.	Not very S.	S. (Acetic acid)	S. in boiling aniline.	S. in c H_2SO_4	Slightly S.	Salts S.; free acids slightly.	S.	?
Aphid polycyclic quinones	Free Q. is S.	S. (variable)	S.	S. (pyridine)	S. in c H_2SO_4	?	Glycosides and salts are S.	S. (stable in NH_4OH).	Unstable
Ciliate polycyclic quinones	Less S. than in S. polar solvents.	S. (some unstable)	S. in acid-alcohol mixtures	S. in pyridine	?	?	?	S.	?
Anthoxanthins	Mostly I.	S.	S. (aqueous CCl_3COOH)	?	?	S.	S.	S.	?
Anthocyanins	Rather I.	Fairly S.	?	?	?	S.	S.	S. (often unstable)	?
Bilins	Bilatrienes S.; others less S.	Bilamonenes S. others less S.	S.	S. (pyridine)	S.	I.	I.; conjugates S.	Salts S.	?
Porphyrins	S.	S.	S.	S. (pyridine)	S.	S. (variable)	Only uro P. is S.	S.	S.
Porphyrans	S.	S.	S.	S. (pyridine)	S. (removes metal)	S. (variable)	(8 COOH groups)S.	S.	S.

Pigment								
Indigotins	S.	S.		S. (pyridine; hot aniline).	S. c H_2SO_4	I.	I.	?
Melanins	I.	Fairly S. in a few special solvents.	I.	I.	Not true Sol.n Erythro M. fairly S.	I.	Erythro and phaeo M. are fairly S.	All S. in 0.5 N NaOH
Ommochromes	I.		S.	I. in lutidine; destroyed by aq. lut.	S.	Variable S.	I.	S. but destroyed
Purines	I.	S. in glycerol; acetone.	I.	S. in aqueous piperazine.	S. in N H_2SO_4 S.	slightly S.	S.	?
Pterins	I.	S. (methanol)	S.	S. (pyridine)	Unstable	S.	Low S. y; conjugates with protein S.	S.	Unstable
Isoalloxazines	Ribitylflavin I. Fairly S. some S. in $CHCl_3$	Unstable		S. (aq. pyridine)	RF unstable	S.	S.	S.	Ribitylflavin unstable

S = soluble
I = insoluble

phors is still uncertain or unknown. In addition there may be representatives of further chemical classes among the many animal pigments not yet analysed. On the other hand some of those represented are interrelated, for instance the porphyrins and bilins and the purines, pterins and isoalloxazines. Plants have certain other classes and subclasses, as well as a greater variety within some classes; this variety has its parallel in other biochemical fields such as the alkaloids, the essential oils, etc. In general biochemical austerity is more evident in animals than in plants and microorganisms, though in one respect biochromes present an exception: there are more classes in animals than in plants. Austerity seems to be greatest in materials with a fundamentally important function and this criterion helps to decide the relative importance of the various classes of zoochrome.

Conjugated double bond systems (p. 12) are the most conspicuous common feature of the molecules of all known zoochromes (Fig. 3.1) and except in the carotenoids these are mainly in the form of condensed ring-systems. The abundance of auxochrome groups based on oxygen and nitrogen also is very evident; both elements form strongly polar groups and so increase water-solubility as well as the chromatic properties. This is important since most metabolic processes take place mainly in aqueous medium; some occur in essentially lipid phases, however, and it is useful to look at the solubility-properties of zoochromes in more detail (Table 3.1).

This shows that apart from the melanins which are very insoluble in all physiological and most other media none are completely insoluble in aqueous media, though the carotenoids, quinones and porphyrins are primarily lipid-soluble; they acquire water-solubility by the use of polar side-chains, often rather numerous, and/or by conjugation with highly water-soluble proteins and glycose derivatives (p. 39). The indigotins are exceptional in not forming water-soluble coloured derivatives but they are scarcely zoochromes in the strict sense (p. 59). Each class tends to have some members soluble in lipids, some in aqueous media and many in media of intermediate polarity (Table 3.1).

It has not been possible to represent on this table all the structural variants which account for the wide spread of solubility-properties within each class but these are important technically as well as theoretically since they provide the basis for extraction and separation of the subclasses. Examples are the various subclasses of carotenoids and porphyrins. Solubility in strong acids and bases has not always been recorded in the table: it usually resembles that at the less extreme pH values but is often complicated by overt chemical reaction with the solute. The metalloproteins and other simple chromoproteins have the usual solubility properties of globular proteins; they are soluble in most aqueous media except in the region of their isoelectric point and insoluble in lipid media.

Zoochromes may be classified first on their relative lipid/water solubility and secondly on the dependent criterion of the abundance of O and N groups in the molecule. Only a few of the carotenoids lack any oxygen so that nitrogen is the more useful for this purpose. There are more nitrogenous than non-nitrogenous zoochromes and this is in contrast to plant chromes: plants economise N much more than animals do.

The molecular structure and chemical properties of the individual classes need some further treatment, in order to understand the basis for their biological functions. The account will be restricted to essential features since the subject has been a centre of interest in a number of previous works (p. VII). *In vivo* many members of most classes are conjugated with protein or other components and therefore the general effects of conjugation will be considered first.

3.1 Chromoproteins and Other Conjugates

In addition to proteins and carbohydrates lipids and occasionally nucleic acids may form chromoconjugates; moreover a chrome may be bound to more than one of these types. Chlorophyll has a lipoprotein conjugant and so have some carotenoids. In the bile itself bilirubin is conjugated not only with lipoprotein but also with cholesterol, glucuronic acid and perhaps other components (GRAY, 1961, p. 47). Occasionally oligopeptides and other N-compounds of low molecular weight are involved though in general a macromolecular protein is essential (SCHEJTER, 1966). Quite often metal ions are components of the complex.

All those biochromes that function as coenzymes (p. 182) are conjugated with protein and the holoenzyme is the functional unit. This is equally true for the oxygen-transporting chromes of the body-fluids and tissues (p. 211). In the simple iron and copper proteins the protein provides the ligands for the metal and so is an essential part of the functional chromophor. Haemovanadin (p. 73) is usually protein-bound, and carotenoids very commonly in animals (CHEESMAN et al., 1967) though not in plants. A number of metal-free porphyrins are conjugated with protein, for instance in the body-wall of the earthworm (FOX and VEVERS, 1960, p. 111) and so are integumental bilins (JUNGE, 1941; LEMBERG and LEGGE, 1949; GRAY, 1961; PASSAMA-VUILLAUME, 1965). Melanins probably are always bonded to protein even where their primary function is not the cross-linking of peptide chains. The structure of indolequinones (p. 61) is such that they can scarcely approach proteins without bonding with them (HACKMAN, 1964) and this is true also for phenolquinones, as in the tanning of leather. There is some indication that the echinochromes are usually complexed with protein (THOMSON, 1957; STERN, 1938), sometimes along with a third component. The ommochromes also are commonly bound to protein (CASPARI, 1949) and so are integumental ribitylflavin (FOX and VEVERS, 1960, p. 158) and pterins (EPHRUSSI and HEROULD, 1945).

In vivo the aphins are present as glycosides and the cochineal of coccids as a simple potassium salt (FOX, 1953, p. 205). In the shells of molluscs and of birds' eggs porphyrins are bonded to the mineral which, unlike protein, does not quench their fluorescence. In plants the flavonoids are nearly always in glycoside form; this is true for the flavonoid of the wings of the Marbled White butterfly (THOMSOM, 1964) and presumably for the water-soluble forms in *Helix* and the sertulariids (p. 82). It is clear that protein is the most common conjugant of zoochromes.

Conjugation is usually stoichiometric, involving specific chemical bonds such as the $4,4'$-keto groups of astaxanthin (CHEESMAN et al., 1967). In the echinochromoproteins the β-OH group of the quinones forms a salt link with basic amino acid groups of the protein, probably of arginine or histidine (THOMSON, 1957, p. 130). In some haemoproteins, as in the simple metalloproteins, bonding is directly and solely with the metal but in others there are also bonds with side-chains of the porphyrin. These various bonds do not represent all the interactions between protein and chrome; the whole primary and tertiary structure of the protein tends to be specific to the particular chrome and only protohaem bonds with the globins of the haemoglobins. Conjugation is usually spontaneous and likewise re-association after separation. The carotenoid and protein of ovoverdin will recombine after careful dissociation (STERN and SALOMON, 1937, 1938). Rhodopsin (p. 163) is a special caroteno-protein which normally dissociates and reconstitutes in the course of its normal cycle of physiological activity; the bond is sensitive to the photo-induced changes in the chrome. Denaturing agents break the bond in most chromoproteins and are used in extraction techniques.

In some cases, however, the bond is very strong and haemocyanin can be denatured without releasing the copper: the affinity of copper for peptides is so great that chemists are surprised that it will accept any other ligand, even O_2, in the presence of protein. The quinone bonds of the melanins also are very strong (HACKMAN, 1964). Metal cofactors usually affect the bonding and this may be one function of such cofactors in enzyme activity. Although a protein-precipitant, ammonium sulphate protects the bonds in ovoverdin (STERN and SALOMON, 1937, 1938) and the alkali metal salts have a weaker action of the same kind; this salting-out phenomenon is reversible.

Strong bonds stabilise the chrome and the whole conjugate; as in many other biological complexes the stabilisation is mutual between the two components (ZAGALSKY et al., 1967). The increased stability is evident particularly towards mechanical (p. 158) and chemical agents and towards light and ionising radiation (p. 156). Most conjugates are more stable than the free components but the converse is true of a number of chromoconjugates, including the green biliproteins of orthopteran insects (PASSAMA-VUILLAUME, 1965). Deoxyhaemocyanin is more sensitive to chemical and other oxidants (SCHUBERT and WESTFALL, 1962) than the free components, and the corrin coenzymes (p. 55) are very sensitive to photodecomposition whereas the free corrins are not (STADTMAN, 1971). Instability is generally proportional to reactivity and is an unavoidable price to pay for the latter. Unlike deoxyHcy the oxygenated form is more stable than the free conjugants so that the whole system favours O_2-loading.

In some instances the biological emphasis is on stabilisation of protein by chrome. This is particularly evident in the integument of animals as PRYOR (1940) first showed for the insect exoskeletal proteins. Dark mammalian hair, with a high melanin/protein ratio, is stronger than blond and stretches less easily (FOX and VEVERS, 1960, p. 30). In the fibres of crimpy (i.e. strongly coiled) wool there is most melanin on the side of maximal curvature of the fibre. Carotenoproteins spread more slowly at interfaces than the free protein (CHEESMAN et al., 1967). Haem tans protein if it is injected under a film of the latter spread on water (LEMBERG and LEGGE, 1949, p. 71). This may be related to the action of haematoporphyrin in depressing the mutual adhesiveness and aggregation of the blood platelets of the rabbit (DAVIS and SCHWARTZ, 1967): it is thought to purloin the sites concerned in normal cell-adhesiveness, and may involve the normal haem-histidine bonds of the haemoproteins (p. 229). In tanning presumably each porphyrin molecule bonds with two or more peptide chains simultaneously. In view of these effects it is not surprising to find indications that chromoproteins play a part in controlling the structure and properties of biological membranes (MANWELL, 1964; CHEESMAN et al., 1967). Planar molecules such as the porphyrins have varied potentialities for membrane-formation (LEMBERG and LEGGE, 1949, p. 172); protoferrihaem, with its vinyl groups as side-chains on one side and polar, acid groups on the other, can pack closely with the molecular plane normal to the surface of the membrane.

There is usually a close correlation between mechanical toughness and insolubility and this is evident in melanoproteins and other exoskeletal proteins. The insoluble chromoproteins of the shell of pulmonates contrast strongly with the easily extracted chromes of the lower gastropods. It is mostly fibrous proteins that are made tougher and more insoluble in this way. At the other end of the spectrum the most soluble globular proteins form water-soluble conjugates even with lipid-soluble chromes and this is put to biological use.

Examples of heat-stabilisation by conjugation include the ovorubin of the eggs of the snail *Pomacea*, haemoglobin as compared with the free globin and rhodopsin as compared with the free opsin (HUBBARD, 1958). The property is rather general in protein-conjugates and is not restricted to chromoproteins. Stability over a wider pH-range is shown in the echinochromoproteins. Increased resistance to proteolytic enzymes is conferred on proteins by ligation with copper (LONTIE and WITTERS, 1966).

Among other properties affected by conjugation are the fundamental optical parameters (p. 5). There is usually a bathochromic shift in the absorption-spectrum of the chrome, so that it is excited more easily; the extent of the shift depends on the particular protein and this has been widely exploited in carotenoproteins. Whereas free carotenoids are nearly all yellow to orange-red the carotenoproteins (Table 3.2) have colours ranging from yellow through orange, red, purple, violet, blue, green, brown, black and even virtually colourless

Table 3.2. Colours and composition of carotenoproteins

Colour	Source	Nature of prosthetic chromes	Nature of the protein
Yellow	Integument of *Asterias*	Astaxanthin or its ester	
	Integument of locust	Astaxanthin	
	Plasma of mammals	β-carotene	
	Egg-yolk of birds	Lutein, zeaxanthin	
Orange	Integument of *Solaster*	Xanthophylls	
	Exoskeleton of *Carcinus*	? Astaxanthin	
Orange-red	Carapace of *Nephrops*	Astaxanthin	
	Blood and eggs of *Idya*	?	Lipoprotein
	Ovary of *Cancer*	Astaxanthin and other xanthophylls	Glycolipo--protein
	Ovaries of *Pecten*	Xanthophylls	Glycolipoprotein
Pink-red	Tissues of *Eylais*	Astaxanthin	
	Ova (ovorubin) of *Pomecea*	Astaxanthin	Glycoprotein
Red	Tissues of *Actinia* (actinioerythrin)	2,2'-bis norastaxanthin di-ester	
	Exoskeleton of Crustacea	Astaxanthin	
	Integument and eggs of *Palinurus*	Astaxanthin and others	Lipoprotein
	Gonad of *Volsella*	Xanthophylls	
	Rhodopsins in general	Retinal I	Opsins
Red-purple	Carapace of *Eriphia*	Astaxanthin	
Purple	Integument of *Crossaster*	Astaxanthin	
	Integument of *Pisaster*	Metridene (xanthophyll)	
	Integument of *Henricia* and *Porania*	Xanthophylls	
	Hind wing of *Oedipoda*	Astaxanthin	
	Ovaries of holothurians	?	Lipoprotein
	Eggs of *Eupagurus*	Astaxanthin esters	
	Porphyropsins of vertebrates	Retinal II	Opsins
Violet	Parasitic Ciliates	?	
	Velum of *Velella*	Astaxanthin	
	Integument of *Rhizostoma* (cyanein)	?	
	Integument of *Fiona*	Astaxanthin	
	Integument of *Asterias*	? Astaxanthin or its ester	
	Blood of *Tanymastix*	? Astaxanthin	
	Iodopsins of vertebrates	Retinal I	Opsins

Table 3.2 (continued)

Colour	Source	Nature of prosthetic chromes	Nature of the protein
Blue	Tentacles of *Porpita*	?	
	Integument of *Diaptomus* and *Heterocope*	Astaxanthin	
	Exoskeleton of *Hippolyte* and *Homarus* (crustacyanin)	Alkali salt of enol form of astaxanthin	Simple protein
	Hind wings of *Oedipoda*	Astaxanthin	
	Blood of *Chirocephalus* and *Tanymastix*	Canthaxanthin	Glycolipoprotein
	Gonophores of *Clava*	Xanthophyll	
	Eggs of *Lepas*, *Daphnia*	? Astaxanthin	? Euglobulin
	Cyanopsins of vertebrates	Retinal II	Opsins
Blue-green	Integument of *Dixippus*	β-carotene and bilin	Albumin
	Blood of *Dixippus*	β-carotene and bilin	Globulin
	Mantle of *Cerithidia*	? Carotenoid acid ester	
Green	Tissues of *Actinia*	Xanthophyll ester	
	Integument of *Asterina* and *Marthasterias*	?	
	Exoskeleton, etc. of *Idotea*	Canthaxanthin, lutein, xanthophylls	Lipoprotein
	Carapace and eggs of *Astacus astacus*	Astaxanthin	Albumin
	Carapace of *Carcinus*	Astaxanthin	
	Wings of locust	β-carotene	
	Integument of *Tettigonia*	Lutein, bilin	Albumin
	Integument of *Sphinx*	Lutein, bilin	Globulin
	Blood of Lepidoptera	β-carotene, lutein, meso-biliverdin (or anthocyanin)	
	Blood of *Plusia*	Lutein- lutein-epoxide, bilin	
	Eggs of *Homarus* (ovoverdin)	Astaxanthin	Glycolipoprotein
Green-brown	Integument of *Gammarus*	Astaxanthin	
Brown	Tissues of *Ficulina*	?	
	Various tissues of *Simocephalus*	? Diketocarotenoid	
Green-black	Eggs of *Astacus astacus*	Astaxanthin	
Black	Eggs of *Cyclops*	? Astaxanthin	
	Exoskeleton of *Homarus*	Astaxanthin	
Colourless	Integument of prawn	? Astaxanthin	

(CHEESMAN et al., 1967). In ternary carotenoid complexes there is no change in absorption unless the carotenoid is bonded directly to the protein moiety; presumably proteins have better auxochrome groups than lipids and carbohydrates.

Many examples are known of proteins quenching the fluorescence of their conjugant chromes and so putting the energy to use within the system (p. 17): the outstanding instances are porphyrins, aphins, flavins and pterins (FOX and VEVERS, 1960, pp. 111, 158). Again there are some paradoxical cases: protein conjugation actually induces fluorescence in the plant biladienes, phycoerythrin and phycocyanin (LEMBERG and LEGGE, 1949, p. 145) which are used by some animals (p. 88).

The protein has a considerable effect on the redox potential of such chromes as ribitylflavin; that of the free chrome is −185 mV whereas those of the flavoproteins can be −60 mV or higher. Collectively therefore the latter catalyse oxido-reductions over a wide redox range. In the O_2-transport chromoproteins the protein determines the essential property:

free haem oxidises to ferrihaem and is unable to bind O_2. In these chemically active chromoproteins the two components interact in the course of the action-cycle, with complex changes in configuration of the whole molecule (p. 230). Essentially the same is true for the sensory carotenoprotein, rhodopsin (WALD, 1959).

Conjugation with carbohydrate or with lipid alone is much more common in plants than in animals, presumably for reasons of N-economy once more. An implication is that these conjugants can deputise for protein, at least for some purposes. Aphins are stabilised and rendered soluble in the blood by their glycose conjugant and flower colours are protected from fading by glycosidation (HARBORNE, 1964). This is not a metabolically useful type of conjugation: from the evidence of cockroach ootheca-hardening system (BRUNET and KENT, 1955) it is a very effective mode of inactivation. There is often a third component in chromoglycosides (MAYER and COOK, 1943, p. 212). Mucoproteins have outstanding adsorbant properties and those of the branchial basket of ascidians take up vanadium (GOLDBERG et al., 1951); the implication is that conjugation probably plays a part in the initial uptake of the chromogenic metals.

In conclusion, conjugation affects the properties of biochromes in many important respects, whether increasing their solubility, lability and reactivity or stabilising and toughening them. It also affects their behaviour in other, more specific ways that will be considered in context.

3.2 Carotenoids

Because of their solubility-properties this class of biochrome was once called 'lipochromes' but this less specific name also is too easily confused with 'chromolipid', once used for lipid-soluble fuscins (p. 35). Carotenoids (Fig. 3.2) are basically C_{40} open chain hydrocarbons with almost complete conjugate double bonding. From this plan some members have radiated in one or more ways: (1) by closure of one or both ends of the chain to form *ionone* rings, (2) by oxidation of one or more substituents on these rings, (3) by shortening of the chain and (4) by special changes in certain substituents and bonds. They are biogenetically related (Fig. 13.1) to a large number of other fat-soluble ethylenic and more saturated hydrocarbons. The chain is usually in the all-*trans* configuration shown for β-carotene (Fig. 3.2, II), giving maximal dipole moment (p. 14) and maximal excitability. Further the ionone rings are coplanar with each other and with the plane formed by the ziz-zag conformation of the main chain and its side-chains, thus giving maximal resonance and related properties (p. 15). The structure of this molecule makes it an ideal donor and acceptor of electrons, i.e. an ideal reversible redox agent (PULLMAN and PULLMAN, 1963).

Animals obtain their carotenoids from plants, directly or indirectly, but they subsequently modify them considerably. They commonly retain carotenes (Fig. 3.2, I, II), the purely hydrocarbon members (FOX, 1953; GOODWIN, 1950a, 1952, 1962a; VUILLAUME, 1969) but especially in the integument there is a preponderance of the partially oxidised subclasses (III–XI) sometimes collectively termed xanthophylls, after the original name for the simplest of them, 3,3′-dihydroxy-α-carotene (III). Since this is now called lutein the generic use of 'xanthophyll' has some merit though it is less informative than *oxycarotenoid* and is best used in the more restricted sense of carotenol (FOX, 1953, p. 64). Within the oxycarotenes the use of informative names for subclasses is now usual, i.e. caroten-ol (III), al, one (IV), olone (alcoholic ketone, V), ol-ester, oxylic acid (VII), oxylic ester, etc. (Fig. 3.2). There are a number of rarer derivatives with epoxide groups (IX), acetylenic bonds (X), allenic bonds (i.e. two double bonds in succession, VIII) and also nor-carotenoids (XI) with the ionone ring contracted to a pentacyclic form (VUILLAUME, 1969; LIAAEN-JENSEN, 1970). Derivatives with the ionone ring fully aromatic have been found in the sponge *Reniera* and conversely there are members with the rings fully saturated or even

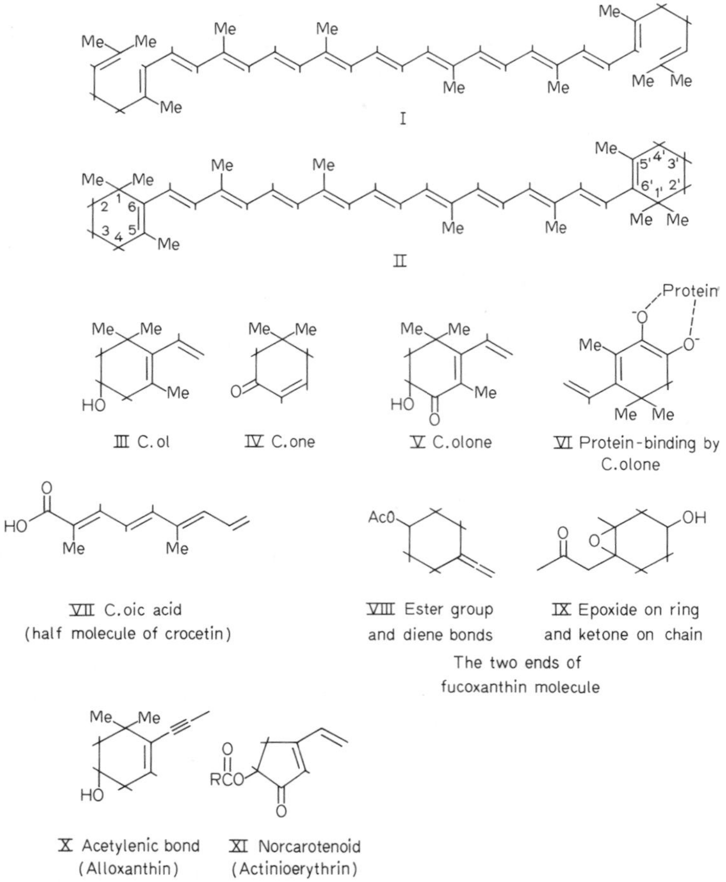

Fig. 3.2. Structural formulae of molecules or parts of molecules of representative carotenoids. C.ol, etc., = carotenol, etc.

reopened (VUILLAUME, 1969). Secondary shortening of the chain also is quite common (GROB, 1960); retinol (vitamin A) is formed from carotene by binary fission, a reaction possibly peculiar to animals. These are all derivatives of the one parent C_{40} compound (p. 281).

Carotenoproteins also are more common in animals than in plants. They are mainly integumental chromes but the photoperceptor retinopsins (p. 163) and probably some enzymes (CHEESMAN et al., 1967) are the exceptions; in addition some carotenoids are transported in this soluble form. The prosthetic group of most carotenoproteins other than the retinopsins is either canthaxanthin (4,4'-diketo-β-carotene) or astaxanthin, its 3,3'-diHO derivative; this austerity contrasts with the variety of free carotenoids. Glycoproteins as well as lipoproteins have been found among the conjugants and in the latter the carotenoid usually bonds to the lipid component, non-stoichiometrically.

3.2.1 Properties

The range of colours in the carotenoproteins (p. 31) rivals that in any class of zoochrome; the proof that this depends entirely on the protein is the common carrot colour of all when denatured by any method. Examples of the absorption-spectra of carotenoids are given by GOODWIN (1952) and VUILLAUME (1969) while more comprehensive tables are available in KARRER and JUCKER (1950), International Critical Tables (1929) and later compendia. Something is known of the relationship between spectral and structural features; the carotenes and neutral oxycarotenoids commonly have a group of three peaks, or two and a shoulder, in the blue to blue-green region of the visible range, the central peak being the highest. By contrast carotenoic acids have a single broad peak in the same general region. The shift of one double bond in the ionone ring, converting β to α carotene removes one of the three peaks. Carotenes have a weak yellow-green u.v.-fluorescence and that of some of the carotenoic acids is blue; the interval between the positions of the u.v. and visible absorptions therefore is not uniform as it is in some classes of zoochrome (p. 69).

The lipid-solubility of free carotenoids is manifested by the frequency with which they colour the triglyceride stores of animals. The carotenes are soluble in paraffins and other apolar lipids which can be used to separate them from the oxycarotenes. Their solubility extends to lipids as polar as acetone and the lower alcohols so that they can be extracted along with the oxycarotenes and then be partitioned off with petrol. The oxycarotenes can be further fractionated according to their polarities by adding water in controlled ratios to the polar phase (Fox, 1953). Fucoxanthin (Fig. 3.2, VIII. IX) remains soluble even in 70% methanol/ 30% water. The oxides and esters may be recovered from petrol by absolute MeOH, giving a useful second step purification. Few classes of biochrome are so amenable to fractionation by partition methods and therefore it is not surprising that they were the subject of Tswet's pioneer studies on chromatographic separation. The range of lipid/water solubilities among carotenoids may have functional significance *in vivo*.

Carotenoids are very sensitive to further oxidation which usually bleaches them; because of this sensitivity they can act as antioxidant 'buffers' for other materials. They are stable to alkalis but sensitive to acids. With concentrated H_2SO_4 they form a transient blue derivative which provides a diagnostic test. The pterins, which are often associated with the carotenoids in animal integuments, give a similar colour-reaction (GUNDER, 1954) but are readily distinguished by solubility and other properties. The $SbCl_3$-$CHCl_3$ test for vitamin A is a variant.

3.3 Fuscins

These are chemically unidentified brown to black chromes, mostly distributed too diffusely and sparingly to be extracted in quantity for chemical identification. From histochemical study it is evident that more than one type of material occurs in this way; the degree of heterogeneity is uncertain but evidently the term fuscin does not represent a discrete chemical class. Some fuscins are soluble in lipids but it seems necessary to recognise aquofuscins as well as lipofuscins and probably each is still a heterogeneous assemblage. Some fuscins are insoluble in both ethanol and xylene, as well as in aqueous media so that at least a third sub-group must be recognised.

In spite of the difficulties VERNE (1926) was able to define a set of properties (Table

3.3) which may refer to a specific chemical class, though it does show some variation in its response to oxidising agents and in other properties. Sometimes the properties demonstrated have been those of a lipid in which the actual fuscin was dissolved. The fuscin of leech botryoidal tissue, often regarded as the most typical invertebrate fuscin, appears to be mainly protoferrihaem (NEEDHAM, 1966b) and that of 'brown fat' in mammals likewise

Table 3.3. Comparison of properties of lipofuscins with those of carotenoids and melanins

Property	Fuscins	Carotenoids	Melanins
Colour	Yellow, orange, brown or black	Yellow, orange, red (Carotenoproteins also purple, violet, blue, green, brown, black, colourless)	Yellow, red, brown, black
U. v. fluorescence	Blue, green, yellow, orange or none (rarely strong)	Blue, green, yellow or orange (rarely strong)	Yellow or none (weak)
Solubility	In lipids only or insoluble	Free carotenoids in lipid only	In a few lipids, and in strong alkalis. Erythromelanins in acid
Conjugation	Not bound to protein	Some bind with protein	All are bound to protein
Stability to dilute acid	High	Low	High
Effect of oxidation on colour	Usually darken but H_2O_2 bleaches them	Bleach	Partially bleach
Action on ammoniacal $AgNO_3$	Do not reduce	–	Do reduce
Reaction with I_2	None: no unsaturated bonds	Form addition compounds: unsaturated bonds	None: no unsaturated bonds
Reaction with $c\,H_2SO_4$	Red solution, stable	Blue solution, transient	No significant change
Reaction with nile-blue in $c\,H_2SO_4$	Blue product, soluble in acetone	–	Green product, insoluble in acetone

appears to be a haem compound (PRUSINER et al., 1970). Some fuscins probably are melanins (VERNE, 1926; BARDEN and MARTIN, 1970) and it is noteworthy that melanins are rather soluble in some lipids. In the mammalian corpus luteum carotenoids may be the origin of fuscous material. The 'fuscin' of the hydroid *Sertularella* may be a flavone (PAYNE, 1931) and that of some nerve cells a pterin (HYDEN and LINDSTRÖM, 1970). If most of these identifications are correct then any idea of fuscins as a discrete chemical class is very suspect.

Most other fuscous materials not only are chemically unidentified but have not even been clearly placed in their solubility group, sometimes perhaps through a desire to believe that all are 'lipofuscins'. VERNE (1926) regarded the dark granules in the cells of the sudorific glands of mammals as lipofuscins but they are freely soluble in the aqueous sweat which can be coloured yellow, orange, red, brown or black, the full range of fuscin colours, and on occasion even green or blue (Fox and Vevers, 1960, p. 165). The dark chrome of the 'wax' of the human external auditory meatus also originates from sudorific glands (Fox and Vevers, 1960) and I find that it likewise is very water-soluble and quite insoluble in all neutral organic media tested, even including methanol; it is soluble in organic acids and bases, however. The colourless component of the 'wax' seems to be mainly scarf-skin, insoluble in the solvents used; it does disperse in water, however, so that the clinical rationale of using oils to remove this material seems questionable. The 'cerochrome' gives a bright yellow solution but its spectrum shows a monotonic decrease in absorption with wavelength. It has some of Verne's fuscin properties.

The 'uranidines' of Krukenberg (1882) also may be regarded tentatively as aquofuscins. They are yellow and water-soluble though unlike cerochrome some are also soluble in neutral polar organic solvents such as ethanol. In solution they have the further fuscin property of depositing completely insoluble dark material on standing. Solutions of urochrome, on which their name was based, behave similarly. There is reason to believe (Rimington and Kennedy, 1962) that the yellow uranidine of holothurians contains ribitylflavin, i.e. the original pigment was a mixture of this with melanogenic material.

Fuscous material has been found in cells of most tissues of mammals, liver, kidney, nerve cells, ependyma, cardiac and smooth muscle, adventitial tissue of blood-vessels, adrenal, interstitial cells of testis and ovary, seminal vesicle, epididymis, lachrymal gland, hypophysis and conjunctival tissue of the nipple. It is not a trivial rarity and merits persevering study. It has been recorded much less frequently in lower vertebrates and invertebrates but this is partly because studies have been fewer. It may be noted that the only integumental site in mammals is the nipple and in invertebrates the wings of some butterflies (Prenant, 1913). In some tissues fuscous material increases progressively throughout life while in endocrine organs it follows a regular short-term cycle (Verne, 1926, p. 46). A physiological significance is implied both here and in nerve cells, where it is depleted by prolonged neural activity. The amount normally present is related to the concentration of redox enzymes (Friede, 1962). In the adrenal, gonads and muscle the amount is inversely related to the content of vitamin E (Deuel, 1957; Weglicki et al., 1968) which is an antioxidant (p. 197). Some lipofuscins at least may be oxidation-products of unsaturated glycerides (Porta and Hartroft, 1969). These are additional motives for studying their chemistry.

3.4 Chroman Biochromes

Plant chromatologists naturally regard the flavonoids as the most important class in this group but relatively few of them are found in animals (p. 82) where also their function is uncertain. The simpler chroman derivatives such as the tocopherols (Fig. 3.3, I), which are vitamins for mammals, therefore deserve prior attention. They are pale yellow oils (Finar, 1964, p. 619) and are obtained from cereal germ oils and other plant materials. There is a long polyisopentenyl side-chain on the ring system, as also in a number of other biochromes, ubiquinone, vitamin K, chlorophyll and cytochrome a; this chain does not have a conjugated double bond system like that of the carotenoids and does not contribute to the chromatic properties.

The tocopherols have redox properties and oxidise to benzoquinones (Pullman and Pullman, 1963, p. 493), i.e. the pyran ring opens reversibly. It is then evident that the tocopheroquinones are closely related to ubiquinone (p. 40) and that the pyran ring of these chromans is formed from the proximal portion of the side chain of UQ; they are probably UQ derivatives and the two classes often occur together in animal tissues.

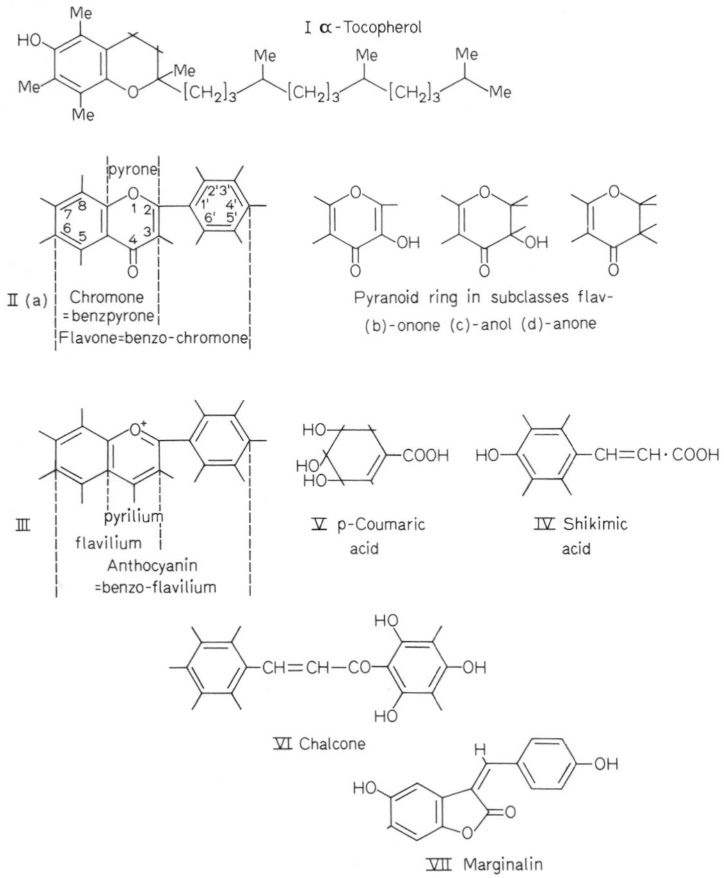

Fig. 3.3. Structural formulae of molecules or parts of molecules of biological chromans and flavonoids, of related chromes and of biosynthetic precursors and intermediaries

3.4.1 Flavonoids

Structurally these are chain-linked benz-chroman derivatives. The structure has radiated rather widely to form several very large classes of plant chrome; in fact these are almost *the* flower chromes, notwithstanding carotenoids and other minor contributors. In one subclass, the flavanones (Fig. 3.3, II d) the pyran ring remains in the same di-H (mono-unsaturated) state as in the tocopherols but in most it has two double bonds. In the main subclasses, the flavones or anthoxanthins (II a) and the anthocyanins (III) it has differentiated in two important ways. In the former it has a ketonic side-chain and so becomes a pyrone ring, and the whole molecule a benzochromone (II); in such a ring the keto group will have quinonoid properties (p. 44). In the anthocyanins (III) the ring-oxygen is in the $\equiv O^+$ oxonium state and forms salts with chloride and other anions. The ring therefore becomes pyrilium and the condensed pair benzopyrilium or flavylium; the complete molecule is therefore a benzoflavylium. The extra double bond associated with the oxonium ion has a strong bathochromic effect so that, as the names imply, the anthoxanthins are mostly yellow but the anthocyanins often blue but also purple and red. The rings are usually heavily hydroxylated, further enlarging the resonance system and increasing the bathochromicity. In the isoflavone subclass the isolated benzene ring is bound at position 3 instead of at 2 on the pyrone ring. Haematein, source of the dye, haematoxylin, and not to be confused with haematin (p. 241) is an anthocyanin derivative which has its lone benzene ring doubly linked to the chroman moiety at the 3 and 4 positions, so adding another ring to the system. The glycose with which plant flavonoids are usually conjugated (p. 29) is mostly glucose but occasionally galactose or rhamnose. The glycosides are termed anthocyans, etc.

3.4.1.1 Properties

Apart from their strong absorption in the visible range the flavonoids have peaks in the u.v., around 257–272 and 320–355 nm, with a shoulder in the region of 279–295 nm. The flavone of the snail *Helix* has a bright yellow u.v.-fluorescence in the same range as its visible colour (p. 17). Unlike that of ribitylflavin (p. 72) it is not quenched by acid but becomes greener. The hydrophilic side-chains and relatively high molecular weights give flavonoids high melting points.

Because of the polar groups and the glycosidation most flavonoids are highly water-soluble and are usually in the cell-sap (BENTLEY, 1960). They are also soluble in acids and alkalis and in polar organic media such as alcohols and acetone; ethyl acetate is a convenient solvent for separating them from the pterins (p. 68) which otherwise they resemble in solubility and some other respects. They are insoluble in apolar lipids.

Although primarily flower colours most flavonoids have redox properties and show how fundamental these are in all chromes. The flavone of *Helix* is reversibly decolorised by the reducing agent dithionite and calcium salts of flavonoids are bleached also by oxidation, whether by O_2 or by light, unless protected by glycosidation.

The salts formed with metals by the phenolic OH groups of the molecule are relatively easily crystallized from acid solution (PAYNE, 1931) and the Ca-salts are still more insoluble. They differ in colour from the free acid and each metal has its characteristic hue (PAYNE, 1931), though a number of them turn hydrangeas blue (p. 14). By virtue of the oxonium ion the anthocyanins form salts also with anions and this is the basis for the strong pH colour-change of familiar members such as litmus; in this case the phenolate anion is blue and the oxonium cation red. The anthocyanin of the weevil *Cionus* has the same pH colour-change. Flavones become colourless in acids which suppress the ionisation of the phenolate groups; they give a characteristic deeper yellow colour with ammonia and other alkalis, such as boric acid. A useful test based on this involves a mixture of boric and acetic acids in anhydrous acetone (MAGER, 1938; WILSON, 1939). The anthocyanins give a characteristic green colour-reaction with NH_4OH. In addition to the use of haematein as a histological dye the yellow luteolin of the plant, *Reseda luteola,* has been used commercially: it was used by the Gauls in Roman times (FOX, 1953, p. 211).

A yellow chrome, *marginalin* (Fig. 3.3, VII), in the pygidial gland of the large water-beetle *Dytiscus* is an aurone (GEISSMAN, 1967), a fairly close analogue of the flavonoids, with a furone instead of the pyrone ring, and therefore it is significant that there is a – CH = group in the link-chain (SCHILDKNECHT, 1971). Aurones occur also in other kingdoms of living organisms.

3.5 Ternary Quinones

Quinones are aromatic diketones (Fig. 3.4, I) usually, but not always (VI) with the two keto groups on the same ring. Ternary quinones, the best known and so usually called 'quinones' without qualification, are certainly typical but it is striking how many biochromes are effecitvely quinonoid, at least in their active state. There are four main subclasses of ternary quinones in animals, benzoquinones (Fig. 3.4, I, II), naphthoquinones (III), anthraquinones (IV) and more highly polycyclic quinones (V, VI, XIV).

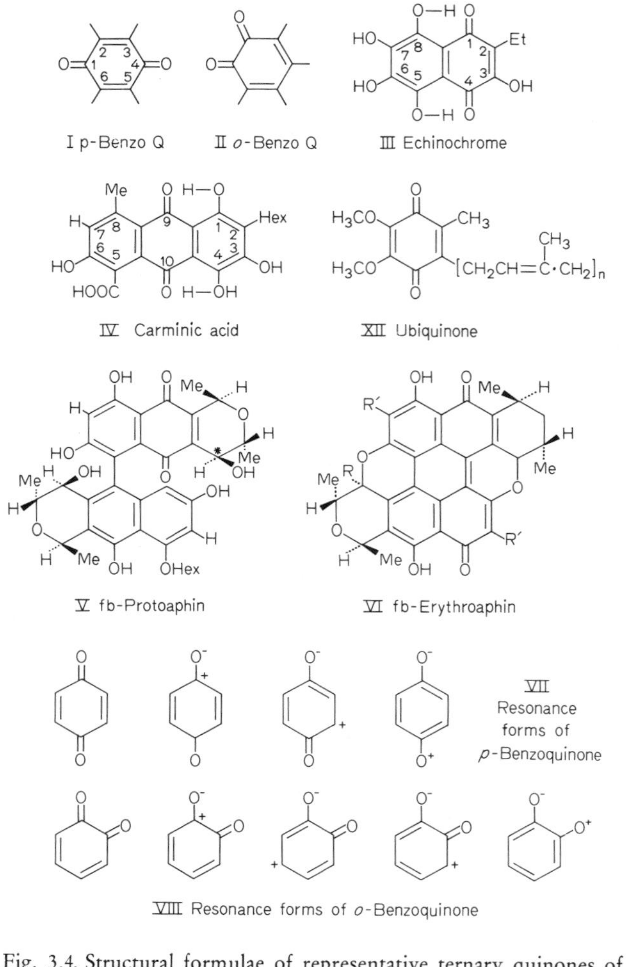

Fig. 3.4. Structural formulae of representative ternary quinones of animals and of resonant forms and intermediaries in biosynthesis and chemical reactions of quinones. In XVI the black disk symbolises the carbonyl carbon of an acetate unit, the methyl group being at the other end of the bond represented as long

Three groups of benzoquinones (BQ) occur in animals; the first is a yellow p-BQ, the respiratory coenzyme Q or ubiquinone (UQ) (Fig. 3.4, XII). Its polyisopentenyl sidechain has a variable number of units, 8 to 10, in the different animal phyla. In the nematode worm *Ascaris* there is a variant, rhodoquinone, with an NH_2 group in place of one of the MeO groups of UQ (OZAWA et al., 1969); this gives it its more bathochromic colour. UQ readily ring-closes the proximal part of the prenyl side-chain to form a chromone (p. 37). The second group, also *p*-BQs, are used for chemical defence (p. 158), also mainly in arthropods. The most common are *p*-BQ itself and its 2-methyl and 2-ethyl derivatives; they are formed at the moment of use from precursor quinols by the action of a polyphenol

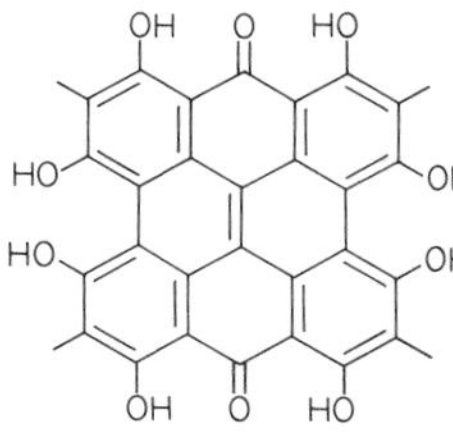

XIII Fringelite

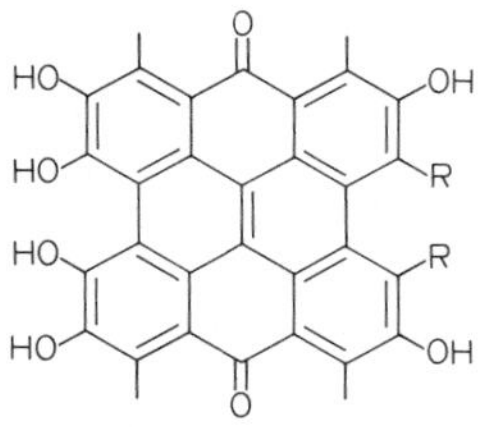

XIV Zoopurpurin

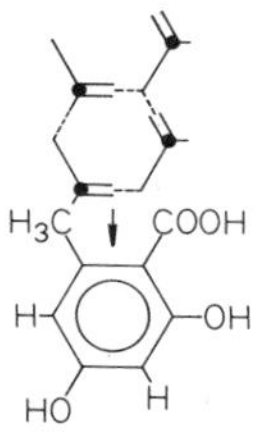

XVI Biosynthesis of orsellinic acid from acetate

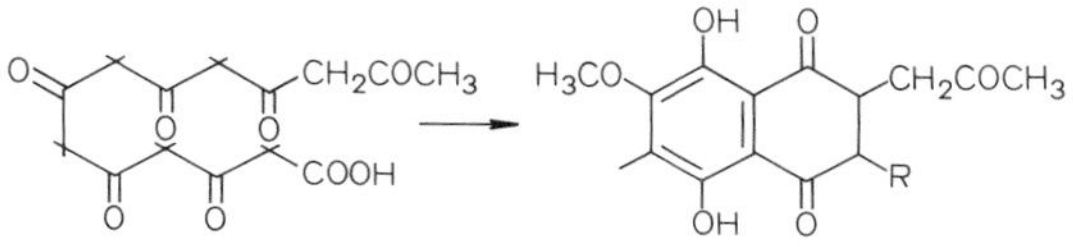

XVII Biosynthesis of NQs from acetate polymer

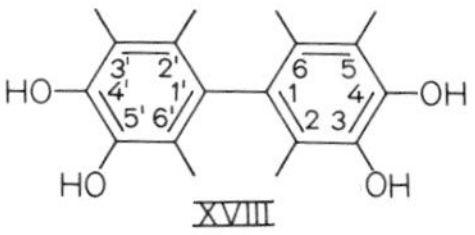

XVIII

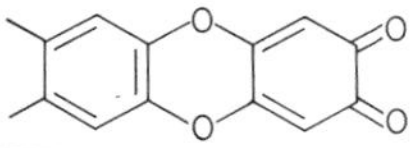

XIX

Dimers in path of phenolic melanin synthesis

IX   X 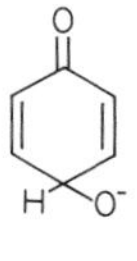   XI

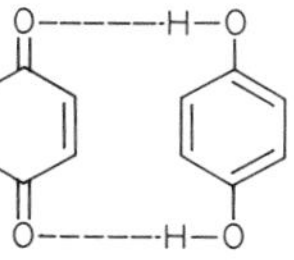

Intermediary radicals in redox cycle of *p*-BQ

p-benzoquinhydrone

Fig. 3.4

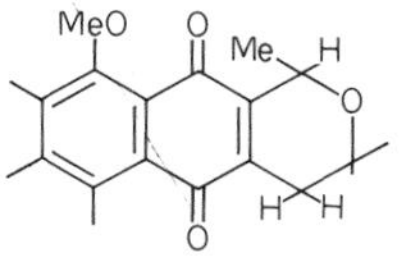

XV Isoeleutherin

oxidase. The third group are used as tanning agents, again primarily in arthropods but also more widely. Already (BRUNET, 1967) four quinol precursors have been identified and all are *o*-BQs, in contrast to the other two groups; they are 4-COOH-catechol (3,4-di-HO-benzoic acid or protocatechuic acid), 4 CH_2OH-catechol, dopamine (3,4-di HO phenyl-β-ethylamine) and N-acetyldopamine. Again the precursor is freed from an inactivating conjugant and oxidised to quinone by a polyphenol oxidase only at the moment of use; the outstanding instability of the *o*-BQs makes this mandatory.

Animal naphthoquinones (NQs) include the vitamin K group and the echinochromes, a large subclass almost entirely restricted to the echinoderms. They are richly and variously substituted on both rings (Table 3.4). It is unfortunate that the trivial name 'spinochrome' was applied to the second echinochrome studied since most of those subsequently discovered also have been called spinochromes, though their distribution is rarely confined to echinoid spines. Structurally all belong to a single subclass and in tribute to MacMunn (1883, 1885) his prior and more generally informative name *echinochrome* is to be preferred. If this were called 'echinochrome$_O$', the O also implying 'original' then the spinochromes could revert to 'echinochrome' with the suffices which by now have become familiar. In any case the precise chemical designation is not excessively cumbersome and is fully informative. MacMunn's echinochrome is 2-ethyl-3,5,6,7,8-pentaHO-1,4-NQ. There is least justification for adapting the trivial names of plant NQs to the purpose (Anderson et al., 1969) until there is good evidence (p. 285) that echinochromes are of dietary origin. In this connection it is noteworthy that MeO as a substituent is rare in animal products (Bergmann and Burke, 1956; Bergmann and Stempien, 1957) but occurs both in UQ (Fig. 3.4, XII) and in some of the echinochromes (of echinoderms other than the sea urchins).

Table 3.4. Substituent groups on echinoderm 1,4-naphthoquinones

Naphthoquinone / Substituent position	2	3	5	6	7	8
Echinochrome	–CH$_2$·CH$_3$	–OH	–OH	–OH	–OH	–OH
Spinochrome A	–CO·CH$_3$	–OH	–OH	–OH	–H	–OH
Spinochrome B	–OH	–OH	–OH	–H	–OH	–H
Spinochrome C	–CO·CH$_3$	–OH	–OH	–OH	–OH	–OH
Spinochrome D	–H	–OH	–OH	–OH	–OH	–OH
Spinochrome E	–OH	–OH	–OH	–OH	–OH	–OH
Namakochrome (*Polycheria*)	–OH	–OH	–OH	–OH	–OCH$_3$	–OH
Acanthaster NQ1	–OH	–OCH$_3$	–OH	–OH	–OCH$_3$	–OH
Acanthaster NQ2	–OH	–OCH$_3$	–OH	–OCH$_3$	–OH	–OH
Ophiocomina NQ1	–OH	–CH$_2$·CH$_3$	–OH	–H	–H	–OH
Ophiocomina NQ2	–OH	–CO·CH$_3$	–OH	–H	–H	–OH
Ophiocomina NQ3	–OH	–H	–OH	–CH$_2$·CH$_3$	–H	–OH
Ophiocomina NQ4	–OH	–CO·CH$_3$	–OH	–H	–OH	–OH
Ophiocomina NQ5	–OH	–CO·CH$_3$	–OH	–H	–OCH$_3$	–OH
Ophiocomina NQ6	–OH	–CH$_2$·CH$_3$	–OH	–H	–OH	–OH
Ophiocomina NQ7	–OH	–OH	–OH	–OH	–CH$_2$·CH$_3$	–OH

Until recently the few known anthraquinone (AQ) zoochromes, from Hemiptera and crinoids, were all 9, 10 derivatives (Table 3.5), i.e. benzene-disubstitued *p*-BQs. However, hallachrome from the polychaete *Halla* proves to be a 1, 2-AQ,- a naphthalene-substituted *o*-BQ. It also has fewer substituents on the rings than the carmines and the crinoid AQs (Table 3.5). The exact positions of the three substituents recognised (Prota et al., 1971) are still uncertain and the associated chrome, hallorange, has not yet been chemically identified. *In vivo* the hemipteran carmines all exist as the alkali salts of the COOH acid group at position 5. The crinoid AQs (Powell et al., 1967; Sutherland and Wells, 1967) have no substituents at 5 and only one has a COOH substituent at any position, so that all three subclasses of AQ zoochromes probably are of independent origin. Their tissue distribution also shows some differences, the carmines being essentially fat-body stores the others mainly integumental.

Those aphid polycyclic quinones (PQs) of which the structure has been determined are all closely related (CAMERON et al., 1964; WEISS and ALTLAND, 1965); all give the same main derivatives when chemically degraded (MACDONALD, 1954). Minor subclasses have been appropriately designated. They appear to be dimers of a pyran-condensed NQ (CAMERON et al., 1964) and it is interesting that such a NQ derivative occurs as a monomer in one echinoid (ANDERSON et al., 1969). Zoopurpurin (XIV) from the ciliate *Blepharisma*, however, is probably more closely related structurally to fringelite (XIII) from fossil crinoids and to the plant PQ hypericin; these are dianthrone derivatives, AQ-dimers. The ciliate PQs are essentially chromes of chromasome-like organelles (p. 118) whereas the aphins are blood-chromes.

The greenish yellow integumental chrome of the Polychaete *Arenicola marina (tinctoria!)* proves to be a polyhydroxy polycyclic quinone of another subclass, based on the benzpyrene nucleus (MORIMOTO et al., 1970). The sulphuric ester form, arenicochrome

Table 3.5. Substituent groups on animal anthraquinones (All have –OH at position 1 and the quinone group at 9 and 10)

Anthraquinone	2	3	4	5	6	7	8
Carminic acid	–CO–(CHOH)$_4$CH$_3$	–OH	–OH	–COOH	–OH	–H	–CH$_3$
Kermesic acid	–CO · CH$_3$	–OH	–OH	–COOH	–OH	–H	–CH$_3$
Laccaic acid	–OH	*	–OH	–COOH	–COOH	–OH	–H
Rhodocomatuline	–OH	–OH	–CO(CH$_2$)$_2$CH$_3$	–H	–OCH$_3$	–H	–OCH$_3$
6-Monomethyl rhodocomatuline	–H	–OH	–CO(CH$_2$)$_2$CH$_3$	–H	–OCH$_3$	–H	–OH
Rhodoptilometrine	–H	–CHOH · CH$_2$ · CH$_3$	–H	–H	–OH	–H	–OH
Isorhodoptilometrine	–H	–CH$_2$ · CHOH · CH$_3$	–H	–H	–OH	–H	–OH
Ptilometric acid	–COOH	–CH$_2$ · CH$_2$ · CH$_3$	–H	–H	–OH	–H	–OH

* HO–⟨⟩– CH$_2$ · CH$_2$ · HN · OC · CH$_3$

(KENNEDY, 1969), appears to be an artefact, the native chrome being sulphur-free. This oxidises to the purple derivative, arenicochromine ($C_{21}H_{12}O_6$), during purification of the extracted material.

As a coenzyme UQ is protein-bound and the insect tanning quinones become protein-bound for a very different functional reason. The BQ secretion of the flour beetle probably binds to the proteins of the flour (MORTON, 1965) and so acquires a pink colour. The echinochromes are mostly conjugated with protein. The precursors of some of the tanning quinones are glucose-bound (BRUNET, 1967) so that they exchange a sugar conjugant for a protein. Glycosides have been recognised also in the secretory glands of some of the insect protective BQs (ROTH and STAY, 1958) so that here again the glycose may be an inactivator. In the blood aphins are glycosides; after shedding they become activated and undergo a series of enzymic oxidations, the yellow protoaphin passing successively through xantho-, chryso- and erythro-aphin, a nice bathochromic series.

3.5.1 Properties

The characteristic quinone properties, which are also the essentially 'chromatic' properties, are most pronounced in the BQs so that rather unusually (p. 14) they are progressively toned down by increasing annelation (ring-condensation). Only the o and p BQs are known because an m-isomer would not be compatible with both the normal 4-valency of carbon and the quinonoid/benzenoid ring-structure (FINAR, 1959, p. 639). The o-BQs (II) are very unstable (THOMSON, 1957) unless further conjugated and no doubt this is why they are rare even as NQs and AQs. Most animal Qs are either p-Qs or have the quinone groups on different rings.

The quinonoid ring is not fully aromatic (Fig. 3.4), in contrast to the quinol, and so it does not form a continuous conjugated double bond system with contiguous benzenoid rings. The chromatic properties of the PQs therefore are due to the quinonoid system alone and, as implied above, the conjugated system of the benzenoid rings condensed on it actually depress the chromatic properties of the quinonoid moiety. This is true also of some of the substituents on the quinonoid ring itself (p. 40). On the other hand the large number of substituents both on this and on the other rings of the polycyclic Qs is due to a more reactive, ethylenic character of a quinonoid than of simple benzene rings (FINAR, 1959); this can be related to the greater polarity and the extra double bond.

The colour of p-BQs is usually yellow but the o-BQs are red; this is because the latter have one more 'resonant species' (Fig. 3.4, VII, VIII) and also greater electrical asymmetry, giving a larger dipole moment (p. 14). As expected they are also more reactive than the p-BQs and the greater instability is another aspect of this. The resonant species of p-BQ are electrically more asymmetrical than indicated by the orthodox formula and it has a much higher dipole moment than benzene. Among NQs vitamin K is yellow in conformity with the above rule but echinochromes range through orange, red, purple, violet and possibly green while the p-AQs of Hemiptera and Crinoidea are all red. From the evidence of the carotenoproteins this bathochromic variety may be due largely to protein conjugates (p. 31) but there is considerable variety in the extracted chromes also and the variety of side-chains must play some part. In the AQs they must be the main factor. It will be noted (V, VI, XIII, XIV) that the PQ zoochromes remain essentially p-Qs in the sense that the operative quinone oxygens are at opposite ends of an axis, even when on different rings. Protoaphin (V) is little more than a substituted p-NQ and is yellow; its more bathochromic series of products (above) and the still darker zoopurpurin (XIV) show clearly that here the increasing aromaticity of the whole molecule becomes the dominant chromogenic factor: the number and type of substituent on the rings does not change significantly in the aphin series of reactions.

The absorption-spectra of the animal Qs (Table 3.6) show that the number of separate chromophor groups giving peaks in the visible range increases progressively with the number of rings and of substituents on them. The simpler Qs are not strongly fluorescent but the PQ subclass are, at least when unconjugated; the products of protoaphin are visibly fluorescent even in daylight.

An effect of the ketone groups is seen in the aquosolubility of Qs, for instance of p-BQ as compared with benzene; it is appreciably soluble in cold, and much more in hot, water. Because of the aromatic foundation solubility is greatest in the polar lipids such as alcohols and ethers and least in petrol. As usual hydrophil groups raise the melting point and even the BQs are solid, though p-BQ is volatile in steam and can be separated from o-BQs through this property. The simplest of the p-NQs and p-AQs also are volatile again showing how the properties of the Q-ring dominate the whole, at this molecular size. The general solubility-properties of all animal Qs are rather similar, so that the hydro-

Table 3.6. Wavelengths (nm) of peaks in absorption-spectra of amimal quinones

Quinone	Solvent			Wavelengths (nm)							Reference
p-Benzoquinone	water	246	298		422						Auct.
p-Benzoquinone	MeOH	243	290	360							Auct.
Rhodoquinone	EtOH	285				515					Ozawa et al., 1969
Echinochrome A	(Et)$_2$O	262	339		462						Millott 1957
Echinochrome A	(Et)$_2$O				467	497	533				Fox and Scheer, 1941
Echinochrome A	(Et)$_2$O					495	531				Kühn and Wallenfels, 1940
Antedon NQ	(Et)$_2$O	262	337		460	(510)					Dimelow, 1958
Spinochrome A	(Et)$_2$O	255	318				525				Fox and Vevers, 1960, p. 204
Spinochrome B	(Et)$_2$O	270	320	390		480					Fox and Vevers, 1960, p. 204
Echinochrome A	EtOH	264	341		467	(520)					Millott, 1957
Echinochrome A	CHCl$_3$	257	339		461	495	520				Millott, 1957
Carminic acid	alk. H$_2$O					492	530				International Critical Tables 1930
Kermesic acid	alk. H$_2$O				493	527	567				International Critical Tables 1930
Carminic acid	H$_2$SO$_4$				474	504	544				International Critical Tables 1930
Kermesic acid	H$_2$SO$_4$				470	501	534				International Critical Tables 1930
Carminic acid	{H$_2$SO$_4$ +				466	498		581	624		International Critical Tables 1930
Kermesic acid	H$_2$B$_4$O$_7$}				463	493		576	623		International Critical Tables 1930
Protoaphin	KH$_2$PO$_4$	273	353		443						Human et al., 1950
Protoaphin	Na$_2$HPO$_4$	234	296	356			518				Human et al., 1950
Xanthoaphin	CHCl$_3$	282	358	378	405	429	459				Human et al., 1950
Chrysoaphin	CHCl$_3$	268	(326)	380	402	(430)	486				Human et al., 1950
Erythroaphin	CHCl$_3$	267			421	446	485	522	560	586	Human et al., 1950
Stentorol	CHCl$_3$				475	520	555				Barbier et al., 1956
Stentorin	in vivo					527	586	618			Møller, 1960
Zoopurpurin (red)	in vivo	247	282	334		490	534	578			Giese and Grainger, 1970
Zoopurpurin (blue form)	in vivo		330			553		602			Giese and Grainger, 1970
Zoopurpurin	90% EtOH				494	540	579				Emerson, 1929
Fabrein	pyridine					566		612			Fontaine, 1934

phily of the polar side-chains is balanced by the hydrophoby of the increasing number of aromatic rings and of alkyl side-chains; this may have biological rationale in permitting variety in colour and other properties without serious variation in solubility. As usual the most water-soluble are those with –COOH substituents or with glycose or protein conjugants.

Quinones have a tendency to return to the fully aromatic form of the ring by reducing to the quinol state; the BQs in particular are strong oxidising agents, in contrast to the aliphatic ketones. The redox potentials, in mV, of a representatives series are: diphenyl quinone + 534, o-BQ + 372, p-BQ + 295, NQs + 156 to − 39, AQs − 266 (MORTON, 1965); again the property is depressed by condensed benzene rings (FIESER and FIESER, 1935). Those with the highest redox values, up to + 950 m V, are more strongly oxidising than any other known biochemical reagent (PULLMAN and PULLMAN, 1963) and the BQs are strongly corrosive (p. 159). Quinones are used as redox electrodes in the laboratory, mainly because their redox reactions are so readily reversible and this is a major virtue biologically also. As indicated by an intermediary colour-change and a paramagnetic state, a free radical (IX or even X) is involved in the reaction. The radical arises by way of a quinhydrone (QH) complex (XI), effectively a quinol H-bonded to a quinone. For steric reasons the o-Qs form QHs more readily than the p-Qs. In some of the PQs it is possible to form QH complexes intramolecularly (III, IV, XIII). In contrast to linear annelation, angular annelation as in phenanthraquinone does not depress the redox potential (PULLMAN and PULLMAN, 1963).

The BQs are very light-sensitive, polymerising rapidly in solution to deposit insoluble melanic products. At the other extreme also the PQ zoopurpurin is destroyed by light, and is very photodynamic (p. 247), to other materials and organisms. This is in accord with the bathochromic colour of the PQs (p. 44) and for both properties it may be relevant that there is angular as well as linear annelation in this subclass.

Their oxidative activity coupled with their avidity for bonding with other molecules largely explains the now familiar role of BQ precursors in hardening the skeletal materials of arthropods and other animals (RICHARDS, 1951; BRUNET, 1967; PRYOR, 1962). They bond peptide chains together making them highly insoluble and strong yet pliable. Darkening is usually but not always involved; it is a measure of the degree of polymerisation of the quinone units or more correctly of the extent of conjugated double bond systems. If aromatic amino acids of the peptides themselves are incorporated into these then dark products should be most rigid but if not then they should be more pliable than colourless bonded peptides. Qs react particularly strongly with − SH and − NH_2 groups and are used in tanning leather. Their effect on metabolically active proteins of course is disastrous.

While the reduced BQs become as colourless as typical catechols the quinols of the other subclasses can be as brightly coloured as the quinone since the Q-ring now joins the conjugate system of the rest of the molecule. There is also the complication of multiple Q/Qol groups. In addition many Q-zoochromes have ionising side-chains and show pH colour-changes (Table 3.7); indeed it is fairly safe to predict that any chrome which shows both redox- and pH-colour changes is an annelated Q. For these reasons also the Qs are outstanding for their variety of brilliant colour-reactions (THOMSON, 1957) and many of these are the basis for diagnostic tests (VOGEL, 1956; FEIGL, 1966).

All Qs form coloured salts with heavy metals, $FeCl_3$ being the most useful reagent. All give triacetoxy-derivatives with acetic anhydride in the presence of a trace of H_2SO_4 or of $HClO_4$. They form brilliant oxonium salts in cH_2SO_4. Hydroxyquinones give coloured products with acetic acid and pyroboracetate and their alkali salts are intensely soloured. There are a number of tests specific to o- as contrasted with p- Qs, for instance reactions with guanidine carbonate, o-phenylene diamine, 2-nitroso-5-dimethyl aminophenol and o-dinitrobenzene plus formaldehyde. The p-isomers of both BQs and NQs give a colour-reaction with ethyl cyanoacetate plus alcohol and ammonia. The BQs and NQs in general are distinguished from the AQs and phenAQs by giving colour-reactions with ammoniacal rhodamine and also with ammoniacal phenyl-3-methyl pyrazole-5-one. The NQs give coloured salts with $CaCO_3$. AQs are distinguished by their slow oxidation and their resistance to cH_2SO_4; in the latter respect they resemble some of the more highly polycyclic Qs. The polyHO-AQs form characteristic coloured lakes with zirconium nitrate in acid or in alkaline solution. The PQs in general reduce to the parent hydrocarbon under the action of zinc dust and they produce mellitic acid by oxidation with HNO_3.

Table 3.7. Colour changes of quinones with pH

Quinone	Colour in acid medium	Colour in alkaline medium	Reference
Naphthoquinones (most)	Red	Purple	–
Antedon NQ	Orange	Red	DIMELOW, 1958
Laccaic acid	Yellow	Red	FOX, 1953, p. 207
Protoaphin	Yellow	Purple	DUEWELL et al., 1948
Xanthoaphin	Yellow	Pink	DUEWELL et al., 1948
Chrysoaphin	Yellow	Crimson	DUEWELL et al., 1948
Erythroaphin	Red	Dark Green	DUEWELL et al., 1948
Stentorin	Red	Green	TARTAR, 1961
Zoopurpurin	Red	Colourless	EMERSON, 1929
Hallachrome	Blue-violet	Green	THOMPSON, 1962
Arenicochrome	Red $\frac{pH}{6.0}$ Green	$\frac{pH}{9.0}$ Purple	VAN DUIJN, 1952
Lithobius violet	Orange-red	Violet	BANNISTER and NEEDHAM, 1971

3.6 Metalloproteins

The remaining known zoochromes all contain nitrogen in the molecule; most are heterocyclic compounds with the N in the ring, since aliphatic N-compounds are colourless. As a member of an open chain system of conjugated double bonds the N-atom causes instability; peptide chains are stable and colourless because they do not form a conjugated system. All protein biochromes therefore owe their colour primarily to non-protein prosthetic groups, though various amino acid side chains are good auxochromes, for instance $-NH_2, -OH, -SH$, guanidine, pyrrolidine, phenol, indole and imidazole. In most chromoproteins the chromophor is a large organic molecule but there is one class in which it is simply a metal atom, in a ligand field provided by some of the above side-chain groups. In some ways this is the simplest type of N-containing zoochrome and that is the reason for considering it at this point; another justification is that functionally the class is related to the metalloporphyrins to be considered in the next section.

The metals concerned are iron and copper, members of the first transitional series, some relevant electronic properties of which have been mentioned (p. 15). Electronic transitions in atoms of these metals are facilitated by innately small differences in energy-levels between d-orbitals of the third, or M, shell and those of s and p orbitals of the fourth, N-shell, as well as by the splitting of levels in the $3d$ group and by the number of unfilled orbitals in all these groups. The splitting is induced by ligands such as those provided by a protein (HESLOP and ROBINSON, 1960, p. 129) and so the ligands contribute to the activating power and bathochromicity of the metal. Ligands bond by donating electrons to the central atom and stabilising it; this does not decrease the metal atom's potentiality for electron-transitions since mainly unoccupied orbitals are used. The sharing of electrons involves charge-transfer from ligands to metal and this affects also the colour and activity of the complex (FEITKNECHT, 1967). Bonding electrons are delocalised over the whole ligand field and the chromophor is emphatically the field as a whole and not merely the central atom.

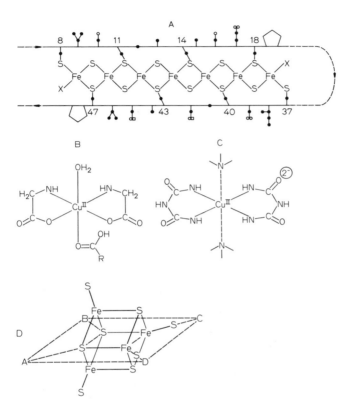

Fig. 3.5. A The prosthetic complex of ferredoxin according to PHILLIPS et al. (1965). Only the relevant part of the apoprotein loop is shown. The symbols for the amino acids are those of WELLNER and MEISTER (1966). The numbers are the positions in the amino acid sequence of the cysteine residues that form part of the prosthetic group. The other S- atoms are inorganic. B Bisglycinato-copper hydrate as a possible model of a Cu^{II} ligand field relevant to copper proteins. After FREEMAN (1966). C Dibiuretocuprate anion, another possible model for the ligand field in Cu-proteins. After FREEMAN (1966a). D Prosthetic complex of the ferredoxin of the bacterium Micrococcus. (After SIEKER et al., 1972)

3.6.1 Copper Proteins

Unfortunately the nature of the actual ligand groups is not known for any copper protein (MASON, 1966) and only partly for one or two iron proteins (PHILLIPS et al., 1965; TANAKA et al., 1965; SIEKER et al., 1972). There is some evidence that imidazole and tryptophan groups are involved in haemocyanin (WOOD and BANNISTER, 1967, 1968; BANNISTER and WOOD, 1971) but copper usually accepts 4 or 5 ligands in all so that probably there is

Table 3.8. Properties of copper-proteins. (Mainly from WOOD, 1968)

Protein	Source	Molecular weight (in 10^4 daltons)	% Cu (Av)	Oxidation state	No. of Cu atoms per molecule	Colour	Peak absorption in visible range (nm)
Oxyhaemocyanin	Arthropoda	40–100	0.17	I	~20	Blue	570
Oxyhaemocyanin	Mollusca	200–900	0.245	I	200–400	Blue	570
Ceruloplasmin	Mammalian serum	16	0.34	I/II = $1/_1$	8	Blue	610
Erythrocuprein	Mammalian erythrocytes	3.35	0.38	II	2	Blue-green	655
Hepatocuprein	Mammalian liver	3.5	0.35	? II	2	Blue-green	660
Cerebrocuprein	Mammalian brain	3.5	0.30	? II	2	Blue-green	660
Lactocuprein	Mammalian milk	?	0.19	?	?	?	?
Tyrosinase (1.10.3.1.)	Fungi	13	0.20	I	4	Colourless	–
Amine oxidases (1.4.3.4.)	Mammalian liver mitochondria	25	0.075	II	4	Colourless or pink	480
Dopamine β-hydroxylase (1.14.2.1.)	Mammalian adrenals	29	0.5	II	2	Colourless	–
Diamine oxidase (1.4.3.6)	Mammalian kidney	18.5	0.07	? II	2	Pink	?
Galactose oxidase (1.1.3.9)	Fungi	7.5	0.09	II	1	Pink-yellow	?
Ascorbic oxidase (1.10.3.3)	Plants and bacteria	14	0.34	I/II = $3/_1$	8	Blue	608
Laccase (1.10.3.2)	Lactree	13	0.33	I and II	4	Blue	615
Uricase (1.7.3.3)	Mammalian liver and kidney	12	0.06	?	1	?	?
Azurin	Bacteria	1.5	0.45	II	1	Blue	625
Plastocyanin	Chloroplasts	2.1	0.58	II	2	Blue	597
Stellacyanin	Lac tree	2.0	?	II	1	Blue	604

some variety (MANWELL, 1964). In galactose oxidase (Table 3.8) and in the iron protein ferredoxin (Fig. 3.5, A) the cysteine SH-group is involved (WOOD, 1968); it has many *a priori* virtues (MORPURGO and WILLIAMS, 1968) and should be a common contributor. As a possible model for the copper proteins it is tempting to consider the biuret type of compound (Fig. 3.5, C) formed by reaction in alkaline medium between $CuSO_4$ and peptides or other molecules with peptide-bond groupings. Where *in vivo* conditions are alkaline special devices would be necessary to prevent Cu from complexing in this way though at physiological pHs it is not the only possibility. The structure of this type of complex (Fig. 3.5, B, C) is essentially also a simple model for a porphyran (p. 53); it has advantages over copper porphyrans which are rare biologically (p. 88). The indication is that as in porphyrans the metal is bonded to four N-atoms and all five atoms are quasi-coplanar. Any further ligands, again as in some porphyran complexes, bond perpendicular to this plane, the whole conforming to the octahedral plan. In the Cu-protein stellacyanin (Table 3.8) there is strong evidence that the ligands are N-atoms (GOULD and EHRENBERG, 1968) and this is the most plausible for haemocyanin (MORPURGO and WILLIAMS, 1968). The complex formed by copper with the simple dipeptide carnosine (IHNAT and BERSOHN, 1970) shows a change of ligands with pH, COOH groups in acid, mainly the imidazole ring N-3 at neutral and amino-N together with peptide-bond N at alkaline pH. In a large protein field at least as much variety may be anticipated, and this could be functionally important.

There is a large number of natural Cu-proteins, all redox in general function but with varied specialisation. Their basic properties are summarised in Table 3.8. To zoologists the O_2-carrier haemocyanin (p. 224) is the best known and it is interesting that mammals and probably other animals have a blue Cu-protein, ceruloplasmin, in their blood, though at a much lower concentration; Hcy may have evolved from such a protein. There are few mammalian tissues which do not have Cu-proteins in amounts which may be metabolically significant. Comparison of ceruloplasmin with Hcy illustrates the great variation in properties among the Cu-proteins. Unlike Hcy ceruloplasmin is paramagnetic, the Cu being in the Cu^{II} form with an unpaired electron in one of the $3d$ orbitals. It is reduced by ferrous iron, ascorbate and p-phenylene diamine and itself reduces O_2 irreversibly so that it could not function as an O_2-carrier. Unlike Hcy it is not oxidised by H_2O_2 and has no catalase activity. It has twice as much Cu per unit weight.

Similar Cu-proteins have been found in various tissues and there are a number with known enzyme activities: tyrosinase, dopamine β-hydroxylase, amine oxidase and uricase in particular. Other specific enzymes are known in plants and microorganisms (Table 3.8). Tyrosinase and the related phenol oxidases are peculiar among Cu-proteins in remaining colourless throughout their cycle of activity while the Cu of Hcy is even more peculiar in its valency-state (p. 235). Cytochrome oxidase is peculiar in absorbing also in the i.r. range though this is not uncommon in simpler Cu-compounds. Clearly this is a versatile class of biochromes.

3.6.2 Iron Proteins

The simple iron-protein zoochromes include haemerythrin (Hery), functionally a fairly close counterpart of Hcy, haemosiderin and ferritin, concerned simply with iron storage and transport in the body, and the ferredoxin group of redox enzymes, more common in microorganisms and plants. This also is a versatile class. The Herys all have relatively small molecules of 10.7×10^4 daltons, like the intracellular haemoglobins, but each has 16 Fe atoms, three times the ratio in the Hbs and Hcys (KLAPPER and KLOTZ, 1968). The prosthetic group has been called 'haemoferrin' (FOX, 1953, p. 304) but there is no evidence of a separable entity larger than the Fe atom. A porphyranlike structure is again implied by the limited knowledge of the ligands (FAN and YORK, 1969). OxyHery is madder red, with absorption-peaks at 280, 330 and 500 nm while deoxyHery is pale yellow with no peaks in the visible range.

Haemosiderin is a yellow to brick red Fe-protein present in the reticulo-endothelial cells of mammals, particularly in tissues forming and destroying erythrocytes, – bone-marrow, spleen and liver. Its significance is to avoid the difficulty in recovering Fe once it has been allowed to revert to the inorganic ferric form. Over half of haemosiderin is amorphous Fe or crystalline FeOOH (BIELIG and WOHLER, 1958) so that there is no question of its being a specific Fe-compound. Only 17.5% is protein, containing eight amino acids which occur also in ferritin. About 25% appears to be a deoxyguanylic acid derivative so that it is a very unusual chromoprotein. The iron is loosely bound and gives a prussian blue reaction without special treatment. Blood-sucking ticks likewise store an iron protein of this kind (DRABKIN, 1951) but the bug *Rhodnius* appears to store inorganic Fe (WIGGLESWORTH, 1943).

Ferritin, the protein that transports Fe round the mammalian body, has a molar weight of 5×10^5 (DRABKIN, 1951), large enough not to be excreted by the kidney. The iron-free protein is called *transferrin* and *siderophilin*. The iron again may be somewhat casually dispersed in the fabric of the protein (GRANICK, 1942; MICHAELIS et al., 1943). Dietary Fe is bound as ferritin in the wall of the gut, transported to the liver and distributed from there. A ferritin has been recognised also in the polychaete *Arenicola* (BRETON GORIUS, 1963).

The *ferredoxins* (MORTENSEN et al., 1962; SAN PIETRO, 1965; MALKIN and RABINOWITZ, 1967) again are yellow in dilute solution and brown or black when concentrated, indicating an absorption peak in the near u.v. and considerable absorption throughout the visible range; they vary greatly in the precise position of the main peak, in redox potential and in other properties. Some are built into larger enzyme-complexes which include a flavoprotein and other components (PULLMAN and SCHATZ, 1967). A possible structure of the essential prosthetic part of the molecule of the ferredoxin of *Clostridium* is shown in Fig. 3.5, A (PHILLIPS et al., 1965; TANAKA, 1965); it shows four rather evenly spaced cysteine residues in each arm of a looped peptide regularly alternating with three inorganic sulphur atoms in each arm and all bonded in a regular steric pattern with 7 Fe atoms. X-ray diffraction studies support the alternative structure shown in Fig. 3.5, D (SIEKER et al., 1972) for the ferredoxin of a second microorganism, *Micrococcus*. Here the steric symmetry is still greater and the stoichiometry simpler. The four inorganic S-groups alternate with the Fe atoms at the corners of a cube. A diagonal set of 4 of these 8 atoms are coplanar with two of the cysteine S-atoms which are bonded to the two Fe atoms of the plane (ABCD). The other two Fe atoms also are bonded each to a cysteine-S but in directions which point to the corners of a tetrahedron completed by the first two cysteines. This is nearer the structure of the porphyrans and other metal-prosthetic groups than the hairpin-model. The iron is half Fe^{III} and half Fe^{II}. *Rubredoxin* differs from typical ferredoxins in having no inorganic S, only one Fe atom and four cysteine residues per prosthetic group.

Adrenodoxin, in the adrenal gland of artiodactyls (SUZUKI and KIMURA, 1965) has a molar weight of 22,000 and resembles chloroplast ferredoxin in spectral absorption. It has two Fe and two S atoms per molecule; in the oxidised state both Fe are in the Fe^{III} state and in the reduced state one only becomes Fe^{II}. The other known Fe-proteins of animals, the *flavodoxins*, are associated with flavin coenzymes and have high molar weights, between 200,000 and 500,000. They have correspondingly more Fe atoms than most of those with small molecules and the Fe maintains its stoichiometry with S. The flavodoxins include milk xanthine oxidase, liver aldehyde oxidase and ox-heart mitochondrial dehydrogenases of succinic acid and NADH (Fig. 8.2). Conalbumen is another Fe-flavoprotein (WAGNER-JAUREGG, 1954). There is evidence for an iron flavoprotein also in the heart-body of terebellid polychaetes (DALES, 1965) so that the type of complex probably is widely distributed.

There is considerable iron in the red and black melanins of mammalian hair (p. 62); in his original extraction of the red 'trichosiderin' SORBY (1878) precipitated it by pH-adjustment which may imply the isoelectric precipitation of a protein and FLESCH (1968) has much evidence that it is a protein containing iron. It would be useful to know if melanin is bonded to an Fe-protein or iron to a preformed melanoprotein. Since FLESCH (1970) postulates a (cysteinyl dopa)$_2$-Fe^{III}-protein complex as an essential agent or intermediary in the biosynthesis of red melanin an Fe-protein might be an essential enzyme, which persists in the final structure. It is noteworthy that ribitylflavin is implicated in melanin-synthesis and also other transitional metals (p. 289). Iron occurs in other zoochromes, in forms as yet unknown; some may prove to be Fe-proteins.

3.7 Pyrrolic Zoochromes

Pyrrole is the simplest stable aromatic N-heterocyclic ring (Fig. 3.6, I) but with only two double bonds it is colourless. Monopyrroles are virtually unknown in living organisms but they form biochromes by chain-linked polymerisation, the links themselves contributing to a conjugate double bonded system involving the whole polymer. There are a few biological di- and tri- pyrroles, an example being the stable red *prodigiosin* (II) of *Serratia marcescens (Bacterium prodigiosum)*. Most of the known pyrrolic biochromes however are tetrapyrroles and have a common biogenetic origin.

The bilins, named after their best known members in the bile fluid of vertebrates, are open chain tetrapyrroles (IV–VI) but these are formed secondarily from porphyrins (III) in which the chain is closed to a 'super-ring' of great geometrical symmetry, with virtually planar form and complete conjugate bonding. The porphyrins therefore are brilliantly coloured (p. 15) with many sharp absorption-peaks, highly sensitive to excitation by light and other forms of energy but at the same time amazingly stable thermodynamically; they have survived virtually unchanged in fossil remains of Silurian age (p. 335) and surely are *the* biochromes par excellence. The simplest natural porphyrins, uro- and copro- porphyrins, have only two types of sidechain on the super-ring, acetic and propionic (Table 3.9), indicating a common origin for all four pyrrole rings (p. 286). In theory these could be arranged in four different ways but in practice only the two series conventionally numbered I and III, and mainly the latter, occur in living organisms. The one shown, *protoporphyrin* (III), is a member of series III but since it has two of its short chains changed to vinyl there are 15 possible arrangements of the side chains. The one shown is conventionally number IX of these and in fact it is the only natural protoporphyrin known. It may be synthesised via uro- and coproporphyrins or independently from an earlier intermediary (Fig. 13.3) and is the parent of the haems and the phaeoporphyrins of chlorophyll.

Free porphyrins occur quite widely in animals but they are photodynamic (p. 247) and usually they are chelated to a central metal atom via the four pyrrole N-atoms. Iron, magnesium and cobalt are the metals most used but occasionally also others of the first transitional series. Manganese has great affinity for the position, which it often comes to accupy *post-mortem* so that *in vivo* it must be specifically excluded. The metalloporphyrins, or porphyrans, containing Fe, Mg and Co are known respectively as *haems, phaeoporphyrans* (VII) and *cobrinic acid. In vivo* they are usually, but not always, conjugated with protein.

Fig. 3.6. Structural formulae of representative pyrrolic biochromes and of intermediaries in the biosynthesis of porphyrins. In VII Am = acetamide; PM = propionamide. For other sidechain symbols see Table 3.9

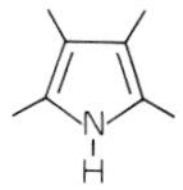

I Pyrrole

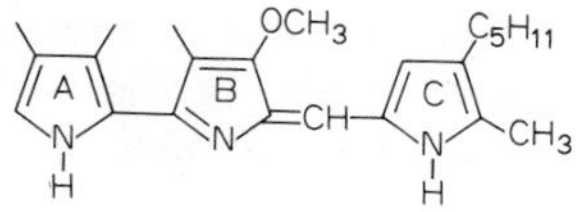

II Prodigiosin

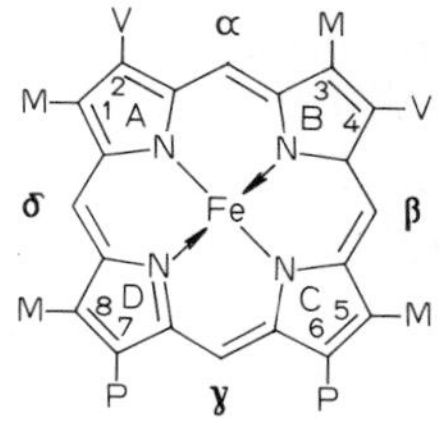

III Protohaem IX

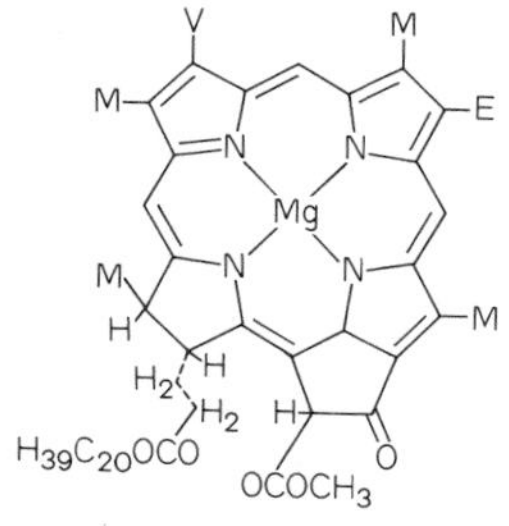

VII Chlorophyll a

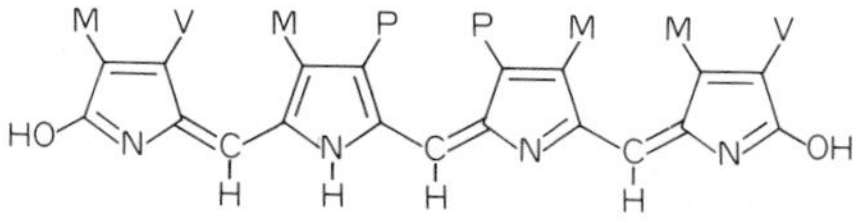

IV Biliverdin IXα

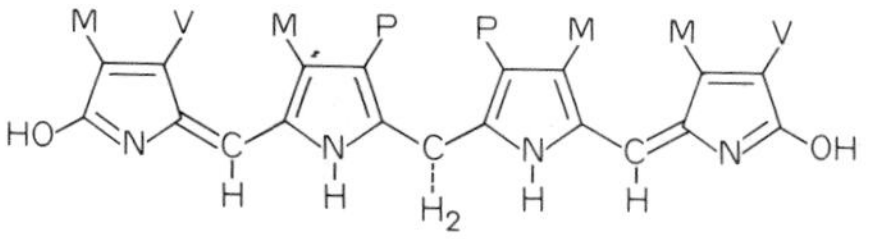

V Bilirubin

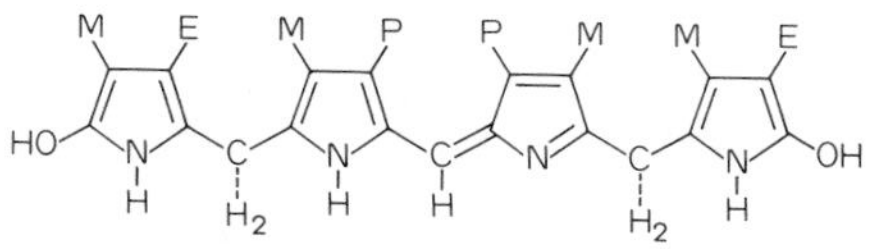

VI Urobilin IXα

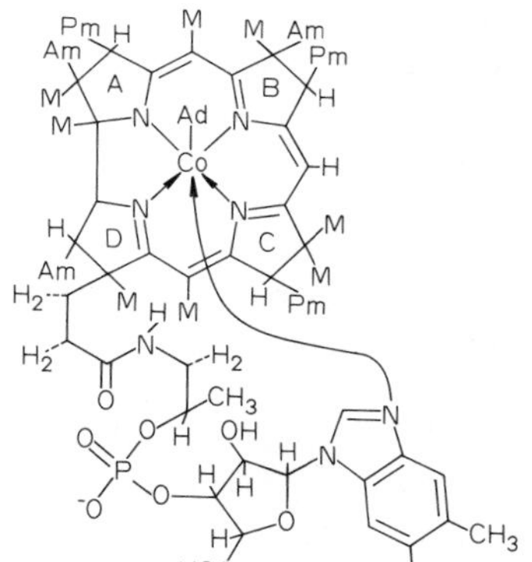

VIII Cobamide B₁₂ coenzyme

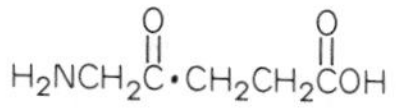

X δ-aminolaevulinic acid

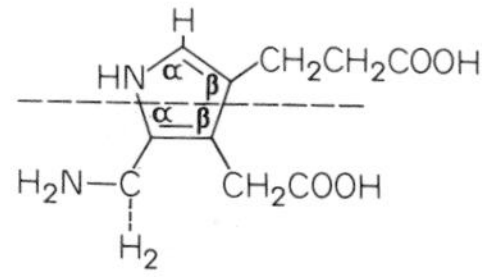

XI Porphobilinogen

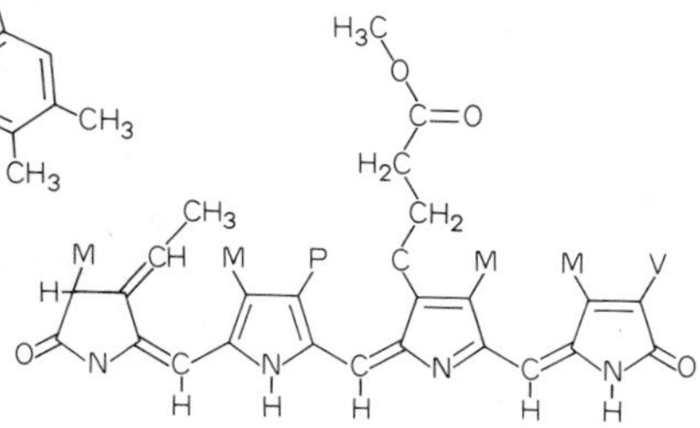

IX Aplysioverdin

Haemoproteins are most common as the *cytochromes* (Cy) of the main respiratory system of all animal cells. The haem varies in its side-chains but is *protohaem* itself in Cy *b* and *c*; that of Cy *a (cytochrome oxidase)* has a farnesyl (isoprenoid) chain bonded at position 2 of the super-ring, and the chain at 8 shortened to formyl (SEYFFERT et al., 1966), and it also has an N-containing group, $C_6H_{12}NO_4$, in the molecule (CAUGHEY,

Table 3.9. Substituents on porphin ring of representative porphyrins.
(A = acetyl, AA = acetamide, E = ethyl, F = formyl, H = hydrogen, M = methyl, P = propionyl, PA = propionamide, V = vinyl)

Porphyrin / Position on ring	1	2	3	4	5	6	7	8
Uro I	A	P	A	P	A	P	A	P
Uro III	A	P	A	P	A	P	P	A
Copro I	M	P	M	P	M	P	M	P
Copro III	M	P	M	P	M	P	P	M
Proto IX	M	V	M	V	M	P	P	M
Chloro	M	F	M	V	M	P	P	M
Meso	M	E	M	E	M	P	P	M
Deutero	M	H	M	H	M	P	P	M
Haemato	M	–CHOH·CH$_3$	M	–CHOH·CH$_3$	M	P	P	M
Cytochrome *a*	M	Farnesyl derivative	M	V	M	P	P	F
Chlorophyll *a*	M	V	M	E	M	–CO·ĊH·COOCH$_3$	H, and propionyl phytate	M, H
Chlorophyll *b*	M	V	F	E	M	–CO·ĊH·COOCH$_3$	H, and propionyl phytate	M, H
Phaeophorbide *a*	M	V	M	E	M	–CO·ĊH·COOCH$_3$	P, H	M, H
Phylloerythrin	M	E	M	E	M	–CO·ĊH·COOH$_3$	P	M
Mesopyrro-chlorin	M	E	M	E	M	H	P, H	M, H
Cobinamide	M, AA	H, PA	M, AA	H, PA	M, M	H, PA	M, PA	H, AA

1967). The *catalases* and *peroxidases* are further haemoprotein redox enzymes. In the O_2-carrier haemoglobins the haem is always protohaem except in several taxa of the polychaetes which have chlorohaem, the vinyl chain at 2 being replaced by formyl. A haemoprotein of uncertain function occurs in the digestive fluid of a number of gastropods (p. 241); its absorption spectrum is very similar to that of Cy *c* and a related haemoprotein is present in the digestive gland (KEILIN, 1956, 1957, 1968).

In the catabolism of porphyranoproteins removal of metal and protein is facilitated by prior opening of the ring to form *choleglobin (verdoglobin)*; *in vivo* it is usually opened at the α-methine bridge but occasionally at γ so that protohaem gives two series of bilins, IX α and IX γ. The intermediate protein-free metallo-bilins, or *verdohaems*, are quite stable and have been isolated from animals. Their side-chains show that the bilins arise mainly from the common haems.

Small amounts of some other porphyrins occur in animals. *Deutero-* and *meso-* porphyrins (Table 3.9) are sometimes formed in the mammalian gut through bacterial action (Rimington and KENNEDY, 1962) but the porphyrin of lactoperoxidase is a mesoporphyrin derivative (MORRISON, 1966). Tricarboxylic porphyrins have been found in polychaetes (KENNEDY and DALES, 1958) and oligochaetes (DELKESKAMP, 1964) and from the Harderian gland of the rat (KENNEDY, 1970). There is a pentacarboxylic form in *Chaetopterus* (KENNEDY and VEVERS, 1954).

Phaeoporphyrins differ in many details from protoporphyrin and they vary in the different chlorophylls (JONES, 1968). Animals degrade dietary phaeoporphyrans which they absorb in a different way from the haems, usually not opening the super-ring at all. Removal first of the phytyl side-chain and then of the Mg atom yields green *phaeophorbides*. Saturation of the two vinyl sidechains to ethyl then gives red *phylloerythrins;* a third stable stage, the green *chlorins*, results from opening the isocyclic ring on position 6. The chlorins therefore are simple mesoporphyrins. In the laboratory further degradation products are possible, still with the porphyrin ring intact (FINAR, 1964, p. 659). Mg is much more easily removed than Fe and this may be why the ring is never opened.

Table 3.10. Known cobamide nucleotides. (From GOODWIN, 1963)

Trivial name	Systematic Name	Base of nucleotide	Source
Pseudovitamin B_{12}	α-Adenyl cobamide	Adenine	Microorganisms of sewage.
Factor A (vitamin B_{12m})	2-Methyl-α-adenylcobamide	2-Methyl adenine	Calf and pig faeces; sewage.
Factor C (Nocardia factor I)	Guanylcobamide	Guanine	*Nocardia;* calf faeces.
Factor F	2-Methylthio-α-adenylcobamide	2-Methyl thioadenine	Chicken faeces; sewage.
Factor G	α-Hypoxanthine cobamide	Hypoxanthine	Pig faeces
Factor H	2-Methyl α-hypoxanthine cobamide	2-Methyl hypoxanthine	Pig and calf faeces
Factor I (vitamin B_{12}III)	5-hydroxy α-benzimidazolyl cobamide	5-Hydroxy-α-benzimidazole	Pig faeces; sewage.

Cobrinic acid is the basis of the *cobamides* or vitamin B_{12} group (VIII) and the metal-free porphyrin is called *corrin*. Again there are many differences from protoporphyrin and in particular the absence of a methine group in the bridge between pyrroles C and D and the extra methyl groups substituted on the α and γ methines may be noted. All the pyrrole rings are partly saturated, all the free side-chain COOH-groups are amidated and that at 7 is bonded via a 1-amino-2-propanyl group to a nucleotide which by its other end shares in the chelation of the cobalt. The degree of saturation of the pyrroles limits the conjugation-resonance to a broken 'inner circle' of the porphyrin skeleton, but extensible via the cobalt to the heterocyclic base of the nucleotide. In isolation procedures the sixth ligand, adenine, is often replaced by a − CN group which is a powerful ligand of these metals. *In situ* the adenine also will extend the resonance system and the colour of the cobamides is dark red. A number of different bases (Table 3.10) have been exploited for the nucleotide loop.

3.7.1 Properties

Most of the porphyrins, porphyrans and their protein-conjugates range from orange-red through red to purple, indicating that the contribution of the porphyrin itself is predominant. The phaeoporphyrins, chlorins and chlorohaem are green but none are blue or violet.

The verdohaems and bilatrienes (IV) show that opening the porphyrin ring has a bathochromic effect, because of the increased dipole moment. Further changes mainly involve saturation of the methine bridges and the colour moves hypsochromically, the extent depending on the positions and number of bridges saturated. Bilirubin is reddish yellow because the middle bridge is affected (V), leaving two separated bipyrryl resonance systems but saturation of an end methine produces biladienes with colours still as bathochromic as red, violet or even green. The bilamonenes have only one bridge still unsaturated and are all orange to yellow. Saturation can be oxidative or reductive and this causes further variation in colour, oxidation tending to add auxochrome groups. This is an ideal group on which to study these and other chromatological properties.

Although closure of the porphyrin ring has a hypsochromic effect it is responsible for the intense Soret absorption peak in the region of 410 nm; this depends on the combined action of the symmetrically placed four N-groups of the pyrroles. It is related also to the intense red fluorescence of the porphyrins. Both intensities are unequalled in any other biochromes. Moreover light at 410 nm has a large quantal value so that the class are most powerful electron-excitants. The fluorescence is quenched by protein-conjugation and by Fe but not by Mg. It is minimal at the isoelectric point of the free porphyrin (LEMBERG and LEGGE, 1949, p. 97) and has a different wavelength-maximum on the alkaline and acid sides of this.

Table 3.11. Reactions discriminating between bilins

Bilins \ Reaction	Gmelin's oxidation	Van den Bergh's diazo	Schlesinger's fluorescence	Colour with methanolic $FeCl_3$	Bingold's pentdyopent
Bilatrienes	+	−	Red (after oxidation)	−	+
Biladienes bilirubin and mesobilirubin	+	+	−	Green	+ (after dehydrogenation)
mesobiliviolin and mesobilirhodin	+	−	Red	Green	+
Bilamonenes urobilin IX α stercobilin	−	−	Green	Violet	−
Bilanes	−	−	−	−	±

On account of the complexity of the system porphyrins absorb also in many other regions of the spectrum and most of the peaks are sharp though not as intense as the Soret. For the group on the protoporphyrin path they lie in the ranges 391–414 (Soret), 487–502, 517–37, 560–76, 613–32 nm when measured in neutral media (LEMBERG and LEGGE, 1949, p. 72). In strong HCl the spectrum is simplified to 400, 550, 600 nm, with slight shoulders at 505, 565 nm. Alkalis cause a 20–30 nm hypsochromic shift but depress the intensity less than acids: anion-formation has less effect than cation-formation. Ligation of Fe adds a new peak in the region of 335–345 nm, removes that at 613–32 and moves the others up by 40 nm (LEMBERG and LEGGE, 1949, p. 94) but it moves the Soret peak down by 20–30 nm (1949, p. 173) and reduces its intensity. There is one peak in the short u.v. range, at 282 nm (NEEDHAM, 1968a) probably due to the individual pyrroles, and this is shifted to 270–275 nm by protein-conjugation, but it only emphasises how essentially geared to absorption in the visible range the porphyrins are. The protoporphyrin group absorb at regular intervals throughout this range, though normally they appear not to be excited by light directly; by contrast the chlorophylls, which are so excited, absorb mainly towards the least energetic end of the spectrum and without the help of other chromes make poor use of it.

The absorption spectrum helps to explain why the free porphyrins are so powerfully photodynamic. Actually they vary considerably in potency, uro- and copro- porphyrins being much more powerful than protoporphyrin. This implies that the action depends on the number of side-chain COOH groups. However the most potent of all is haematoporphyrin which like protoporphyrin has only two of these groups (RIMINGTON and KENNEDY, 1962) so that there must be other relevant factors. It is tempting to believe that proto has been naturally selected for relative innocuousness.

Phaeophorbides retain the complex porphyrin spectrum but that of the bilins is much simpler. They have only one peak in the visible range and it is very flat-topped; also they are not fluorescent. Conjugation with metals sharpens the peaks, adds new ones and may induce fluorescence, providing a useful diagnostic test for some subclasses (Table 3.11). The absorption spectrum of the IX α series differs slightly from that of the IX γ counterpart (RUDIGER, 1970):

Solvent Bilin	MeOH-NH$_4$OH	MeOH-HCl
Biliverdin IX α	261,376,653	261,308,376,700
Biliverdin IX γ	266,372,657	261, 360,700

Prodigiosin has peaks at 290, 370, (500) and 550 nm (LEMBERG and LEGGE, 1949, p. 538).

As a class the pyrrole chromes are relatively insoluble in neutral aqueous media (Table 3.1): when uncharged the N-atom leaves the pyrrole ring almost as hydrophobe as a purely carbocyclic ring. The super-ring shields the N-atoms and makes the porphyrins even more hydrophobe than the bilins. However this is counteracted by such polar groups as COOH on the side-chains and in general the biological porphyrins are more water-soluble than the bilins. Uroporphyrin, with eight COOH groups, is quite insoluble in lipid solvents, including partially polar types such as ether. Turacin, the copper-uroporphyran in the feathers of the turaco bird, is soluble even in neutral and salt-free rain water. Esterification of the COOH groups with alkanols decreases the water-solubility once more. In strong acids such as HCl some of the pyrrole N-atoms in the free porphyrins become active bases and this increases the aquo-solubility. No doubt these various properties are biologically significant and certainly they are invaluable in isolation and purification. As implied above porphyrins are amphoteric and their isoelectric point lies between pH 3.0 and 4.5, so that acetate buffer systems hold them in this region, where their water-solubility is minimal. With the exception of uroporphyrin they can be transferred into ether in this range and then after washing they can be transferred back to HCl, of strength inversely related to the number of side-chain COOH groups. The HCl concentration or 'number' required for this is diagnostic for each porphyrin and is used in their identification. To some extent bilins also can be separated by the method (LEMBERG and LEGGE, 1949, p.

115). Their amphoteric properties make the porphyrins ideal material for chromatographical separation and a wide range of eluents is available for those bilins that retain polar sidechains.

The best extractants of the tetrapyrroles from biological materials are (1) a mixture of ethyl ether and acetic acid (2) methanol-HCl mixtures and (3) 5% aqueous HCl. These denature and break the bonds of large conjugants, replacing them with small substituents of appropriate polarity for solution in the extractant. Details of the methods and of further treatments are given in LEMBERG and LEGGE (1949), GRAY (1961), LASCELLES (1964) and FALK (1964).

3.7.1.1. Chemical Reactions

Because of the stability of the porphyrin ring there is a lack of chemical reactions that diagnose all porphyrins. A difference in absorption-spectrum between cationic and anionic forms (p. 77) is not peculiar to this class though the precise differences can be used to identify individual members. Protoporphyrin is distinguished from all other porphyrins by giving a green and not a purple colour with HCl in $CHCl_3$ (CHU, 1946). The superring can be resolved into constituent pyrroles by reducing agents such as $SnCl_2$ in acid medium or by strong oxidisers such as chromic acid but tests for free pyrroles are not very satisfactory. The very characteristic spectral and solubility properties therefore are most fortunate.

Porphyrans are identified by removing the protein conjugant and substituting a simpler N-compound such as pyridine, which produces a 'haemochrome' with a characteristic absorption-spectrum. Oxidation and reduction reactions of the metal, again with characteristic spectral changes, also can be useful. Most dramatic is the liberation of the free porphyrin and its brilliant fluorescence by strong acids.

Once the super-ring is opened the pyrrole rings as well as the methine bridges (p. 52) are more easily saturated and desaturated and even the side-chains are more easily modified; consequently the bilins are as remarkable for their chemical reactivity as the porphyrins for their inactivity, particularly as many of the reactions are reversible. They also show 'prototropic' or intramolecular reactions; for instance the spontaneous change of bilirubin to biliverdin is not a true oxidation but two of the methylene bridge H-atoms move to positions on the pyrrole side-chains (GRAY, 1961, p. 17).

In the gut of vertebrates the vinyl side-chains are hydrogenated to ethyl, forming meso-bilirubin, and further hydrogenation gives fully saturated bilanes (bilinogens) which are then dehydrogenated again as far as the bilamonene stage (LEMBERG and LEGGE, 1949, p. 137). Alternatively saturation can be effected oxidatively as in the classical Gmelin test for bilins. Treated with a mixture of nitrous and nitric acids a bilin passes hypsochromically in turn through all the intermediate stages and colours to colourless bilanes; whatever the initial stage it completes the series: green → blue → violet → purple → red → yellow → colourless. The actual reactions are somewhat complex and are detailed by Vuillaume (1969). The verdohaems do not give this series of reactions (FOX. 1953, p. 275) so that the metal stabilises the molecule even when the super-ring has been breached. Not surprisingly there are a number of diagnostic chemical tests for bilins and for their subclasses. The most common are given in Table 3.11. Bilins also interact with their protein-conjugants more than porphyrins and the protein affects their colour and other properties.

3.8 Indolic Zoochromes

The indole nucleus (Fig. 3.7, I) is a condensed benzopyrrole but the simple indoles such as tryptophan are colourless. In the laboratory porphyrin-analogues, the phthalocyanins, have been synthesised (p. 201) and are among the most brilliant chromes known but they have not been found in living organisms. The only known class of brightly coloured indolic biochromes are the *indigotins* (II), mere dimers but linked by a double bond which gives both coplanarity of the monomers and a maximal extent of conjugate double bonding. Only one of these, 6,6′-dibromoindigotin (II), is actually produced by animals, and even this is produced from its precursor only after secretion from the body.

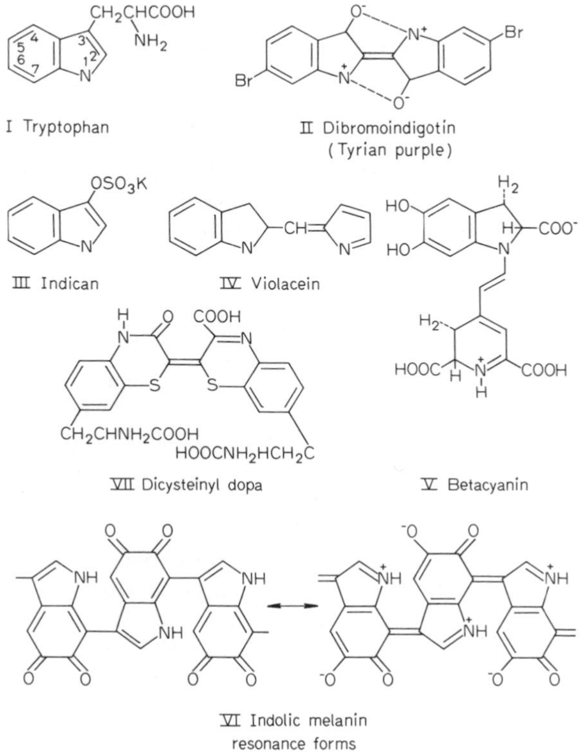

Fig. 3.7. Structural formulae of representative indolic biochromes and of intermediaries in their biosynthesis

I Tryptophan
II Dibromoindigotin (Tyrian purple)
III Indican
IV Violacein
V Betacyanin
VII Dicysteinyl dopa
VI Indolic melanin resonance forms

3.8.1 Indigotins

Dibromoindigotin is formed in the secretion of the adrectal (hypobranchial) gland of *Murex*, *Nucella (Purpura)* and a few related whelks. The unbrominated indigotin, together with indigo red or indirubin, a 2–3′ linked member, and indigo green, 1,1′-dimethyl indigotin, are formed in voided mammalian faeces and urine, and occasionally in the bladder and intestine, but always by the action of microorganisms, on indolic monomers. A number of plants of several taxa synthesise indigotins, by enzyme-action, in extracts of crushed tissues, i.e. again not *in vivo*. The product of *Isatis* was used as the body-paint, woad, by the Celts of Britain and that of *Indigofera* was used as a fabric dye. Dibromoindigotin likewise, as Tyrian Purple, was used for this purpose. Possibly this is why they are not synthesised *in vivo;* in secretions they may have a contingent protective action. There is no definite knowledge of their function. This class is the least water-soluble of all biochromes (Table 3.1) and would be difficult to handle in vivo.

Indigotin has an absorption peak between 600 and 630 nm and therefore has some absorption in the near i.r.; in consequence it could have had some heating effect as a body-paint. Indirubin absorbs between 520 and 540 nm and the colour transmitted indicates that this is the only peak in the visible range. The brilliance of the colours of the indigotins is due largely to their quinonoid structure. Like the ternary quinones some of them, including dibromoindigotin, sublime without melting. They have high melting points so that they are strongly polar without being hydrophil. The trans form is the more stable isomer, owing to the attraction between the side-chain quinone oxygens at 3,3' and the N-atoms at 1,1' respectively and the replusions between the like atoms. H-bonding at these points further stabilises the planarity of the molecule and facilitates electronic excitation.

3.8.2 Other Indolic Oligomeric Chromes

There are two classes of indolic biochromes somewhat analogous to the indigotins, though neither is known from animals. One is the indolyl-pyrryl-methines (Fig. 3.7, IV); it is probably the basis of the well known *violacein* of *Bacterium violaceum* (BEER et al., 1954) which may have a second indol substituted on the methine carbon. It is synthesised from tryptophan (McMoss and HAPPEL, 1955).

The second class includes the *betacyanins* (V) responsible for the familiar colour of the beetroot. Long regarded as anthocyanins they do prove to be fairly near analogues of the flavonoids (p. 38) since a lone aromatic ring is chain-linked to a condensed pair. They have the distinction of being one of the few types of biochrome with a pyridine ring; moreover this is in the very unstable partially unsaturated state.

3.8.3 Indolic Melanins

The term *melanin* was first applied, without chemical implications, to the dark pigments of hair, feathers and other structures of vertebrates (FURTH and SCHNEIDER, 1902). When it was shown (RAPER, 1928) that this is essentially an indolic polymer the tendency was to limit the term to this chemical class (CASPARI, 1953; THOMAS, 1955). However it seems reasonable to retain the trivial but meaningful term melanin for all black pigments, since probably all are quinone polymers, and to distinguish subclasses by their main monomer: the present subclass therefore is the *indolic melanins*. Another reason for retaining melanin as the generic term is that they do not classify into 'species' at all sharply. In mammalian melanin there are several different indolic monomers and probably also some phenolic monomers (NICOLAUS, 1968) and this is true for some other 'typical' melanins. At the other extreme the typical phenolic melanins contain as much as 5% nitrogen (NICOLAUS, 1968), possibly as indole. Nicolaus uses 'allomelanin' for the predominately phenolic melanins of plants, fungi, bacteria and some arthropods but taxonomically also the distinction is not absolute and some arthropod melanins are mainly indolic (NICOLAUS, 1968, p. 179). Moreover mammalian blond melanin *(phaeomelanin)* and still more the trichosiderin *(erythromelanin)* of red hair and feathers (PROTA et al., 1968; FLESCH, 1970) may have a preponderance of non-indolic monomers. The melanins of the various tissues of mammals differ considerably (NICOLAUS, 1968; LILLIE, 1969). Animals may use phenolic monomers only for special purposes since tyrosine is their usual source of all monomers, but in their usual policy of N-economy plants mostly have relatively pure phenolic melanins. That of lignin may be a polymer mainly of phenylpropane (BOLKER and BRENNER, 1970) while that of the fungus *Ustilago* gives mainly the dihydroxyphenol, catechol, on degradation. However, the melanin of the skin of ripe bananas contains mainly DOPA (dihydroxyphenylalanine) which partly at least may be ring-closed to indole (GRIFFITH, 1959; SWAN, 1963).

The melanin of the ink of cephalopods is indolic (NICOLAUS, 1968) but nothing certain is known about the chemical nature of the monomers in other invertebrates. In the melanogenic tissues and body-fluids of most of the higher invertebrates tyrosinase or a related phenolase has been found and no doubt melanins are widespread. On the other hand since the original discovery of BECKER (1938) a large number of dark invertebrate chromes have proved to be ommochromes (p. 63). Like the latter the black pigments of planarians and leeches are soluble in acid methanol (NEEDHAM, 1965a, 1966b) but they are more bathochromic and lack typical ommochrome properties, and conceivably are oligomeric melanins.

Melanins are so refractory to most solvents and reagents that it has not been possible to determine their molar weights or exact structures (NICOLAUS, 1968). They may be heterodisperse as well as having the great variety of monomer units. The extent of branching in the polymer also is unknown, but is probably small. Melanins are always bound to protein which, as in most chromes of the integument, does not confer water-solubility. In any case this is precluded by the large size of the melanin polymer and because of the large percentage of the polar sidechains of both components taken up in the bonding. Also because of the size and shape of the melanin polymer the protein must be in effectively fibrous, extended form and this also depresses water-solubility. The insolubility of arthropod sclerotins is relevant here.

It is probable (CROMARTIE and HARLEY MASON, 1957; PULLMAN and PULLMAN, 1961; ROBSON and SWAN, 1966) that indole-5,6-quinone, the main monomer of the indolic melanin pathway (Fig. 13.4), bonds mainly between position 3 of one and position 7 of the next (Fig. 3.7, VI) though in some indolic melanins there is evidence of 4–7 and 2–7 bonds (NICOLAUS, 1962; HEMPEL and DIENNEL, 1962). The protein is bound by its – SH and – NH_2 side-chain groups (MASON and PETERSON, 1965) mainly to position 2 (SWAN and WRIGHT, 1956). The portion of the melanin shown in Fig. 3.7, VI implies that the polymer is a giant system with complete conjugate double bonding, resonating with at least one charged species and involving a complete series of bound semi-quinone radicals. There is evidence also for free radicals, therefore non-stoichiometric in amount but always most abundant in the eumelanins (NICOLAUS, 1968). X-ray diffraction studies (THATHACHARI, 1970) indicate that the indole residues form stacks with their planes almost parallel. It is therefore possible that sterically affine monomers build an ordered structure while the others act as free radicals and in other ancillary capacities.

3.8.3.1 Properties

Like all black materials eumelanin absorbs strongly, without any peak, throughout the visible range but whereas the absorption by graphite is uniform throughout the range that by melanin increases progressively as the reciprocal of the wavelength down to 300 nm. As its colour implies erythromelanin has a distinct absorption-peak in the visible range and this is at 530–560 nm (NICKERSON, 1946). Both erythro- and phaeo-melanins have a dull u.v. fluorescence but eumelanin has none, probably because there is complete internal quenching.

Eumelanin is insoluble in water and dilute acids though quite soluble in hot alkalis and in hot strong H_2SO_4. It is soluble in acetic and formic acids and in a number of neutral organic solvents such as diethylamine and ethylene chlorhydrin (LEA, 1945). From solution in alkali the material precipitates at its isoelectric point. Erythro- and phaeo-melanins have wider solubilities, for instance in dilute HCl.

Eumelanin is very resistant to both oxidation and reduction but prolonged oxidation with H_2O_2 turns it pseudo-blond as hairdressers well know; in bright sunlight oxygen itself bleaches melanin quite strongly, a mainifestation of its photosensitivity (p. 172). Reducing agents turn it a reddish brown, quite distinct from the red of erythromelanin. Melanin from the choroid of the eye can be reduced all the way to a leucocompound, following hydrolysis (STOVES, 1953).

Both eu- and erythro- melanins have a high content of heavy metals, particularly iron (FLESCH, 1968). Copper has been found in hair-melanin, the ink of *Sepia* and other melanins BOWNESS and MORTON, 1953). Zinc is present in high concentration in melanosomes (FITZPATRICK, 1966) and in the melanin of the vertebrate eye (BOWNESS et al., 1952). Calcium and barium also often are present. Some of these appear to be trapped in the course of melanogenesis (p. 289) but Zn is bound at a late stage to the completed melanoprotein; it increases light-absorption by melanin and therefore may play a positive role in the function of the material. Iron appears to be bound stoichiometrically and so also may have more than casual significance. Melanin may be a good cation-exchanger in fact (WHYTE, 1958). The high iron content may be relevant to the strong tendency of melanin itself to produce paramagnetic free radicals and mobile electrons (SWAN, 1963; PULLMAN and PULLMAN, 1963, p. 497). The number of these is increased by irradiation of black hair and depressed by chemical reduction. Calculations of the expected charge-distribution over an indole quinone monomer and dimer indicate that melanin should be an exceptionally good electron-acceptor and that the power is likely to increase with polymerisation. The difference in energy-level, between the highest occupied and lowest unoccupied molecular orbitals should be as low as 3.2 kcal per mole, indicating that it should be very readily excited by light. PULLMAN and PULLMAN also consider that the geometry of the polymer should enable it to act as a semiconductor; at least the extensive delocalisation of π-electrons should permit charge-transfer. There is as yet no experimental evidence for semiconduction (BLOIS, 1969) but if it should occur it would be adequate to excite quinone units at a distance (p. 13).

Dark hair is stronger mechanically than blond so that eumelanin may there function like the phenolic melanins in sclerotin but it must be remembered that the melanin of hair remains mainly in the melanosomes of the cells (p. 118) and the way in which it strengthens the keratin proteins is not evident. Aromatic compounds in general do bind strongly to hair, through dipole-induced dipole interactions (BREUER, 1971).

Some histochemical tests for melanins are listed in Table 3.3.

3.8.4 Other Zoochromes of the Indolic Melanin Pathway

A variety of coloured oligomeric intermediaries in the melanogenic pathway have been demonstrated in the laboratory (p. 288) and whereas they appear too reactive to become definitive zoochromes erythromelanin is an indication that relatively hypsochromic stages can be stabilised. BU'LOCK (1960) identified one intermediary as the dimer, 5,6-dihydroxyindol-indol-5,5-quinone and this quinhydrone type of dimer might provide a basis for stabilisation. Erythromelanin is pink in acid and yellow in alkaline medium (FOX and VEVERS, 1960, p. 187) and the violet chromes of planarians and leeches (p. 61) show a similar colour change in alkaline medium. They also precipitate at neutrality and show other melanoid properties. They give a positive pinesplint reaction for pyrroles, as do indoles in general. Their colour, a dusky violet, indicates that they are higher polymers than erythromelanins and it may seem curious that the absorption peaks of the leech chrome, at 502 and 541 nm are not at a longer wavelength than that of erythromelanin; there is in fact a long shoulder in the violet-transmitting region (Fig. 2.1). BEER et al. (1954) found that the oxidation-products of 5,6-diHO-indole have a peak at either the 502 nm or the 540 nm region depending on the oxidant used; in the leech chrome the peaks probably are due to oligomer units and the shoulder to more highly polymeric structural features.

Some coloured oligomers are obtained when melanins are degraded in the laboratory (NICOLAUS, 1968; NICHOLLS, 1969) and it seems likely that the bright chromes produced by oxidising tryptophan and other indoles in acid medium (DALGLEISH, 1955) may be related oligomeric quinones. The violet product from tryptophan itself has an absorption-peak again at 540 nm (NEEDHAM, 1968b). It turns yellow in alkaline medium and precipitates just on the acid side of neutrality. If it is true that tryptophan and other indoles readily oxidise in the same way as the tyrosine-derived indoles then it is particularly curious that so many animals have exploited a different chromogenic pathway from tryptophan: the ommochromes are the most important class in this pathway.

3.9 Ommochromes (Phenoxazones) and Related Biochromes

The trivial name was applied by BECKER (1938, 1941) to dark chromes extracted from the ommatidia of arthropod compound eyes and previously thought to be melanins. They proved to be phenoxazones, i.e. substituted phenoxazines (Fig. 3.8, I). The simplest and most common is *xanthommatin* (II), bright yellow as the prefix indicates. The aspartyl side-chain at position 5 and the fourth ring formed by closure of a second aspartyl at 4 (Butenandt's numbering) are clues to the origin by condensation of two hydroxykynurenine (HOK) residues (p. 289) derived from tryptophan (Fig. 3.8). Some chromes of collateral paths are indicated on this figure; kynurenine itself (REBELL et al., 1957) and HOK (YAMASHITA and HASEGAWA, 1966) are already pale yellow, presumably due mainly to the double bond of the aspartyl γ-CO group and to the increase in dipole moment due to the aspartyl chain. The ommochromes are dimers (*ommatins*, II) or oligomers (*ommins*, III) of kynurenine derivatives probably synthesised via distinct branches of a common pathway.

Since BECKER's death work on the class has been due mainly to the school of BUTENANDT, who has reviewed the field at intervals (BUTENANDT, 1957; 1960). A recent review by his pupil LINZEN (1967) also contains much information. Becker had already distinguished between the lower molecular weight and more reactive ommatins and more bathochromic, non-dialysable ommins. The actual molecule of ommin A is smaller than the decrease in dialysability implies. It is further interesting for the addition of a thiazine and not a second oxazine ring. Both oxazine and thiazine dyes are extensively used commercially, as also are many other analogously bridged aromatic systems, for instance in the anthraquinones and the isoalloxazines. In the broadest sense most classes of commercial dyes and not a few biochromes are based on this plan.

Two further zoological ommatins have been recognised and identified, *rhodommatin*, a 2-glucoside of dihydrogenoxanthommatin (diHX) and *ommatin D*, a 2-sulphuric ester of diHX. Five further ommins have been distinguished but their structures are not yet known; they vary in degree of polymerisation. LINZEN (1966) has recognised a third subclass, *ommidines*, intermediate between the other two in properties but nearer to the ommins in having sulphur in the molecule. This is the subclass earlier called 'acridioxanthin' by CHAUVIN (1938) and 'insectorubin' by GOODWIN and SRISUKH (1950). In locusts there is evidence for a fourth subclass, with a very unstable molecule. *Lacciferinic acid*, in the scale insect, may be an ommochrome (SINGH et al., 1966). Other subclasses of phenoxazone biochromes occur in microorganisms (LINZEN, 1967); the antibiotic *actinomycin* is a member of one of these.

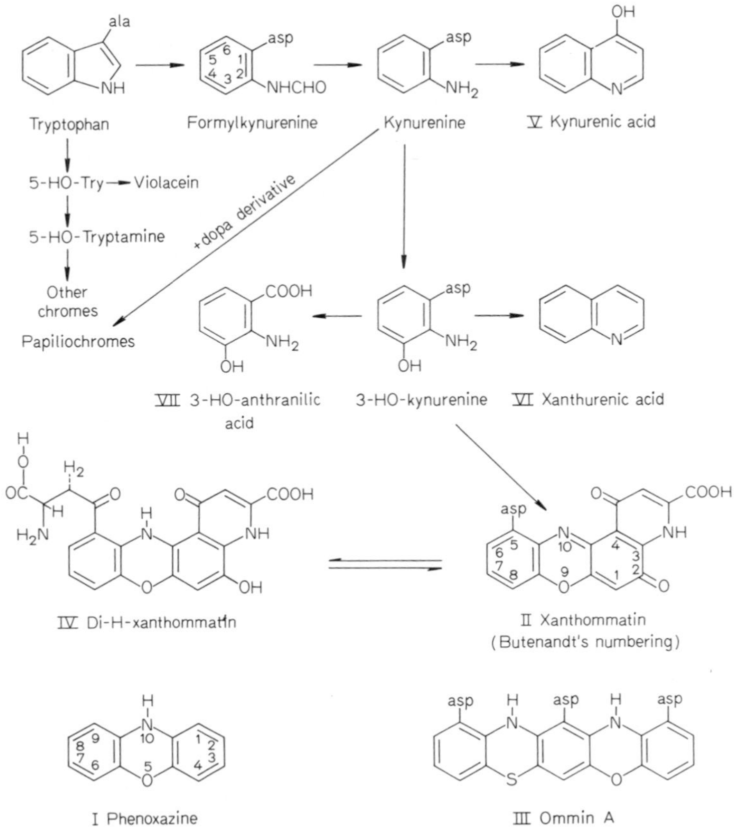

Fig. 3.8. Structural formulae of representative ommochromes; some of the biosynthetic pathways from tryptophan

Animal ommochromes are mainly integumental and screening eye-pigments and so are bound to a protein matrix in intracellular granules (p. 122) but it is not clear if there is also stoichiometrically bound protein. Denaturing agents are the best extractants but these would help to remove the chrome from matrix-protein, as well.

3.9.1 Properties

Ommochromes range in colour from the gold of xanthommatin (X) through red, purple and violet to almost black. Most of them are dark *in situ* where even X can appear dark brown (NEEDHAM and BRUNET, 1957); it is not surprising that a number of them were assumed to be melanins. A peculiar feature of the class is that the reduced form is more bathochromic (red) than the oxidised (yellow). This is the more surprising since X becomes hydrogenated at positions 2,10 (IV) and so both loses a quinone double bond and suffers an interruption of bond-conjugation at the oxazine ring (IV).

The absorption-spectrum of simple phenoxazines in $5N$ HCl is characterised by four well-spaced peaks, A, B, C, D at 225–40, 265–90, 365–75 and 435–85 nm, only the last being in the visible range. That of the ommins lacks the C-peak and differs in other respects. In contrast to the kynurenine monomers none of the class is strongly u.v. fluorescent, except in strong mineral acids; internal quenching is implied.

All are very insoluble in neutral organic solvents, as would be expected from the many polar groups in the molecule. They are soluble in acids and alkalis and particularly in acidic organic solvents or mixtures, such as formic acid and MeOH-HCl respectively. In alkaline media they are unstable, the ommatins in particular. In cH_2SO_4 there is a bathochromic change in colour again particularly evident in the ommatins; BECKER called the reaction 'halochromy'. The ommins give their maximal shift in colour in 75% H_2SO_4; in higher concentrations the shift is reversed and in cH_2SO_4 the product is yellow, with some resemblance to X (NEEDHAM, 1970a). Presumably the thiazine bridge is broken.

In situ the ommatins occur in either form, oxidised or reduced. In solution they spontaneously oxidise in air so that there is a provisional case that they can act as redox agents *in vivo*. Eggs of the echiurid, *Urechis caupo*, contain X and it can be reversibly reduced *in vivo* simply by lowering the O_2 pressure (HOROWITZ and BAUMBERGER, 1941). The ommins are usually in the reduced form *in situ* and they do not oxidise so readily, even in alkaline medium: the reduced form is more stable than that of the ommatins. The redox property of the ommatins depends not only on the $=O$ and $-N-$ groups actually hydrogenated but also on the N-group at position 3, the $>C=O$ and $-OH$ groups of the aspartyl side-chains and on the whole of the pyridine ring formed by ring-closure of the aspartyl at 4. This is a striking demonstration of the extent of electronic delocalisation in coloured molecules (p. 12). Ommin A lacks the $-OH$ group at 2 but nevertheless has some redox properties due to the oxygen of the oxazine ring having oxonium properties. This gives conjugation-continuity across that ring and may explain the above paradox that the reduced form is the more bathochromic. In the ommatins effectively every polar group is implicated and it would be interesting to know if the thiazine S of the ommins is in the sulphonium form: this would further extend the conjugated system and might be the main reason for the very bathochromic colour of reduced ommins.

3.9.2 Other Zoochromes of the Kynurenine Pathway

A yellow chrome related to kynurenine (K) has been extracted from the hair of rodents. It is particularly evident in the white strains of rat and has the strong u.v. fluorescence of K and its derivatives. The white and yellow *papiliochromes* of the wings of the Papilionidae may be fairly closely related (UMEBACHI and TAKAHASHI, 1956). The coloured members may be complexes of K with a quinone, which would imply something similar to the precursors used by BUTENANDT et al. (1954b) in their model of ommatin biosynthesis (p. 290).

BRUNET (1967) finds that 3-HO-anthranilic acid (VII) is oxidised to a product which colours and hardens the silk of the Robin-moth, *Samia cecropia*; this therefore is a further alternative (p. 62) to phenolic tanning agents. The colour is pale so that the conjugated bond systems are not extensive. There is some evidence that the cocoon of oligochaetes may be hardened and darkened by tryptophan derivatives rather than by phenolic materials (NEEDHAM, 1968b). The cocoon of *Eisenia* is initially lemon yellow and eventually a purple-brown; the yellow stage is consistent with a kynurenine pathway and there is considerable indolic material in the clitellum, and in the cocoon itself.

3.10 Other N-heterocyclic Zoochromes: Purines, Pterins and Flavins (Isoalloxazines)

There are no biochromes in which the imidazole ring (Fig. 3.9, I) is an essential chromophor though it is part of the benzimidazole moiety of one of the cobamides (Fig. 3.6, VIII) and of the purine nucleus (Fig. 3.9, VI). The simplest fully unsaturated N-hexacyclic ring, pyridine (Fig. 3.9, II) also is rather rare in biochromes. Even the pyridine nucleotide coenzy-

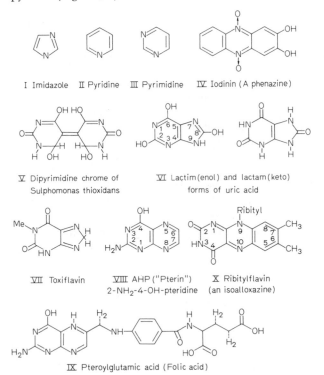

Fig. 3.9. Structural formulae of members of various classes of N-heterocyclic aromatic biochemical compounds including purines and biochromes of the pterin and isoalloxazine classes

mes have no absorption peak in the visible range though in high concentration they are pale yellow owing to the spread of an absorption centred at 340 nm. A pyridine ring is present in the molecule not only of the ommatins (p. 63) and the betacyanins (p. 60) but also of the plant chrome, *berberine* (MAYER and COOK, 1943). The mono- and dibenzologues of pyridine, i.e. quinolines and acridines, have coloured members many of them useful as dyes but they are not known as biochromes. Structurally xanthurenic acid is a quinoline derivative.

The common diazine hexacylic aromatic rings, pyrimidine (III) and pyrazine (IV) are more common in biochromes; the third isomer, pyridazine, with the two N-atoms ortho-related is very rare in nature. As isolated rings all these are colourless and so are the purines, imidazole-condensed pyrimidines. Enlargement of this imidazole to a pyrazine ring has a marked bathochromic effect and many of the resulting pterins (VIII) are visibly coloured, though rarely beyond yellows and reds. Further condensation of a benzene ring with the pteridine nucleus to form flavins or isoalloxazines (X) leaves the colour still in the yellow region, yet ribitylflavin is one of the most important of all biochromes. By contrast the phenazines or dibenzpyrazines (IV) include biochromes with brilliant bathochromic colours, in spite of having only a single N-heterocyclic ring. In having two polar atoms bridging between two benzene rings they are analogous to thiazines such as methylene blue, to phenoxazines (p. 63) and more distantly to the anthraquinones (p. 42). The critical chromogenic feature is the strongly polar axis established transverse to the long apolar axis of the molecule. A pyrimidine ring in place of one of the benzenes, as in the flavins, weakens this polarity. Two of the best known phenazine biochromes are the deep blue antibiotic *iodinin* (IV), of *Chromobacterium*, and *pyocyanin* (p. 200), produced by *Bacterium pyocyaneus* in pus and sometimes in sweat.

There are a few other relatively simple pyrimidine biochromes, for instance the substance V (Fig. 3.9), produced by *Sulphomonas thioxidans* (MAYER and COOK, 1943, p. 264). It is another cuff-link dimer, analogous to the indigotins (p. 59) and to pterorhodin (XII). A somewhat similar dimer, except that the link is a $= N -$ bridge, is the product *purpuric acid* of the murexide test for acidic pyrimidines such as alloxan and for purines and pterins. The term 'murexide' was applied because the colour is so similar to that of the dibromoindigotin of *Murex* (p. 59) and it is interesting that purpuric acid is a close structural analogue also. *Thiochrome*, the oxidation product of vitamin B_1 is a pale yellow pyrimidine-pyrimidine-thiazole condensed ring compound which may play a part in the function of this vitamin (RISINGER, DURST and HSEIH, 1966).

3.10.1 Purines

A number of purines are often regarded as biochromes because in many animals they confer structural whiteness (p. 136). This is usually due to reflection from crystalline deposits, all relatively insoluble in aqueous media. Iridescent colour-effects also are frequent. The most common of these purines are uric acid and guanine but xanthine (PURRMANN, 1939) and isoguanine (GILMOUR, 1965, p. 148) also occur. With the help of appropriate substituents the purine molecule does absorb in the visible range: as the name indicates *toxiflavin* (VII) is yellow; it is a bacterial product (VAN VEEN and MARTENS, 1934). It appears that the saturation of this molecule at position 8 improves the polar axis of the quinone, a m-Q (p. 44) but stabilised by suppression of the lactam-lactim tautomerism characteristic of most purines (VI). The purines absorb powerfully in the u.v. and in the broad sense are biochromes for this reason also (p. 9).

3.10.2 Pterins

This class of biochrome was so named because the first natural source discovered was the wings of pierid butterflies (HOPKINS, 1895; WIELAND and LIEBIG, 1944). Like the ommochromes it has the distinction of being more common and varied in animals than in plants. The condensed pyrimidine-pyrazine ring-system (VIII) now proves to be synthesised *in vivo* in the way its structure might imply, by inserting a C_1 unit into the imidazole ring of a purine. All the biological members are structurally derivable (but not biosynthetically derived, p. 290) from the 2-NH_2,4-OH pteridine (VIII), AHP (aminohydroxypteridine) and the term pterin refers to this particular subclass.

Table 3.12. Substituent groups at positions 5–8 in pterins (2-amino- 4-hydroxy pteridines) of animals

Pterin \ Position	5	6	7	8
Neopterin (Bufochrome)	--	$-[CHOH]_2 \cdot CH_2OH$	–H	--
Biopterin	--	$-[CHOH]_2 \cdot CH_3$	–H	--
Tetra-H-biopterin (? = Hynobius blue)	–H	$-[CHOH]_2 \cdot CH_3$; –H	–H; –H	–H
Ichthyopterin (7-OH-biopterin)	--	$-[CHOH]_2 \cdot CH_3$;	–OH	--
Sepiapterin	--	$-CO \cdot CHOH \cdot CH_3$	–H; –H	–H
Isosepiapterin	--	$-CO \cdot CH_2 \cdot CH_3$	–H; –H	–H
Drosopterin (?)	–H	$=COH \cdot CHOH \cdot CH_3$	–H	--
Neodrosopterin	–H	$-CO \cdot CHOH \cdot CH_3$	–H	–H
Folic pterin	--	$-CH_2OH$	–H; –H	–H
Ranachrome 3	--	$-CH_2OH$	–H	--
AHP-COOH (6-Carboxypterin)	--	–COOH	–H	--
6-Carboxyisoxanthopterin	--	–COOH	=O	--
DiH-AHP	--	–H	–H; –H	–H
AHP ("pterin")	--	–H	–H	--
Xanthopterin	–H	=O	–H	--
Leucopterin	--	–OH	–OH	--
Chrysopterin	–H	=O	$-CH_3$	--
Isoxanthopterin	--	–H	=O	–H
Ekapterin	--	–H	$-CH_2CHOH \cdot COOH$	--
Erythropterin	--	–H	$=CH \cdot CO \cdot COOH$	--
Lepidopterin	--	–H	$=CH \cdot C(=NH) \cdot COOH$	--
6,7 dimethyl lumazine (2-OH)	--	$-CH_3$	$-CH_3$	--

The number of known animal pterins is large (MATSUMOTO et al., 1971) and increasing. Table 3.12 gives the known or probable ring-substituents in the more familiar members; virtually all variants concern positions on the pyrazine ring. In addition to monomers a cuff-link dimer, *pterorhodin* (XII) has been found in insects (GILMOUR, 1965, p. 153). In *folic acid*, or *pteroylglutamic acid* (IX), and its derivatives a pterin is conjugated with p-aminobenzoic acid and glutamic acid. Some pterins are complexed with tryptophan and this has a bathochromic effect (ZIEGLER, 1961). *In vivo* most of them are conjugated with protein, lightly or more firmly. In *Bacillus subtilis, neopterin* is conjugated with glucuronic acid (KOBAYASHI and FORREST, 1970). Unlike porphyrins and flavins pterins do not usually bond with heavy metals (ZIEGLER-GUNDER, 1956) though *xanthopterin* forms an interesting compound with titanium. Some pterins show the keto-enol and lactam-lactim tautomerism of such purines as uric acid (VI).

3.10.2.1 Properties

Some of the pterins are colourless *in situ* and act like the purines, as structural whites (ZIEGLER-GUNDER, 1956; FORD, 1947; HOPKINS, 1942). Most of them range from pale yellow, through orange to red (Table 3.13) and therefore in situ can be confused with carotenoids flavones and flavins. As extracted pterins seem to be more hypsochromic than when *in situ* (MATSUMOTO et al., 1971) where conjugation or stabilisation in the quinonoid (keto) tautomeric form may be the bathochromic agencies. The difference may explain

Table 3.13. Absorption and fluorescence of Pterins

Pterin	Colour	Absorption Peaks of absorption (nm)		Colour	Fluorescence Peaks of emission (nm)		References
Neopterin (Bufochrome)	None			Blue			BAGNARA and OBIKA, 1965
0.1 N HCl		248	322				
0.1 N NaOH		255	363				
Biopterin	Pale Yellow			Blue	438		BAGNARA and OBIKA, 1965; VUILLAUME, 1969
0.1 N HCl		250	325				
0.1 N NaOH		260	365				
Tetra H biopterin	None			None			ZIEGLER, 1961
Hynobius blue	None			Blue			BAGNARA and OBIKA, 1965
7-hydroxybio-pterin (Ichthyopterin)	Pale Yellow			Blue	432 460	690	JACOBSON and SIMPSON, 1946; GOTO and HAMA, 1958
0.1 N HCl		289	344				
0.1 N NaOH		256	343				
Sepiapterin	Yellow			Yellow-green			FORREST and MITCHELL, 1954; VUILLAUME, 1969
0.1 N HCl		279		429			
0.1 N NaOH		268		440			
Isosepiapterin	Orange			Yellow			VUILLAUME, 1969
Drosopterin	Red			Orange			VISCONTINI et al., 1957
0.1 N HCl		265	475				
Neutral		265	485				
0.1 N NaOH		265	505				
Neodrosopterin	Red			Orange			VISCONTINI et al., 1957
0.1 N HCl		257					
0.1 N NaOH		245					
Isodrosopterin	Red			Orange			ORTIZ and WILLIAM-ASHMAN, 1963
0.1 N HCl		260	477				
0.1 N NaOH		260	501				
Ranachrome 3	Pale Yellow			Blue			BAGNARA and OBIKA, 1965
0.1 N HCl		246	322				
0.1 N NaOH		255	364				

why the colours usually attributed to the various pterins are more bathochromic than their absorption peaks imply (Table 3.13). Thus the red pterins have their peak of longest wavelength at 450–465 nm, the orange xanthopterin at 380–390 nm and the pale yellow *ekapterin* at 360 nm: with these values the colours should be golden yellow, very pale yellow and colourless respectively. They are usually less bathochromic also than would be expected from their structural formulae because of competition between sidechains for keto-enol transitions (MILLAR and SPRINGALL, 1966, p. 826) with consequent breaks in the conjugate bonding. Most recent records of absorption-spectra show three peaks but rarely more than one is in the visible range. All three move in the same direction under the influence of substituents, change of medium etc., and the pterins show these regularities as well as any class of biochrome.

In contrast to the purines most free pterins are u.v. – fluorescent. The number of emission-bands corresponds to the number of absorption bands and again the wavelengths are fairly consistently 150–200 nm greater than the corresponding absorption bands (Table 3.13). The resultant fluorescence-colour has a somewhat shorter wavelength than the visible colour transmitted so that it cannot be responsible for the above discrepancy between visible colour and position of absorption peaks though it may explain the apparent hypsochromic shift in colour upon extraction since fluorescence is quenched *in situ* by protein-conjugation. Fluorescence is quenched also by acids and is most enhanced in $5N$ NaOH. Some pterins show a change in the colour of the fluorescence with pH: that of xanthopterin

Table 3.13.(continued)

Pterin	Colour	Absorption Peaks of absorption (nm)			Colour	Fluorescence Peaks of emission (nm)				References
6-Carboxypterin (AHP-6-COOH)										Viscontini et al., 1957; Bagnara and Obika, 1965.
0.5 N HCl		265		320						
0.1 N NaOH		264		365	Blue					
Xanthopterin	Orange yellow									Jacobson and Simpson, 1946; Hama et al., 1960.
c Acetic		270		380						
0.1 N HCl					Yellow	460	523	592	620	
Neutral		275		375	Blue	455		592	620	
0.1 N NaOH		255		391	Blue-green	463	516	592		
Pterin (AHP)	None				Light Blue					Forrest and Mitchell, 1955; Vuillaume, 1969.
0.1 N HCl		245		312						
0.1 N NaOH		252		360						
Leucopterin	None									Jacobson and Simpson, 1946.
0.1 N NaOH		240	285	340	Blue		462	516		
Solid						436	462	535	620	
Chrysopterin	Yellow				Blue-violet					Vuillaume, 1969
pH 11.0		252		385						
Ekapterin	Pale Yellow				Yellow-green					Viscontini and Stierlin, 1961
pH 1.0		230	265	360						
pH 13.0		255	290							
Erythropterin	Dark red									Viscontini and Stierlin, 1961, 1963; Vuillaume, 1969.
pH 1.0		240		320	445					
pH 7.0			300	420	450 Violet					
pH 13.0		240		310	475 Orange-red					
Lepidopterin	Red				Yellow					Viscontini and Stierlin, 1961, 1963.
pH 1.0		238		310	450					
pH 13.0		238		310	465					
Isoxanthopterin	Pale Yellow				Violet	410				Mori et al., 1960; Vuillaume, 1969.
0.1 N HCl		220	286	340						
0.1 N NaOH			254	339						

is yellow in acid, blue in neutral media and blue-green in alkalis. This hypsochromic shift by acids in extreme cases will be interpreted as quenching. Other quenchers include oxidising agents such as $KMnO_4$; this provides a technically useful distinction from the fluorescence of the flavins which on the other hand is quenched by alkalis. Pterin fluorescence is quenched also by reducing agents such as dithionite and this is reversible in parallel with the redox state of the chrome. It is quenched also by light, the pterins in general being very photolabile.

Like the purines the pterins are sparingly soluble in water but mostly soluble in dilute acids and alkalis. They are amphoteric but with weak acid and base groups, again like the purines. They therefore form salts only with the monovalent metal cations and with strong anions. As would be expected they are soluble in polar but not in apolar organic solvents (Table 3.1).

Most pterins are redox agents but with a low oxidation-potential so that they are fully oxidised at physiological O_2-pressures. This is confirmed by calculations of charge distribution, using molecular orbital methods (PULLMAN and PULLMAN, 1963); these indicate that the reduced pterins are good e-donors but the oxidised forms only moderate acceptors. For this reason they do not bond with the transitional metals (p. 69). Folic acid is a still poorer oxidiser than the unconjugated pterins. The oxidised form of a pterin has a much higher resonance energy than the reduced form, due mainly to the $-$ OH and $-$ NH_2 substituents: without these even the 5, 6, 7, 8 tetra-H-pteridines do not oxidise readily. The 2-NH_2, 4-OH subclass, the pterins, have the highest resonance-energy per π-electron of all pteridines, both positions and substituent groups being optimal; in fact the value per electron is higher than in ribitylflavin and it seems likely that there has been very strong natural selection for the class and subclass. Resonance-energy is greatly depressed in the metabolic degradation products of the pterins. The pteridines and especially the pterins usually have their oxygen substituents in the more stable quinonoid, $=$ O, form although resonance is greatest when they are in the enol, $-$ OH, form.

The murexide reaction distinguishes pterins from all other classes of true biochrome. The blue reaction-product with cH_2SO_4 has been mentioned (p. 35). Few other reactions diagnostic of the pterins have been developed and no suitable reagents for demonstrating them *in situ* (FOX and VEVERS, 1960, p. 153).

3.10.3 Flavins (Isoalloxazines)

These are essentially benzene-condensed pteridines and the pteridine is not a pterin. Only one isoalloxazine is at all common biologically, *ribitylflavin* (RF), 6,7-dimethyl-9-ribityl isoalloxazine (X); it is quite as unique in importance as this distinction might be taken to imply. With the open-chain ribitol substituted at position 9 it is a modified nucleoside and it normally exists in nucleotide form, as coenzyme for a large number of respiratory enzymes (p. 189). Initially different names were given to the flavins from various sources, hepatoflavin, ovoflavin, lactoflavin, uroflavin and lyochrome but all proved to be RF. Flavins are defined as isoalloxazines with methyl substituents at 6 and 7 and an alkyl chain at 9 (PENZER and RADDA, 1967); the subclass therefore includes also two derivatives of RF in living organisms, *leucoflavin*, or 1,10-diH-RF, the reduced form of RF, and *lumiflavin*, 6,7,9-trimethylisoalloxazine, produced by photolysis of RF in alkaline medium. Photolysis in neutral or acid media produces *lumichrome* (XI) which has the pyrazine ring fully unsaturated and so is not an isoalloxazine but an alloxazine.

As a coenzyme RF occurs in two nucleotide forms, flavin mononucleotide (FMN) and flavin-adenine dinucleotide (FAD). In D-amino acid oxidase and perhaps in general the prosthetic group is bonded to the protein apoenzyme via the 3-N of the flavin (WELLNER, 1967). Some of the enzymes have a complex prosthetic group: that of cytochrome b_2 is an equimolar mixture of RF-5′P and protohaem plus a chain of 15 orthodox deoxynucleotides (i.e. DNA) per haem (HIROMI and STURTEVANT, 1966). One of the metals of the first transitional series is usually associated as a cofactor and this is in sharp contrast to the pterins.

3.10.3.1 Properties

RF is a bright chrome yellow in solution and almost orange in the solid state, It has absorption peaks at 223, 267, 373, 445 and 475 nm (LAMBOOY, 1963); the 445 nm peak corresponds to the lowest singlet excited state, $\pi \rightarrow \pi^*$, and the 373 peak to $n \rightarrow \pi^*$ transitions (SONG, 1971). It has a brilliant u.v. – fluorescence, maximal emission being in the range 520–565 nm, varying according to conjugants and other conditions. The hue is almost identical with that maximally transmitted in visible light (WAGNER-JAUREGG, 1954). If free RF were primarily an integumental chrome it might have seemed that it had been exploited for this useful synergism; it is possible to envisage other respects in which the coincidence could be useful. Fluorescence is quenched 90% when FMN is conjugated with the adenine nucleotide to form FAD; this is due mainly to the adenine and is a good example of a useful chromatic function of a purine. Most other aromatic structures that will conjugate also have a marked quenching effect (WEBER, 1966) with the same implication. Acids and alkalis outside the narrow physiological range of pH 4–8 all quench and so do reducing agents; in contrast to their effect on the pterins (p. 70) oxidising agents do not quench. The two photolytic products of RF are an index of its outstanding photosensitivity. It also has a powerful photodynamic effect on other molecules (p. 249). Crystals of a stoichiometric complex with a phenol show increased electrical conductance on illumination (PENZER and RADDA, 1967), indicating some form of charge-transfer to the phenol.

Conjugation with ribitol makes isoalloxazine very water-soluble and it is insoluble in organic solvents such as $CHCl_3$. Conjugation with phosphate, a second nucleotide and a protein all further increase water-solubility. The free RF is soluble in the more polar organic solvents, including such non-ionic types as alcohols and acetone. Pyridine is a very good solvent and will elute it from fuller's earth, a useful purification step. Aqueous pyridine also forms a complex with FMN (WEBER, 1966) and the fluorescence of RF is maximal in pyridine-water (FOX, 1953). These features may have some relevance to the usual sequential action of pyridine and flavin nucleotide enzymes in the respiratory system (p. 185). Complex-formation by RF, and the properties of the complexes, are very dependent on the solvent (WEBER, 1966) and water plays a unique steric role; this is related as much to the hydrophobic aromatic nucleus as to the hydrophil components of the molecule (PENZER and RADDA, 1967).

Redox changes dominate the chemical behaviour of RF in all its conjugates. The E'_0 of the free RF is − 185 mV, rather strongly reducing, and those of FMN, FAD and lumiflavin are in the same general region. Protein conjugation raises the value (p. 32) as far as the region of redox neutrality. In contrast to the pterins RF in its enzymes donates and accepts electrons with equal facility, an ambivalence to which the structure of the flavin itself also contributes materially (SZENT-GYORGYI, 1960). There is a particularly small change in resonance-energy between oxidised and reduced forms (PULLMAN and PULLMAN, 1963, p. 396) and in this connection it is interesting that the whole prosthetic group of cytochrome b_2 is isomorphous and equally stable in the two states (LABEYRIE, 1971).

When reduced *in vitro*, using powdered zinc, RF passes through green *(verdo-* and *chloro- flavin)* and red *(rhodoflavin)* stages to the colourless *leucoflavin*. The coloured intermediaries are free radicals as shown by their paramagnetic properties. The green colour is due to a blue radical Fl· RH mixed with unchanged flavin and the red to Fl· R⁻ or to [MFl· R]⁺ ions, where Fl· is the flavin free radical, R a relevant anionic group and M one of the metal cofactors (MÜLLER et al., 1971). The flavin radical (SLATER, 1966) is a semiquinone (quinhydrone) with the high redox power of such intermediaries (p. 46).

It will be noted that in this ring system again (p. 67) an m-Q is stabilised (X) and that two molecules can form a quinhydrone without steric hindrance.

Oxidised RF compounds have a planar ring-system whereas the reduced form is bent around the axis formed by the two nitrogens, N-5 (N-9) and N-10, of the pyrazine ring (KIERKEGAARD et al., 1971) and this plays an important part in the redox cycle. Free energy is minimal in the planar configuration, when adenine and other aromatic systems can fit most closely. Further details are given by SLATER (1966; PENZER and RADDA, 1967; WELLNER, 1967; YAGI, 1969 and KAMIN, 1971).

3.11 Other Zoochromes

A number of zoochromes that have been examined have not been identified, even as far as their chemical class (Table 3.14). No doubt further work will show that some of these do belong to known classes of biochrome, even if they are rather unusual members, or belong to new subclasses but it seems equally likely that some may represent classes as yet unrecognised in living organisms. In view of the small percentage of coloured materials of animals that so far has been studied it is possible that still further new classes are present, though probably not in proportion to the number still to be examined; there is enough taxonomic uniformity to be sure that representative species give much of the information but there can be surprisingly sharp chromatological differences between closely related lower taxa (p. 93). It is among integumental chromes that variety is most evident but in the deeper tissues collectively there are as many known classes (Table 5.1) and these tissues have been little studied. In the body-fluids coloured materials other than the oxygen-carriers (p. 211) have received little attention.

Space does not permit a detailed account of what is known about the chromes in Table 3.14 but this is available in the references given and shorter accounts are given by Fox (1953) and Fox and VEVERS (1960). These authors also describe a number of unidentified metal-based chromes not included in the table. Four only of the partially characterised chromes will be considered, *haemovanadin* because it is the greatest classical enigma and three in which I have been interested recently, *rufin*, the orange secretion of the slug, *Arion rufus*, a brown chrome in the chromagogen tissue of the earthworm *Lumbricus* and a yellow one in the Cuvierian Organs of the holothurian, *Holothuria forskalii*. These therefore are from blood, epidermis, mesoderm and gut tissues respectively.

Haemovanadin: In situ this chrome is green and in some ascidians it is free in the plasma; in others it is in special intracellular bodies of cells which also contain H_2SO_4 at 9% concentration or $1.4N$ (WEBB, 1939). There are also cells of other types and colours in ascidian blood; conceivably some may contain other V-chromes but most do not and some certainly have carotenoids. Vanadochrome is conjugated with protein which can be separated from the chrome by dialysis or by precipitation, the bonding being rather loose. The chrome has a molar weight of 900 (WEBB, 1939) so that it must be a relatively complex molecule. HENZE (1927) thought that it was a pyrrole compound but this is now disproved (BIELIG, 1967); it cannot be a simple metalloprotein (p. 47) but the intriguing *ferroverdin* of *Streptomyces* (CANDELORO et al., 1969) might provide a clue. The properties of vanadochrome have been further investigated by CALIFANO and colleagues (1947, 1950). *In vitro* it can be changed from the green V^{IV} to the brown V^{III} state. In some ascidians there appears to be an analogous, or perhaps more correctly homologous, niobium compound (CARLISLE, 1958) and in others there is an iron analogue (ENDEAN, 1955). Other properties of vanadochrome are considered later (p. 205).

Table 3.14. Partially characterised zoochromes

Source of chrome and trivial name	Colour and absorption peaks	Other properties	Reference
Porifera			
Halichondria: tissues	Blue	Redox and pH changes	ABELOOS and ABELOOS, 1932
Aplysina: tissues	Yellow	Oxidising black at alkaline pH; soluble in H_2O and organics.	KRUKENBERG, 1882
Cnidaria			
Pelagia: tissues	Magenta	Chromoprotein	FOX and MILLOTT, 1954b
Adamsia: acontia	Purple: (435, 450, 465, 555nm)	Blue u.v.f.; pH-change	CHRISTOMANOS, 1953
Anemonia: ectoderm	Green: (512nm)	Green u.v.f.	FOX and VEVERS, 1960, p. 169
Tealia, Anemonia, ectoderm: actiniochrome	Purple: (572nm)	Destroyed by heat	FOX and VEVERS, 1960, p. 161
Calliactis: tissues: (calliactin)	Red-purple	pH and rH changes	LEDERER et al., 1940
Cerianthus: tissues: purpuridin	Red, purple or violet	Soluble in alkaline H_2O; stable	KRUKENBERG, 1882
Alcyonaria: skeleton	Red	Iron pigment	DURIVAULT, 1937
Turbellaria			
Polycelis: tissues	Violet	Soluble, N HCl	NEEDHAM, 1965a
Rhynchodemus, 2 spp.: rhabdite cells	Blue sp.	pH-change; soluble in acid alcohol	MOSELEY, 1877
	Red. sp.	No pH-change; insoluble in acid alcohol	
Annelida			
Hirudinea: integument	Violet (502nm)	Soluble, NHCl; pH, and slight redox, changes in colour	NEEDHAM, 1966b
Lumbricus: chloragogen	Red-brown; no peaks in visible range	Soluble in MeOH-HCl	NEEDHAM, unpublished.
Eulalia, Phyllodoce: integument	Green	Two or three distinct chromes, non-fluorescent; soluble in polar solvents.	LEDERER, 1940
Neanthes: egg	Yellow: (400, 450–480nm)	Soluble in MeOH	SUZUKI, 1968
Lumbriconereis: egg	Pink (365–85, 405, 450–80nm)	Soluble in MeOH	SUZUKI, 1968
Chaetopterus: integument	Red-brown	Melanoid; soluble in dilute alkali	KENNEDY and NICOL, 1959
Pomatoceros: gills.: 2 chromes	1. Pink: (500, 535nm)	Water-soluble	DALES, 1962
	2. Blue: (272, 405, 505, 542nm)	Insoluble in water	DALES, 1962
Arthropoda			
Triops,Cypris: connective tissue	Green	pH and rH colour changes	FOX, 1955
Geometrinae: wings	Green	pH change	FORD, 1955, p. 5
Daphnis: wings	Green	pH change	FORD, 1955, p. 5
Psylla mali: integument	Green or red	Variable with host plant	MAHDIHASSAN, 1947

Table 3.14. (continued)

Source of chrome and trivial name	Colour and absorption peaks	Other properties	Reference
Lampyrid beetles: tissues: lampyrine	Pink: (565nm)	U.v.f.; pH colour-change; soluble in acid and alkali, insoluble, neutral H_2O and organics	METCALF, 1943
Anoplodactylus (pycnogonid) blood corpuscles	Pink, purple or blue	Redox change	DAWSON, 1934
Mollusca			
Patella: ova	Green (620–630nm)	Chromoprotein, water-soluble,	BANNISTER et al., 1969
Pigment Y	Yellow (380–390nm)	Prosthetic group, soluble in polar organic solvents	GOODWIN and TAHA, 1950
Janthina: secretion (? and shell): janthinin	Violet (app. 590, 530, 490nm)	Red fluorescence in u.v. and visible; pH colour-change; soluble in polar organics;	MOSELEY, 1877
Doris: digestive gland	Blue	pH and rH colour-changes; soluble in dilute acid	ABELOOS and ABELOOS, 1932
Chromodoris	Blue (620–622nm)	pH and rH colour-changes	PREISLER, 1930
Arion ater (var. *rufus*): integument	Orange-yellow	Soluble in H_2O (420, 440nm) and MeOH-HCl (470, 500 nm)	NEEDHAM, unpublished.
Octopus: branchial heart appendage: adenochrome	Red (505nm)	Redox change; soluble in H_2O, not in organics.	FOX and UPDEGRAFF, 1944
Ectoprocta			
Bugula: Body wall	Purple (545nm)	Redox colour-change; water soluble; insoluble in organics	KRUKENBERG, 1882; VILLELA, 1948
Echinoderma			
Molpadia intermedia: coelomic corpuscles	Red (569, 613nm)	Redox change (588nm, reduced)	CRESCITELLI, 1945
Holothuria forskali :integument ?Ribitylflavin	Yellow (442, 475nm)	Fluorescent in U.V. and visible; soluble, water and polar organics.	KRUKENBERG, 1880 MACMUNN, 1890 RIMINGTON and KENNEDY, 1962, p. 594
Holothuria forskali :cuvierian organs	Yellow (450, 474nm)	Very stable and insoluble; soluble in c H_2SO_4 and in c NH_4OH	NEEDHAM, unpublished
Tunicata			
Ascidian vanadocytes vanadochrome	Green,	Chromoprotein water soluble	HENZE, 1911 WEBB, 1939
	Red-brown	Free chromophor water soluble	CALIFANO and BOERI, 1950
Ascidia fumigata :blood plasma	Yellow	Blackens in alkaline aerated medium; soluble in H_2O and organics	KRUKENBERG, 1880
Aves			
Psittaci: feathers	Pale yellow	U.v. fluorescent: gold to green (505, 540nm to 479, 514nm); soluble in hot alcohol	VÖLKER, 1937

Table 3.14 (continued)

Source of chrome and trivial name	Colour and absorption peaks	Other properties	Reference
Psittaci: feathers	Red to yellow; yellow in solution (433, 459nm to 420, 440nm)	Non-fluorescent; soluble in pyridine-HCl, in petrol	VÖLKER, 1942
Mammalia Urochrome	Yellow (no peak in visible range)	Soluble in water, insoluble in apolar lipids.	RANGIER, 1935
Cerumen chrome	Yellow (no peak in visible range)	Soluble in aqueous media, insoluble in neutral organics	NEEDHAM, unpublished

Rufin: At first this was thought to be a bilin (DHÉRÉ and BAUMELER, 1928; DHÉRÉ et al., 1928) and to be related to the red-brown *rufescin* of *Haliotis rufescens* (TIXIER, 1945). The differently coloured pigments of the shells of other species of *haliotis* are indeed bilins (p. 88) so that *rufescin* probably is a bilin derivative but *rufin* has none of the positive properties of bilins (Table 3.11) and DHÉRÉ himself appeared less convinced of this relationship than later reviewers. Unlike the more primitive gastropods the pulmonates in general do not have bilins in their shells and this seems likely also for the integument (COMFORT, 1951). Rufin resembles the flavone of the snail *Helix* in being the secretion of the epidermal glands, in reducing to a colourless product and in deepening in hue at alkaline pH. It resembles the *Helix* flavone also in being very water-soluble and in being secreted on irritation. However the colour is more orange and it lacks the typical flavone properties. The aqueous solution has twin peaks in the visible range, at 420 and 440 nm, the single peak of the *Helix* chrome being around 385 at neutral pH, i.e. outside the visible range. In the u.v. rufin has peaks at 234 and 267 nm. By extracting with acid alcohol rather than H_2O DHÉRÉ missed these features and risked extracting other chromes in addition. In HCl the colour is red and the absorption peaks move up to 476, 499 nm, near the positions DHÉRÉ found in alcoholic HCl. Rufin gives no clear pyrrole reaction. Strong acids cause a strong yellow u.v.f. and, under conditions not yet specified, a violet visible colour. I have not obtained DHÉRÉ's green product through the action of H_2SO_4 but this also depends on rather critical conditions. Rufin contains no iron.

Chromagogen Brown: This probably is the 'fuscin' recorded by VAN GANSEN (1956) and ROOTS (1960). It is extracted along with a small amount of protoferrihaem after ribitylflavin has been extracted by water and the carotenoid by methanol. The brown chrome is extracted by methanolic HCl and from this it will only partially partition into $CHCl_3$ or into CCl_4. The absorption-spectrum has a featureless decline with increase in wavelength; alone among the reagents tested H_2SO_4 brings up a shallow peak around 405 nm. A violet u.v.f. is associated with the chrome but $cHCl$ and cH_2SO_4 turn this yellow; the visible colour is deepened by these acids and when partitioned into $CHCl_3$ it becomes almost purple. There is a positive reaction for iron which remains preponderately in the MeOH-HCl phase when partitioning with $CHCl_3$. The iron is released only by ashing so that the chrome is not related to haemosiderin or to the ferredoxins. It precipitates on neutralisation, possibly indicating a protein component. Chromatographical development in a 5:3:2 mixture of collidine: ethanol: water gives a mobile and an immobile component of the same colour. There is virtually no redox change in colour.

Cuvierian Yellow: This bright yellow chrome has some unusual properties. It is inextractable by any of the common organic and inorganic solvents and even by commercial (90%) formic acid and by cHCl. Only cH_2SO_4 and cNH_4OH (0.88 s. g.) dissolve appreciable amounts. In the latter it has absorption-peaks at 226, 290, 450 and 474 nm and in 50% H_2SO_4 (for technical convenience) at 208, 288, 456 and 523 nm. In 50% H_2SO_4 therefore it is red though as extracted by cH_2SO_4 it has the *in situ* yellow colour. Both *in situ* and in these solvents it has a brilliant yellow u.v.f. Spectrum and fluorescence show considerable resemblance to those of ribityflavin, as also does the yellow chrome of the integument (p. 37), but the fluorescence of RF is quenched at extreme pHs and its visible colour does not persist unchanged in cH_2SO_4. Although it does not extract the chrome cHCl produces a green derivative, whereas it has little effect on the colour of RF. The whole substance of the Cuvierian organs is very resistant to all four of the powerful reagents mentioned so that the chrome may be stabilised by this to an extraordinary degree.

3.12 Zoochromes in Relation to pH

A number of zoochromes show a striking change in colour at a certein pH value (Table 3.15) and this is one criterion for their identification. It is also a familiar feature of some classes of organic chrome synthesised in the laboratory and is used in the simple method of measuring pH by 'indicators'. It depends on the change from unionised weak acid or base to ionised salt, with the associated changes in the various parameters considered in chapter 2. The properties and functions of biochromes therefore are very dependent on pH. Some natural and laboratory chromes show two pH colour-changes, indicating two relevant ionising groups, separated by an appreciable pH range and therefore usually on opposite sides of neutrality. The intermediate colour is due to an amphoteric salt, therefore. Examples are the echinochrome of *Arbacia* (VLÉS and VELLINGER, 1928), the Calliactis chrome (FOX and PANTIN, 1944) and arenicochrome (VAN DUIJN, 1952).

It is noteworthy that most of the identified zoochromes of Table 3.15 are quinones, as are the commercial indicators and pH colour-change is less characteristic of some other classes of biochrome. However many of them have quinonoid members (p. 40) and certainly most classes have at least one polar group which ionises within the central pH-range. The porphyrins (LEMBERG and LEGGE, 1949, p. 68), pterins (ZIEGLER-GUNDER, 1956) ribitylflavin and the anthocyanins have amphoteric ions over a certain range. One example of a pH-indicator chrome *in vivo* is that which registers a spatial gradient in pH in the egg of the polychaete *Perinereis* (p. 274). Examples are rather common in plants.

No doubt biochromes are greatly affected by the normal pH buffering *in vivo*, the rate of redox reactions varying considerably with pH. There is a difference of as much as 6pK units between the ground state and the first excited state of some chromes (GLASS, 1961, p. 831). Squid metarhodopsin (p. 164) is very sensitive to pH changes (p. 167) and so are the O_2-transport zoochromes (p. 214). Cytochrome c is high-spin (p. 192) below pH 2.5 and low spin above pH 4.7 (SCHEJTER, 1966). Helicorubin (p. 241) oxidises to a ferric complex in acid medium but not in the normal alkaline crop-fluid of the snail (KEILIN, 1957).

A number of zoochromes are both pH and redox colour-change indicators: examples

Table 3.15. Zoochromes that change colour according to pH

Zoochrome	Colour on acid side of transition	Transition – pH (or colour)	Colour on alkaline side of transition	Reference
Carotenoids				
Mytiloxanthin	Red		Yellow	SCHEER, 1940
Metarhodopsin	Red	7.0	Yellow	PROSSER and BROWN, 1961 p. 357
Flavonoids				
Helix flavone	Colourless		Yellow	KUBISTA, 1950
Sertularia argentea flavone	Colourless	8.0	Yellow	PAYNE, 1931
	Yellow	8.5	Orange	PAYNE, 1931
	Orange	10.0	Brown	PAYNE, 1931
Sertularia pumila flavone	Colourless	6.5	Yellow	PAYNE, 1931
	Yellow	7.0	Brown	PAYNE, 1931
Bombyx, chrome of silk (?flavone)	Green		Yellow	JUCCI, 1932
Cionus (weevil), anthocyanin	Red		Blue	HOLLANDE, 1913
Quinones				
Arbacia, echinochrome	Orange	4.0–6.5	Violet	VLES and VELLINGER, 1928
	Violet	7.0–8.0	Yellow	
Paracentrotus, NQ	Yellow		Green	GLASER and LEDERER, 1939
Echinarachnius, NQ	Purple		Green	JUST, 1939
Antedon NQ	Orange		Red	DIMELOW, 1958
Antedon (? AQ)	Yellow	Red	Violet	ABELOOS and TEISSIER, 1926
Laccaic acid	Yellow		Red	FOX, 1953, p. 207
Protoaphin	Yellow	7.0	Magenta	DUEWELL et al., 1948
Xanthoaphin	Yellow		Pink	DUEWELL et al., 1948
Chrysoaphin	Yellow		Crimson	DUEWELL et al., 1948
Erythroaphin	Red		Green	DUEWELL et al., 1948
Zoopurpurin	Red	7.0	Colourless	VUILLAUME, 1969
Arenicochrome	Red	6.0	Green	
	Green	9.0	Violet	KENNEDY, 1969
Hallachrome	Red	8.3	Green	RAZZA and STOLFI, 1931
Lithobius, violet	Red-Orange	5.0	Violet	BANNISTER and NEEDHAM, 1971
Haemerythrin	Yellow		Red	MARRIAN, 1927
Porphyrin derivatives				
Protoporphyrin	Green	very low	Purple	CHU, 1946
Haliotis, shell chrome (? coprophorphyrin)	Blue		Violet	TIXIER and LEDERER, 1949; RÜDIGER, 1970
Aplysioviolin	Blue		Violet	LEDERER and HUTTRER, 1942
Aplysiorhodin (? urobilin)	Violet	Red	Yellow	LEDERER and HUTTRER, 1942; RÜDIGER, 1970
Aphysioazurin	Green		Blue	RÜDIGER, 1970
Melanins				
Erythromelanin	Pink		Yellow	FLESCH, 1968

Table 3.15 (continued)

Zoochrome	Colour on acid side of transition	Transition – pH (or colour)	Colour on alkaline side of transition	Reference
Unidentified chromes				
Halichondria, tissue chrome	Blue		Yellow	ABELOOS and ABELOOS, 1926
Velella chrome	Red	Yellow	Blue	HAUROWITZ and WAELSCH, 1926 KROPP, 1931
Adamsia, acontiachrome	Colourless Purple	1.9 12.4	Purple } Yellow }	CHRISTOMANOS, 1953
Calliactis, tissue chrome	Orange-yellow Blue	Brown Violet	Blue } Red }	FOX and PANTIN, 1944
Rhynchodemus tissue chrome	Red		Blue	MOSELEY, 1877
Haliotis, foot chrome	Yellow-green		red	ABELOOS and TEISSIER, 1926
Janthinin	Blue	Violet	Purple	MOSELEY, 1877
Doris, chrome of digestive gland	Blue		Yellow	ABELOOS-PARISE, and ABELOOS-PARISE, 1926
Chromodoris chrome of tissue and blood	Pink-orange	4.3–5.6	Blue	CROZIER, 1922 PREISLER, 1930
Rufin of *Arion*	Red	Yellow	Violet	DHERE et al., 1928
Rufescin of *Haliotis*	Green Red	3.0 7.0	Red } Yellow }	DHERE and BAUMELER, 1930
Eulalia, two integumental chromes	1. Green 2. Green		Brown } Violet }	LEDERER, 1940
Phyllodoce, integumental chrome	Green		Red	LEDERER, 1940
Perinereis, chrome of egg	Violet		Yellow	SPEK, 1930
Hirudinea, 'black' integumental chrome	Violet		Yellow	NEEDHAM, 1966b
Triops, chrome of connective tissue	Green		Brown	FOX, 1955
Lampyrine	Rose red		Dull blue	METCALF, 1943
Geometrinae, chrome of wings	Brown		Green	FORD, 1955, p. 5
Daphnis nerii chrome of wings	Green		Brown	FORD, 1955, p. 5
Styelopsis	Green		Red	ABELOOS and TEISSIER, 1926

(Table 3.15) are the tissue-chrome of the sea-anemone *Calliactis*, the green chrome of the crustacean *Triops*, the blue of the nudibranch *Chromodoris* and the violet of the centipede *Lithobius*. The dual property is typical of many quinones and is one measure of their versatility. Most of the commercial quinonoid pH-indicators show a redox colour-change also. The transitional metals owe their chromatic excellence to the combination of, and labile transition between, ionic and covalent properties, between low-spin and high-spin states (p. 228). It seems likely that the secret of 'chemi-osmotic' coupling (MITCHELL, 1966) lies in this field.

The auxochrome groups (p. 11) of the dyeist (Gurr, 1967) are mostly ionising substituents and their ionic bonding properties are as useful in this industry as their bathochromic action. Such ionising groups make some biochromes very good dyes. Dyes thus are large anions or cations which use the electrovalent bond for binding 'fast' to their substrate (Conn, 1953; Gurr, 1967). Biochromes use the property *in vivo* for chemical conjugation and for binding to matrix material. In fact the property probably is much curbed *in vivo* since many zoochromes bind to muscle and some other tissues in the course of extraction from their normal site. The violet chrome of leeches (Needham, 1966b) seems to bind specifically to the nucleus, chromaffin *par excellence;* presumably this is a cationic dye. Extractants probably expose normally protected binding groups on the matrix proteins as well as on the chromes themselves and this is an indication of what could occur normally if there were no safeguards. The use of tyrian purple and other biochromes as commercial dyes has been mentioned (p. 59). The coccid anthraquinones are still used for various purposes and a variety of terms, 'cramasie', carmine, crimson, cochineal, etc. bear witness to a long tradition in many fields. Carmine and haematoxylin continue to hold their own as dyes for microscopical biological preparations.

Also relevant here are ionic-exchange properties. The melanins are cation-exchangers (Whyte, 1958) and so are the Fe and Cu protein enzymes (Bersin, 1957). This is in keeping with the predominance of anionic ('acidic') dyes commercially.

3.13 Conclusion

There are relatively few known chemical classes of zoochrome, some of which are obtained from dietary sources. Within each class only a restricted subclass usually is exploited. All classes have molecules with the characteristic chromatic properties considered in the previous chapter; they are photo- and chemo- excitable and in turn excite chemical reactions essentially of a redox type. In detail there are many significant differences in properties between the classes, useful technically for separation and identification as well as being important biologically. In general all are most soluble in polar lipid media but most have derivatives that are soluble in aqueous media. Few are soluble in apolar lipids. The melanins are sparingly soluble in all media and this is true also of some unidentified pigments.

Carotenoids have a particularly low water-solubility and are exceptional also in having an open chain molecule. The others are ring-condensed, polycyclic and heterocyclic, and mostly with N atoms in the rings and oxygen groups as side-chains. The quinonoid structure is universal in one main class and common in a number of others: this is in keeping with the outstanding chemical reactivity of this structure. Many zoochromes form chromoproteins and other conjugates which modify their properties in various useful ways. Most zoochromes have weakly acidic or less commonly weakly basic auxochrome groups so that they change colour with pH and can act like commercial indicators and dyes.

A considerable number of zoochromes studied have no evident relationship to the chemical classes already recognised among biochromes. These constitute an urgent and exciting challenge to research.

Chapter 4

The Taxonomic Distribution of Zoochromes

It is not possible to discuss the biological significance of zoochromes without knowing something of their taxonomic distribution, so far as it is known. A detailed survey would be tedious, since it is already clear that nearly all classes of chrome are widely distributed. For the same reason it is unfortunately difficult to make many clear-cut generalisations, and attempts are likely to fall from, as well as between, the two stools. Probably most rewarding is an initial comparison between animal and plant chromes.

4.1 Distribution between Animals and Plants

Animals obtain their carotenoids from plants, but they produce a number of derivatives which are virtually or completely unknown in plants themselves; vitamin A and the carotenoproteins are the best examples. Animals also have relatively more oxycarotenoids. Fuscins have not been recorded from plants. As already emphasised flavonoids are *the* flower-chromes, but are used by very few animals, which obtain them from plant-food. Although 80% of the known biological ternary quinones come from plants alone, most of the remainder are equally restricted to animals. Benzoquinones are more common in animals than in plants (THOMSON, 1957). The important benzoquinone, ubiquinone (UQ), occurs in all animals and is one of the quinones common to both animals and plants. It shows other major taxonomic distributional variations, being present in Gram-negative but not in Gram-positive bacteria, which have the naphthoquinone, vitamin K, as functional alternative; some bacteria have both, and some neither. The lower fungi have vitamin K, and other fungi UQ; plants have both.

It is debatable if the phytoflagellates are true animals, in which case chlorophyll occurs only adventitiously in animals (1) in the gut, (2) in unicellular symbionts and (3) in chloroplasts salvaged in functional condition from dietary plant material (HINDE and SMITH, 1972). The breakdown-products, phaeophorbides, phylloerythrins and chlorins, are used by animals as integumental chromes and are rare in plants themselves. Haemoglobin and the other oxygen-carrier zoochromes are rare in plants though the cytochromes and the simple iron and copper metallo-proteins are common in both kingdoms. Free porphyrins and bilins are much rarer in plants than in animals. Nevertheless the bilins used by plants, namely, phycocyanin, phycoerythrin, and phytochrome, are used for much more fundamental physiological purposes than any known animal bilin by animals; even so the name phytochrome, which could have been used as the counterpart of 'zoochrome', is a little pretentious. Amusingly enough the sea hare, *Aplysia* uses plant phycocyanin and phycoerythrin, but only for 'visual effect' purposes.

Plant melanins are phenolic rather than indolic and indolic chromes are rare in plants; ommochromes also have not been found, though related phenoxazones occur in microorganisms. Pterins are rare in plants, though folic acid is synthesised by microorganisms. The free purines of plants differ, methylated derivatives such as caffeine being common and uric acid rare. Plants have few nitrogen-containing chromes, apart from the porphyrins. The betacyanins (p. 60) one of the few other N-containing classes in plants is not known in animals! Since plants squander nitrogen in alkaloids and some other products, the paucity of N-containing classes of plant chrome is intriguing.

4.2 Distribution between the Phyla and Classes of Animals

An outline classification of animals (Table 4.1) is given for reference.

Since all taxonomic groups have exploited a large proportion of the chemical classes of chrome and since there are more phyla than classes of chrome it is preferable to survey according to the latter rather than phylum by phylum. For instance it is possible to make several useful generalisations about *carotenoids:* they are surprisingly abundant in spite of their extraneous origin and further they are not necessarily much poorer, in quantity or variety, in carnivores than in herbivores, from which the former derive them at third hand. Cephalopod molluscs and whalebone whales have very little but there are a number of herbivorous mammals which also have virtually none (Table 14.1) and actively exclude carotenoids from the tissues (p. 302). The mud-burrowing interrelated phyla Priapula, Sipuncula and Echiura have very little in the integument, unlike most other phyla.

Carotenoproteins are found in parasitic Protozoa (CHATTON, LWOFF and PARAT, 1926) and in a sponge (CHEESMAN et al., 1967) as well as in most of the other phyla examined (Table 3.2) so that the ability to conjugate the two components may have been acquired early in animal evolution. Carotenoproteins have been found mainly in Crustacea and insects, and next in abundance in echinoderms and Cnidaria. Vitamin A has not been found in Protozoa, Porifera or Coelentera (FISHER and KON, 1959) and so may be a later evolution. On the other hand, some of the most peculiar of animal carotenoids have been isolated from the sponge *Reniera* (VUILLAUME, 1969, pp. 45–46).

The Porifera, Echinoidea, Polychaeta, Gastropoda and Insecta have β-carotene as their main carotenoid (Table 14.1) usually together with echinenone, whereas most of the large groups of marine invertebrates, Cnidaria, Asteroidea, Bivalvia and Crustacea have mainly oxycarotenoids (GOODWIN, 1962). Among vertebrates, the Amphibia and mammals have mainly carotene, while more brightly coloured groups, the teleosts and birds, have oxycarotenoids at least in their integument. It is not clear that this difference will explain the distribution among invertebrates, though it is generally true that the oxycarotenoid/carotene ratio is highest in the integument and lowest in the deep viscera. The γ-carotene isomer is rare in invertebrates (FOX, 1953, p. 82).

Fuscins: Recent records might imply that fuscins are common in mammals and rare elsewhere but this may be more an index of research interest than of differential distribution. They have been recognised in lower vertebrates (VERNE, 1926; SMITH, 1955) and among invertebrates, in the wings of butterflies (PRENANT, 1913), the nerve cells of sipunculids and gastropods (SCHREIBER, 1930), the ciliated cells of the gills of the mussel, *Mytilus* (BENAZZI-LENTATI, 1939), the chromagogen tissue of the polychaete, *Thoracophelia* (FOX et al., 1948), the coelomocytes and connective tissue of the feather-star, *Antedon* (DIMELOW, 1958) and the axial organ of some echinoids (VEVERS, 1963). The records are thin but the taxonomic distribution potentially wide.

Flavonoids: There are few well established instances of significant amounts of flavonoids in animals (FORD, 1944; MORRIS and THOMSON, 1963; FELTWELL and VALADON, 1970). These are mainly in larval and adult Lepidoptera and in plant-feeding Hemiptera and Coleoptera. In adult Lepidoptera they must derive from larval food. Pugmoth caterpillars feed specifically on the flowers of their host-plant and their colour varies with that of the flower. The only two plausible records in other phyla are the yellow secretion of the foot and mantle of the garden snail, *Helix* (KUBISTA, 1950) and the yellow to brown tissue-pigments of the marine hydroid, *Sertularella* (PAYNE, 1931). Even so it is curious that a marine carnivore should have this class of chrome; endogenous synthesis is implied.

Table 4.1. The major taxa of animals in relation to chromatology

Phylum (Ph), Class (Cl) and subclass (S. Cl.)	Any outstanding chromatological features
1. *Unicellular Grade*	
Ph. Protozoa	Mostly transparent, colourless animals
Cl. Mastigophora (Flagellata)	Includes photosynthesing types
Cl. Rhizopoda (Sarcodina)	Radiolaria and Foraminifera often coloured
Cl. Sporozoa	Parasitic, rarely pigmented but may affect host's chromes.
Cl. Ciliophora	Polycyclic quinones as chromes
2. *Multicellular Grade*	
Ph. Mesozoa	Microscopic parasites, unpigmented
2.1 *Coelodiploblastic Metazoa*	
Ph. Porifera	Various bright colours, mainly carotenoids, some peculiar to phylum
Ph. Cnidaria (Coelentera)	Most sub-taxa have coloured members
Cl. Hydrozoa	Carotenoid-accumulations in gonads, buds and regenerates
Cl. Scyphomedusae	Large medusae, often brightly coloured
Cl. Anthozoa	Large polypoid Cnidaria with variety of colour
Sub. Cl. Alcyonaria (Octacorallia)	Bright iron chromes, carotenoids and bilins
Sub. Cl. Zoantharia	Great variety of chromes in anemones and corals
Ph. Ctenophora	Mainly delicate structural colours
2.2 *Triploblastic, bilaterally symmetrical Metazoa*	
Ph. Platyhelmia	
Cl. Turbellaria	Bright chromes, particularly in marine polyclads and terrestrial triclads
Cl. Trematoda	Parasitic, rarely coloured but sometimes induce coloration in host
Cl. Cestoda	
Ph. Gnathostomulida	Microscopic, poorly known; chromes rare
2.2.1 *Pseudocoelate Triploblasts*	
Ph. Acanthocephala	Parasitic, sometimes brightly coloured
Ph. Nematoda	Little integumental colour; haemoglobin in some; some internal chromes identified
Ph. Gastrotricha	Microscopic, little colour
Ph. Kinorhyncha (Echinodera)	Microscopic, little colour
Ph. Rotifera	Small, little colour except in cold lakes and deep water
Ph. Endoprocta (Kamptozoa)	Small, little colour
2.2.2 *Spiralian, protostome line of Triploblasts*	
Ph. Nemertina	Colourful; Hb in blood corpuscles
2.2.2.1 *Coelomate spiralian Triploblasts*	
Ph. Priapula	Little integumental colour; haemerythrin (Hery) in corpuscles of body-fluid
Ph. Sipuncula	Little integumental colour; Hery in coelomic corpuscles
Ph. Echiura	A number of bright chromes, sometimes in most organs; Hb in coelomic corpuscles
Ph. Mollusca	Very varied integumental and shell colours; haemocyanin (Hcy) in blood of many; some have Hb

Table 4.1 (continued)

Phylum (Ph), Class (Cl.) and Subclass (S. Cl.)	Any outstanding chromatological features
Cl. Monoplacophora	
Cl. Polyplacophora	Hcy in blood; myoglobin (Mb) in some muscles
Cl. Gastropoda	Great variety of shell and integumental chromes, some unidentified; Hcy in blood
Cl. Scaphopoda	
Cl. Bivalvia	Shell chromes largely unidentified
Cl. Cephalopoda	Ommochromes in integument; melanin in ink

2.2.2.1.1 *Polymeric spiralian Coelomates*

Ph. Annelida	Mostly well pigmented but not often brightly; emphasis on porphyrins and bilins; chromagogen tissue with varied chromes
Cl. Polychaeta	Most colourful Class; Hb, chloro Hb and ? Hcry in blood or coelom
Cl. Oligochaeta	Not bright; usually Hb in blood plasma
Cl. Hirudinea	Various chromes in integument and botryoidal tissue
Ph. Arthropoda	Varied chromes in many sites. Ommochromes and pterins predominate, porphyrins rare
S. Ph. Onychophora	Coloured integument
S. Ph. Tardigrada	Microscopic, brown
S. Ph. Pentastoma	Parasitic, little colour
S. Ph. Chelicerata	
Cl. Merostomata	
S. Cl. Xiphosura	Coloured integument; Hcy in plasma
Cl. Arachnida	
S. Cl. Scorpionida	Coloured integument; Hcy in plasma
S. Cl. Aranida	Integumental chromes unidentified; some have Hcy in plasma
S. Cl. Acarida	Heavy carotenoid-accumulations in integument
S. Cl. Opilionida	Coloured
S. Cl. Solpugida	Coloured
Cl. Pycnogonida	Blood corpuscles of varied colour (*Anoplodactylus*)
S. Ph. Mandibulata (Antennata)	
Cl. Trilobita	Brown chromes still visible in many
Cl. Crustacea	Colourful: carotenoids, bilins, melanins, ommochromes pterins, ribitylflavin; Hb, Hcy
S. Cl. Entomostraca	Small, relatively little integumental colour; Hb in blood of some, in all orders
S. Cl. Malacostraca	Larger, strong and varied integumental colours. Hcy in blood, never Hb
O. Isopoda	Ommochromes in integument; Hcy in blood
O. Amphipoda	Carotenoids and ommochromes in integument; Hcy in blood
O. Decapoda	Large and brightly coloured; Hcy in blood
Macrura, Anomura	No melanin in integument, many other chromes
Brachyura	Melanin in integument and other chromes
O. Stomatopoda	Hcy in blood
Cl. Myriapoda Chilopoda	Violet to red connective-tissue chrome
Cl. Myriapoda Diplopoda	Ommochromes or porphyrins in integument; defensive benzoquinones
Cl. Insecta	Many chromes: carotenoids, flavonoids, phenolic and indolic melanins, bilins, ommochromes, pterins, ribitylflavin and unidentified
S. Cl. Apterygota	
S. Cl. Pterygota	
O. Ephemeroptera	

Table 4.1 (continued)

Phylum (Ph), Class (Cl) and Subclass (S. Cl.)	Any outstanding chromatological features
O. Odonata	Structural colours; ommochromes in muscle
Orthopteroid group	Large insects, varied colours
O. Isoptera	Subterranean, little colour
O. Hemiptera	Variety of chromes, including polycyclic quinones
O. Coleoptera	Colourful, chemical and structural colours
O. Hymenoptera	Bright and dull colours
O. Lepidoptera	Most colourful insects; colour mainly in wings, varied chromes and structural colours
O. Diptera	Mainly dull colours, structural and unidentified chromes.

2.2.3. *Lophophorate oligomeric Coelomates*

Ph. Brachiopoda	Mostly dull, little known chromes; Hery in blood of lingulids
Ph. Ectoprocta (Polyzoa)	Colonial, variously coloured; gut forms brown bodies
Ph. Phoronida	Small, tubicolous, little colour; Hb in blood corpuscles
Ph. Pogonophora	Tubicolous, little colour; Hb in blood corpuscles

2.2.4 *Orthoradial deuterostome line of Triploblasts*

2.2.4.1 *Oligomeric Deuterostomes*

Ph. Chaetognatha	Small; pelagics transparent and colourless; deeper types coloured
Ph. Echinoderma	Very colourful, mainly carotenoids and quinones
Cl. Crinoidea	Naphtho- and anthra-quinones; no carotenoids
Cl. Asteroidea	Oxycarotenoids mainly, chloroporphyrin, echinochromes
Cl. Ophiuroidea	Oxycarotenoids, echinochromes, melanin
Cl. Echinoidea	Echinochromes and a few quinones of other classes, carotenoids, melanins
Cl. Holothuroidea	Melanins; ?ribitylflavin, obscure chromes in viscera; Hb in coelomic corpuscles of some
Ph. Hemichorda	Chromes in integument and stolons, unidentified

2.2.4.2 *Polymeric Deuterostomes*

Ph. Chordata	Most chromes except ?ommochromes
S. Ph. Urochorda (Tunicata)	Colourful, varied chromes, often in blood corpuscles, carotenoids; vanadochrome in blood, plasma or corpuscles
S. Ph. Cephalochorda	Little colour
S. Ph. Vertebrata (Craniata)	Most classes of chrome, especially melanin (indolic); Hb in blood-cells of virtually all
Cl. Agnatha (Cyclostomata)	Monomeric Hb molecule
Cl. Chondrichthyes (Elasmobranchii)	Dull melanic colour
Cl. Osteichthyes (Teleostomi)	Colourful: melanins, carotenoids, pterins, ribityl-flavin (RF)
Cl. Amphibia	Colourful: melanins, carotenoids, pterins, RF
Cl. Reptilia	Some colourful: chromes as in Amphibia
Cl. Aves	Colourful: feathers, wattles, legs, beak: melanins, carotenoids, unidentified chromes, structural colours
Cl. Mammalia	Only melanins, in hair and skin; some structural colours; duller than other classes of vertebrates

Ternary Quinones: Apart from ubiquinone (UQ) all sub-classes of ternary quinones are rather sporadically distributed in animals. Free benzoquinones are largely restricted to arthropods though sclerotisation by *o*-BQ intermediaries may be quite widespread, particularly among invertebrates (BROWN, 1950; FOX and VEVERS, 1960, p. 50) and lower chordates. They may be involved also in hardening the egg case of the dogfish (KRISHNAN, 1959) and the pecten of the eye of some kinds of lizards (SANNASI, 1969). One BQ has been found in echinoderms (MOORE et al., 1966) and a quinol in the scent gland of the beaver (THOMSON, 1957, p. 7). There are taxonomic variations in the number of isopentenyl units in the side chain of UQ, viz., 8 or 9 in Protozoa, 9 or 10 in Insects and mammals and 10 in Echinoderms and Annelids (LESTER and CRANE, 1959; CRANE, 1965) – unfortunately, not a tidy phylogenetic trend.

Among naphthoquinones (NQ), vitamin K has not been recognised in animals other than the mammals and birds. As chromes of the integument, and other tissues the NQ's are common only in the echinoderms, and particularly in the class Echinoidea (SINGH et al., 1967; ANDERSON et al., 1969). One has been recognised in brachiopods (OHUYE, 1936), and the aphins are probably dimerised NQ's.

Until recently the coccids, homopteran bugs, were the only animals known to have significant amounts of anthraquinones (AQ) but these have now been recognised in crinoids and in the polychaete, *Halla* (p. 42). If the aphins are dimerised NQ's, then both homopteran bugs and echinoderms have both subclasses. Moreover the echinoderms appear, on the evidence of fringelite (Fig. 3.4), to be able to dimerise AQ's. The aphins are found only in that one small group of Homoptera and the AQ dimers only in the fossil crinoids and in ciliate protozoa; 'zoopurpurin' (p. 43) occurs in *Blepharisma* (GIESE, 1946, 1953; GIESE and GRAINGER, 1970) and two were found in the blue *Stentor*, 'stentorin' (LANKESTER, 1873) and 'stentorol' (BARBIER et al., 1956). Another is in *Fabrea* (DONNASSON and FAURÉ-FREMIET, 1911; FONTAINE, 1934.)

Metalloproteins: Of the simple metalloproteins the oxygen-transporting copper protein, haemocyanin (Hcy), again is rather restricted in distribution. It has been found in the blood of cephalopod, gastropod, some polyplacophoran and possibly some bivalve molluscs, of decapod, stomatopod, isopod and amphipod malacostracan crustacea and in *Limulus*, scorpions, some spiders (REDMOND, 1968) and the pedipalpid, *Mastigoproctus* (LOEWE et al., 1970) among the Chelicerata. By contrast the enzymic copper-proteins (Table 3.8) and particularly tyrosinase and cytochrome oxidase, are very widely distributed, though as yet investigated in few groups.

The simple iron-protein O_2-carrier, haemerythrin (Hery), is more narrowly distributed than its copper analogue. It has been found only in cells of the body-cavities of sipunculids, priapulids, some lingulid brachiopods and possibly of the blood of the polychaete, *Magelona*. The simple iron-protein enzymes of the ferredoxin type also may be less widely distributed than their copper analogues but again are probably much more common than the oxygen-carrier. They may be less common than copper proteins because iron-porphyranoproteins are so efficient whereas copper forms no counterpart.

Haemoproteins: It is extremely improbable that any animal has no haemoproteins. The cytochrome enzymes (p. 191) are almost universal: even the nematode, *Ascaris,* has normal respiratory enzymes at one phase of its life-cycle. Catalases and peroxidases are widespread but not universal (p. 193), and their distribution is not known in detail. The haemoglobins (Hb) or oxygen-carrier chromes of circulatory systems (p. 211) are the most widely distributed O_2-carriers, particularly if the intracellular version, myoglobin, is included. The distribution of Hb nevertheless is somewhat sporadic (Table 4.2) except in the vertebrates where only a few species of antarctic fishes are known to be normally without it. The only major taxa not yet known to have Hb in any species are the Porifera, Coelentera, Rotifera, Sipuncula, Polyzoa, Brachiopoda, Chaetognatha, Onychophora, Chelicerata, Tardigrada and Protochordata (FOX and VEVERS, 1960, p. 86). The haem of the Hb's is protohaem in all instances except the serpulimorph, flabelligerid (chlorhaemid) and amphareted polychaetes, which have chlorhaem (p. 223); this may be present also in *Peobius*, a small planktonic worm possibly linking the annelids with the echiurids (PICKFORD, 1947; MANWELL, 1964). After the vertebrates the Annelida is the phylum which has most exploited Hb's.

The helicorubin type of haemoprotein has been found only in the gut-fluids of some molluscs, crustaceans, (FOX and VEVERS, 1960, p. 108) and polychaetes (PHEAR, 1955). These are all members of the 'spiralian' group of phyla.

Table 4.2. Distribution of protohaemoglobins

Blood corpuscles of	Blood plasma of	Coelomic corpuscles of
Vertebrates	Annelids (most)	*Glycera, Terebella,* and some other polychaetes
Some bivalves (*Ensis, Arca*)	Entomostracan Crustacea (some of every subclass)	Echiurids
Phoronida		
Nemertines (most)	Larvae of chironomids	*Thyone, Caudina,* and *Cucumaria*
Few Turbellaria	*Planorbis*	
	Some nematodes (pseudocoele)	
	Pogonophora	
Muscle of	Nervous system of	Other cells
Vertebrates: red muscle	*Daphnia*	*Thalassema:* most cells
Daphnia: most muscles	*Aplysia:* giant neurons	*Gastrophilus* ⎫ tracheal
Gastropods ⎱ radula	*Lymnaea:* nerve sheath	*Anisops* ⎭ end cells
Amphineura ⎰ muscles	*Aphrodite*	Turbellaria: parenchyme
Aphrodite: pharyngeal muscle	*Thalassema*	Ciliate protozoa
Ascaris and some other nematodes: bodywall musculature	Nemertines	*Chironomus* ⎫
		Daphnia ⎬ ova
Arenicola, sabellids: bodywall musculature.		*Thalassema* ⎪
		Scolopus ⎭

Haems and Porphyrins: Free haems and porphyrins and their secondary protein-conjugates also are most common in molluscs and annelids (FOX and VEVERS, l.c. p. 112) and particularly in polychaetes (MANGUM and DALES, 1965). They are more often of the III than of the I series (p. 53) and would appear to be diverted from the main biosynthetic path of the haemoproteins (Fig. 13.3). This could explain why the corresponding haem and porphyrin are so often found together. An exception to both these rules is that uroporphyrin I, without the haem, is the main porphyrin of the shells of the lower gastropods and of the skins of the opisthobranchs, *Aplysia* and *Tritonia,* and the pulmonate, *Arion ater.* It is also the porphyrin of planarians (MACRAE, 1956, 1961, 1963) and of the bones and teeth of the fox-squirrel (TURNER, 1937).

Two further generalisations are possible: first that porphyrins and haems are rather common in two other spiralian phyla, the nemertinea and platyhelmia. They are rare in arthropods, however, and at present the only instances are coproporphyrin and its haem in the millipede, *Polydesmus angustus* (NEEDHAM, 1968a), and protohaem in the egg-shells of the phyllopod crustaceans, *Artemia* (J. and D.M. NEEDHAM, 1930) and *Triops* (FOX,

1955), and in the bodies of blood-sucking insects (WIGGLESWORTH, 1943). Secondly the breadth of distribution of the III isomers is in the order proto > copro > uro, i.e., in inverse order to the biosynthetic sequence; this again supports the idea of simple diversion from the path of biosynthesis which normally halts at protoporphyrin. Protoporphyrin is common in some asteroid echinoderms, in the egg-shell of birds and in some mammalian organs but even so it is most common in the spiralian phyla and so are both copro and uro pairs. Both protoporphyrin and protohaem are present in the integuments and other tissues of oligochaetes, polychaetes and possibly leeches, but there is mainly the haem in the guts of trematodes (LLEWELLYN, 1954), leeches (NEEDHAM, 1966b) and polychaetes (PHEAR, 1955). Protohaem is common also in the secreted tubes of polychaetes, even including serpulimorphs, which have chloro Hb in their blood. The latter do have chlorohaem in most of their tissues, however, but the free chloroporphyrin is recorded only from the integuments of two asteroid echinoderms, which do not have chloro Hb (KENNEDY and VEVERS, 1954).

Free copper porphyrans have been identified in two instances. The striking water-soluble red chrome of the feathers of the touraco (CHURCH, 1870) is Cu-uroporphyrin III (WITH, 1957; KEILIN and McCOSKER, 1961), and there is a Cu-phaeophorbide in *Aplysia* and in the related opisthobranch, *Akera* (KENNEDY and VEVERS, 1956; RIMINGTON and KENNEDY, 1962). Cu thus can replace both Fe and Mg.

With so much haem material available in animals it is surprising that some, and mainly annelids, should use phaeophorbides and the further degradation-products of chlorophyll as integumental chromes. Even herbivorous mammals absorb large quantities from the gut, though they do re-excrete them in the bile. It is more understandable that insects should use these (FOX and VEVERS, 1960, p. 117–8) and equally there is reason why *Aplysia* uses plant-bilins in large amounts; these animals have only the enzymic haemoproteins.

Bilins: Probably because the haemoproteins collectively are so widespread, their degradation-products the *bilins*, are very common as integumental chromes. They have been identified in all the major groups of animal except the Protozoa, Porifera, Platyhelmia, Echinoderma, Tunicata (RÜDIGER, 1970) and Chelicerata. However, bilins of the γ-series (p. 54) are confined to the wings of pierid butterflies. Most records of the α-series concern the green and blue trienes; violet dienes have been identified only in the gastropods *Aplysia* and *Haliotis* but there are also other dienes in molluscs (BANNISTER et al., 1970; RÜDIGER, 1970). Bilirubin has rarely been found except in the gut-fluids of vertebrates and some other animals (p. 241) and bilamonenes except in the excreta of vertebrates. One integumentary monene is 'aplysiorhodin' RÜDIGER, 1970) and in the integument of some orthopteroid insects there is some indication (PASSAMA-VUILLAUME, 1966) of choletelins (diketonic b-monenes). Dipyrryl fragments, the bilifuscins (LEMBERG and LEGGE, 1949), including the 'copronigrin' of mammalian faeces are not often retained in the body but 'lanaurin' of yellow wool is an example. Equally, at the other end of the haem-degradation path, the verdohaems (p. 54) are not commonly retained (FOX, 1953, p. 274).

Indigotins: Dibromoindigotin (p. 59) is narrowly restricted to the muricid gastropods. Other indigotins may be more widespread than in mammalian urine and faeces and conceivably may have more significance for some animals.

Melanins: Phenolic and indolic melanins together are very widespread and it will be useful to know more about their respective distributions. It seems certain that vertebrate melanin has a predominance of indolic monomers and similarly the 'ink' of cephalopods and some arthropod melanins. Equally certainly other arthropod melanins are based mainly on phenolic amines, or even on N-free quinols. The phaeo- and erythro-melanins of birds and mammals have not yet been found elsewhere but since they must have evolved independently in these two classes, other taxa also may have exploited them. The discovery by BECKER (1938, 1941) followed by later work (p. 63) that many invertebrate dark pigments are ommochrome has added further uncertainty and few genuine melanins have been definitely identified except in molluscs and arthropods (NICOLAUS, 1968).

Invertebrate brown to black pigments still believed to be melanins include those of the sponge porocytes (Fig. 5.1), the tissues of the sea anemone *Metridium* (FOX and PANTIN, 1941, 1944), the chaetae of the polychaete *Chaetopterus* (KENNEDY and NICOL, 1959), the body-wall of the black slug, *Arion ater*, the snail *Limnaea* and other gastropods, the integuments of the sea urchin, *Diadema* (MILLOTT and JACOBSON, 1952), the sea-cucumber *Holothuria* and other echinoderms and the shed body-fluids of echinoderms and arthropods. The black chromes of planarians and leeches may be oligomeric melanins (p. 62).

Ommochromes: This is another class of chrome most characteristic of the spiralian group of phyla and until recently believed peculiar to them. Ommochromes are probably universal in the eyes of arthropods and very widespread in the integuments of this phylum (LINZEN, 1967). They are the main chromes of the integument of cephalopods (SCHWINK, 1953, 1956) and have been found in the eggs of the echiurid *Urechis* (HOROWITZ, 1939; LINZEN, 1959), the branchial crown of sabellid worms (DALES, 1962) and the integument of the nemertine *Lineus* (VERNET, 1966). Outside the spiralian group they have been found only in the eyes of the Medusa *Spirocodon* (YOSHIDA, 1968) but they may occur in some vertebrates. VERNE (1926, p. 209) mentions acid and alkali-labile yellow, red and violet granules in the cells of the integument of reptiles, while a violet pigment in the skin of the fish *Lepadogaster* is definitely thought to be an ommochrome (CASPARI, 1953; NEUBERT, 1956). The first enzyme of the biosynthetic pathway, tryptophan pyrrolase, is very active in the liver of vertebrates and a close by-product of the path, kynurenic acid (Fig. 3.8, V), was first discovered in the urine of dogs, as its name indicates. Rat liver extracts containing mitochondrial cytochrome c and presumably also more specific enzymes, will synthesise xanthommatin from added 3-hydroxykynurenine (YOSHI and BROWN, 1959). Normally the path may be blocked in most vertebrates but becomes open (BUTENANDT et al., 1960a) in the pathological condition alkaptonuria (p. 249). Collateral paths, of which some are chromogenic, also may function in vertebrates; for instance in the hair of some rodents there are yellow chromes derived from kynurenine. Papiliochromes (p. 65), also products of a collateral path from kynurenine, have been found only in the wings of Lepidoptera (UMEBACHI, 1962; UMEBACHI and TAKAHASHI, 1956). Most of the other tryptophan-derived chromes which have been recorded are from members of the spiralian group of phyla.

Pterins: Pterins again were first found in butterfly wings (HOPKINS, 1895) and, also like ommochromes, are rare in plants and most common in insects and crustacea. They are abundant also, and again as integumental chromes, in poikilothermic vertebrates. Some are known only from arthropods or from these vertebrates but a high percentage are common to the two groups (p. 330). Pterins are not uncommon as reflecting chromes in the eyes of vertebrates and are usual in those of insects and other arthropods. As a vitamin, pteroylglutamic acid is known to be exogenous yet essential for mammals, birds, fish, insects, crustacea, nematodes and protozoa. Ascaris-blue (TORRI, 1956) is a derivative of this pterin-conjugate. The unconjugated pterin-coenzymes are essential for the lower invertebrates (KIDDER, 1967) and also are probably very widely distributed. Pterins have been detected in the internal organs and urine of mammals.

Ribitylflavin: As the essential component of the coenzymes FMN and FAD (p. 185) ribitylflavin must be present in every normal cell of every animal. In mammals, birds, fish, insects, Crustacea, nematodes and Protozoa it is known to be demanded in the diet and so may be exogenous in all animals. This may be why so many animals store such large amounts (p. 305), whether in organs such as the heart and kidney, which need so much for their normal metabolism, or in purely storage sites such as the chromagogen tissue of annelids.

Ribitylflavin (RF) is quite widely distributed also in the integuments of animals and, like the pterins, particularly in Crustacea and poikilothermic vertebrates. Teleosts with small scales, or none, have mainly RF whereas those with large scales have pterins (Fox, 1953, p. 287). If the synthetic analogue, 9-phenyl-5,6-benzisoalloxazine, is fed to, or injected into, white rats it is incorporated into the hair (HADDOW et al., 1945) though RF itself is not known to do this under normal conditions. It does not have a strong colour for visual effect purposes but it does seem to contribute in the leech, *Erpobdella* (NEEDHAM, 1966b) and possibly in the planarian *Polycelis* (NEEDHAM, 1965a). The yellow bands of the brandling earthworm, *Eisinia*, are definitely due to the RF of the chromagogen tissue showing through transparent regions of the body-wall. It may be the major yellow chrome of the integument of *Holothuria forskalii* (p. 37) so that representatives of most of the main phyla have it as part of the superficial coloration.

Again like the pterins, RF is common in the retinal tissues of vertebrates but, unlike the pterins, not in those of arthropods. It was found in the choroid and pigment-epithelium of three genera of teleosts, *Prionotus*, *Centropristis* and *Stenotomus* (WALD, 1936,) *Beryx*, (LEDERER, 1938), man and some other vertebrates (KEILIN and SMITH, 1939) and reptiles (Fox, 1953, Table I). It is in the tapetal layer of the bush-baby, *Galago* (PIRIE, 1958). It is present also in the bioluminescent organs of lampyrid beetles (p. 208) and is rather widely associated with such organs. Its distribution therefore is similar to that of the pterins in some respects but not in others.

Other Chromes: Vanadochrome is restricted to particular sub-groups of ascidians (WEBB, 1939), though relatively high concentrations of vanadium have been found in the opisthobranch gastropod, *Pleurobranchus plumula*. The distribution of partially characterised chromes necessarily must remain as uncertain as their identity.

4.2.1 Conclusion

Some zoochromes, such as haemovanadin and dibromoindigotin, are very restricted in distribution whereas others, including carotenoids, ribitylflavin and the haemoproteins are universal. It would be possible, if more were known of the chromes as yet unidentified, particularly in the less well known phyla, to make quantitative estimates of the relative breadths of distribution of the various chemical classes. An intuitive ranking in order of decreasing breadth, would seem to be: ribitylflavin, porphyrin derivatives, carotenoids, metalloproteins, pterins, melanins, fuscins, quinones, ommochromes, flavonoids, haemovanadin, dibromoindigotin and papiliochromes. It is noteworthy how many of the more widely distributed are exogenous in origin. The spiralian group of phyla appear to have the greatest variety of chromes, including some virtually peculiar to them; in general they probably have the largest amounts per unit weight. Most phyla have a wide variety and what evidence is available indicates that this is true also for the less well know phyla, mainly phyla of few species, with relatively small individuals; these are considered in the next section.

4.3 Distribution in the More Obscure Phyla and Classes

Available information on the lower Metazoa comes largely from the treatise of HYMAN (1940 onwards). It shows that apart from the Gastrotricha, which are in the protozoan range of size, all have some pigment. The planktonic rotifers also are mostly transparent and colourless but some rotifers are highly coloured and much the same differences are found among the Chaetognatha. In the following groups there is a wealth of colour, which merits detailed study: Nematoda, Nemertina, Brachiopoda, Ectoprocta, Hemichordata and Tunicata.

In some of these a number of the chromes have been identified, for instance in nematodes: carotenoids (FRANK and FETZER, 1968), the benzoquinone, rhodoquinone 9 (OZAWA et al., 1969), and ubiquinone itself (SATO and OZAWA, 1969), porphyrins (LEDERER and TIXIER, 1947), haemoglobin in various tissues and in the body-fluid, cytochromes and bilins (SMITH, 1969), cobamine (WEINSTEIN, 1966), the pterin Ascaris-blue (TORRI, 1956), ribitylflavin (SMITH, 1969) and possibly melanin. This is an impressive list from such a superficially dull group. In nemertines carotenoids (VUILLAUME, 1969, p. 58), protoporphyrin (FOX and VEVERS, 1960, p. 112), haemoglobin and ommochromes (VERNET, 1966) have been recognised. Echiurids have xanthophyll carotenoids, haemoglobin – sometimes in most organs of the body (FOX and VEVERS, 1960, p. 84), chlorophyll derivatives, including the well known green 'bonellin' (LEDERER, 1939), and xanthommatin in the eggs of *Urechis* (LINZEN, 1959). In brachiopods carotenoids are abundant (VUILLAUME, 1969, p. 59), particularly in the gonads (HYMAN, 1959, pp. 562, 593), one naphthoquinone has been found (OHUYE, 1936), haemerythrin transports O_2 in some lingulids (KAWAGUTI, 1941), porphyrins occur (FOX and VEVERS, 1960, p. 112) and there is iron in some fossil shells (ROWLEY and WILLIAMS, 1933; CLOUD, 1941). The Ectoprocta have mostly carotenes, with some xanthophylls (VUILLAUME, 1969, p. 59), and an unidentified purple chrome in *Bugula* (VILLELA, 1948a). The chromes of hemichordates have not been identified but FOX (1953, Table I) shows that most known classes are present in ascidians. In virtually all groups carotenoids have been detected.

Among arthropods much is already known of the Crustacea and insects but recently considerable knowledge of the chromes of myriapods and chelicerates has been acquired, again showing variety and some peculiar features. In myriapods little carotenoid has been found. Some groups of millipedes use benzoquinones for defence and the centipedes may have a hydroxyquinone in their connective tissue (BANNISTER and NEEDHAM, 1971). Polydesmic millipedes have coproporphyrin and coprohaem as integumental chromes (NEEDHAM, 1968a), while iulids have ommochromes (NEEDHAM and BRUNET, 1957) and so have glomerids.

Some mites and other acarines have vast amounts of carotenoids (MANUNTA, 1948; GREEN, 1964) and they have been found also in the nerves of *Limulus* (MONAGHAN and SCHMITT, 1931). A number of spiders produce defensive quinones. The 'myohaematin' which MacMunn found in spiders (VERNE, 1926, p. 445) presumably is cytochrome or myoglobin but ticks also acquire Hb from their food and modify it in various ways (FOX, 1953, p. 264; FOX and VEVERS, 1960, p. 171). Ommochromes are present in the eye of the king crab, *Limulus*, and of spiders (VUILLAUME, 1969, p. 82) and in the integument of some spiders. Guanine is common in the integument and excreta of spiders and as reflecting material in the eye of *Limulus*. Clearly there is a wide range of chromes in both of these groups of arthropods.

4.3.1 Conclusion

When the full account is available it will probably be found that nearly all phyla and classes have much the same wide range of zoochromes as the large and well-known groups. Among the latter there are some peculiarities of particular phyla but relatively few; and almost none are absolute (section 4.2). Such groups as the myriapods indicate that sharper distinctions may be evident at lower taxonomic levels and this is worth examining in more detail. A number of examples at the various levels have already been considered in section 4.2 and need not be repeated.

4.4 Distribution between Lower Taxonomic Units

There are some distinctions already at the class-level, for instance between the integumental chromes of chilopod and diplopod myriapods. Among molluscs ommochromes as integumental pigments appear to be peculiar to the Class Cephalopoda. None of the Orders of the malacostracan crustacea have haemoglobin in their blood but a number have haemocyanin whereas some members of every main subdivision of the Entomostraca have Hb and none have Hcy. In the Order Decapoda the Brachyura have melanin but not ommochromes as the dark pigment of the integument, while the converse is true for the Macrura and Anomura. There are porphyrins and bilins in the shells of the lower Orders of gastropods but more insoluble, enigmatic chromes in those of pulmonates (COMFORT, 1951). The yellow of the hen's iris is due to carotenoid but that of the pigeon is different and unidentified.

The integumental chrome of gammarid amphipods is carotenoprotein but those of the jassid amphipods are ommochromes (NEEDHAM, 1970a). The polydesmid-iulid contrast is perhaps at a somewhat higher taxonomic level. The IX γ isomer of biliverdin is peculiar to the family Pieridae (RÜDIGER, 1970) and so are the pterins, as wing pigments (FORD, 1947). Papiliochromes (p. 65) are restricted to the family Papilionidae. The adults of the amblystomid urodeles have virtually all the pterins known among Amphibia whereas the salamandrids have none (BAGNARA and OBIKA, 1965). *Loligo* has virtually no carotenoids except in the retina, whereas the octopus has much (WALD, 1941; FOX and CRANE, 1942). There are carotenoids but no ribitylflavin in the blood of some snakes but conversely in others.

Some zoochromes appear to be sporadically present, in taxa of varying size. Protohaemoglobin is present in all vertebrates except a few species of fishes, in most sub groups of all classes of annelids, in many nemertines, in some members of every group of entomostracan crustacea, in a smattering of holothurians, and finally in isolated genera and species of some arthropod and molluscan groups. There is much the same variably patchy distribution of the other oxygen-transport chromes. For other functional types an apparently patchy distribution could be due merely to patchy investigation so far, for instance the distribution of pterins as reflecting materials in the vertebrate eye (Fox, 1953, p. 298). In easily investigated cases, however, for instance the absence of carotenoids from the body-fat (Table 14.1), there is no doubt that the distribution is very sporadic.

When exact comparisons are possible, it is clear that chromatic contrasts can be quite sharp down to generic and even specific levels. PAYNE (1931) found that some sertularid hydroids have carotenoids and no flavonoids but other genera reciprocally. Free chloroporphyrin is restricted to the starfish genera *Luidia* and *Astropecten* (p. 54). There is protoporphyrin in the integument of the genus *Asterias*, but not of *Marthasterias* (KENNEDY and VEVERS, 1953). *Lumbricus* has a high ratio of protoporphyrin to its haem in the integument while the converse is true for the related earthworm *Allolobophora* (NEEDHAM, 1968b). Biliverdin is present in the eye of the cladoceran, *Polyphemus*, but not of the related *Daphnia* (GREEN, 1961).

The yellow body-colour of the wood-wasp *Urocercus* is different from that of *Vespa* and within the latter genus *Vespa vulgaris* has xanthopterin whereas *Vespa rufa* has a redder unidentified chrome (LeCLERQ, 1951; BECKER, 1937). The yellow skin-chrome of *Rana esculenta* is almost certainly carotenoid while that of *Rana temporaria* is probably a pterin (ZIEGLER-GUNDER, 1956). No doubt the most striking and complex example concerns the distribution of proto- and chloro- Hb's in the serpulimorph polychaetes described in detail by FOX and VEVERS (1960, p. 100).

4.5 Conclusion

It seems clear that chromatic contrasts can be as great between species and genera as between the higher taxa. Indeed they are more clear-cut because so few species are involved. Evidently chromatic evolution is very labile and every phylum has by now exploited all classes of chrome in most feasible capacities. The generic and specific, and no doubt intra-specific, contrasts show the process still in action. Some new mutations permitting new syntheses, for instance of the IX γ bilins in pierid butterflies (p. 88), well may have great selective advantage and since zoochromes are so often terminal members of their biosynthetic pathways such mutations can be relatively common.

Contrasts at the lower taxonomic level mainly concern integumental chromes where colours rather than chemical properties are important. By contrast, the exygen-transport chromes, and stored materials in general, are common throughout larger taxa and some of the metabolic enzymes (chapter 8) are invariant through all phyla. Among integumental chromes there must have been endless parallel, convergent and reticulate evolution; a multivariate type of analysis might be applied to this, when knowledge is adequate, and the results may also clarify the general evolution of the taxa concerned. Particular chromes common to widely diverse taxa must be good examples of the 'isologous' materials of FLORKIN (1966).

It may be stressed that quite a number of the classes of zoochrome have been able to produce a wide range of colours, so that in principle any one class could meet all the visual-effect requirements of any one taxon. Yet in fact most species use quite a wide range of chemical classes, though in different permutations from those of species in other phyla. Presumably the explanation is that this spreads the metabolic load and makes best use of available materials. Where haem is used extensively there will be plenty of bilins, which otherwise would go to waste. Another advantage of the variety is that no biosynthetic pathway is called upon to produce a large variety of products and so control of production is relatively simple.

Chapter 5

Distribution and State of Zoochromes in the Body

Distribution in the body should reflect the functional significance of various chromes very closely. It is useful to distinguish three successive levels for study, (1) distribution between the various organs, tissues and fluids of a typical higher animal, (2) location within organs and tissues and (3) location and physical state within the cell.

5.1 Distribution of Zoochromes between Organs, Tissues and Fluids of the Body

In principle a preliminary comparison between the lower Metazoa, which have little tissue –, or organ –, differentiation and the phyla with maximal differentiation (Table 5.1) should be very instructive. For instance, haemoglobin has been found in the parenchyme of some Platyhelmia and this supports orthodox views on the origin of the vascular system of the higher phyla. As yet unfortunately the phylogenetic outlook has scarcely motivated chromatological studies on the Porifera and Coelentera, which are the most critical phyla in this context. In these animals a chrome is visible anywhere in the body and so all are 'integumental' in this sense. In the Cnidaria some appear to be distributed rather indiscriminately in both cell-layers (Fox, 1953, p. 91), perhaps for this reason. On the other hand the carotenoids of *Tubularia* and *Hydra* appear to be mainly in the endodermis, and so may have some other significance. It may be noted that the green *Hydra* has its symbionts in this layer. In the Alcyonaria the bright chromes are mainly in the corallum, an exoskeletal material in origin, and often there is little colour in the actual tissues. In most other cnidarian corals conversely the corallum is white and the tissues coloured; *Stylaster* resembles the Alcyonaria in having a coloured corallum. Already in this phylum the gonads are often brightly coloured, for instance red or purple in *Aurellia*; these organs are ectodermal in the Hydrozoa but endodermal in the other two classes, so that the chromes are specific to reproductive tissue, irrespective of origin. Carotenoproteins accumulate in the eggs (p. 265). Ocelli are deeply pigmented and are ectodermal.

The bright chromes of sponges are mainly in mesogloeal cells (Fig. 5. 1, C) often designated chromocytes (HYMAN, 1940, p. 288, 295) though it is not certain that they represent a single, discrete line of cells. As in the Cnidaria, symbiotic algae are in the inner (choanocyte) layer of those sponges which exploit the phenomenon. The porocytes of some ascon grade sponges contain many insoluble melanic granules (Fig. 5.1, A, B) and these cells originate from the outer, 'dermal' layer. Spongin is yellow, probably due to the halogenated aminoacids in it; these occur also in the more compact organic axes, usually much darker in colour, of some anthozoan Cnidaria. Sponges differ from Cnidaria in having less organ-differentiation and therefore less localisation of chromes, but they appear to be more advanced in having an abundant mesogloeal cell-population, including definite chromocytes. This may be because the inner, choanocyte layer are more specialised cells than cnidarian endodermal cells, the Porifera in these respects being more advanced than the Cnidaria. The chromocyte could be the precursor of the parenchymal pigment-cells of planarians, which then differentiated into both sub-epidermal and coelomic chromatic tissues of higher phyla.

Organogenesis and chrome-segregation are considerably more advanced in the Turbellaria, and such terricolan triclads as *Bipalium* have distinctive superficial colour patterns. The distribution in the deeper organs is not readily displayed until the coelomate stage is reached, when most organs can be exposed by simple dissection. Few zoologists forget their first impression of the chromatological display on dissecting the common frog; there are many equally striking cases among both vertebrates and invertebrates, e.g., *Holothuria*,

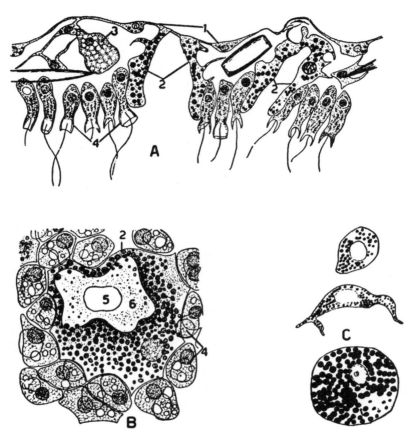

Fig. 5.1. A, B Section through, and surface view of, small portion of body-wall of sponge Clathrina showing melanin granules in porocytes (2). C Chromocyte of mesogloeal layer of Clathrina. (All from HYMAN, 1940)

the bug *Rhodnius*, and the larvae of some simulid flies (demonstrated to me by Dr. P.L. MILLER). By contrast, in some animals one particular chrome seems to run riot through all the organs of the body. Without any necessarily pathological implications this phenomenon might be termed a 'jaundice'. Haemoglobin is widely distributed in the polychaete *Aphrodite*, the nematode *Ascaris*, the crustacean *Daphnia* (in anoxic water), and particularly the echiurid *Thalassema* (FOX and VEVERS, 1960, p. 84). Biliverdin and the phaeophorbides are rather widely distributed in the polychaete, *Nereis diversicolor*, and naphthoquinones in echinoid organs. In the echinoid, *Diadema*, there is melanin round most organs (MILLOTT and JACOBSON, 1952). Even in these cases, however, there is a variety of other chromes in the body and it is worth examining the individual tissues and organs in turn. The distribution of partially characterised and unidentified chromes should be recorded since it might help in their identification but space is not available here.

5.1.1 Exoskeletal Structures and Other Secretions

Cuticles, thicker exoskeletons, shells and secreted tubes are common among invertebrates; they are organic, inorganic or various combinations of the two. Most are coloured, at least in the outer layers, and for 'visual-effect' purposes have become the 'integument'. Inorganic materials in general have brighter colours than the organic. Many are unidentified. Carotenoids have been recognised in the exoskeleton of some insects (Fox, 1953, p. 132) and Crustacea (Needham, 1970a). They are conspicuously absent from mollusc shells. The dark colours of insect and probably many other, cuticles are polymerised benzo quinones, modified sometimes by structural colours (Neville, 1967). Porphyrins occur in the mineral prismatic layer of the shells of prosobranch molluscs and some bivalves (Comfort, 1951), and in pink pearls (Fox, 1953, p. 264). Bilins colour the shells of other primitive gastropods and also green pearls. The brown of the tubes of some sabellid polychaetes is due to protoferrihaem but that of *Myxicola* to chloroferrihaem (Smith and Lee, 1963), – the only record of this as a free haem except in chromagogen tissue (p. 106). There is ribitylflavin in the exoskeleton of the centipede, *Lithobius forficatus*.

Turning to other secreted materials, carotenoids are present in beeswax and in the silk of some insects. The mantle and foot of the snail *Helix* secrete a flavone and again some insect silks contain this class of chrome. Various arthropods secrete benzoquinones. There are two porphyrins in the Harderian secretion of the rat (Derrien and Turcini, 1925; Kennedy, 1970) and haemoglobin itself in the radula and teeth of the winkle *Littorina* (Fox and Vevers, 1960, p. 85). The 'ink' of the sea hare, *Aplysia,* is a bilin. The muricid gastropods secrete dibromoindigotin and there are often indigoid chromes in the sweat of mammals (Fox, 1953, p. 215; Fox and Vevers, 1960, pp. 164–5). Cephalopod 'ink' is an indolic melanin. The nemertine *Lineus* secretes some of its integumental ommochrome.

5.1.2 Epidermis

Unfortunately the epidermis has not always been distinguished from the rest of the integument. One recent report on the 'cuticular' chromes of an insect was based on the whole carcass after removing merely the viscera! The situation is not improved by the use of the term 'hypodermis' for epidermis by some arthropologists, but in any case (Fig. 5.2) the epithelium is invaded by the deeper tissues, including chromatocytes (Green and Neff, 1972).

Carotenoids are found in the epidermis of some asteroids (Hyman, 1955, p. 381), Crustacea (Fox, 1953, p. 123, 135) and locusts (Goodwin, 1934), in the face of the pheasant and in the feathers of many birds (Fox, 1953, p. 168). Flavones are recorded from the epidermis of the bug *Leptocoris* (Fox, 1953, p. 212). Naphthoquinones are common in that of echinoids (Fox, 1953, p. 195).

There is uroporphyrin in the epidermally derived rhabdites of Turbellaria (MacRae, 1961), coproporphyrin in the feathers of thirteen orders of birds, and phaeophorbides *a* and *b* in the epidermis of the squash-bug *Anasa* (Metcalf, 1945). Hedgehog spines also contain porphyrins (Derrien and Turcini, 1925). The feathers of the touraco contain the unusual copper uroporphyrin, 'turacin'. Haemoglobin is present in the epidermis of

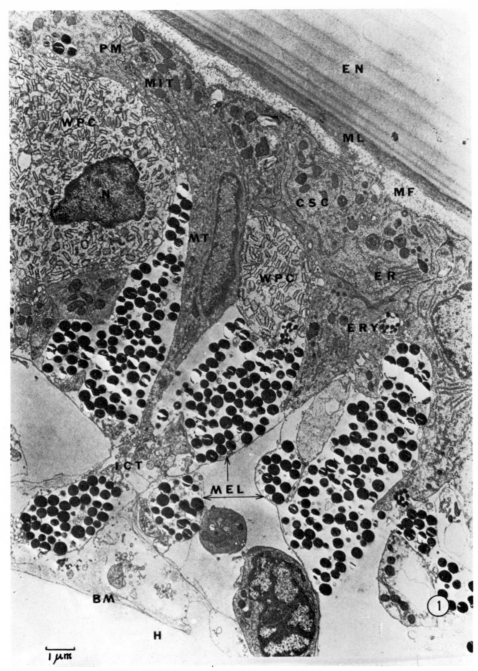

Fig. 5.2. Electron-micrograph of section normal to surface of fiddler crab to show relationship of chromocytes to epidermis, three types of chromocyte and chromasomes in situ. BM = basement membrane of epidermis; CSC = cuticle-secreting cells; EN = endocuticle; ER = endoplasmic reticulum; ERY = erythrophore; H = haemocoele; ICT = intraepidermal connective tissue; MEL = melanophore; MF = moulting fluid; ML = membraneous layer of exoskeleton; MIT = mitochondria; MT = microtubule; N = nucleus; PM = plasma membrane; WPC = leucophore. (From GREEN and NEFF, 1972)

the nematode, *Ascaris*, and probably of the echiuroid, *Thalassema*. Bilins are in the epidermis of the polychaetes *Nereis* and *Chaetopterus* (DALES and KENNEDY, 1954).

Melanins occur in the epidermis of the bivalve *Dreissenia*, brachyuran Crustacea (FOX, 1953, p. 133), birds and mammals and probably other vertebrates. All three varieties are common in hair and feathers. Ommochromes have been detected in the epidermis of the locust (FOX and VEVERS, 1964, p. 54; GOODWIN, 1962). Both pterins and ribitylflavin occur in the epidermis of Crustacea (FOX, 1953, p. 298).

5.1.3 Respiratory Epithelia

Protected respiratory epithelia are usually unpigmented so that any colour in the blood shows through. The external gills of Amphibia, polychaetes and some other groups have the chromes of the general integument, probably without any specific respiratory significance. A number of unusual chromes have been found in Sabellid crowns (DALES, 1962), which are feeding, as well as respiratory organs. It would be interesting to know if any of these are in the respiratory epithelium itself. The only clear record of a chrome in this epithelium is the haemoglobin in the tracheal end-cells of the larva of the bot-fly, *Gastrophilus*, (KEILIN and WANG, 1946) and of the adult water-boatman *Anisops* (MILLER, 1964). The violet chrome of the centipede *Lithobius* is in mesenchymal cells surrounding the tracheal epithelium.

Melanic material is common in the lungs of tetrapods (Fig. 5.3) and is always in mesenchymal scavenging cells called 'dust cells'; it originates outside the organs though not only from the air: injected carotenoids have been found to accumulate there (DRUMMOND et al., 1934). Nematode parasites increase the amount in the frog. Acanthocephala increase the amount of ommochrome in the operculum of the isopod *Asellus* (Fig. 11.2) but this is not a respiratory structure proper. The haemoglobin of the insect tracheal end-cell seems to be the only true chrome of a respiratory epithelium, therefore.

5.1.4 Neural Tissue

Probably in all animals neural and epidermal tissues have a common origin. Carotenoids are highly concentrated in the ganglion-cells of *Anodonta* and other molluscs and are present in the axons of both the nonmedullated nerves of the lobster and king-crab, *Limulus*, and the medullated nerves of the frog (MONAGHAN and SCHMITT, 1931). A progressive accumulation of fuscous material is common in the neurons of vertebrates, gastropods, sipunculids and other invertebrates (FOX and VEVERS, 1960, p. 164). There is haemoglobin in the giant nerve-cells of the sea hare *Aplysia* (CHALAZONITIS and ARVANITAKI, 1950) and probably in the cell-bodies and/or axons of nemertines, the polychaete *Aphrodite* and the echiurid *Thalassema* (FOX and VEVERS, 1960, p. 84–5).

The greyness of nerve cell-bodies is partly a simple contrast to the reflecting whiteness of the lipids of the myelin round the axons but there is enough cytochrome and other mitochondrial chromes (FOX and VEVERS, 1960, p. 104) to give a distinct coloration, -greyish because of the mixture. Some neurons, usually in particular 'nuclei', or centres are frankly pigmented. The dark colour of the substantia nigra and substantia ferruginea (locus coeruleus) of the midbrain of primates and some other mammals is probably melanin, formed by the polymerisation of quinonogenic neurotropic agents such as epinephrine and dopamine. The pigment is distributed more widely, in more or less continuous longitudinal columns (MARSDEN, 1969). The chrome of the red nucleus and other coloured nuclei is not known.

Much of the vertebrate glial tissue has a common origin with the neural tissue and the cells often are shaped and pigmented rather similarly to the chromatocytes; these again have the same origin, and form a sheath around some peripheral nerves (Fig. 18.1). There

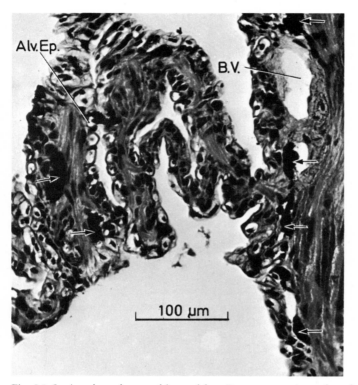

Fig. 5.3. Section through part of lung of frog Rana temporaria to show location of black pigment (arrows) in connective tissue scavenging cells ('dust cells').In Amphibia the alveolar epithelium (Alv. Ep.) has conspicuous deep cells. B.V. = blood vessel

are naphthoquinones in the sheath of the radial nerve-cords of echinoids (FOX, 1953, p. 197) and melanins in vertebrates. There is haemoglobin in the sheathing tissue of the snail *Limnaea*, and free porphyrins in the spinal region of homoiothermic vertebrates (KLÜVER, 1944; 1951). Some poikilothermic vertebrates have a green chrome here, probably a bilatriene. In holothurians (HYMAN, 1955, p. 132) and other animals there are unidentified chromes. The exact site of the copper protein, cerebrocuprein, is uncertain (PORTER, 1966).

5.1.5 Sense Organs

Light-perceptor organs are easily recognised, in most animals which possess them, by their dark masking pigments and sometimes by the perceptor chrome itself, a retinopsin (carotenoprotein) in nearly all cases so far examined (p. 163). Melanin is the masking pigment in vertebrates, and is developed in the wall of the optic cup itself (iris and pigment-epithelium) as well as in the choroid, i.e. in cells of neural crest origin. Arthropods have ommochromes as main masking pigments; there may be different ommochromes in the different ommatidia of the same eye (BUTENANDT, 1957) implying some function more specific than mere masking. In addition ribitylflavin, pterins and purines serve in a reflecting, and perhaps in other, capacities (p. 170) in the eyes of vertebrates and arthropods (Fox, 1953, p. 115). There are carotenoids in some other types of sense-organ (p. 176).

5.1.6 Endocrine Organs

Of the vertebrate endocrines the posterior pituitary, adrenal medulla and pineal are neural in origin, the remainder of varied homology. The pineal is grey, partly due to melanic material (MARSDEN, 1969) and the adrenal medulla also accumulates melanin (Fox, 1953, p. 226). The latter has a common origin with melanocytes, from the neural crest, but epinephrine (adrenalin) itself, the main product of the medulla, can oxidise to a polymerising quinone; this is related to the 'chromaffin' reaction of the medulla cells with chromic acid and other oxidising agents (CARLETON and LEACH, 1938).

Chromes of the non-neural endocrines include carotenoids, in the anterior pituitary (Fox, 1953, p. 175), adrenal cortex (VERNE, 1926, p. 227), corpus luteum and interstitial tissue of testis. The parathyroids are yellow-brown (MAXIMOV and BLOOM, 1942, p. 316) possibly indicating carotenoids and fuscins and the latter accumulate in the pituitary and the adrenal cortex. The thyroid is orange in the frog and red brown in man; the secretion in man is yellow and the colour is attributed to breakdown products of haemoglobin (GRAY, 1944, p. 1433). The hypophysis accumulates injected porphyrins but does not normally contain any. Other vertebrate endocrines are colourless, apart from a rich blood-supply, and there is little pigment in invertebrate glands: the branchial heart-appendage of cephalopods contains adenochrome (Table 3.14) and the nephridium of sipunculids a brown chrome, but these are not proved to be endocrine organs.

These organs and particularly the non-neural group therefore are not rich in biochromes. Carotenoids are most common, as in so many tissues. The neural glands show their origin in their pigments.

5.1.7 Tissues Derived from the Neural Crest of Vertebrates

The crest is remarkable for the number of different tissues it produces (HORSTADIUS, 1949), including the chromatocytes of the integument and viscera. In the lower vertebrates they are 'chromactive', functioning in colour-changes (chapter 6) and contain carotenoids, melanins, purines, pterins and ribitylflavin (GOODRICH et al., 1941; Fox, 1953, pp. 145, 287). In birds and mammals they are not chromactive and are restricted to melanin and carotenoids (birds) or to melanin only (mammals). Here one type of melanocyte, the 'epidermal melanocyte', passes melanosomes (p. 118) into the basal cells of the epidermis (BREATHNACH, 1969). In all groups of vertebrates melanocytes and sometimes other sub-types invade the connective tissue sheaths of blood vessels, nerves, peritoneum and various organs.

5.1.8 Connective Tissues

For simplicity this may be taken to include all the less specialised skeletal tissues and all the cells in them although some, such as the vertebrate chromatocytes, may originate independently of the general mesenchyme. Certainly most invertebrates have a population of dermal chromatocytes similar to those of vertebrates topographically and in other respects. Sometimes they intrude between epidermal cells, as in the vertebrates, and they have radiated into an equal or greater variety (p. 138). They make the main contribution to integumental colour (VERNE, 1926, p. 242) and probably evolved from the chromocyte of the porifera (p. 95). In leeches they contain porphyrins, porphyrans, bilins, purines, ribitylflavin and unidentified chromes (NEEDHAM, 1966b) and in Crustacea carotenoids, melanins, ommochromes, purines, pterins and ribitylflavin. Echinoderms collectively have carotenoids, porphyrins, naphthoquinones and melanins, and some polychaetes, and other animals, have a comparable variety. By comparison the connective tissues proper, i.e., those actually forming skeletal structures, are relatively free of chromes. The special skeletal tissues also are relatively free (see below).

5.1.9 Fat Tissue

Vertebrate fat is deposited in modified connective tissue cells. The fat-body of the frog is heavily loaded with carotenoids and there is visible coloration in the fat of most vertebrates. Experimentally administered lipid-soluble dyes also tend to accumulate in the stored triglycerides, which normally are fluid at body-temperature. The fat-bodies of some invertebrates originate from the coelomic epithelium and so are not strictly homologous with that of vertebrates; again their triglycerides commonly contain carotenoids but, as in the vertebrates, not universally (Table 14.1). The fuscous chrome of mammalian brown fat may be protoferrihaem (p. 53). No doubt any of the lipid-soluble chromes (Table 3.1) may tend to accumulate in fat, though relatively few have been identified here.

5.1.10 Internal Skeletal Structures

As already indicated (above) these special connective tissues are mostly colourless though elastic fibres and ceratotrichia are yellowish. The brown pigment in the enamel of the incisor teeth of some rodents is of course epidermal in origin. Most other instances of colour in internal skeletons of vertebrates are abnormalities but this is not true for the echinoderms, virtually the only other triploblastic phylum with an extensive internal skeleton. This is strongly coloured with naphthoquinones in most echinoids, – as are so many of the tissues. If the sea otter feeds on the urchins it acquires naphthoquinones in its bones and this is the most common way abnormal coloration is acquired by vertebrate skeletons.

Various species of teleost deposit bilatrienes in this way, quite sporadically, but it is not certain that the origin is exogenous. In some bilin spills over into neighbouring organs (VERNE, 1926, p. 207) and in *Cottus* nearly all organs are affected. Gallstones show how great is the affinity of bilins for calcium salts. Under pathological conditions porphyrins are deposited in the growing regions of the bones of the cat and guinea pig (DERRIEN and TURCHINI, 1925) and porphyrins again have a high affinity for calcium salts, as shown by their abundance in the egg-shells of birds. Experimentally the anthraquinone alizarin is easily introduced into mammalian bones and the surprise is that the whole phenomenon is not more common. Normally the skeletal structures must be protected, or regularly depigmented (see section 5.2).

5.1.11 Muscle

Most muscles when cleared of blood are virtually colourless though there are commonly some pigment granules in the sarcoplasm (MAXIMOW and BLOOM, 1942, p. 165) and in very active muscles such as the flight muscles of insects there can be a distinct pink colour from the cytochrome of the densely crowded mitochondria. It was here that KEILIN (1925) first studied this important haemoprotein and MACMUNN (1886) first discovered it in the muscle of the sea anemone *Actinia;* he recognised its haemoprotein nature and called it 'actiniohaematin'. In heart muscle there may be as much as 10 µg of ribitylflavin per gram (FOX, 1953, p. 287).

Equally functional is the myoglobin present in 'red' muscles of various groups of animal (Table 4.2). Some smooth muscles also contain it. The fuscous material which accumulates in the heart muscle of some vertebrates (VERNE, 1926, p. 223) probably comes from myoglobin. More enigmatic is the sporadic occurrence of a number of other chromes in the muscles of restricted groups of animals. The green bilatriene found in one species of earthworm, *Allolobophora chlorotica*, is said to be in the body-wall muscles (KALMUS et al., 1955). Phaeophorbides *a*, and *b* are concentrated in the muscles of the polychaete *Chaetopterus* (VERNE, 1926, p. 224). Clearly this is of exogenous origin and the same is true for the carotenoid, astaxanthin, in the muscle of the salmon and some other fishes. The ommochrome of dragon-fly flight-muscles (PRYOR, 1962, p. 381) may be endogenous and this chrome very readily binds to muscle tissue in the course of laboratory extraction procedures (NEEDHAM and BRUNET, 1957). As in internal skeletons the sporadic presence of adventitious chromes is puzzling.

5.1.12 Blood System

In the fluid itself the oxygen-transport chromoproteins, simple metalloproteins or porphyranoproteins, are most important (chapter 9). In addition there are smaller amounts of many other chromes. Some of these, such as ceruloplasmin (SARKAR and KRUCK, 1966), are permanent constituents while others are materials in transit. Of the latter, carotenoids and ribitylflavin are rather widespread, whereas others are related to tissue-chromes peculiar to the species. In addition to those already mentioned (p. 87) there are various products of Hb in the blood of polychaetes (MANGUM and DALES, 1965) and ferrihaemochrome in *Rhodnius* (WIGGLESWORTH, 1943). Bilins are very common in vertebrates, where normal levels can be rather high; in addition to bilirubin in the horse (p. 241) there are red and blue bilins in labrid and cyclopteran teleosts, possibly related to phycoerythrin and phycocyanin (FOX, 1953, p. 278). Precursors of clotting chromes are frequently present also (NEEDHAM, 1970 c); melanin is the usual pigment of the clot but in one silkworm ommochromes are involved (INAGAMI, 1954). In addition to major chromes of known function there is the vanadochrome of ascidians (p. 73) and protoaphin in aphids.

Blood Cells: In vertebrate erythrocytes, in addition to Hb, there is erythrocuprein, another copper-protein. In ascidians most of the cell-types in the blood are coloured, rather variously; only one is known to contain the vanadochrome. The cells of nemertine blood seem to be more varied in colour than can be due to Hb alone. In most animals some types of blood-cell ingest experimentally injected dyes (VERNE, 1926, p. 222; WIGGLESWORTH, 1959) and presumably they normally ingest coloured metabolic products.

Endothelial Tissues: Some of the cells lining the blood vessels and sinuses also have this phagocytic function, the classical examples being the Kupffer cells of the vertebrate

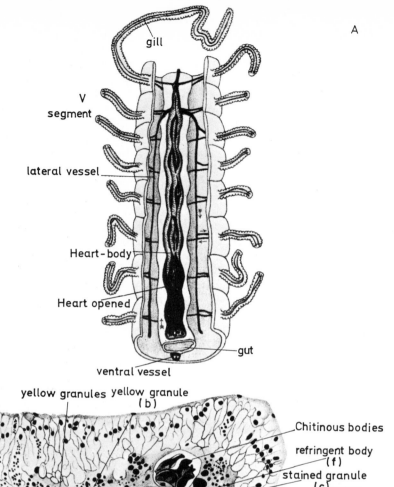
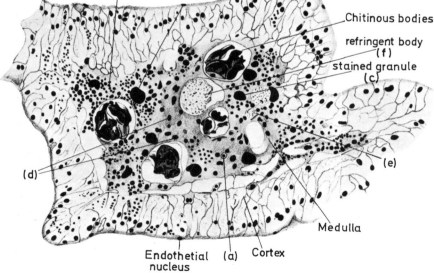

Fig. 5.4. Heart-body of polychaete Audouinia filigera
A Diagram of a dorsal dissection of relevant part of body with the heart opened longitudinally to show the heart-body. Arrows indicate direction of circulation.
B T.S. of one strand of the heart-body showing the endothelium of the heart surrounding a 'cortex' derived from the heart-wall invaginated and enclosing a 'medulla' derived from the coelomic lining. The latter contains yellow pigment granules (b) as well as various other inclusions. (a) = granule in vacuole; (d) = nuclei of medullary cells; (e) = mass of granules. (From PICTON, 1899)

hepatic sinusoids. Cells of the spleen accumulate haemosiderin and ribitylflavin and store as well as scavenge. Phagocytes carrying scavenged pigment tend to accumulate in this and in the lymphoid organs. The endothelial cells of the blood vessels of phoronids vigorously take up dyes and other material (HYMAN, 1959, p. 268). The heart-body of some Polychaeta (Fig. 5.4) is an intrusion of the vessel wall into the lumen and contains a core of coelomic epithelial cells (FAUVEL, 1959). It is a storage and productive tissue for blood chromes. The fat-body and pericardial tissues of insects (RICHARDS and DAVIES, 1957), and no doubt of other arthropods, represent the remains of the coelomic epithelium and not of the vascular endothelium.

5.1.13 Coelomic System

The functions of the coelomic system are so often similar to those of the vascular system that its relations with zoochromes are inevitably rather similar. The two are confluent in Hirudinea and possibly also in brachiopods and echinoderms. In various invertebrates the coelom becomes the main circulatory system and the polychaetes *Nephthys*, *Terebella* and *Travisia* have Hb in both coelomic and vascular systems. O_2-carrier chromes are rarely free in the coelomic plasma (p. 211) though echinochromes are often in both cells and plasma (FOX and VEVERS, 1960, p. 140). Precursors of melanin are free in the coelomic plasma of echinoderms (MILLOTT and JACOBSON, 1951; HYMAN, 1955, p. 411), a parallel to the system in the blood of arthropods.

Coelomocytes: Some types of coelomocyte in virtually all coelomate invertebrates are coloured. Haemoglobin is sporadically present (Table 4.2), as O_2-carrier, and haemerythrin is only in this site. Among other chromes in these cells, carotenoids as usual are most common, for instance in echinoids, in *Sabella* and *Amphitrite* (DALES, 1964) and probably in brachiopods (HYMAN, 1959). Protoferrihaem has been found in *Amphitrite* (DALES, 1964) and either bilins or phaeophorbides in *Nereis* and *Sabella* (DALES and KENNEDY, 1954). *Diadema* has melanin (VEVERS, 1963), making three classes in echinoids collectively. A pigment in the coelomocytes of *Antedon* is regarded as a lipofuscin (DIMELOW, 1958). Fuscous chromes in those of articulate brachiopods are not all lipid-soluble (HYMAN, 1959, p. 557). The enigmatic purple chrome of the ectoproct *Bugula* (p. 75) is mainly in its coelomocytes.

Some at least of the free coelomocytes are phagocytic scavengers and accumulate chromes, for instance, in phoronids (HYMAN, 1959, p. 268), ectoprocts (ib p. 415), brachiopods (ib p. 593), asteroids (HYMAN, 1955, p. 387), echinoids (ib p. 568) and holothurians (ib p. 145). Many holothurians have in the peritoneum curious ciliated and cup-shaped groups of cells, called *urns*, which take up such material and eventually break free. The coelomic urns of sipunculids may have a similar significance.

Peritoneum (Coelomic Lining): In invertebrates the peritoneal cells are often heavily pigmented. By contrast the vertebrate peritoneum is a squamous epithelium and is usually pigment-free; chromes apparently in the coelomic wall are retroepithelial. The whole of the invertebrate peritoneum is phagocytic and has been shown to take up injected dyes in sipunculids (HYMAN, 1959, p. 680), annelids and molluscs (POTTS, 1967), *Phoronis* (HYMAN, 1959, p. 268), Ectoprocta (ib p. 415), Brachiopoda (ib p. 593), Hemichordata (ib p. 119), and Asteroidea (HYMAN, 1955, p. 388). The cells of the walls of excretory organs opening from the coelom behave similarly, e. g., in phoronids, brachiopods (HYMAN, 1959, p. 268, 593) and annelids (BAHL, 1945).

Certain special regions of the peritoneum are very heavily pigmented. The typical example is the chromagogen tissue (NEEDHAM, 1970b) of annelids, sipunculids (HYMAN, 1959, p. 680) and related phyla. The extra-vasal tissues and the core of the heart-body belong to the same extensive system (FAUVEL, 1959, p. 70) and in the Hirudinea it is represented by the botryoidal tissue. Owing to the extensive occlusion of the coelom in much of the body of the Hirudinea the tissue has become a complex system of strands interspersed between the other tissues. There are pigments in the pericardial and some other coelomic epithelial tissues of molluscs, – Keber's Organ of bivalves, the branchial heart appendage of cephalopods, the nephridial gland of gastropods, and the large dorsal coelom peculiar to the Monoplacophora (POTTS, 1967). As indicated, the arthropod fat-body and related tissues also are coelomic in origin. Further examples are the vasoperitoneal tissues of phoronids and the glomerulus of hemichordates. All contain large amounts of a variety of zoochromes; they are the chromatic stores par excellence.

The annelid chromagogen contains Hb, ferrihaem, free porphyrins, bilatrienes, ferritin, ribitylflavin, lipofuscins (NEEDHAM, 1970b) and carotenoids. Collectively the extravasal tissues of polychaetes have all the relevant porphyrins of the biosynthetic pathway (Fig. 13.3), together with their ferrihaems and haemoglobin itself (MANGUM and DALES, 1965). There are sometimes bilins and also ribitylflavin. Heart-bodies likewise contain these chromes, together with other precursors and products of the porphyran pathway, including ferritin (DALES, 1965). In sabellids, such as *Megalomma*, there is chloroferrihaem and chloroferro Hb itself: in fact this is the only known intracellular site of Chl Hb, and almost certainly it is being stored here. Even in serpulimorphs, the tissue has protoferrohaem, in accordance with the biosynthetic path (p. 286, Fig. 13.3).

The various molluscan chromagogen tissues are red, orange, brown or black, depending on the prevailing chromes which include melanin, bilirubin (POTTS, 1967) and adenochrome (Table 3.14). In insects collectively (HOLLANDE, 1923) the pericardial cells contain carotenoids, melanins, ommochromes and biliverdin, which becomes bilirubin in the presence of glucose (WIGGLESWORTH, 1943). The fat-body, again in insects collectively, contains carotenoids (HOLLANDE, 1914) anthocyanins (HOLLANDE, 1913), anthraquinones (FOX, 1953, p. 205) and bilins (METCALF, 1945).

5.1.14 Excretory Tissues

Those excretory organs opening from the coelem take up various chromes (p. 243) and in the annelids, at least, these are much the same as those in the chromagogen tissues. In the nephridium of *Arenicola* there is coproporphyrin III and its ferrihaem, while in *Myxicola* there are both these and chloroferrihaem and coproferrihaem I (MANGUM and DALES, 1965). Excretory organs without an opening into the coelom nevertheless behave very similarly. The malpighian tubules of insects accumulate a large variety of biochromes, in large quantities: carotenoids, fuscins (VERNE, 1926, p. 229), ribitylflavin, pterins (FOX, 1953, p. 298) and ommochromes (KIKKAWA, 1953). There are relatively high concentrations of ribitylflavin in the vertebrate kidney (FOX, 1953, p. 286).

Urine: The chromes actually excreted by this route are virtually unknown except in the vertebrates. Mammals pass bilirubin, urobilin IX and traces of uroporphyrins. The most abundant, urochrome, is still unidentified (p. 76). Pterins and ribitylflavin are excreted at a low, steady rate and indigotin, indirubin (and sometimes urorosein) in proportion to bacterial action, in the gut and urinary system. Melanin also may be present. The urine of most animals is visibly coloured.

5.1.15 Reproductive Tissues

The ovaries and eggs of animals are usually coloured, to some degree at least. By far the most common chromes here (Table 12.1) are the carotenoids, usually in massive quantities (Goodwin, 1950, 1952, 1962). They are present in most, but not all vertebrates and in echinoderms, arthropods, molluscs, annelids and Cnidaria. In most other phyla, also, carotenoids are implied by the colour. The colours of eggs are more varied because the carotenoids there frequently become protein-bound (p. 31). The ovary and eggs of echinoids often contain naphthoquinones. Haemoglobin occurs in the ovary of *Chironomus* (Fox, 1955), and is passed into the eggs, and this is true also for the louse, *Pediculus*. In *Rhodnius* it is derivatives of Hb which are transmitted in this way. There is haemoglobin round the anterior uterine horns and vitellaria of trematodes (Lee and Smith, 1965) and in fact where separate vitellaria occur they are usually coloured. The oviducts of the herring and some other teleosts are silvery with purines. Accessory reproductive organs such as the nidamental glands of cephalopods are often pigmented. The placenta of mammals may become coloured with bilins, haemosiderin and other haem products (Diwany, 1919; Verne, 1926, p. 228).

Testes usually have little pigment, the small amounts of chromes in the endocrine cells being inconspicuous, but there are significant amounts of carotenoids in some molluscs, annelids and other worms. Both carotenoids and fuscins occur in the endocrine cells (Fox and Vevers, 1960, p. 163). There are echinochromes in echinoids and Hb in *Ascaris:* in both of these cases the chrome occurs also in most other tissues of the body. In fact there is a general tendency, particularly in arthropods, for the predominant integumental chrome to be present also in the testis, testis-sheath or vas deferens. There are phaeophorbides a and b in the testis-sheath of the squash bug, *Anasa tristis*, (Metcalf, 1945) and haemoglobin in the accessory reproductive organs of the water bug *Macrocorixa* (Brindley, 1929). Ommochromes, also present in the integument, surround the vas deferens of woodlice (Needham, 1970a). The chromes of the testis-sheath of the various genotypes of *Drosophila* are closely related to those of the eyes of the same mutant (Ziegler, 1961). Lampyrine, an unidentified chrome of the luminescent and other organs of fireflies, occurs also in the testis (Metcalf, 1943). Melanising activity in the semen of rabbits is proportional to the amount of melanin in the pelage (Beatty, 1956). This general parallelism is valid also for the female system but less striking because of the large amounts of chromes, mainly carotenoids, in transit to the eggs.

5.1.16 Alimentary System

Including materials in the food, the alimentary system contains more biochromes even than the integument and the storage tissues. Even in the fasting animal there is a considerable variety of chromes both in the gut-wall and in the tissues of the digestive glands and other gut-appendages. Chromes absorbed from the food are often stored in the alimentary system itself.

Gut-wall: Dietary carotenoids are stored in the wall of the pyloric caeca of asteroids (Fox, 1953, p. 97) and holothurians, and in the gut-wall of many other groups; as usual they are the most common class. Next in abundance are the various haem derivatives, of which almost every conveivable member is found in annelids (Dales, 1957; Mangum and Dales, 1965). The gut-wall of *Rhodnius* has a protoferrihaemochrome and protoverdohaem (Wigglesworth, 1943). *Nereis* has biliverdin IX, while the chlorophyll derivatives, phaeophorbides *a* and *b*, are common in the polychaetes *Chaetopterus* (KENNEDY and NICOL, 1959), *Nereis* and *Owenia* (DALES, 1957). Presumably, *Nereis* is not purely carnivorous. Both phaeophorbides occur also in the salivary glands of the squash bug, *Anasa*, (METCALF, 1945). Haemerythrin is present in the gut-wall of sipunculids (SMITH and LEE, 1963) and there is a purple to brown chrome in *Priapulus* (FÄNGE, 1969) which likewise is probably haemerythrin or a derivative. Echinochromes are concentrated here by *Echinus* (MOORE, 1937) and pterins occur in the ascidian gut (KARRER, 1948).

Livers: The vertebrate liver contains carotenoids (FOX, 1953, pp. 141–80), and particularly vitamin A, which is said to be in such high concentrations in the liver of polar bears as to make the tissue toxic as a food (HEILBRUNN, 1952 p. 193). There are fuscins (FOX and VEVERS, 1960, p. 163), hepatocuprein (PORTER, 1966), breakdown products of haem and ribitylflavin. The Crustacean liver likewise has mainly carotenoids (FOX, 1953, pp. 121, 123, 128) and some bilins (BRADLEY, 1908). Molluscan digestive diverticula contain carotenoids (FOX, 1953, p. 107, 115), phaeophorbides (ib p. 107–8), bilins (ib p. 276) and ribitylflavin (ib p. 110).

Secretions from these organs into the gut-lumen include carotenoids (VERNE, 1926, p. 224) and haem derivatives. In *Aphrodite*, Hb itself is passed from the gut caeca (BLOCH-RAPHAEL, 1939), and helicorubin (p. 241) in *Helix* and other pulmonates. Similar haemochromes are secreted in serpulimorphs and terebellids (PHEAR, 1955). Bilins are secreted not only by vertebrates but also by Crustacea (BRADLEY, 1908) and cephalopods (FOX, 1953, p. 276). Thus most of these chromes are present also in the cells of the organ; in addition, however, the vertebrate bile is the normal outlet for chromes and other large aromatic molecules.

Gut Lumen: Studies on fasting animals distinguish secreted material from direct food-constituents, which have trivial interest. Secreted materials come via minor glands and epithelial cells of the gut-wall as well as from the major glands. As usual carotenoids are most common, – mainly carotenes in the polychaete *Chaetopterus* (KENNEDY and NICOL, 1959) and in the guinea-pig (FOX, 1953, p. 180); astaxanthin in whales (*ib* p. 178); and carotenoprotein in the crab *Carcinus* (*ib* p. 121). In its proventriculus the fulmar petrel accumulates massive quantities of carotene and vitamin A, in solution in oil (ROSENHEIM and WEBSTER, 1927); it regurgitates this foul-smelling oil when threatened by a predator.

In quantity haem-derivatives come second but in variety first, particularly among polychaete worms (PHEAR, 1955; MANGUM and DALES, 1965). Methaemoglobin is found in the sea mouse *Aphrodite* (PHEAR, 1955), protoferrohaemochromes in *Aphrodite* (Pantin, 1932), *Hermione* and *Harmathoe* (PHEAR, 1955) and chloroferrohaemochrome in those which have chlorohaemoglobin in the blood. Free chlorohaem is not found, but always protohaem, usually in the ferric form, whether in polychaetes, in leeches (NEEDHAM, 1966 b) or in the bug *Rhodnius* (WIGGLESWORTH, 1943). Some free ferrohaem is occasionally found (MANGUM and DALES, 1965). Uro- and copro-porphyrins and haems also are common in polychaetes; free porphyrins have been found also in larval mites (FOX, 1953, p. 264) and in human pernicious anaemia patients.

Phaeophorbide *a* has been found in the gut-lumen of the polychaetes *Marphysa* and *Sabellaria* (MANGUM and DALES, 1965), and of the silk-worm (FOX and VEVERS, 1960, p. 118), and both *a* and *b* in the sheep (FOX, 1953, p. 264). Other herbivorous mammals excrete at the phylloerythrin stage; it is interesting that chlorophyll-derivatives are taken into the body, at any rate temporarily, by animals already having abundant porphyrins of their own.

Coccids inevitably pass anthraquinones this way (FOX, 1953, p. 205), though there is none in the gut-wall itself. Ommochromes are excreted in the meconium of newly emerged nymphaline butterflies (p. 245). In all, the gut handles most zoochromes, but particularly carotenoids in both directions, and members of the haem path in the excretory direction.

5.1.17 Conclusion

It is clear that no main tissue of the body is invariably devoid of pigment and that in animals collectively most tissues contain a variety of chromes, representing most of the known chemical classes and some others. This extensive distribution between tissues is as impressive as the breadth of taxonomic distribution. It could be interpreted in two quite different ways, either as indicating manifold functions for most zoochromes or as evidence that most are casual products of metabolism liable to be formed, or to accumulate, anywhere. The bilins (VERNE, 1926) and some others do seem to leak indiscriminately from one tissue to those which happen to be topographically nearest, supporting the second alternative, but for the most part there is a sharp change in colour at the boundary of each organ, which therefore appears to select the chromes it contains.

While making due allowance for present unevenness of information it is possible to make a semi-quantitative analysis of the tissue-distribution of the main classes of chrome (Table 5.1). Fuscins have been omitted because of their probable heterogeneity (p. 35), the flavonoids because of their rarity in animals, and the simple metallochromes because these are probably much less investigated in invertebrates than the other known classes. All haems and haemoproteins have been pooled because further subdivision would add numbers rather than information. This leaves nine classes, all of which occur in the epidermis, and in neural crest plus connective tissue and all but one in blood-plasma. Next in variety of content come exoskeletal materials and the tissues associated with blood-vessels and coelomic walls. In no main tissue have fewer than two of the classes been found.

Although the gut system is most consistently pigmented, not all the classes have been found there. If integumental structures are taken collectively there is no doubt that they have the greatest variety both between and within chemical classes. Fat tissue naturally has fewest classes and next in order come nerve-cells, endocrine organs, internal skeletons, gonads and, perhaps surprisingly, excretory organs.

Analysis at right angles to this (Table 5.1) shows that carotenoids are the most widespread class, with haem and its complexes second, followed closely by the haem-derivatives, – free porphyrins and bilins. Quinones and indolic melanins have been found in nearly half of the tissues. Surprisingly, perhaps, ribitylflavin is widely distributed in quantity: in smaller amounts of course it is present in virtually all tissues (p. 89). Even ommochromes and pterins are not narrowly restricted in sites.

Apart from the melanogenic bloods, coelomic fluids and semen, only integumental and connective tissues contain melanin; no other class is so restricted. So far pterins have rarely been demonstrated in blood and coelomic systems, and in other tissues deep in the body (FOX and VEVERS, 1960, p. 157) but they are probably common, at least in coenzyme quantities.

Table 5.1. Distribution of main classes of zoochrome between sites in the body (x indicates one or more records)

Class of chrome / Site in body	Carotenoids	Quinones	Haem and derivatives	Porphyrins	Bilins	Melanins	Ommochromes	Pterins	Ribityl flavin	Total
Exoskeleton	x	x		x	x	x	?		x	7
Epidermal secretion		x		x	x	x	x			5
Epidermis	x	x	x	x		x	x	x	x	9
Respiratory epithelia	x		x			x	x			4
Neurons	x		x			x				3
Glia & white matter	x	x		x	x	x		?	?	5
Sense organs	x					x	x	x	x	5
Endocrine organs	x			x		x				3
Vertebrate neural crest	x					x		x	x	4
Connective tissue	x	x	x	x	x	x	x		x	8
Internal skeleton		x		x	x					3
Fat tissue	x		x							2
Muscle	x		x	x	x		x		x	6
Blood-plasma	x	x	x	x	x	x	x		x	8
Blood cells	x		x			x				3
Blood vessel tissues	x		x	x	x	x	x		x	7
Coelomic fluid		x	x			x				3
Coelomocytes	x	x	x						x	4
Peritoneum	x	x	x		x					5
Excretory organs	x		x							3
Urine				x	x			x	x	4
Testis sheath of arthropods			x	x			x	x		4
Ovary	x	x	x							3
Gut-wall	x	x	x	x	x			x		6
Liver	x		x	x	x				x	5
Gut-secretions	x		x		x					3
Gut-lumen	x	x	x	x	x		x			6
Totals	22	13	20	16	15	14	10	6	12	128

5.2 The Location of Chromes within Organs and Tissues

Some organs consist of a variety of tissues, in addition to the universal connective tissues, blood vessels, etc., and so it may be important to know which tissue actually contains a particular chrome. Within a tissue there may be more than one type of cell, as well as intercellular material and tissue-spaces and it is useful to know the exact location.

A few examples will make this clear. There is quite a sharp distinction between the chromes in the chief ('parenchyme') cells of the liver and those of the Kupffer (reticuloendothelial) cells. The former contain bilirubin (JIRSA, 1969), the 'lipomelanin' of the Dubin-Johnson syndrome (BARONE et al., 1969) and haemosiderin (STURGEON and SHODEN, 1969), while the latter contain ferritin (STURGEON and SHODEN, 1969), copper complexes (GEDIGK, 1969) and the fuscins of 'ceroid storage disease' (FISHER, 1969). The fuscin formed in gut and uterus as a result of vitamin E deficiency (p. 197) is deposited mainly in the smooth muscle cells. Mammalian cutaneous melanin is formed in the epidermal melanocytes but is then passed on to all the keratinocytes of the vicinity (BREATHNACH, 1969). In the lung, inhaled soot is sequestered (Fig. 5.3) almost entirely in phagocytes *behind* the bronchiolar and alveolar epithelia (HEPPLESTON, 1969). The adrenal cortex and medulla (section 5.1) could be cited as an example of a significant contrast between the tissues of one organ. There is very evident chromatic differentiation between the various integumental chromatocytes, but elsewhere relatively little knowledge of the chromatic differences between topographically associated cells. Electron microscope studies such as that of GREEN and NEFF (1972) are valuable here.

The evidence from living Porifera is that extracellular chromes within the body were rare in primitive metazoa but already in the scyphomedusa *Pelagia* the visible coloration is due to extra-, as well as intra-, cellular chromes (FOX and MILLOTT, 1954a). Secondly, zoochromes in solution in the body fluids may have evolved as soon as they could serve a useful purpose: there is haemoglobin in the pseudocoele of some nematodes and possibly even in the parenchymal spaces of platyhelmia. The intracellular site of the oxygen-carrier chromes of vertebrates and some other groups may be secondary, and it certainly has advantages (p. 215), but it is common in metazoa as primitive as the nemertines and may have evolved in parallel with plasma O_2-carriers.

It might seem that chromes in solution should be more useful for rapid colour-change than solid material inside cells but in fact the most rapid changes, by cephalopods (p. 138), use intracellular chromes and no doubt these are best for rapid changes with a detailed patterning. For such purposes as storage, an intracellular site is best on almost all scores; it ensures tidiness and segregation into small units so that contamination and infection cannot spread and it permits sequestration in concentrated, inert form, which nevertheless can be mobilised rapidly when necessary. Ferritin in the Kupffer cells is a good example of the facile mobilisation of a difficult chrome: the iron is changed from Fe^{III} to soluble Fe^{II} enzymically, passed through the cell-membrane into the blood and, in Fe^{III} form, rebound to the plasma β-globulin as transferrin.

It appears that in fact there are severe restrictions on the amount and variety of chromes permitted in extracellular sites. In the circulatory system only oxygen-transport chromes, certain enzymes and materials in transit are found and even these are largely excluded from the tissue-spaces of those animals with a 'closed' blood-vascular system. If the digestive diverticula of the isopod *Asellus* are isolated into Ringer's solution containing methylene blue, this chrome is rapidly transferred to the lumen; presumably *in vivo* such chromes would be passed into the gut lumen and so eliminated from the body. The excretory organs and other systems discussed in section 5.1 share in this type of control. From the tissue spaces of animals with a closed circulation the material is first passed into the vascular system. This is seen when bruising in humans causes leakage of blood into the spaces; the 'black and blue' display of bilins and other breakdown-products of the haemoglobin are relatively rapidly removed. Jaundice involves the escape of bilirubin or, in the pike, biliverdin (Fox and Vevers, 1960, p. 120) from blood to tissue-spaces and as a further complication it adsorbs strongly to elastic tissue (Gray, 1961, p. 80; Rook et al., 1968,

Fig. 5.5. Three dimensional section of skin of lizard Anolis showing three types of chromatophore and extracellular yellow oil droplets in the dermis. (From Fox, 1953; after von Geldern, 1921)

p. 1150). Since the sclerotic of the eye particularly is affected, collagen also must adsorb the bilin. Nevertheless removal can be complete in a matter of weeks. In *xanthaemia* or pseudojaundice due to excess carotenoids, again the tissue-spaces are affected; in this instance the sweat is found to be an auxillary route for elimination (Fox, 1953, p. 174). *Ochronosis*, another yellow discoloration associated with alcaptonuria (p. 249), affects the connective tissue matrices rather generally, i.e., cartilage as well as tendon, sclera, dermal collagen and the basement membrane of the epidermis (La Du and Zannoni, 1969); again the chrome is removed if the condition is corrected, and again the sweat glands assist.

The therapeutic use of the tetracycline antibiotics leads to chelation with the salts of the teeth and bones as a yellow chrome (SAXEN, 1969), and invertebrate skeletal materials are affected similarly.

Heavy metals such as silver are deposited as dark material among the collagen fibres of the oral mucosa (FISHER, 1969; GEDIGK, 1969) but this is in insoluble, granular form. The fate of such materials is well known from the barbarous practice of tattooing: here cinnabar, (red mercuric sulphide), prussian blue (which deposits insoluble iron salts) and other heavy metal salts, as well as india ink (finely particulate carbon), are directly introduced into the dermis. They are taken up by phagocytes (GEDIGK, 1969). The experimental use of colloidal ferric hydroxide shows that all the metal salts are bound at a common site on the intracellular carrier-molecules of the phagocyte, an all-purpose protein-carbohydrate-lipid complex (GEDIGK, 1969). Whereas the iron can join its normal metabolic pool as transferrin, however, the salts of most other metals are indigestible and most of the phagocytes appear to become embalmed and move no more, though some reach the spleen and lymph nodes to perenniate there. It is known that amoebocytes *can* be eliminated through various epithelia and the stasis in this case is an intriguing mystery. There might be a temptation to regard the cells as artificial chromatocytes on the understanding that the stasis also is very unnatural: natural chromatocytes are much more mobile and labile.

A few biochromes are normally deposited extracellularly, for instance (Fig. 5.5), the yellow oil-droplets in the skin of the lizard *Anolis* (VON GELDERN, 1921) and the guanosomes of the integument of some fishes (HOAR, 1966, p. 592). Lipofuscins sometimes are deposited extracellularly (PORTA and HARTROFT, 1969) but this may be pathological rather than physiological. Carotenoids and purines probably can be re-mobilised as necessary.

5.2.1 Conclusion

Chromes are probably usually specific to particular cell-types. The Metazoa have evolved some extracellular chromes within the body, mainly soluble chromoproteins of the circulatory system and chromes in transit via that system. Other chromes which by some means get into extracellular sites are removed via gut, circulatory system and various excretory organs, provided they can be metabolised or dissolved. Completely insoluble materials are phagocyted but not further eliminated. Such materials are normally rare, all biochromes being physically and chemically tractable when necessary.

5.3 The Location and State of Zoochromes within Cells

Collectively zoochromes exist in several different states within the cell, dispersed throughout the cytoplasm, in spherical 'vacuoles' of varying size, or in solid 'granules' of varying, but usually angular, form. To some extent at least the form is related to the physical properties of the material, whether soluble in water, lipid, or neither. Until the advent of electron microscopy little further significance in the particular state was envisaged, though for their possible classificatory value further minor variants were recognised. It is now clear that the chromatic 'inclusions' of the integumental chromatocytes are organelles almost as highly organised as mitochondria and there is much scope for further studies at this level of magnitude. As for materials in general, inclusions in the cisternal-golgi compartment of the cell are usually destined for secretion from the cell while those free in the 'supernatant', or cytoplasm proper, are permanent organelles, of appropriate functional significance.

Variations in size, shape and development of pigment-inclusions are becoming more meaningful and more diagnostic at the electron-microscope level.

The oxygen-carrier chromes are almost the only type which is sometimes found free in aqueous solution in the cytoplasm; in the mammalian erythrocyte there is no nucleus and the haemoglobin (Hb) occupies the whole volume, apart from a skeletal stroma. Moreover the Hb solution is so concentrated that it crystallizes out on slight provocation, for instance, when taken into the gut of the blood-sucking tick, *Ixodes,* (DIWANY, 1919), and even *in vivo* in the erythrocytes of some vertebrates, and in the tracheal end-cells of the larva of *Gastrophilus*. The high concentration has a number of virtues (p. 211) not feasible for chromes free in the plasma. Little is yet known about the state of Hb in those invertebrates which have it in corpuscles, except that the molecule is as small as that of vertebrate Hb.

The haemerythrin of sipunculids and priapulids (p. 224) is restricted to one or more small vacuoles in the cell so that mere microscope-examination might imply a lipid solution. On cytolysis the chrome in fact proves freely water-soluble, so that *in situ* the vacuole may be part of the cisternal system, i.e., a segregated part of the cell: this is an obvious subject for E.M. work. Water-soluble dyes injected into animals are sometimes sequestered in this way, inside coelomocytes (HYMAN, 1959, p. 594).

Some of the chromes of the epidermis, and of the connective tissue parenchyme cells, are thought to be freely dissolved in the cytoplasm (HYMAN, 1951, p. 78) but there are few clear records of this. A number of the chromes of eggs of animals are in cytoplasmic solution, for instance haemocyanin in the egg of *Carcinus* (BUSSELEN, 1971). The ovorubin of the egg of the snail, *Pomacea,* is a carotenoprotein in solution in the jelly around the egg (FOX and VEVERS, 1960, p. 70–71). On the other hand the melanosomes of the frog's egg and the yolk-platelets of birds' eggs are large organelles. In view of the way in which the chromatophores of cephalopod molluscs operate in chromomotor changes (p. 151), it might be assumed that their chrome is in solution but of course a dispersion of small granules is equally suitable for that mechanism.

Biochromes in fluid vacuoles usually prove to be free carotenoids, in lipid medium; as they grow such vacuoles tend to run together to form a single, large globule. Carotenoids in lipid vacuoles occur in both ectodermis and endodermis of the sea anemone *Metridium* (FOX and PANTIN, 1941), in the integument of asteroids (HYMAN, 1955, p. 381), and in the ovaries and eggs of many groups (FOX, 1953, p. 68); no doubt free carotenoids are usually in this state. In birds' feathers this is the initial state but later the chrome becomes dispersed throughout the keratinocytes (DESSELBERGER, 1930), and a similar dispersion occurs in the exoskeleton of insects and Crustacea. Coloured globules in the retinae of some vertebrates (WALLS, 1942) are mainly carotenoids in lipid medium. However, in some of the lower vertebrates integumental carotenoids are thought to be in granular form (BAGNARA, 1966) so that they may share in the usual chromomotor responses and those of the sea anemone *Actinia* also appear to be granular (M. and R. ABELOOS-PARISE, 1926). The carotenoids of plants (VERNE, 1926, p. 262) and of lizards (ib. p. 451) sometimes crystallize in their vacuoles and this may be the explanation of other granular carotenoids.

A brown chrome in some of the coelomocytes of brachiopods is in lipid vacuoles (HYMAN, 1959, p. 557) and so may be a lipofuscin (p. 35): other lipofuscins appear to be in granular form (FOX and VEVERS, 1960, p. 162) so that this may provide a basis for separating two sub-groups. Other zoochromes said to be in fluid vacuoles include the anthraquinones of the coccid, *Dactylopius,* (FOX, 1953, p. 205), the bilatriene of *Allolobophora chlorotica* (KALMUS et al., 1955), adenochrome in the branchial heart-appendage of the octopus (Fox, 1953, p. 307) and the chrome in the epidermis of Turbellaria (Hyman, 1951a, p. 78). Many bilins certainly are soluble in lipids (Jirsa, 1969) but *in vivo* the coccid anthraquinones are in water-soluble form (p. 45).

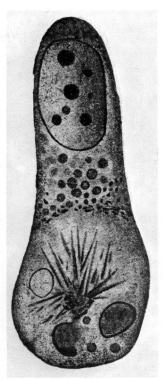

Fig. 5.6. Crystals of bibiverdin inside cell of chorionic villus of placenta of dog. (From Verne, 1926)

Among solid states the crystalline is most easily recognized (Fig. 5.6); it is not uncommon for chromes in situ. In addition to the free carotenoids mentioned, some carotenoproteins also appear to be crystalline, for instance, 'crustacocyanin' of the crustacean integument (Verne, 1926, p. 262). In the cells of the body-cavity of *Priapulus*, haemerythrin is sometimes birefringent, and therefore effectively crystalline (Fänge, 1969), like the examples of vertebrate Hb quoted. Hb is crystalline also in the botryoidal tissue cells of leeches (Harant and Grassé, 1959, p. 538); this may be a prefunctional store, corresponding to those of polychaetes (Breton-Gorius, 1963; Roche, 1965). Here the crystals are built of units of macromolecular dimensions, and the electron microscope resolves both the molecules and their main sub-units (Fig. 5. 7). Serpulimorph chlorohaemoglobin crystallizes similarly. Biliverdin is sometimes crystalline (Fig. 5. 6) in the chorionic villi of the placenta (Diwany, 1919) and similarly phaeophorbides *a* and *b* in *Anasa tristis* (Metcalf, 1945). Blue crystals have been recognised in the cells of the adrectal gland of muricid gastropods (Fox, 1953, p. 219) and so may be precursors of the dibromoindigotin (p. 59). Pterins are crystalline in the cells of the scales of the wings of pierid butterflies (Fox and Vevers, 1960, p. 153). The reflecting properties of purines, pterins and ribitylflavin depend essentially on orientated crystalline states. The structure is particularly regular in the iridosomes of fishes, where (p. 151) it is associated with fascinating and enigmatic motility (Denton, 1971).

An intriguing instance is the material in an orange-coloured haemocyte of ascidians

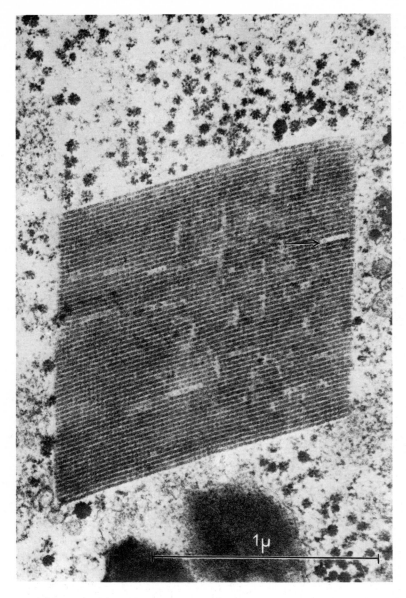

Fig. 5.7 A. Electron-micrograph of crystal of haemoglobin in a chromogogen cell of the polychaete Arenicola. The arrow points to one of the 'faults' in the crystal lattice

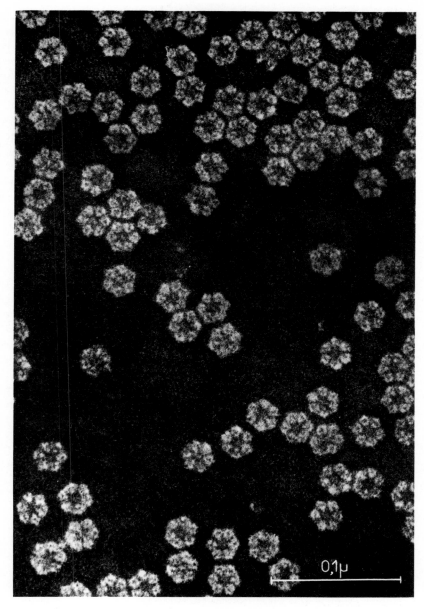

Fig. 5.7 B. Electron-micrograph of molecules of haemoglobin of Arenicola showing six of the sub-units. There are two decks of these in the molecule as indicated by two of the molecules seen in profile near the top of the picture. (From BRETON-GORIUS, 1963)

(WEBB, 1939). This cell is packed with eight to ten elliptical plate-like inclusions filled with the chrome, which is crystalline or is bound to a crystalline matrix. Each plate has a spherulitic structure, the crystallites radiating from a focus at one end of the major axis of the ellipse. The crystallites have negative birefringence with respect to their length.

The term granular has been applied to virtually all kinds of discontinuous distribution of biochromes in cells, though some have observed the convention that 'granules' are visibly distinct under the light microscope, and are both solid and angular, and that spherical and ovoid inclusions must be fluid. This last assumption is not necessarily true, as shown by such ovoid yet structured organelles as mitochondria, lysosomes and chloroplasts. Both shape and texture of chromatic inclusions vary greatly and electron-microscope studies of a wide range of these should be very rewarding. The vanadocytes (p. 73) of ascidians contain eight to ten pigment-inclusions, collectively filling the cell but retaining their integrity although forced into sector-shaped masses (WEBB, 1939). These may be fluid and the angularity entirely due to close packing. Under certain conditions the cell has a morula form so that its membrane may be weaker than that of the inclusions. The purple zoochrome of *Bugula* (Table 3.14) occurs either as ellipsoid, or as angular, microscopic granules (FOX, 1953, p. 309).

Most of the known types of biochrome, including carotenoids (p. 33), have been found at least at times in granular form. Lipofuscins have this form also in heart-muscle (VERNE, 1926, p. 223) and in the mesenchyme cells and coelomocytes of *Antedon* (FOX and VEVERS, 1960, p. 164). Under the light microscope, melanins are nearly always described as granular and this appears to be true of ommochromes in insects (FOX and VEVERS, 1960, p. 52-3) and isopods. Pterins are granular in *Drosophila* and in the skin of the salamander (FOX and VEVERS, 1960, p. 165). Ribitylflavin is in this form in the cells of insect malpighian tubules. Most unidentified zoochromes also are in this general form, for instance, the dermal pigments of sipunculids (HYMAN, 1959, p. 624) and arenicochrome (DALES, 1963). In the ectodermis of *Metridium*, melanin appears to be more finely dispersed than in other animals (FOX and PANTIN, 1941).

The porphyrins of the triclad Turbellaria are adsorbed to the rhabdites, which are formed in sunken epidermal gland cells and passed up into the epidermis (MACRAE, 1961). This is one of the few instances where a chrome is deposited on a cell-organelle with an independent function.

5.3.1 The Chromasome

The typical integumental pigment-granule or chromasome is about 1μ in diameter, spherical or ellipsoidal. The term should not be confused with 'chromosome' unfortunately already prë-empted, not very aptly, for the chromophil nucleic acid helices of the cell-nucleus. 'Chromatosome' would be confused less easily but also has been prëempted (p. 122) for a consortium of several chromatocytes (FINGERMAN, 1963). The chromasome is in the mitochondrial range of size and shape and therefore has been considered a modified mitochondrion, phylogenetically if not ontogenetically. Claims have been based mainly on the

---

Fig. 5.8 A. Electron-micrograph of longitudinally sectioned premelanosomes from human foetal retinal melanocytes. When matrix-proteins sheets are cut normally to their surface (13) they show a characteristic periodicity. In some sections normal to the surface the periodicity appears to be due to a helical structure of the protein filaments (15). In sections parallel to the surface of the sheets (14) the periodic structures are seen to be in dress in all the filaments giving a lattice-like picture. (From BREATHNACH. 1969)

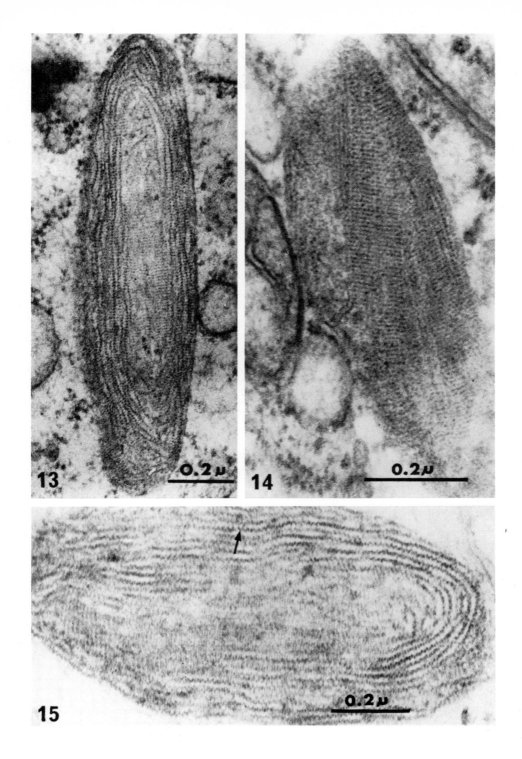

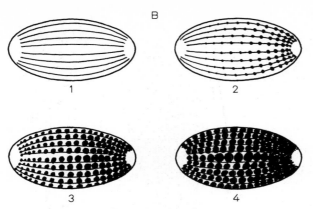

Fig. 5.8 B. Diagrams showing successive stages in the deposition of melanin on the protein-matrix of the melanosomes of the retinal melanocytes of a mammal. (Based on MOYER, 1963)

melanosome but VOINOV (1928) suggested this also for the biliverdin granules in the cells of the leech, *Glossiphonia*. Further, the stentorin granules of *Stentor coeruleus* behave like mitochondria in taking up the dye, Janus Green (WEISZ, 1950); the mitochondrial relationship of the melanosome continues to receive support (WOODS et al., 1963; DU BOY et al., 1963), because of this implied cytochrome oxidase activity, and some other features. However, the structure resolved by the electron-miscroscope is very different and it would be difficult to derive the plan of the matrix of the melanosome from that of the cristae of the mitochondria. Mitochondria multiply when melanosomes (BREATHNACH, 1969) and other chromatic inclusions do (GEDIGK, 1969) but this is in parallel and not in series! There is no detected ontogenetic relationship and any phylogenetic affinity has been masked by divergence.

Melanin is deposited at a large number of loci on the protein matrix of a preformed organelle, the *premelanosome* (SEIJI et al., 1963) and it seems that this body can have two very different forms and modes of formation in mammals (MOYER, 1963, 1966; BREATHNACH, 1969). In the cells of the pigment layer of the mouse's retina, derived from the optic cup, the matrix is in the form of a sheaf of filaments 5 nm thick, running from one point to the opposite end of the somewhat elongated premelanosome, and caught up at each end. The whole (Fig. 5.8) has some resemblance to the inclusions in the special type of orange cell in tunicates (p. 115). The filaments are coiled coils (DROCHMANS, 1966) and are assembled free in the cytoplasm before a membrane is built round them. Melanin is then deposited at many loci along the filaments, but earliest near one end of the body (Fig. 5.8, B). The deposits increase in size until they fill the whole melanosome (MOYER, 1963).

In melanocytes of neural crest origin the premelanosome develops in the endoplasmic-golgi cisternal systems (for export, p. 101) and the membrane is laid down before the matrix-filaments. These become aggregated into sheets which, in transverse section of the melanosome appear concentrically or spirally wound (Fig. 5.9). MOYER (1966) finds that some integumental phaeomelanosomes, presumably the dermal melanosome type (BREATHNACH, 1969) which is not extruded from the cell (p. 101), are formed in the retinal melanosome manner. Some variation in structure, if not in site of formation, seems consistent with the variation in gross size and shape of melanosomes between species and even between the sexes. For instance there are rodshaped bodies in black, but spherical ones in 'blue',

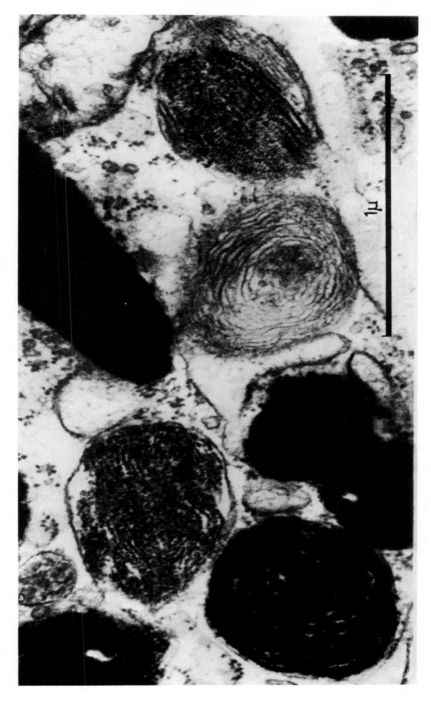

Fig. 5.9. Electron-micrograph of transversely sectioned premelanosomes of the neural crest melanocytes of a mammal. In this type the protein-matrix sheets are spirally wrapped. (From BREATHNACH, 1969)

feathers, while the phaeomelanosomes of buff, and the erythromelanosomes of red, feathers are ovoid (BOHREN et al., 1943). In human skin there are rods and spherules, and in mouse melanoma ellopsoids (FOX, 1953, p. 226). Mitochondria vary in much the same ways.

There appears to be no significant difference between the melanosomes of the lower vertebrates, which are motile in their asteroid melanphores (p. 148) and those of the mammals which are generally thought to be immobile except for the intercellular transfer from epidermal melanocytes. In fact they may still have some power of dispersion and aggregation within the mammalian melanocyte; the difference between dark and fair-skinned human races is probably due not to differences in density of cell-population but to differences in degree of display of the melanosomes within the individual cell (SZLABÓ, et al., 1970). The darker races also have more melanosomes per cell (SZLABÓ, 1957).

The other vertebrate chromatocytes, yellow, red and white, variouly known as lipo-, xantho-, erythro-, pterino-, irido-, and guano- cytes (or phores) have chromasomes essentially with the same structure as melanosomes (BAGNARA, 1970; ALEXANDER, 1970) notwithstanding the variety of chromes involved; this supports the general view that all have a common origin from the neural crest. Xanthophores may contain carotenoids or pterins, or both (ALEXANDER, 1970) and perhaps ribitylflavin also (BAGNARA, 1966). Erythrophores are probably always purely pterinophores (OBIKA et al., 1970). Iridophores and guanophores contain purines and/or pterins.

All vertebrate types come to contain tyrosinase and indeed this is usually more active in mature cells of the other types than in mature melanocytes, where perhaps the enzyme has been exhausted. It would seem that either the melanocyte is the ancestral type or that all have the potentiality for conversion into melanocytes. Most of the other chromes have been implicated in melanin synthesis (p. 289) and the present fact gives some support for these claims. The two types of melanosome-matrix occur also among pterinosomes (BAGNARA, 1963, 1966; MATSUMOTO, 1965b; 1970); here the concentric, lamellate type has mainly drosopterins while the fibrous type has sepiapterins and other pterins (MATSUMOTO, 1970). The two may have different taxonomic distributions, though both occur in teleosts and amphibia. The pterin is attached to the matrix in the same way as melanin, and moreover the same mode is seen in the pterinosomes of *Drosophila* (ZIEGLER and JANICKE, 1959). The iridosomes of fishes are highly structured (BAGNARA, 1966); those of amphibia have as matrix a stack of platelets, each bounded by a double membrane as described for the lamellar type of melanosome (TAYLOR, 1970). Crustacean leucosomes are prisms, 0.5×0.15 µm (GREEN and NEFF, 1972). Carotenoids have been detected both in the endoplasmic reticulum and in the cytoplasm (MATSUMOTO, 1970); again the former may be for export from the cell but this chrome presumably is only processed (e. g., protein-conjugated), and not synthesised here (p. 312). Crustacean carotenosomes have a diameter of 0.23 µm (GREEN and NEFF, 1972).

The ommochromasomes of insects also have a structure similar to vertebrate chromasomes (ZIEGLER and JANICKI, 1959), so that this may be quite universal. The gross uniformity of the classical 'granule' therefore is now magnified into the highly structured uniformity of the chromasome. The highly organised structure of all chromasomes renders more credible the fact that several types of the organelle can be maintained and independently operated within the same cell, as seen most strikingly (Fig. 6.1) in *Carausius* epidermal cells (GIERSBERG, 1928). Some vertebrate chromatocytes also contain more than one type of chromasome. However, what has sometimes been interpreted in Crustacea as a polychrome cell or a syncytium of several types of cell is now regarded as a consortium of separate cells with thin party-walls (FINGERMAN, 1963). In fact, as long ago as 1923 (VERNE, 1926), Millot separated one cell from another. For the most part metazoan chromatocytes have become monochromatic specialists.

5.3.2 Protozoan Chromes

Protozoa such as *Euglena* certainly have more than one type of chromasome, for instance, chloroplasts and carotenosomes ('haematochrome' granules). In *Euglena* the latter, but not the former, are motile (JOHNSON, 1939; JOHNSON and JAHN, 1942). In the phytoflagellates the chloroplasts contain the whole set of biochromes required for photosynthesis and this may account for most of the twelve chromes, in all, found in Dinoflagellates (STRAIN et al., 1944). In this organelle the organisation of polychromacy lies essentially at a still lower level of magnitude, which is effectively the macromolecular level. The mitochondrion is another organelle with polychromatic organisation at this level; the variety is less than in chloroplasts but since it is common to all cells, in all animals, it is the more important in the present context.

In principle the Protozoa might be expected to reveal a good deal about zoochrome-structure at the sub-cellular level but apart freom the phytoflagellates relatively few Protozoa have large quantities of chromes, detectable at the light microscope level; their respiratory co-enzymes probably are organised much as in a metazoan cell and there is rarely enough to produce a definite coloration. Eye-spots are well structured organelles (p. 171), but again are largely confined to the phytoflagellates. A red chrome is distributed in granular form throughout the cytoplasm of the foraminiferan, *Myxotheca*. Radiolaria often have a pigmented body termed a phaeodium, and some parasitic amoebae and ciliates (p. 256) are pigmented. The only other significant protozoan zoochromes appear to be the polycyclic quinones of *Stentor, Blepharisma* and *Fabrea* (p. 86). They are in structured chromasomes and are arranged in very regular longitudinal tracts; this is in keeping with the high degree of organisation of the whole microanatomy of ciliates (TARTAR, 1961). They are in the cortex, the anatomically analogous site to that of the integumentary pigments of the metazoa. It would be useful to know the location of the haemoprotein of *Glaucoma* and *Paramecium* (p. 87). More details of the exact state of the various chromes in eggs (p. 269) also would be very helpful here; they may be considered to recapitulate a protozoan stage, but many have become heavily pigmented for developmental reasons.

5.3.3 State of Zoochromes in Storage and Scavenging Cells

Here present knowledge is even more limited. VERNE (1926, p. 259) commented on the irregularity of pigment fragments inside amoebocytes and presumably characteristic of excretory chromatic material. Storage granules are more likely to be structured, like starch-grains, etc. It has already been shown that a number of pathological chromes are stored in lysosome-like bodies, sometimes containing the enzymes typical of these organelles. Examples are heavy metal compounds (GEDIGK, 1969), bile pigments (JIRSA, 1969) and lipofuscins (PORTA and HARTROFT, 1969). The bodies are called phagosomes (Fig. 5.10) or phagolysosomes and the implication is that pathological chromes are usually scavenged by amoebocytes (p. 113). An increase in number of lysosomes in the cell is induced and again there is a parallel multiplication of mitochondria (GEDIGK, 1969). Scavenged fat forms intracellular bodies as large as 10 μm (THOMPSON, 1969). The enigmatic 'lipomelanin' of the Dubin-Johnson syndrome develops as bodies in the cisternal system of the liver parenchyme cell (BARONE et al., 1969) so that the inclusions are either abnormal secretory, or excretory, material of this cell.

None of these are normal storage bodies and information on such bodies would be

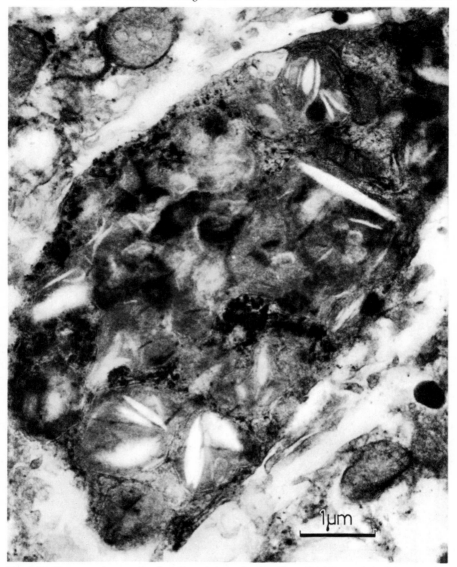

Fig. 5.10. Electron-micrograph of section of part of a Kupffer cell of liver of rat fed a 'necrogenic' diet (deficient in Vitamin E and selenium) for four weeks. The cytoplasm is packed with conglomerates of ceroid (fuscin) material, electron-translucent and of varied crystalline form. The dark granules between conglomerates are due to the standard test for acid phosphatase. (From PORTA and HARTROFT, 1969)

welcomed. The chloragosomes of earthworm chloragogen cells are almost spherical and 2 μm in diameter (see NEEDHAM, 1970b, Fig. 7), i.e., within the chromasome and mitochondrial ranges. Clearly the form and site of chromatic inclusions inside cells can help to diagnose their significance. It may be presumed that the melanin of melanosis coli (FISHER, 1969) is in phagosomes and not in free melanosomes.

5.3.4 Conclusion

In body-fluids zoochromes are usually in true solution but within cells this is uncommon except in O_2-transport cells of the circulatory systems, in some ova and in the phase-separate vacuoles of some other cells. Some intracellular solutions are so concentrated as to be virtually crystalline or liquid-crystalline. A number of zoochromes are in an overtly crystalline state. The state in the various type of 'granule' of other chromatocytes is not easily determined at the level of the light microscope. As revealed by the electron-microscope the most common condition is that the chrome is attached to a structured matrix inside ellipsoidal *chromasomes* of the 1μm order of size. Mitochondria and chloroplasts may be other specialised variants of this type of organelle. By contrast scavenged chromes are found in irregular masses in lysosome-type organelles which again are just visible in the light microscope. The intracellular state shows considerable relationship to function.

5.4 Biochromes and the Solid State

Since *in situ* most chromes are in a solid state this merits further consideration. For camouflage and storage purposes, where physical factors are more important than chemical reactions, this is usually an advantage. The stabilisation conferred by the solid state is well demonstrated by some fossil biochromes (p. 336), and a number of chromes are effectively preserved by simply allowing carcasses to dry. The raw material of commercial carmine is in this form and lampyrine was first extracted from firefly material which had been stored dry for sixty years (METCALF, 1943). It is noteworthy that the *Cypridina* bioluminescent system (p. 207) preserves well in the same way.

Chemical activity is not precluded by the solid state and indeed can be as high as 50% of that in free solution (YONETANI et al., 1966); the main shortcoming is rather the necessarily higher activation-energy (WILLIAMS, 1966). Possible compensating advantages are semiconduction (p. 13) and other properties for which a regular geometry is a great asset. Most of the evidence on reactivity in the solid state has been obtained from the familiar metabolic chromoproteins, some of which are normally in solution but others, e.g., mammalian Hb, are effectively in a solid state. The molecules of Hb do interact, even in solution (p. 231) but possibly more effectively in the solid state. In crystallised Hb the individual molecules will undergo their repeated cycles of oxygenation and deoxygenation just as in solution (WATSON, 1966). Similarly if crystals of deoxyhaemerythrin are exposed to air they turn pink, indicating O_2-uptake; at the same time they fragment owing to the strain due to changes in molecular conformation involved (MANWELL, 1964) and this shows one of the drawbacks of the solid state. Myoglobin also can be chemically changed in the solid state (GURD et al., 1966) and another haemoprotein, cytochrome c peroxidase, reacts half as fast in the solid state as in solution (YONETANI et al., 1966). The reaction is accompanied by the normal colour-change from brown to red (p. 194) and in this case there is no change in crystal form. The cytochromes themselves, bound to the *oxysomes* of the mitochondria, are effectively in a solid form and the oxysomes function quite normally down to oxygen pressures of only 1 mm of mercury (CHANCE, 1964). At these low pressures adsorption to a solid could be at least as effective as absorption by a solution. The complete intracellular respiratory system (p. 185) will operate in the solid state (MORTON and ARMSTRONG, 1963).

This system includes the normally dissolved pyridino-, and flavo-, proteins and it has further been shown that ribitylflavin experimentally incorporated into a solid matrix of methyl cellulose will undergo reversible oxidation-reduction in the normal way. Moreover the molecules are able to rotate in this gel, though not as rapidly as in solution (PENZER and RADDA, 1967). These authors consider this medium to be nearer *in vivo* conditions than a free solution and equally it is nearer to the normal state of the organelle-bound respiratory chromes than are the more familiar and rigid solid states. Phthalocyanins, laboratory analogues of the porphyrans (p. 59) likewise transfer electrons in the solid state (ELVIDGE and LEVER, 1959; BRATERMAN et al., 1964).

The frozen-dried (lyophilised) state is even more active than that of the simple solution, but probably is not a model for any natural state. It loses in stability what it gains in reactivity. Both deoxyHb (DRABKIN, 1966) and some carotenoproteins (CHEESMAN et al., 1967) are more stable in solution than when lyophilised in the usual form of a thin film. In this state deoxyHb takes up O_2 explosively (DRABKIN, 1966); moreover it is converted to metHb and not oxygenated as in its normal functional cycle. To be stable the solid state must be orderly and not produced under stress. The aqueous medium surrounding large solute molecules is relatively orderly and stabilising and no doubt this more common biological state is the best compromise between reactivity and stability. It is more reactive than the orderly solid and more stable than lyophilised material.

Large molecules are already ordered solid systems and their reactivity and other properties are less affected than those of small molecules by the transition from a state of solution to one of the macro-solid states. Large biochrome molecules may have their essential properties enhanced, as for instance in the large insoluble macromolecules of melanin (p. 61). This process may continue through the further extension of the solid system by the matrix of the chromasome (p. 120). Further possibilities emerge if the chromasome is motile within its cell (p. 138): indeed it may be capable of automobility! There are also transport and chemical possibilities.

Conclusion: Not only the integumental chromes but most zoochromes are effectively in a solid state. This does not seriously depress their activity and greatly increases their stability. A consequent possibility is that integumental and other essentially solid chromes may show electrical and chemical activities *in situ*. Many further indications of this will be given. Both stability and reactivity demand an ordered solidification, from the molecular level upwards.

5.5 The Stability of Zoochromes

The view that stabilisation is one biological reason for the solid state in biochromes raises wider issues. Stabilisation is one important result also of conjugation with proteins and other components (p. 30). Further, the basic molecular structure of chromes (p. 15) itself tends to make for thermodynamic stability, so that stability is already inbuilt. Porphyrins have survived in fossil form possibly even since the Precambrian (MEINSHEIN et al., 1964).

Paradoxically (p. 15) the molecular structure at the same time confers reactivity, i.e., kinetic instability, so that the reactivity/stability compromise also is inbuilt. Even the porphyrins are unstable to light. This is true also for carotenoids and ribitylflavin (FOX, 1953, p. 284). Even melanin is bleached in bright light, and even in such desiccated material as human hair. Rather naturally chromes are more unstable towards light than towards any other agent but there are some exceptions: the anthraquinone, carmine, and the flavonoid, haematoxylin, are rather fast to light even when free and in solution.

It seems clear that there has been natural selection for the various stabilisation devices. Free *p*-benzoquinones are more common than free *o*-benzoquinones which are very unstable. The pteridines in general are rather unstable in spite of their heterocyclic aromatic structure but this can be rectified by $-$ OH and $-$ NH$_2$ side-chains (MILLAR and SPRINGALL, 1966, p. 826), another reason (p. 71) why all the biological pteridines are pterins (2-NH$_2$-4OH-pteridines). These are very resistant to boiling and to acids and alkalis. Ligating metals (p. 188) also give stability, and oxygen stabilises the O$_2$-carrier zoochromes (p. 229). This is reminiscent of the stabilising effect of many substrates on their enzymes. A large number of enzymes are chromoproteins (p. 184) but the phenomenon probably has more general biological significance: biological systems *in toto* are most stable when most active and rapidly decay after death. They have dynamic, steady-state stability.

Most stabilising agents are effective against all relevant destructive agents, electro-magnetic radiation of the u.v. to i.r. ranges, and oxygen and other chemical agents, including enzymes. This is logical since all tend of affect the same basic chromatic properties (chapter 2).

Conclusion: The compromise between stability and reactivity in zoochromes is manifest at both molecular and higher levels of magnitude. There has been natural selection for the compromise. Stabilisation is usually valid against all relevant destructive agents.

5.6 General Conclusion

The distribution of zoochromes between organs and tissues is a problem of general biological significance but their distribution between cells and extracellular compartments becomes essentially a physiological problem of what distributions and physical states are feasible and tolerable. Distribution within the cell is inseparable from a consideration of precise physical states at lower levels of magnitude which eventually make connection with the basic molecular properties of biochromes, already considered.

III. Physiological Functions of Zoochromes

Traditionally Physiology means the study of all activities in living organisms so that strictly speaking all the activities of biochromes are physiological. However, since the emergence of Biochemistry as a discrete science there has been a tendency to consider that physiology is concerned more specifically with biophysical, including biomechanical, aspects of function. The recognition of a subject Physiological Chemistry shows that the physical and chemical aspects rarely can be separated profitably and in the broad outlook of the general biologist there is not even technical justification for separating them. In the present study of zoochromes such a schism would be maiming.

Since colour is based on molecular energy-transitions (chapter 2) the biochemical functions of zoochromes in fact are more fundamental than their biophysical or 'true' physiological functions, apart from photoperception (chapter 7). The main biophysical function in animals, the dynamic type of integumental colour-change (chapter 6) uses zoochromes quite secondarily and incidentally, as materials in an effector organ of which they possibly do not even provide the motor energy, still less the controlling forces. The biophysical functions will be considered in this section and the more typical biochemical functions in section IV. The role of zoochromes in reproduction and ontogenesis will be treated in section V since, although these are still generally regarded as 'true' physiological aspects they involve biochemical more than biophysical processes, both in relation to zoochromes and more generally. Moreover logically they lead on to other ontogenetic aspects of chromatology in section VI and so to the final section on evolutionary evidence.

Chapter 6

The Functions of Integumental Zoochromes

The main, but not the only, functions of integumental chromes are in the contrasting roles of crypsis (camouflage) and semasis (advertisement). These involve general biological, ecological, behavioural and other aspects peripheral to physiology proper. Again these are essential aspects for the zoologist and they will be given appropriate, but not excessive, emphasis. The biological aspects were discussed by POULTON (1890) and more recently by COTT (1940); physiological aspects have been reviewed by PARKER (1948), CARLISLE and KNOWLES (1959), WARING (1963), FINGERMAN (1963), and others. Table 6.1 is a modified summary of POULTON's classification of the subdivisions of these two main *visual-effect* functions. POULTON made apatesis (deceitful coloration) his first main subdivision, so that mimicry of other animals was grouped with the cryptic mimicry of static surroundings. This was unfortunate since all animals mimicked are sematic types; this type of mimicry (pseudosemasis) more logically should be classified with the sematic group of functions.

Table 6.1. Classification of integumental colour schemes of animals. (Modified from POULTON, 1890)

A. *Cryptic* (camouflage)	1. To escape detection by predators 2. To avoid detection by prey and to facilitate approach	
B. Sematic (Advertising, signalling)	1. Aposematic (warning)	
	2. Episematic (attracting)	(a) Epigamic: directed towards the opposite sex (b) Episomatic: serving non-sexual purposes such as luring prey
	3. *Pseudosematic* (mimicking sematic schemes)	(a) *Pseudaposematic* (b) Pseudepisematic

Crypsis is most crucial for animals preyed upon but is common also in predators, both those which lie in wait for the prey and those which approach by stealth. At its best crypsis is very effective, extending to countershading, shape-disruption, shadow-elimination, etc. (THAYER, 1918; COTT, 1940). Advertisement involves bright, conspicuous colours, in large patches, with sharp boundaries and contrasts, and is interpreted by the 'target' animal in the light of appropriately vivid, previous experience, as either warning or welcoming. The contrasting transverse colour bands of wasps are the best known warning pattern and the bright colours of male insects, teleosts and birds the most evident examples of sexually attracting displays. The latter must often compromise with camouflage requirements and in many cases the colours and bold patterns are hidden except during the actual process of courtship. It is for this reason, also, that the female of most species in all these groups is inconspicuous. In some species the male also is normally cryptic and assumes advertisement devices only in the breeding season. Actual light-emission, bioluminescence, is often used where courtship occurs in dark conditions; it has the advantage of competing only very briefly with crypsis. Moreover it is a biochemical chromatic function also (p. 206).

The episomatic subdivision, patterns attracting for purposes other than reproduction, is the smallest and least familiar. Perhaps the best known are the lures of some deep sea fishes but still better use is made of colour-pattern and movement by certain trematodes (p. 257). The bright colours of sea anemones may have this significance (p. 348). Another minor sematic variant is the sudden display of a conspicuous pattern, usually when in motion, so as to startle, dazzle or distract an observer (COTT, 1940). The display is usually brief so that the displayer has disappeared by the time the image has faded from the observer's retina. The device is particularly effective as a bioluminescent pulse by deep sea animals (HOAR, 1966, p. 587). The purpose is escape, i.e., crypsis rather than advertisement.

It is clear that collectively the various members of the classes of zoochromes found in integuments are adequate for all visual effect requirements. Some animals do little more than match shades of grey while others match particular coloured backgrounds, and some a patchwork of colour. Some copy a relatively nondescript background with a 'heather mixture' of chromes, i.e., *a pot pourri* of bright colours when viewed closely but becoming a mere olive to russet when viewed from a distance. In animals which also need to change their colour on occasion, effective monotony of this type has an advantage over true monotony.

It is now generally accepted (COTT, 1940; KETTLEWELL, 1961) that integumental chromes do serve these and some other useful biological purposes and most of the difficulties which troubled earlier generations, because they envisaged a more restricted number of functions, are now resolved. Even POULTON (1890), in spite of his recognition of many types of visual effect and his intense conviction in the general validity of DARWIN's still novel doctrine, considered many examples of animal colours and patterns to be useless. In fact he concluded hat all zoochromes were originally functionless (POULTON, 1890, p. 13), merely incidental by-products of metabolism, and that only subsequently some of them were put to camouflage and other use by the opportunist action of natural selection. The latter in short "created significant colours". Being essentially a naturalist he rather overlooked the already well known metabolic functions of at least haemoglobin and chlorophyll.

The camouflage function certainly is an evolutionary aspect which only natural selection can fully explain: no obvious form of use and disuse theory will meet the case. It is not surprising, perhaps, that scepticism about the adaptive value of integumental coloration has been most evident among LAMARCK's followers and compatriots. As late as 1926, VERNE (pp. 484–95) was arguing thus with considerable warmth and the outlook is still wistfully evident in VUILLAUME (1969). VERNE contended that resemblances between supposedly cryptic animals and their surroundings, for instance fish and chamaeleons, were rarely very close. He also cited the insect, *Tylopeis thymifolia,* which is green but sits on substrata of various colours and frequently crosses from one to another. Similarly, the cricket *Oedaleus nigrofasciatus* has green and brown varieties but both can be found on the same background. This is true also of the spider, *Thomisia,* while the shrimp, *Palaemon,* is brown on black backgrounds and white in blue surroundings, – only very approximately camouflaged, therefore. Reindeer and sable are not white in the Arctic, while the polar bear is white in regions with no snow. In general these exceptions serve to emphasize the rule. Like other critics of the time Verne did not attempt a quantitative assessment of the evidence from all sources, or distinguish all possible functions.

Animals committed to moving around, on a variable background, cannot be perfectly camouflaged at all times and in any case survival is the fortune of the species rather than of particular individuals. Most species living on a very varied terrain, for instance, *Carcinus* in the early crab stages, carry a rich colour polymorphism; each colour-morph matches some one type of background very closely but inevitably each is found at times on a non-matching type. It is less easy to measure the adaptive and survival value of this polymorphism than of monomorphism on a uniform background but results showing positive values for all the morphs of *Cepaea* have been obtained and statistically tested (CAIN and SHEPPARD, 1950; SHEPPARD, 1958; CLARKE, 1960; BANTOCK, 1971).

It was already known to VERNE (1926, p. 486) that animals of a number of cryptic species deliberately seek a matching background, a phenomenon called 'chromatropism' or, perhaps better, *chromotaxis*. Dark individuals of the fish, *Ericymba*, select a dark background more often than do light coloured fishes (MAST, 1916; BROWN and THOMPSON, 1937) and crayfish behave similarly (BROWN, 1939). Verne discounted the value of such choice because, in an experiment by KRAUSE on green and brown Mantids, given the choice of backgrounds of these colours, the choice was imperfect. The results, 12 on homochromous and 4 on heterochromous background would now be regarded as having some significance, though the sample should have been larger. The results of another experiment criticised by VERNE were almost exemplary on this score; this was by DI CESNOLA on the relative predation rates upon mantids of these same two colours sitting on backgrounds of the two colours:

Number of Mantids initially	Colour of Mantids	Colour of background	Number of Survivors	After (days)
20	Brown	Brown	20	17
45	Brown	Green	6	17
20	Green	Green	20	17
25	Green	Brown	0	11

Since HALDANE (1927) and FISHER (1931) showed how small a selective advantage can be critical for the differential survival of one type relative to others the significance of this phenomenon of chromotaxis and of the whole property of background-matching is scarcely open to question. SUMNER (1935) showed that fishes adapted to background colour were less predated by birds than groups of non-adapted individuals and ERGENE (1952) found the same for the insects, *Acrida*, and *Oedipoda*. In recent years the whole phenomenon has been fully validated by KETTLEWELL, (1956, 1961, 1965) and his colleagues, with statistically adequate data along various lines of approach.

This demonstrates the evolution of crypsis actually in progress in the so called "industrial melanism" of *Biston betularia* and other moths. As many as 70 of the 780 species of British Macrolepidoptera are at present replacing the light coloured 'typical' forms of the pre-industrial era by melanic forms, in industrial areas, and to a diminishing extent to leeward of these areas, but not to windward (southwest). In the industrial areas tree-trunks and other resting places, which previously were light grey from the growth of lichens, are now blackened and the melanic mutant is well camouflaged, while the peppered 'typica' is conspicuous. The converse is true in pristine areas such as Dorset. For a number of species there are historical records showing that the percentage of the melanic form in the population has increased progressively over the past century. For *Biston*, it is now 85% compared with less than 1% in Dorset (see Table below). The curve has been sigmoid, as mathematically predicted. It will be noted that the population also includes a small percentage of a third form, 'insularia' which does not concern the present theme.

By the release and subsequent recapture of survivors of marked batches of moths of both typical (typ.) and melanic (carb.) forms of *Biston* in each type of environment, KETTLE-WELL found that the percentage of the inconspicuous form recovered was almost exactly twice that of the conspicuous form, in both localities (columns 3 and 4):

Locality and Background	Normal %s in the 'wild'		%s of released moths recaptured		%s choosing backgrounds		%s seen to be taken by birds	
	Typ.	Carb.	Typ.	Carb.	Typ.	Carb.	Typ.	Carb.
Rural (lichen)	94.6	1.0	12.5	6.3	66.1	35.6	13.7	86.3
Industrial (blackened)	9.5	86.9	25.0	53.2	33.9	64.4	74.1	25.9

Robins, Hedgesparrows and Great Tits were seen to take a higher persentage of the more conspicuous form on both types of tree trunk (columns 7 and 8). The differential is higher than in the recapture experiment since the mode of supply of moths by the observers was a periodic replenishment, therefore differentially penalising the conspicuous form and enhancing the selectivity of the predators. Appropriate tests showed the two forms of *Biston* to be equally palatable.

The moths themselves selected the background which matched their own colour again about twice as frequently (columns 5 and 6) as the other. The choice is far from perfect but is significantly better than chance. The predators' skill also was not exemplary though it amounts to a powerful force of natural selection. *In toto* the evidence is irrefutably in favour of the evolution of cryptic coloration under natural selective forces.

There is now equally good evidence for the adaptive significance of the conspicuous types of coloration (POULTON, 1890; COTT, 1940) and this need not be recapitulated in detail here. In almost every case the bold warning patterns have been found correlated with defensive properties which make the warning useful to both parties. Animals with such patterns do not attempt to hide or escape, are rarely attacked and still more rarely killed and eaten by predators, – which usually give visual evidence of repenting any such attempt. Using the monkey, *Cercopithecus*, as test-predator, CARPENTER (1921) found that 84% of aposematic species of insects were distasteful but only 18% of cryptic species. This extended and confirmed a great deal of earlier work by POULTON (1887) and others. COTT (1940, p. 274) found that only 0.17% of the 11,585 insects taken from the stomachs of tree-frogs in the wild were of aposematic species; similar results were obtained by others using as predators bats, which do not usually rely on the visual cues.

Coloured lures are used by various predators and are very ingeniously supplemented by movements (COTT, 1940, pp. 383–6). Direct evidence that they do lure prey to be captured was obtained by DIXON and DIXON (1891) using the spider-crab, *Hyas*. This, like many other such predators, is otherwise very well camouflaged.

Circumstantial evidence that some bright colours play a significant role in reproduction is abundant and impressive (HUXLEY, 1938). If only the male is brightly coloured then he alone displays but where the sexes are alike both do so. Display is clearly directed towards an individual of the opposite sex, though sometimes a threat-display is used towards another individual of the same sex. There is reasonable evidence that display accelerates gonad-maturation, readiness to mate and increased fertility. Coloration and display are the consequence of natural selection, rather than of inter- or intra-sexual selection, as DARWIN envisaged.

In view of such conclusive recent work it is surprising to find VUILLAUME (1969) still echoing VERNE almost verbatim in the charge of 'finalism' or teleology, which others now prefer to call *teleonomy* (PITTENDRIGH, 1958), – recognising that the 'final' cause may be nothing more abstruse or debatable than natural selection. Verne's 'homophany', the idea that animals match only the general shade and illumination of the background, rather than any precise colour-pattern, also is sustained by Vuillaume, with the same implication that somehow the charge of finalism cannot be levelled against this more dilute and less efficient version of camouflage by animals.

Certainly there remain some apparent, and perhaps some real, anomalies and puzzles, but most of these may be due to imcomplete knowledge of the situation or to unrecognised additional factors distorting the observations and results of experiment. Verne raised the important point, later stressed by COTT (1940), that an animal must be viewed as its potential predator or prey would see it, though he well might have concluded that the experiments above and those which he quoted, showed that the visual pictures of other animals are usually rather similar, and rarely very dissimilar, to that of man. Perhaps it is more important to ensure that experiments as far as possible are done under natural conditions; the splendid conspicuousness of the peacock on the lawn or the display-ground can be quite concealing in its natural jungle haunts (HUXLEY, 1934). There are other good examples of this, including the kingfisher and the zebra (STEPHENSON and STEWART, 1946, p. 18). In this context it is amusing to re-read W.H. DAVIES' poem, "The Kingfisher":

> It was the rainbow gave thee birth
> And left thee all her lovely hues;
> And, as her mother's name was Tears,
> So runs it in thy blood to choose
> For haunts the lonely pools, and keep
> In company with tears that weep.
> Go you and with such glorious hues
> Live with proud Peacocks in green parks.

One example of an unexpected complication is the observation of POULTON (1890, p. 179–180) that some distasteful species of insects are sematically coloured in years when they are plentiful but cryptically coloured when few in number; the sematic colouring can only be adaptive for the species when there are enough individuals to benefit from the sacrifice of those necessary to educate the population of predators. The numbers of locusts in migration-years certainly are adequate and the *migratoria* phase has an essentially sematic colouring; nevertheless the scheme is a positive signal to other locusts, rather than a warning pattern (p. 311), and so it may have other advantages. Here numbers are adequate to dispense with the advantages of both crypsis and warning.

Another complication which is now fairly well understood is that since on land temperature and humidity can affect background-colour and show a regular correlation in their effects, some terrestrial animals under certain conditions are adapted to these factors more strongly than to background colour itself. This is irrelevant to aquatic animals, and in terrestrial forms it is complicated by colour schemes which help to control body-temperature. These may be in the same, or in the opposite direction to the background adaptation. In poikilotherms they are usually in the same direction since soil tends to be warm, dry and light-coloured in low latitudes and conversely in high latitudes and so the background-adaptation helps the animals to absorb heat in cooler latitudes and reflect it in the tropics. Homoiotherms, however, need to radiate their own heat in low latitudes and conserve it in high latitudes and they do tend to be oppositely coloured to poikilotherms, i.e., dark in the tropics and light in high latitudes. These colours are not universally anticryptic, as might be supposed: in tropical forests and beyond the snow-line they provide very good camouflage.

For thermoregulatory reasons, poikilothermic terrestrial animals tend to be dark at high altitudes, even in low latitudes, and this can be anticryptic. The over-riding need to absorb energy applies also to freshwater aquatic animals at both high altitudes and high latitudes (BREHM, 1938) and again the animals tend to be very conspicuous. These examples are enough to show that other adaptations interact with, complicate, and may even dominate, cryptic-requirements. They may modify other visual-effect schemes, also.

6.1 Integumental Colour-Change

Many of the colour-patterns considered above are genetically determined (p. 323) and fixed, so that their significance is not physiological except 'in the eye of the beholder', or when associated with activities such as selecting the appropriate background, searching for prey or displaying to a mate. This is an adequate reason for not considering the general biology of adaptive coloration and other visual effects in further detail here. Their 'function' is biological rather than physiological.

On the other hand a considerable number of the higher metazoa have a labile coloration, responsive to relevant environmental changes and this is essentially physiological. There are two types of lability, changes in amount of chromes and changes in the display of the amount already present. These have been called 'morphological' and 'physiological' respectively but the first more correctly is morphogenetic and therefore developmental-physiological. The distinction between the two perhaps is best recognised by the respective terms *chromogenic* and *chromomotor* colour-change. Chromogenic changes are the more primitive and widespread, and all animals capable of chromomotor change are capable of the former also; the converse is not true: even man shows integumental chromogenic change. When both are present they are usually under a common control-system and this is another reason for regarding chromogenic change as truly physiological.

The useful distinction has been made (FITZPATRICK et al., 1965) between *chromatophores* which are capable of chromomotor change, and *chromatocytes* which at most have chromogenic lability. Unfortunately this still leaves the problem of a generic term to cover both, and possibly a third type of cell in which the chrome is purely static apart from the replacement of wear-and-tear.

6.1.1 Chromogenic Change

In this response there is usually a change both in the number of chromatophores and in the amount of chrome per cell. In the simplest case, of black-and-white adaptation, the number of melanocytes and amount of melanin in the dorsal integument increases while those of the light chromes, purines and some pterins, decrease when the animal is moved to a darker background, or is illuminated more strongly; the converse changes occur on a lighter or more weakly illuminated terrain. The response is a resultant of one to incident illumination and one to reflection from the background. SUMNER (1944) found a roughly straight-line relationship between the amount of guanine in the skin of the fish, *Girella*, and the logarithm of the albedo, that is, the ratio of the amount of light reflected from the background to the incident illumination. Melanin was inversely related to log. albedo. The same relation was found also for the number of melanophores in *Lebistes*

(SUMNER, 1940); the relation roughly follows the Weber-Fechner relation between sensitivity of a photoreceptor and light intensity-level.

In animals which have a variety of differently coloured chromatophores all of these may show chromogenic responses. In adaptation to simple changes in the albedo all the darker colours, including even yellows, tend to behave like melanin while purines and pale, reflecting pterins are virtually the only chromes in the other group. In adaptation to changes of hue of the background the cells of the different colours behave appropriately and differentially; in the most adept animals the range of combinations is considerable but it is not yet fully known for any animal.

The chromogenic type of response is most rapid in the young animal, indicating some relationship to the once-for-all type of chromogenesis, but it continues throughout life in many species and remains reversible at all ages. Decreased display of a chrome is accompanied by a destruction of its cells and a removal of chromasomes from the remainder: chromogenesis therefore can take both positive and negative signs, in what corresponds to functional hypertrophy and atrophy in other cells and organs.

The reciprocity in response between light and dark groups of chromes extends to countershading, i.e., the strong colour-difference between dorsal and ventral surfaces, with intervening gradation. Experimentally, and presumably sometimes in nature, countershading can be reversed, by illumination from below, and in fact this was the way in which chromogenic change was first demonstrated (CUNNINGHAM and MACMUNN, 1893). OSBORNE (1940) and others have since confirmed this reversal and other manifestations of the phenomenon. Some lepidopteran larvae show a chromogenic response to variations in the hue, as distinct from the intensity of incident light (DURKEN, 1916, 1923); this may be mainly a background response since the background colour will be changed in the same direction but no doubt in nature the larvae sometimes receive such colour-filtered light.

Like the chromomotor response (p. 139), chromogenesis is evoked also by other factors and in relation to requirements other than camouflage. It shares in colour-changes in response to temperature or to humidity. In insects and semi-terrestrial Crustacea, low temperature promotes melanin-synthesis (VERNE, 1926, p. 336; FOX and VEVERS, 1960, p. 41; VERNBERG, 1962; NICOLAUS, 1968). The growing goldfish does not change from black to gold unless the temperature is above a critical value (ARCANGELI, 1926); of course this is a once-for-all developmental change. The ommochromes of insects also are produced maximally at low temperatures (GOODWIN, 1952). In solitary locusts (NICHOLAS and FUSEAU-BRAESCH, 1968) and in larval salamanders (VERNE, 1926, p. 327) the temperature response dominates that to albedo. Red pterins show the same response to temperature (EPHRUSSI and HEROLD, 1945); while carotenoids, being exogenous (p. 301), are transported to the integument in larger quantities at low temperatures. Temperature also controls the seasonal chromogenic change in mantids from green in spring to brown in summer (PASSAMA-VUILLAUME, 1964). Increased humidity increases the amount of dark chromes in the butterfly *Hestina* (FOX and VEVERS, 1960, p. 42).

Some birds and mammals are capable of a special type of chromogenic change, namely the moulting of their plumage or pelage and its replacement by differently coloured feathers or hair. The Inca dove darkens in response to humidity: this is not a rapid change and moulting can be as protracted as the more orthodox chromogenic changes. More commonly, however, it is used for seasonal colour changes whether related to camouflage or to reproduction and then it is often relatively rapid and dramatic. The winter whitening of the ptarmigan, mountain hare, etc., are good examples of the former. In the varying hare

(LYMAN, 1943) and in the male of some species of bird there are three colour-changes per year, two for camouflage and one for the breeding coloration.

These seasonal moulting changes like the more orthodox chromogenic changes are triggered mostly by light or by temperature changes. Day-length or photoperiod control the autumn change in Bonaparte's weasel, *Mustela cicognanii*, (BISONETTE and BAILEY, 1944), and in the willow grouse, *Lagopus lagopus* (HOLST, 1942), whereas in *Mustela ermina* it is controlled by the fall in temperature (ROTHSCHILD, 1944; SCHMIDT, 1954). In the mountain hare the spring moult is induced by the change in photoperiod, but temperature controls the *rate* of the change (FLUX, 1970).

In principle, arthropods and other animals which moult also can effect chromogenic changes at that time but since in most there is a regular developmental sequence of moults, the known colour-changes are mostly developmental in significance. Birds and mammals often have at least one such developmental colour-moult but they are so commonly perennial that the seasonal cycle attracts most attention. The developmental changes are once-for-all or 'irreversible' and not cyclic. They are sensitive to environmental modification, however; certain phenocopies of the Himalayan rabbit and Siamese cat develop black tips to the ears, nose, feet, etc., only if reared at a low external temperature, though the pattern is then fixed and not abolished by any subsequent moulting.

6.1.2 Chromomotor Colour-Change

In this type of change, colour is displayed and obscured by two main methods, either by movements of chromasomes within the chromatocytes, or by a change in shape of the whole cell. The latter occurs only in cephalopods and some pteropod molluscs (NICOL, 1964), the former in annelids, Crustacea, insects, echinoderms, cyclostomes, elasmobranchs, teleosts, Amphibia and reptiles. In the phantom larva of the fly, *Corethra*, the chromatophores themselves disperse and aggregate by amoeboid movement but this seems to be a rare method. For details of the mechanism, see PROSSER and BROWN (1961), HOAR (1966), and FINGERMAN (1963, 1970); KNOWLES and CARLISLE (1959) and WARING (1963) give accounts on Crustacea and vertebrates respectively. Some recent developments are recorded in Annals of the New York Academy of Sciences, vol. 100 (1963) and American Zoologist, vol. 9 (1969).

The cephalopod mechanism can be extremely rapid; in one response waves of darkening and pallor sweep over the surface like the image of moving ripples on the surface of the water. These animals also have a very varied repertoire of responses, relevant to crypsis, warning, courtship and other functions (HOLMES, 1955; CARLISLE, 1964; NICOL, 1964). Movement of chromasomes within the cell is necessarily a slower mechanism and the frog requires several days to complete an adaptation to changed background; at its best, however, as in some teleosts, the response is complete within a minute, i.e., within two orders of magnitude of the speed of cephalopod reactions. At the other extreme, the frog's response is not much more rapid than chromogenic changes.

Again the dark chromatophores have been most studied though most of the others are known to show chromomotor responses. They behave differentially and with the same adaptive rationale as in chromogenic changes. The outstanding and logical difference is that chromomotor responses play no part in seasonal and other leisurely changes. In the cephalopods the white pigment is static and the various colours in the active chromatocytes are all due to ommochromes. In Crustacea, by contrast, there are almost as many classes of chrome as colours of cell, and there is nearly as much variety in some of the poikilothermic vertebrates, and in leeches and some other annelids. Most of these use the variety to good effect in background adaptation; the shrimp *Crago* (KOLLER, 1927), the crab *Portunus* (ABRAMOWITZ, 1935) and the teleost *Fundulus* (FOX, 1953, p. 147) can match black, white, grey, yellow and red backgrounds, while the shrimp *Palaemonetes* (BROWN, 1935) and the flounder *Paralichthys* (KUNTZ, 1916) can adapt to blue and green also. Some can copy details of background-pattern with fair precision.

Chromomotor changes preserve coutershading but unlike chromogenic changes they cannot be used to reverse this pattern, though presumably they can operate a pattern which has been chromogenically reversed. Of course the motor type of change has compensating advantages over the chromogenic type. Any changes which can be completed within an hour or so can be programmed to perform a regular diurnal cycle. Indeed, *Nereis* (HEMPELMAN, 1939) and leeches (JANSEN, 1932) have such a diurnal change although they do not adapt to background changes. The diurnal colour-change programme is geared to an internal 'clock' (NEVILLE, 1967), which is checked and reset by the external solar clock since its period is not exactly 24 hours, when 'free running' in constant light or constant darkness. It is evident that the internal clock would not be more efficient if its period were exactly 24 hours; even in the tropics the daylight period varies significantly over the year and with the amount of cloud, so that 'entrainment' by actual light-conditions is the only alternative to an extremely sophisticated endogenous system.

A diurnal chromomotor cycle has been detected in the Crustacea, *Idotea*, and *Uca* the stick insect *Carausius*, the sea urchin *Diadema*, the lamprey *Lampetra*, and the minnow *Phoxinus*, in salamander larvae and some Anura and in the lizard *Phrynosoma*, i.e., both aquatic and terrestrial animals of a number of the higher phyla. In the isoped *Idotea* the intrinsic cycle persists for as long as 8–9 weeks in total darkness but under a changed external light-cycle it is promptly reset to synchronise with this. In some animals the internal cycle tends to drift back to its original period (BROWN and WEBB, 1949), even under the continued operation of the new external clock, and in the absence of detected clues from the original entrainer. Conceivably this is an indication that there is also an inbuilt knowledge of the expected programme of the normal solar cycle, or at least an inbuilt conservatism. Although in some localities on land the diurnal temperature-cycle in principle could be an effective entrainer, no record of this is available for chromomotor changes. Light is always more reliable.

The macruran Crustacea are dark at night, while *Uca* and other Brachyura are pallid, and other animals have an intermediate shade. The biological explanation of this variability is not evident: if it were immaterial what shade of coloration the animal has in darkness there would be no selective advantage in including a chromomotor response in the circadian cycle at all. In fact, however, the inclusion appears to be so important that the response is innate to every part of the integument; in isolated legs of *Uca* the melanophores behave differently according to the hour at which they are removed from the body (FINGERMAN, 1963, p. 71). The internal clock presumably is intrinsic to every chromatocyte. Further, in the swimming crab, *Macropipus* (*Portunus*), only the black and white cells show this diurnal change (FINGERMAN, 1963, p. 69) and this is biologically rational since there is colour-vision only above a certain light-intensity. The diurnal change is probably relevant mainly to the nightmare predation periods of dusk and early dawn and this may also hold the clue to the variable night-time shades noted above.

Among marine littoral animals there is also an intrinsic and entrained chromomotor cycle corresponding to the local tides. It is doubled where there is a double tide, in certain inshore waters; on very few shores is the tide closely synchronised with the moon's transit, and for one at least of the two tides per day no visual lunar clue is available so that tidal movements themselves are the entrainers (WILLIAMS and NAYLOR, 1969). At the same time it is noteworthy that the circadian cycle is modulated by a 'carrier rhythm' with a period of a fortnight, evidently a lunar cycle. Similarly the tidal cycle is modulated by the diurnal rhythm.

To littoral and sublittoral animals a tidal cycle of colour-change has evident value during daylight. A complication is the variable amount of light received from the moon itself during the night. Over the course of each lunar month the tidal cycle shifts progressively relative to daylight so that interaction between the setting mechanisms of the two cycles is very useful.

As already indicated (p. 136), chromomotor and chromogenic responses to a large extent have common control mechanisms. This was first recognised by BÁBÁK (1913) and is sometimes called Bábàk's Law. It is actually a special case of a wider phenomenon: in muscles and many other systems the factors which evoke physiological activity also induce growth, repair, functional hypertrophy, etc. (NEEDHAM, 1964). The adaptive value, in all cases, is evident enough. There are chromomotor as well as chromogenic responses to the signals other than light itself. A motor response to humidity-changes has been found in the stick-insect, *Carausius*, (GIERSBERG, 1928), and the frog, *Rana*, (ROWLANDS, 1950, 1952); both darken maximally at high humidity and blanch in dry conditions. *Rana* needs water in liquid form but many arthropods are very sensitive even to water vapour, i.e., to relative humidity.

The response to temperature-change is generally consistent with the chromogenic response. *Phrynosoma* (PARKER, 1938) and the frog *Hyla* (EDGREN, 1954) darken at low temperature and both *Hyla* and the grasshopper, *Koscuiscola*, become pale under hot conditions (FINGERMAN, 1963, p. 11). For *Carausius* the high-temperature colour is green and the 'cold' colour black: for this stick insect green (foliage) is the background colour. Even some aquatic animals show a chromomotor response in this same direction to temperature, for instance the crab, *Callinectes*, the shrimp *Palaemonetes* (FINGERMAN, 1956) and the phantom larva of the midge *Chaoborus* (FINGERMAN, 1963, p. 11); most do not, however, for the reasons given (p. 135). There are some other anomalies: the shrimp *Hippolyte* (GAMBLE and KEEBLE, 1900), and *Uca* (BROWN and SANDEEN, 1948) become pale at both high and low temperatures while *Idotea* (MENKE, 1911; FINGERMAN, 1956) and the shrimp, *Macrobrachium*, (SMITH, 1930), darken at both extremes. No doubt the habitat of each animal must be considered: *Uca* is almost a terrestrial crab, but *Hippolyte* and *Idotea* are aquatic, and the reason why they show any responses at all to temperature-change needs investigation.

The copepod, *Hyperia galba*, lives as an ectoparasite on a typical transparent medusa; it is dark when free-swimming but retracts its chromes on making contact with its host. Retraction makes it transparent, like the host, and thus well camouflaged. It also retracts on contacting any other surfaces, whatever their colour (SCHLIEPER, 1926), and again transparency will make it inconspicuous.

As in chromogenic change, motor responses to temperature are not always concerned with camouflage (p. 137) and altogether there are a number of chromomotor reactions with other significance. In cephalopods they play a part in courtship, while in the three-spined stickleback, the cuckoo wrasse, and the black sea bream, the nuptual colour change is chromomotor (CARLISLE, 1964). Again the male of the lizard, *Anolis*, changes from green to brown during mating and in cichlid fishes there are chromomotor changes in relation to egg-laying and egg-tending (FINGERMAN, 1963, p. 16). In vertebrates and some other animals the coloured blood is often used for these and other purposes, but by a very different type of chromomotor mechanism, as for instance in blushing and flushing in light-complexioned humans. There are chromogenic and static counterparts to this in the wattles of gallinaceous bird and in the nose and buttocks of the mandrill (Fox, 1953, plate II). Emotional colour changes are sometimes effected also by the orthodox chromomotor mechanism, in cephalopods, lizards and the clawed frog *Xenopus* (PROSSER and BROWN, 1961, p. 507).

6.1.3 The Mechanism of Chromomotor and Chromogenic Responses

The account will refer almost entirely to the chromomotor response, particularly at the level of the chromatocyte itself, where the two responses are so different, but it can be assumed that in general the paths of mediation and control are identical. Knowledge is mainly of responses to light but the indication is that stimuli via other afferent pathways connect centrally with a common efferent system. Some chromatocytes respond directly to relevant stimuli and so are independent sensori-motor systems. They are said to give a *primary* response, as opposed to *secondary* responses through the neural and humoral systems. Some animals use both types of response.

6.1.3.1 Primary Responses

These are not more rapid than the indirect type, quite the contrary; they represent a primitive relic the main advantage of which should be to facilitate very local patterning. In fact, however, teleosts match a complex background pattern by secondary mechanisms; this is logical since the dorsal surface comes to resemble a background facing the ventral surface, which maintains a very different coloration, in aquatic types very often that of the 'background' facing the dorsal surfaces, i.e., the sky or the upper waters. No direct light-induced response is feasible here.

A primary response has been found in the young of the teleosts: *Perca, Salmo, Macropodus* and *Hoplias*, and of *Xenopus* (FINGERMAN, 1963, p. 129), but not later in life; a secondary mechanism is not feasible until the nervous system is functioning. Other fish, *Scyliorhinus, Mustelus, Fundulus, Lebistes* and *Gambusia*, from the outset have only secondary mechanisms; fish with viviparity, or with a long embryonic phase, have the nervous system functioning before camouflage is necessary. On the other hand the zoea larva of *Crago* has only the primary mechanism although its neuromuscular mechanism is well developed. A number of fishes revert to a primary mechanism if they are blinded and so does the lizard, *Anolis* (KLEINHOLZ, 1938); the implication is that the primary mechanism generally persists after it has been superseded and masked by the secondary type. Even in adults of the 'horned toad', *Phrynosoma*, and of the chamaeleon, responses to temperature-change (FINGERMAN, 1963, p. 10) and to incident light are primary responses; they are demonstrable in regions denervated and tested after complete degeneration of the peripheral supply (PARKER, 1938). ERGENE (1950–6) and KEY and DAY (1954) showed that adult orthoptera are able to copy their background using only a primary mechanism, and there is no doubt that in some animals the latter does function alongside fully developed secondary systems, at all stages of the life cycle.

In *Diadema* (YOSHIDA, 1956), in prawns (CARLISLE, 1964) and in *Xenopus* (VAN DER LEK, 1967), a local primary response of chromasome dispersion has been demonstrated by training a microbeam on a single chromatocyte. The most effective wavelengths are 425–500 nm for *Diadema* and *Xenopus*, but between 300 and 450 nm for *Crago* (PAUTSCH, 1953) and *Uca* (COOHILL et al., 1970).

It is possible that the chromatocytes serve as light-perceptors also for other responses, such as the spine-movements in *Diadema* (MILLOTT and YOSHIDA, 1956, 1957), since the action-spectrum for the two responses is very similar. If so, then it is further possible that the chromatocytes mediate the general dermal 'light-sense' (p. 173). These further functions would demand some nervous connections with other systems, if only in the form of a superficial nerve-plexus, and then there might be some doubt if the chromatocyte is more 'independent' as an effector than it is as a receptor. There is evidence that the movement of the proximal masking pigment of the crustacean eye may be caused by direct photomechanical transduction (LOCKWOOD, 1968, p. 90) and this would support the probability of a similar mechanism in primary chromatophores of the general integument.

6.1.3.2 Secondary Responses

Secondary or indirect responses to light naturally are mediated mostly through the eye and optic tract but instances of other light-perceptor paths have been detected (ZOOND and EYRE, 1934). In some cases general dermal photo-perceptors may be responsible or, in vertebrates, the pineal (YOUNG, 1935) or even the hypothalamus itself (SCHARRER, 1928; MENAKER 1972). These paths feed into the C.N.S. and evoke systemic responses, as the optic information does, and so are distinct from the primary mechanisms. Responses to them are abolished by transection of appropriate nerves but not by tying off the blood supply. They are therefore neurally rather than hormonally mediated, and they interact with the other responses mediated through the N.S.

Although the afferent paths for secondary responses to temperature, humidity, tactile and other stimuli have been little studied (STEINACH, 1901), the interactions just mentioned provide circumstantial evidence that they do all feed centrally into a path common to that for light-signals. Central control-ganglia for chromatocytes have been found in the brain of cephalopods; it so happens that they are known as the 'central' ganglia. The arthropod centres likewise are in the brain and those of vertebrates specifically in the hypothalamus. Sub-centres with excitatory and inhibitory actions respectively on the chromatocytes have been demonstrated in cephalopods (BOZLER, 1928, 1929); there is also an efferent relay station in the sub-oesophageal ganglion. In fishes there is a relay-centre in the medulla oblongata which in some species causes an interesting minor oscillation in the response of the effectors (WILHELM, 1969).

The efferent limb of the response is neural in cephalopods, in leeches and in some teleosts and reptiles. It is purely hormonal in Crustacea, Insecta and most Elasmobranchii and Amphibia. An hormonal efferent path is more common in reptiles than originally supposed (FINGERMAN, 1963; HADLEY and GOLDMAN, 1969; BARTLEY, 1971). Some teleosts and elasmobranchs have mixed efferent systems, one phase only of the response of the chromatocyte being under neural control; in *Mustelus* and *Squalus* this is the blanching phase (PARKER, 1943). Primitive teleosts such as the eel have bineural control but bihumoral responses also can be demonstrated so that there is dual control of both phases. Some degree of humoral sensitivity, and so presumably of normal humoral control, seems rather general not only in vertebrates but even in cephalopods, where the nervous mechanism is particularly sophisticated. By contrast, arthropod chromatocytes have no innervation, and no nerve-supply has been detected to those of some vertebrates.

With this variety of efferent control it is far from clear which is the more primitive, neural or humoral (WARING, 1942, 1963). Among vertebrates it once seemed that humoral control was characteristic of more primitive groups and the coleoid cephalopods certainly seem a good example of a group in which colour-change is a recent acquisition (HOLMES,

1955). Moreover it is unlikely that a nervous mechanism once acquired would be subsequently lost. In evolution, nervous systems antedated circulatory systems with systemically transported hormones, but the latter were probably well established by the time chromatocyte control became essential, i.e., in the higher phyla, and would be suitable for controlling simple responses, as well as being more economical than neural control for sustained responses and slow responses. There is certainly a strong correlation between rapid responses and neural control.

The vertebrate neural chromomotor system is part of the autonomic, the fibres for darkening being essentially parasympathetic and cholinergic, and those for blanching sympathetic and adrenergic. Presumably the outflow for the former is entirely craniosacral and the latter thoraco-lumbar, though this remains to be clarified for the lower vertebrates (HOAR, 1966, p. 26ff); chromatophore-control should be a convenient function on which to test it. Unfortunately peripheral, postganglionic fibres of the two modalities tend to run together so that local differential denervation and stimulation are difficult; systematic high level transections and micromethods seem necessary. Gross nervous stimulation usually causes pallor, i.e., that component dominates. Following denervation the converse effect is found (HOGBEN and MIRVICH, 1938) and initially this was interpreted as indicating mononeuronic innervation only, by the blanching component. However, a cold-block distally to a nerve-transaction prevents local darkening so that this darkening must be due to positive activity in an antagonistic nerve. Differential stimulation of this component, without cutting the blanching fibres, has been achieved in the catfish tail (PARKER and ROSENBLUETH, 1941). During the regeneration of transected nerves the fibres of one or other modality may grow the more rapidly so that stimulation in intermediate stages may give variable results (ABRAMOWITZ, 1936). It is fairly easy to induce a state in which the chromasomes rhythmically disperse and aggregate (FUJII, 1969).

The dual control has been demonstrated not only for melanophores but also for the erythrophores of the squirrel fish, *Holocentrus* (PARKER, 1937) and the xanthophores of *Fundulus* (FRIES, 1942). Consequently there is a neurological basis for the observation that all types of chromatophore may respond together, as they do in the chromogenic response.

In taxa with an hormonal efferent path the hormones all appear to be secreted by cells of neural phylogeny and this favours the view that nervous control is the more primitive. The vertebrate hormonal system is best known and shows the same simple antagonistic pair of controls as the neural system. A melanin-display hormone, or 'B-substance' (LUNDSTROM and BAND, 1932; YOUNG, 1935; HOGBEN, 1936), now usually called the melanophore stimulating hormone (MSH) is secreted from the intermediate lobe of the pituitary, and produces darkening in cyclostomes, elasmobranchs, teleosts, amphibia and reptiles. It also causes dispersion of the erythrosomes, in the minnow. An antagonistic hormone, 'W-substance', also was isolated from the pituitary but this has not the universal distribution of MSH. In most vertebrates it seems more likely that a pineal hormone, melatonin, N-acetyl-5-methoxytryptamine, which is highly concentrated there, is the normal blanching hormone (YOUNG, 1935; HOAR, 1955; BAGNARA, 1960, 1963; WARING, 1963; CHARLTON, 1966; FUJII, 1969). It also induces aggregation of xanthosomes and erythrosomes (FUJII, 1969), but it is not effective in all fishes with hormonal control. Adrenalin also usually causes blanching, as would be expected of the sympathetic mediator, but this again is often ineffective. It may be that different hormones are used, very haphazardly, in the various subgroups of lower vertebrates, or that frequently there is a hormonal dispersor but a neural aggregator.

MSH also promotes chromogenesis, e.g., melanisation in man, and in melanomas (Fox and Vevers, 1960, p. 43) and pterin-synthesis in the larvae of *Rana sylvatica* (Bagnara, 1966). It depresses the synthesis of purines in the white iridophores, notwithstanding the biosynthetic connection between purines and pterins (p. 290). Experiments increasing (Vorontzova, 1929) or decreasing (Peredelsky and Blacher, 1929) the amount of pituitary tissue give results in accordance with this, and hypophysectomy increases the production of purines (Verne, 1926, p. 354; Bagnara, 1966). Chromogenically the W-hormone of the pituitary again acts antagonistically to MSH and it promotes purine synthesis (Bagnara, 1958). Similarly pineal melatonin depresses melanin synthesis (Lerner and Case, 1959) both in the frog's skin and in human melanoma. It is noteworthy that the adrenocorticotropic hormone (ACTH) of the pituitary, which is structurally related to MSH, has a similar melanogenic action (Hama, 1963; Kim et al., 1963) increasing both the number of melanophores and the number of melanosomes per cell (Chavin, 1956). It was thought to be responsible for the bronzing of the skin in Addison's disease of the adrenal cortex, but in this disease MSH output as well as that of ACTH is increased (Geschwindt, 1966).

Because chromogenesis is required in so many biological contexts it is affected also by various other vertebrate hormones such as thyroxin, prolactin and the reproductive hormones.

Darkening and blanching factors have been extracted from the nervous system of the starfish, *Marthasteria*, (Ungar, 1960) but these may prove to be orthodox synaptic transmitter-agents rather than systemic hormones. In the stick-insect *Carausius*, a darkening hormone is produced by neurosecretory cells of the brain (Giersberg, 1929), and in the fly, *Corethra*, this is probably released via the corpus allatum (Prosser and Brown, 1961, p. 514). The hormone from this gland, neotenin, also promotes chromogenesis of the bathochromic group, for instance in the grasshopper *Acrida* (Fingerman, 1963, p. 91) and in both solitary (L'Helias, 1970) and gregarious nymphs (Joly, 1951) of locusts. Under experimental conditions, analogues of neotenin cause excess pigmentation in embryo locusts (Novak, 1969). Ommochrome synthesis certainly is increased, in *Carausius* (Dustmann, 1964).

The hormone of the prothoracic gland, ecdysone, which generally acts conversely to neotenin, has been found to depress melanin synthesis (Jones, 1956) and Girardie (1962) found that a blanching hormone was produced by the intercerebral region of the brain which might be expected to pass the hormone via the corpus cardiacum to act on the prothoracic gland. Certainly the cardiacal hormone seems to have a reciprocal effect to that of neotenin on the colour of Phasmids (L'Helias, 1970). However, the position is far from clear and probably not simple. In the insect metamorphic changes ecdysone controls all changes including those of pigments (Karlson and Brückmann, 1956; L'Helias, 1970). Moreover another hormone, *bursicon*, of the intercerebral region, now appears to have a direct action in the metamorphosis of diptera and this is melanogenic (L'Helias, 1970). It seems that hormonal action during metamorphosis may be very different from that in camouflage chromogenesis and that analyses must soon attempt to distinguish between all the different biological situations. It will also be noted that most of the evidence is on chromogenesis, since few insects have good chromomotor changes.

The Crustacea provide a strong contrast in this respect and in consequence their chromogenesis has been almost completely neglected. Perhaps this is because their chromomotor mechanisms fully occupy for an indefinite period anyone who becomes interested. Results to date present a formidably complex, incomplete solution to a complex problem (Knowles and Carlisle, 1959; Prosser and Brown, 1961; Carlisle, 1964; Fingerman, 1963, 1970),

Table 6.2. Crustacean chromomotor hormones

A. *Displaying chromes*

Colour and site of chromatophore	Hormone and/or endocrine tissue giving extract having appropriate action	Reference
1. *Black*		
Uca	Uca-darkening hormone (UDH), alcohol insoluble; brain and thoracic nerve cord; Sinus gland;	SANDEEN, 1950 { BOWMAN, 1949 ENAMI, 1951
	C. N. S. of *Limulus*	BROWN and CUNNINGHAM, 1941
Crago, body (Crangon)	Post-commisural organ (PCO): Crangon-darkening hormone (CDH), alcohol-insoluble.	KNOWLES, 1953 BROWN, 1946
	Posterior thoracic nerve cord of Anomura Abdominal N. cord of Reptantia	BROWN and SAIGH, 1945 BROWN, 1952
Crago, tail-fan	Sinus gland (alcohol insoluble fraction) Post-commisural organ (CDH) and other CNS hormones affecting body	BROWN, 1952 BROWN, 1946 BROWN and KLOTZ, 1947
Squilla	PCO hormone and Hs. of other parts of CNS	KNOWLES, 1954
Ligia and other isopods	Brain H	{ ENAMI, 1941 FINGERMAN, 1956
	"B substance"	SMITH, 1938
Eye, masking chromes, distal set	Sinus gland, alcohol-insoluble factor; Brain and circum-oesophageal commisure (COC) factor	FINGERMAN and MOBBERLY, 1960
2. *Red*	In general respond as black	
Palaemonetes	Sinus gland, alcohol-soluble H.	BROWN, et al., 1952
Cambarellus	Sinus gland factor; positively charged protein	FINGERMAN, 1957 FINGERMAN and AOTO, 1958
Uca	Sinus gland alcohol-soluble H	BROWN, 1950
Palaemonetes	Brain and COC factor (different from SG factor)	FINGERMAN, 1957
Leander	PCO: alcohol-insoluble, electronegative, "B substance"	CARLISLE and KNOWLES, 1959
Uca	CNS	BROWN, 1952
3. *Yellow*	In general respond as black	
Uca	Sinus gland factor	STEPHENS et al., 1956
4. *White*	In general respond conversely to black	
Palaemonetes Crago, Leander	Sinus gland factor	PERKINS and SNOOK, 1932
Orconectes	Sinus gland factor	BROWN and MEGLITSCH, 1940
Uca	Sinus gland, acetic-insoluble factor	STEPHENS et al., 1956
Crago, Uca	CNS factor	{ BROWN, 1952 PROSSER and BROWN, 1961 p. 518

Table 6.2 (continued)

B. *Retracting Chromes*

Colour and site of chromatophore	Hormone and/or endocrine tissue giving extract having appropriate action	Reference
1. *Black*		
Crago, body; Macrura in general	Sinus gland of all decapods; alcohol-soluble factor, *Palaemonetes*-lightening hormone (PLH)	BROWN, 1952 { BROWN and EDERSTROM, 1940 FINGERMAN, 1957
Crago, tail	Sinus gland factor of Macrura (*not* of Brachyura), *Crago* tail-lightening hormone (CTLH)	BROWN and EDERSTROM, 1940
Uca	Sinus gland factor	ENAMI, 1943, a, b
Crago, body and tail	PCO, alcohol-soluble factor; *Crago* body-lightening hormone (CBLH)	BROWN, 1952
Crago, body	Posterior thoracic nerve cord of Anomura	BROWN and SAIGH, 1946
Uca	Brain and abdominal ganglia, COC	{ ENAMI, 1943, a, b FINGERMAN, 1966
Sesarma	Brain and medulla terminalis	ENAMI, 1950, 1951
Isopoda	Organ of Bellonci (? = X organ): ? W-substance of Smith (1938)	OKAY, 1946 McWHINNIE and SWEENEY, 1955
Eye, masking chromes, distal and proximal	Sinus gland factor, and factor(s) from PCO, brain and other parts of CNS	{ KLEINHOLZ, 1936 BROWN et al., 1952 FINGERMAN, et al., 1959
2. *Red*		
Palaemonetes, Leander, Crago, etc.	Sinus gland PLH; Peptide of 1000 daltons (*Pandalus*).	KOLLER, 1928 PERKINS and SNOOK, 1932 FERNLUND and JOSEFSSON, 1968
Sesarma	Sinus gland, eyestalk ganglia, PCO, thoracic ganglia	ENAMI, 1950, 1951
Uca	Eyestalk H	BROWN, 1950 CARLISLE and KNOWLES, 1959 p. 53
Leander	PCO: MeOH soluble, electronegative peptide (A-substance)	KNOWLES, 1956
Cambarellus	Brain and COC: positively charged protein	FINGERMAN and AOTO, 1958
Pandalus	*Eriocheir* brain	OTSU, 1965
Leander	*Squilla* PCO	CARLISLE and KNOWLES, 1959
3. *Yellow*		
Uca	Sinus gland H	STEPHENS et al., 1956
Palaemonetes	Sinus gland	PROSSER and BROWN, 1961, p. 516
4. *White*		
Leander	Sinus gland H.	BROWN, 1952
Palaemonetes and other Macrura		CARLISLE and KNOWLES, 1959 p. 50
Uca	Sinus gland; acetic-soluble fraction	STEPHENS et al., 1956
Macrura	PCO of *Leander* and *Squilla*: A-substance, electropositive peptide. Other parts of CNS	CARLISLE and KNOWLES, 1959, p. 50 KNOWLES, 1952; FINGERMAN, 1957

The most widespread group is the *p*-benzoquinones (p. 40) used by various classes of arthropods (BURTT, 1947; ROTH and EISNER, 1962). They cannot have evolved directly from the materials used to harden the exoskeleton since these are *o*-BQs. As free quinones they are strongly corrosive oxidants, and the volatility (p. 45) also is exploited by those which eject the material forcibly. The bombardier beetle *Brachynus* secretes a quinol precursor mixed with one-third if its weight of hydrogen peroxide and with a heat-stable catalase. This releases O_2 from the H_2O_2 at a rate which becomes audibly explosive through the acceleration due to the heat of the reaction itself. The quinol is oxidised to quinone which is carried by the aerosol jet as far as 45 cm (SCHILDKNECHT, 1971). From its tracheal glands the cockroach *Diploptera* secretes a mixture of *p*-benzoquinone, 2-methyl-1,4-BQ (*p*-toluiquinone) and the ethyl analogue (ROTH and STAY, 1958) and the mealworm *Tribolium* produces the last two (ALEXANDER and BARTON, 1943; LOCONTI and ROTH, 1953). These turn the flour pink, probably through conjugation with $-$ SH or $-$ NH_2 groups of its proteins (MORTON, 1965).

Iulid millipedes also have mainly *p*-BQ derivatives (CASNATI et al., 1963), and likewise South American chelicerates (VUILLAUME, 1969, p. 97). In the latter di-, and tri-, methyl substituted *p*-BQs predominate, with some toluquinone and unsubstituted BQ (BARBIER and LEDERER, 1957). Millipede quinones blacken the human skin and the lesion heals very slowly (BURTT, 1947). The alkyl-substituted BQs are among the most powerful bacteriostatic agents: Maesoquinone is a vermifuge and embelin both anthelminth and fungicide (THOMSON, 1957, p. 9). Other quinones are natural antibiotics (MORTON, 1965). Naphthoquinones also are effective fungicides (THOMSON, 1957) and include some natural antibiotics (MORTON, 1965). It is suggested (VEVERS, 1963) that echinochromes (p. 42) may be used for this purpose, against bluegreen algae, which otherwise would be able to coat the echinoid test disastrously; they do grow on pale specimens of *Echinus esculentus*. NQs have been used experimentally to prevent proliferation of the Cyanophyceae.

Poisonous properties are implied, though not proved, for the dibromoindigotin of muricids (p. 59), or for its precursor. DUBOIS (1903) found a colourless poisonous material also present in the purple gland of *Murex* but this was not shown to be chemically related to the chrome or to its precursor. The various species of the sea hare, *Aplysia*, are interesting in this connection (RÜDIGER, 1970). *A. depilans* secretes only a colourless, milky, toxic fluid on irritation, but *A. punctata* secretes both this and the violet and red plant bilins (p. 1). *A. limacina* secretes only the bilins which are non-toxic but provide the smoke screen type of defence (p. 306). If analogy is a safe guide the implication is that dibromoindigotin may provide a second type of defence, an important part of which may be the objectionable odour which accompanies the reactions by which it is produced from its precursor. The colour and odour may also be memorable warnings about the toxin to individuals which have previously attacked these gastropods.

A considerable number of other external secretions contain chromes and in no case has a function been established. Examples are the ommochrome in the mucus secretion of the nemertine *Lineus* (p. 97), arenicochrome (p. 43) which stains the hands so strongly that the species was once called *Arenicola tinctoria* (FOX and VEVERS, 1960, p. 173), ribitylflavin by *Eisenia* and probably by some leeches (p. 90), flavone by *Helix* (p. 82) and rufin by *Arion rufus*. The uroporphyrins of planarians are highly concentrated in the rhabdites and in the gland cells which produce these (MACRAE, 1961) and may add to the defensive effect of the rhabdites. Turacin (p. 88) very readily washes out of the turaco's feathers and conceivably may be unpleasant in the mouth of a predator; it is not a free porphyrin and so will not have a serious photodynamic effect. Conceivably the chrome in cerumen (p. 37) may contribute to its defensive properties, but those in sweat are too varied and too sporadic in incidence to have more than incidental significance.

A number of the antibiotics of microorganisms are coloured. Actinomycin, one of the best known because of its inhibition of DNA transliteration (p. 64) is a phenoxazone, related to the ommochromes. Another quite different form of chemical defence is by anti-oxidants (p. 197) most of which are chromes.

6.2.4 Thermal Regulation

It has been seen (p. 137) that temperature may affect integumental colour either for camouflage purposes or to assist in temperature-regulation. At high altitudes and in high latitudes (p. 135) both terrestrial and aquatic poikilotherms tend to be heavily coloured, either red or black. The latter absorbs the more strongly at all wavelengths in the solar spectrum but red is less transparent to heat than it is to shorter waves (BURGEFF and FETZ, 1968). These differentials may meet the requirements of different species. Black has the highest *calorescence* (EDSER, 1902) and some inorganic black materials such as charcoal, platinum black and appropriately oriented crystals of tellurium, are particularly efficient, transmitting virtually no light or i.r. Some melanins are almost as efficient, at least in the near i.r. but they are u.v. and visible absorbers rather than long wave heat-absorbers (TREGEAR, 1966). Individuals of *Uca* with their melanosomes displayed were found to be 2°C warmer than pallid individuals (BROWN and SANDEEN, 1948). Normally this crab and many other poikilotherms are maximally dark at low external temperatures, so that they absorb solar heat maximally, and pallid at high temperatures so that they reflect heat maximally (PARKER, 1906; BAUER, 1914). Migratory phase locusts each morning bask in the sun, exposing their dark colours maximally, and so expedite the attainment of an adequate take-off temperature. The 'horned toad', *Phrynosoma*, is maximally dark in the morning and evening, and pallid at mid-day (KRUGER and KERN, 1924); other reptiles also show this response. Melanin-display is maximal at 5°C (PARKER, 1938). The seasonally polymorphic butterfly *Colias* is lightest in colour in the warmest season (WATT, 1969); the dark form of early spring basks in the sun, exposing the dark undersides of the wings where the blood is most easily exposed to direct as well as to pigment-trapped heat. It absorbs heat faster than the pale-winged type. Normally the frog's skin reflects differentially the longer-wave heating radiation (DEANIN and STEGGERDA, 1948), but under cold, wet conditions this is partly offset by the usual darkening of colour.

The apparent paradox that homoiothermic animals tend to be darkest in low latitudes (p. 135) and white in the arctic probably is due to the fact that dark chromes are also best radiators of internal heat. Where the main consideration is the dissipation or conservation of internal heat this converse to the poikilotherm colouring is adaptively rational. Obviously when the external temperature is higher than the body-temperature an homoiotherm must on balance absorb heat and the darkest types will be most at a disadvantage; a dark colour is useful only where there is a fairly consistent need to dissipate internal, or absorb external, heat. There are situations in higher latitudes where homoiotherms can benefit from absorbed solar energy. Even in temperate regions zebra finches were found to require 25% less food when their feathers were painted black (HAMILTON and HEPPNER, 1967); the paint did not affect any other properties of the feathers. Equally, dark poikilotherms will radiate more heat than lighter types if the heat-flow is in that direction: POULTON (1890, pp. 17, 19) described an interesting experiment of Lord WALSINGHAM showing that dark butterflies melt snow faster than white individuals of the same species. There are few large poikilotherms in high latitudes and the smaller types mostly hibernate; they could not obtain enough heat with a black integument nor conserve significantly with a white one.

6.3 Conclusion

Integumental zoochromes serve many functions, *crypsis* (camouflage), *semasis* or *advertisement* (alluring and warning), protection against radiation, mechanical and chemical protection and heat-control. There is often a wide variety of chromes, each in a separate *chromatophore*. The biological value of crypsis has been proved by innumerable observations, by field and laboratory experiment and by the sophisticated adaptations in the mechanisms which control the colour patterns. Some species have a once-for-all *chromogenesis* of a static pattern but many have a *chromogenic* mechanism adaptable to a variety of backgrounds and some have both this and a *chromomotor* colour-change. The two share a control mechanism, nervous or neuro-hormonal.

There is strong circumstantial evidence for the various advertisement functions and some experimental proof. Some of these advertisement patterns can be varied but not so commonly as camouflage; the control mechanisms share pathways with the crypsis system and this may lead to confusion between functions and to apparently anomalous patterns. Again light and other signals can evoke more than one of the functions of integumental chromes; equally a variety of signals may evoke a particular function, e.g., crypsis-responses. Crypsis is distinguished from other patterns by its strong motif of countershading. Spontaneous but externally 'entrained' *circadian* and tidal cycles of colour-change, for crypsis and other purposes, are common in the appropriate environments.

In colour-change the chromatophores may respond directly to external signals by a primitive *primary* mechanism or indirectly via the nervous and hormonal systems *(secondary response)*. In all taxa the afferent path is nervous but the efferent side is nervous, hormonal or both. Each colour-type of chromatophore has some independence of response but also conforms to one of two main responses, darkening and blanching, due to the display respectively of the dark and bathochromic chromes and the light and hypsochromic group; the display of one group is accompanied by the eclipse of the other. Cephalopods have a neuromuscular effector organ and have fully exploited this for speed and versatility. Other taxa operate a mechanism of dispersion and aggregation of *chromasome*-granules within their richly arborescent chromatophores; at best this is considerably slower than the fastest responses of the cephalopods. The precise motive force of the chromasomes is uncertain. The complete mechanism by which complex colour patterns change to match a variety of backgrounds also cannot yet be explained in detail.

Chapter 7

Zoochromes and Sensory Perception

It seems probable that chromes play a part in the function of most sense cells and the primary function of chromatophores (p. 141) in response to various types of external signal is relevant here. At present little is known of the rôle except in photoperceptors but chromes are present in other sensory epithelia.

7.1 Light Perception

It has been known for a century that carotenoid derivatives play the crucial rôle in the transduction of light-stimuli into electrical stimuli in the vertebrate eye, eventually initiating impulses in the optic pathway and there is now considerable knowledge of the mechanism of this (PIRIE and VAN HEYNINGEN, 1956; WALD, 1960; DARTNALL, 1972). The carotenoid derivatives are *retinals* (Fig. 7.1 A), aldehydes of the alcohols, vitamins A_1 and A_2, which are haplomers of β-carotene (Fig. 3.2), hydroxylated at the point of fission. Retinol A_2 and its retinal have a second double bond in the ionone ring and so are more bathochromic than the A_1 molecules. Retinal$_1$ (R_1) performs the same function in the eyes of cephalopods (HUBBARD and ST. GEORGE, 1958, 1959), Crustacea (KAMPA, 1955; WALD and BERG, 1957), insects (WOLKEN et al., 1960; MARAK et al., 1970) and planarians (PIRENNE and MARRIOT, 1955). The wavelength sensitivity-curves of the eyes of various other invertebrates imply the mediation of similar chromes, in some cases extracted and partially characterised. In the starfish, *Marthasterias*, the agent may be a complete C_{40} carotenoid (VEVERS and MILLOTT, 1957). Only the vertebrates are known to have both R_1 and R_2. In the unstimulated retina the retinal is conjugated with a lipoprotein, opsin, of molar weight 40,000, forming retinopsin. Absorption spectrum and other properties vary with the opsin, and both R_1 and R_2 form two distinct types of conjugate, scotopsins for vision at low light-intensities, and photopsins for full daylight-vision. The absorption maximum for the latter is in the ranges 528–600 nm and 600–650 nm respectively for R_1 and R_2, compared with 480–520 nm and 512–533 nm respectively for their scotopsins (DARTNALL and TANSLEY, 1963). These values bear a relation to the difference in solar intensity spectrum under the two conditions. The scotopsins are usually in the rods and the photopsins in the cones of vertebrate retinae; there are about 4.2×10^6 molecules of scotopsin per rod cell.

The effect of light on these retinopsins is best known for *rhodopsin*, the R_1 scotopsin of vertebrates. The retinal moiety undergoes a series of isomeric changes (Fig. 7.1, B), with considerable changes in colour and associated properties. The initial intermediaries are more bathochromic and subsequent stages more hypsochromic. In unstimulated rhodopsin the retinal has the *cis*-conformation about the 11 C–12 C double bond and probably also about the 12 C – 13 C single bond (HONIG and KARPLUS, 1971); in its physiologically active phase it isomerises in stages back to the all-*trans* form characteristic of carotenoids. The intermediaries are unstable and therefore after the first critical photoactivation step, the process needs only thermal energy. The series of small energy-steps enables the energy released to be harnessed to the generation of a potential in the perceptor cell, the basal end of which becomes more electropositive relative to the distal end. The opsin is split off in the process, since sterically it fits the *cis* form only. Both moieties are re-used, and are reconjugated after the retinal has been re-isomerised to the *cis*-form. Actually, 70% of the all-*trans* retinal is first reduced to vitamin A, which is *cis*-isomerised and then reoxidised to the aldehyde (Fig. 7.1, B). The reason for this complication is not very clear but since the replacement of any loss or shortfall of retinal is met by retinol from storage-

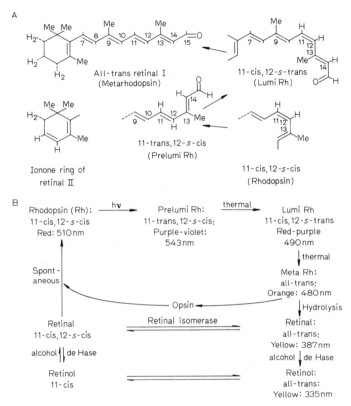

Fig. 7.1. The functional cycle of rhodopsin in vertebrates. A The series of changes in the retinal itself. B The complete cycle of rhodopsin, including regeneration phase. (After WALD, 1959 and HONIG and KARPLUS, 1971)

sources this path must operate in any case. The 'regeneration' of vertebrate scotopsins is possible only at low light-intensities so that in broad daylight they are out of action. Their functional levels of illumination, together with a varied external field and slight eye-movements, permit the process to keep pace with the photic phase of the cycle. The conjugation of opsin with *cis*-retinal is exergonic, and therefore spontaneous, and moreover it promotes the oxidation of further retinol molecules (GLASS, 1961, p. 901); conceivably this is related in some way to the predominance of the retinol path over the more direct one.

The 11-*cis* form is sterically hindered, unlike some of the other *cis*-isomers; thermodynamically it is therefore the most 'improbable' isomer but once formed it is kinetically very stable, except to light, which is the most effective agent for causing *cis-trans* changes (PULLMAN, 1972). This isomer is therefore the ideal triggering agent for the purpose, notwithstanding the sacrifice of resonance-energy common to all the *cis*-isomers, because of their shorter dipole (p. 14). The classical view is that rhodopsin has the simple 11-*cis* isomer but HONIG and KARPLUS (1971) maintain that it has the *cis*-conformation also around the 12–13 bond (Fig. 7.1, A). This isomer is particularly unstable to light; it also provides the best explanation of the number of intermediaries, including prelumirhodopsin, the small energy-steps between these and the need only for thermal energy, 4 kcal per mole, for all steps but the first.

A total of six quanta, absorbed by six neighbouring rods, is the thereshold stimulation for detection centrally (GLASS, 1961), so this implies summation at some point in the path. A quantum absorbed at one point in a rod can affect chemical activity up to 100 μm away so that the summation may be peripheral, by interaction between neighbouring rods. There is little knowledge of the precise mechanism of photo-chemo-electrical transduction; ERHARDT et al. (1966) suggest that electron-transfer along the side-chain of the retinal plays a part. Also, since rupture of the bond with opsin exposes − SH groups of the latter, these may be involved in the transduction. A retinal-phosphatidylethanolamine complex has been found in the eyes of vertebrates; it is suggested that this is a store of *cis*-isomer for rapid resynthesis of rhodopsin (ANDERSON, 1970) rather than functioning in the active phase of the cycle. The activity cycle of *porphyropsin*, the scotopsin based on R_2, is less well known (BAUMANN, 1971) but differs only in detail from that of rhodopsin. It is less easily bleached.

Table 7.1. Distribution of Retinopsins

Rhodopsins	Porphyropsins
Mammals	Larval Amphibia
Birds	Aquatic adult Anura
Crocodiles	Water-breeding stage of Amphibia
Snakes	*Protopterus*
Pseudoemys	A few marine teleosts
Adult Anura	About half of stenohaline freshwater teleosts
Urodeles: terrestrial stage	*Petromyzon:* freshwater phase
Marine elasmobranchs	*Lampetra*
Most marine teleosts	
About half of euryhaline freshwater teleosts	Both Rhodopsin and Porphyropsin
Marine cyclostomes	Adult *Xenopus* (8:92%)
Petromyzon: marine phase	About half of all freshwater teleosts
Cephalopods	A few marine teleosts
Some insects	
Crustacea	Animals unexamined
Limulus	Marine and freshwater mammals
Nereis	Diving birds
Anthomedusae	Most lizards and Chelonia
	Freshwater elasmobranchs
	Many invertebrates

Why vertebrates collectively have both R_1 and R_2 retinopsins is still not quite clear. Certainly the ranges of their absorption maxima are significantly different but these overlap considerably, owing to the variation in the opsin moieties and it would seem that either should be able to cover the full range of wavelengths needed by the exploitation of further opsins. Possibly the phenomenon is connected with the need to maintain a wavelength gap between the scotopsins and the photopsins of any particular animal. There is a definite ecological basis for the distributions of R_1 and R_2 (Table 7.1) but this is not so simple as at first supposed. All terrestrial and nearly all marine vertebrates have R_1, whereas many fresh-water types, e.g., cyclostomes, teleostomes, dipnoi, and the amphibia in their aquatic phases, have R_2. Newts revert to R_2 when they return to the water to breed and the sea lampreys and amphidromic teleosts change between R_1 and R_2 when they move between the sea and fresh water.

The permanently aquatic 'clawed toad', *Xenopus laevis*, has both R_1 and R_2 and so have a large number of fresh water teleosts (MUNZ, 1971). Further, many other fresh-water fishes have only R_1. It helps to recognise three groups of freshwater fishes: (1) 'primary' taxa, which are intolerant of sea-water, (2) 'secondary', euryhaline types, which can spread from one river system to another via the sea, and (3) 'peripheral', amphidromic types which regularly breed in the one medium and spend much of their life in the other. In the first group nearly a half of the species have R_2 only and nearly a half have both retinals, the remainder having R_1 only (Table 7.2, A). In the other two groups essentially the converse holds: nearly half the species examined have R_1 only, a half have both and very few have R_2 alone. Among marine teleosts 92% have R_1 only and the remainder either R_2 or both. For both groups of euryhaline fresh-water fishes therefore R_1 is as characteristic as for marine fishes and R_2 is specific only to stenohaline freshwater types. In both main osmotic categories of freshwater fishes a mixture of the two is slightly more common than the one alone and this perhaps is the most significant feature. There is regularly a difference in wavelength of maximal absorption between the pair in any species.

Table 7.2. A. Distribution of Retinals 1 and 2 in Retinae of Teleost Fishes. (Based on MUNZ, 1971)

Ecological type	Number of species known to have		
	Retinal 1	Both R1 and R2	Retinal 2
Marine	58	3	2
Freshwater			
(a) Stenohaline	8	28	25
(b) Euryhaline: Coasters and Amphidromics	28	31	3

B. Positions of Absorption Peaks of Rhodopsins of some Marine Teleosts in relation to average depth at which they live. (WALD et al., 1957)

Species	Flounder	Cod	Redfish	Lancet fish
Depth (m)	4–20	10–150	80–390	400
λ max (nm)	503	496	488	480

The Cyprinid, *Scardinius*, has more R_2 in winter and in other conditions of lower light-intensity than in summer and other conditions of bright light. Similarly, tropical fresh-water fish have a higher R_1/R_2 ratio than those of higher latitudes, with the sun at a lower altitude. River waters and other fresh waters tend to have more clay, etc., in suspension and so transmit less light than salt water, particularly at the short-wave end of the spectrum. The longer wavelength of the absorption-maxima of the porphyropsins therefore is appropriate to the environmental conditions of fish which use R_2, relatively low total illumination coupled with differential impoverishment in the short-wave light.

Fresh-water fish living in deep water mostly have R_2 and surface forms R_1 and this has the same rationale. Conceivably those which have both R_1 and R_2 move rather frequently and rapidly through a large range in depth.

By contrast marine fishes have little horizontal restriction and have tended to become depth-specialised. They need only one scotopsin therefore and have modified the opsin to give an absorption maximum appropriate to their normal depth. Salts flocculate down any suspended matter and in fact it is the longer wave visible rays which are differentially absorbed by pure water (p. 7). Consequently it is R_1 alone which has been adapted and the absorption maximum of its scotopsins decreases progressively with depth (Table 7.2, B). These 'rhodopsins' are increasingly yellow in colour and so have been called *chrysopsins*.

It is now believed that the gnathostomes originated in the sea and therefore R_1 was the original retinal in vertebrates. R_2 evolved when fishes moved into fresh waters and since then various groups have moved so frequently between the two environments that a full explanation of all details of the distribution of R_1 and R_2 will take time: reciprocally it should help with the general evolutionary problem. Some apparent anomalies may be due to the recapitulation of recent evolutionary changes (WALD, 1958).

The photopsins of vertebrate cones have the same retinal as the rod scotopsin of the same species, though, as already seen, the range of absorption maxima of the R_1 photopsins alone is far greater than that of either set of scotopsins. The R_1 series are violet and the R_2 series blue so that they are called iodopsins (iodine vapour is violet) and cyanopsins, respectively; the latter therefore again are more bathochromic than their R_1 counterparts. Some tortoises (GRANIT, 1943; KIMURA and HOSOYA, 1956) have cyanopsins but the same general ecological rules apply as for the scotopsins. The activity-cycle of both series is essentially similar to that of the scotopsins.

Cephalopod rhodopsins have absorption maxima between 475 and 495 nm and those of Crustacea between 462 and 515 nm. Those of insects, at 515 nm (MARAK et al., 1970) and 535 nm (GOLDSMITH, 1960), and that of *Limulus*, at 520 nm, somewhat overlap the porphyropsin range. In their behavioural responses the Crustacea collectively show sensitivity-maxima over the whole range, 470–545 nm (WATERMAN, 1960). Since only R_1 has been identified in invertebrates they bear out the contention that in principle R_1-retinopsins can cover the absorption range of the R_2 series also. The retinopsin of the polychaete, *Nereis*, absorbs at 500 nm, in the rhodopsin range, but that of the starfish has a maximum at 580 nm (PERSHIN, 1951), in the iodopsin range. It seems likely that some invertebrates have a scotopsin-photopsin differentiation. Some insects have values around 437–440 nm, far below the ranges of all retinals (PROSSER and BROWN, 1961) and these are due to a different class of chrome (p. 172), which is sometimes present with the carotenoid (MARAK et al., 1970). In euphausid Crustacea 90% of the large stores of vitamin A itself are already the 11-*cis* form and this should expedite the generation of active retinal, provided the alcohol dehydrogenase is available; this essential enzyme apparently is not present in all invertebrates (GLASS, 1961, p. 901). The active form always is the aldehyde so that there may be a different path for its generation in some phyla.

Light isomerises squid 11-*cis* retinal to the all *trans*-form, just as in the vertebrates. However, the metarhodopsin differs in being stable at room temperature (PROSSER and BROWN, 1961, p. 361) though it is very sensitive to pH-changes. Moreover, rhodopsin is regenerated in the light (HUBBARD and ST. GEORGE, 1958; KROPF et al., 1959), a great advantage over the vertebrate system. Many of the vertebrates have compensated for their disadvantage by their separate photopic and scotopic systems, only the latter suffering complete bleaching in bright light and the former being active only above a certain threshold light intensity. The generator potential of those invertebrates studied, arthropods, cephalopods and starfish, resembles that of vertebrates in that the end of the receptor cell nearest the first relay with the optic tract becomes electropositive to the other end, upon excitation. Because of the inverted retina of vertebrates, however, the relation to the direction of incident light is opposite to that in invertebrates. On illumination the resting potential rises in vertebrates while that in invertebrates falls, so that in all cases there is increased positivity at the end towards the incident light. The large eccentric retinular cells of *Limulus* can produce a propagated spike from their generator potential change. In vertebrates the propagated impulse is generated only in the next cell of the path, a neuron proper, the rods and cones having no functional axon (Fig. 7.2).

7.2 Colour Vision

It would be very surprising if the ability of some animals to distinguish colours, and other differential wavelength effects, did not depend on the same basic property, -the differential absorption of spectral light by appropriate molecules. This has been the guiding principle in all studies on colour vision but research has not been as straightforward as might have been anticipated. The principle has been justified but the mechanism seems to be a good example of the inability of natural selection to do more than improvise with what was already available. In vertebrates the scotopic-photopic system was available and it was adapted for colour vision.

Man does not discriminate wavelength-differences in dim light though when the intensity is adequate he can distinguish hues of considerably shorter wavelength than the scotopic sensitivity maximum. This is compatible with the scotopsin acting as one wavelength-discriminator since the spectrum is treated subjectively as a circle (p. 6); however, its slow rate of regeneration in bright light makes this improbable. According to LAND (1959) all spectral hues can be discriminated with only two differentially sensitive chromes, say a rhodopsin and an iodopsin, but the consensus still favours the classical trichromatic theses of YOUNG and HELMHOLZ. Certainly at least two retinopsins are active in bright light since the absorption-spectrum of iodopsin alone does not correspond very closely with the action-spectrum of the intact eye (HOAR, 1966, p. 493).

It is widely believed that all the wavelength discriminators are in the cones (HENDRICKS, 1968) and that only one is present in any one cone. Three different chromes have been extracted from cones (HENDRICKS, 1968) but with absorption maxima at 450, 525, 535 nm, respectively, they do not cover the visible range at all well and only the last is even within the iodopsin range. To trace individual chromes to particular cones is a formidable problem but it is now possible to obtain action-spectra from individual cones (MUNZ, 1971) or from small groups of the same modality. Individual ganglion cells, receiving from such groups show action-spectra which vary considerably, probably because the action-spectra are composite already at this point. However, action-peaks cluster around three modes, 450–470 nm, 520–540 and 580–600 nm (GRANIT, 1955); these values cover the visual range rather well, bearing in mind that because of the colour circle none need be very near either end of this range. In the retinal projection-system there is provision for mixing these three 'modulator' components to give a 'dominator' action spectrum, with a maximum at 562 nm, responsible for our perception of whiteness in broad daylight.

BROWN and WALD (1964) introduced the technique of recording the action-spectrum of individual cones of the human retina by a valuable absorption-difference method. This again indicates three wavelength discriminators, with absorption maxima at 450, 525 and 555 nm; the last, in particular, is at a significantly shorter wavelength than that of Granit. It is noteworthy that none of the values quoted could be due to rhodopsin, which in man absorbs maximally at 495 nm. KEILIN and SMITH (1939) thought that the 455 nm peak might be due to ribitylflavin; the value is rather high for the main peak of ribitylflavin in the visible range, but JUDD (1951) found a value as low as 440–450 nm, near the main peak of ribitylflavin. Granit's more recent values for the other two chromes agree quite well with the averages quoted by JUDD for work up to 1951: 540–550 and 590 nm. Direct measurements of absorption in the foveal region of the human eye also gave peaks at 540 and 590 nm (RUSHTON, 1958), so that these two, at least, seem well established. Dichromat colour-blind persons lack the 590 nm (red) peak and deuteranopes have a subnormal absorption at the 540 nm (green) peak (RUSHTON, 1958). The rare tritanopic condition, of blue-yellow confusion, presumably is due to a defect in the short wave modulator.

There is some evidence that the frog has colour vision, but based on chromes in two types of rod and one type of cone; the sensitivity maxima are at 440, 502 and 560 nm respectively (MUNTZ, 1964). In the goldfish, microspectrophotometry indicates three types of cone, each with a single retinopsin, their sensitivity-mexima being at 455, 530 and 625 nm respectively (MUNZ, 1971). The numbers of the three types are in the ratios 1:4:2. There is a different chrome in each member of every twin pair of cones and in 97% of these the two are red and green, the remaining 3% being green and blue pairs. This may be relevant to the fact that red-green confusion is the commonest type of colour-blindness in man.

By no means all vertebrates give evidence of having colour vision but there is reason to anticipate a trichromatic basis where the property has been evolved; the evolution must have occurred independently many times so that uniformity of mechanism cannot be assumed without direct evidence in each case. It appears that relatively little use is made of the longer visible rays and this is indicated also by the wide range of wavelengths which give a subjective impression of redness (Fig. 2.1). This may be due to the limitations of carotenoids as perceptor chromes but alternatively, it may be related to the lower quantal values of radiation in this range and its relatively low intensity in the solar spectrum, features which may have led to the exploitation of carotenoids in the first place. The indifference to long waves is even more evident in those insects with colour vision.

In insects the individual wavelength-discriminators appear to be in separate retinulae (BURKHARDT, 1962). These are large enough to be stimulated individually by microbeams of monochromatic light and to permit recording peripherally to the 'mixing' in the projection-path. This elegant method has shown that the fly, *Calliphora*, again has three discriminators in the visible range, with sensitivity maxima at 470, 490 and 520 nm, much more bunched than in the vertebrates. All the retinulae have in addition a second maximum, at 350 nm, which could be due to a pale yellow chrome, most probably a pterin, common to all retinulae (MOTE and GOLDSMITH, 1971). The bee has a similar quartette of discriminators with absorption-maxima at 300–340, 400–480, 480–500 and 500–650 nm (CARTHY, 1964). These insects therefore can see 'colour' in the long u.v. and this has been confirmed by behavioural studies.

The visual properties of the photoperceptor cells of the barnacle, *Balanus amphitrite* are very intriguing. They indicate a perceptor chrome with two physiologically active stable states, responsible for two alternative 'early' perceptor potentials. The decision which of these is set up depends on the light waveband to which the cell was previously adapted: if this was red then a purely depolarising potential is set up but if blue then a largely hyperpolarising potential-change (HILLMAN, HOCHSTEIN and MINKE, 1972). A simple form of colour vision is implied. Following adaption to blue light red has an effect by causing prolonged depolarisation as the 'late' perceptor potential and this is switched off by a further exposure to blue. The colour-adaptation memories for both early and late potentials have a long persistence.

7.3 Other Zoochromes of the Retina

There is no doubt that the melanin and other dark chromes of eyes screen the perceptor cells from misdirected and excessive light, by surrounding the visual unit; this is the individual ommatidium of arthropods and both the whole eye and the individual perceptor-cells of other phyla. For protection against an increase in light-intensity either the chromasomes move within pigment cells, much as in dermal chromatocytes, with which they are usually homologous, or alternatively the perceptor cells themselves move into a static screen. Both methods are developed in some taxa. In many arthropods the chromasome-movements have a further function; they permit the eye to function with great form-, and movement-, discrimination in bright daylight, by means of an 'apposition' image, and with great sensitivity but low acuity ('superposition image') in dim light. The screening cells form a distal and a proximal cuff round the optical system of each ommatidium so that when the chromasomes are fully dispersed each ommatidium is completely isolated from all neighbours and the deeply placed retinulae are stimulated only by light directly along their axis (Fig. 7.2. A). In dim light the granules aggregate distally and proximally, exposing the whole of the retinular region to any available light from almost any angle. In *Limulus*, even the individual rhabdomeres of an ommatidium can be optically isolated by a further, transverse movement of granules.

The inverted retina of vertebrates allows the perceptor ends of the rods and cones to be always partially embedded in the screening layer, the chromasomes moving out, in bright light, to surround the perceptors more completely; this occurs in all groups but the snakes and the mammals (HOAR, 1966, p. 498). The effect is very similar to that of the movements in arthropods. In addition many vertebrates move the outer segment (perceptor limb) and the 'ellipsoid' of the rods and cones into and out of the pigment screen by means of a *myoid*, or contractile portion, of the cell itself (Fig. 7.2, B). This lies between the ellipsoid and the nucleus of the cell. At night the rods of the salmon move out of the screen and the cones in, and conversely by day (ALI, 1959; NICOL, 1963). The exact mechanism of the movement each way is not clear.

The use of tapetal or reflecting layers of pigment to direct any light reaching the screen back to the perceptors can account for the purines and some of the pterins and ribitylflavin present in many eyes. This reflection again increases sensitivity in dim light (Fig. 7.2 A) and decreases discrimination so that it is advantageous not to expose these materials in bright light. A number of Elasmobranchii have a very efficient reflector for this purpose; the reflecting material is guanine and it fills a number of plate-like cells stacked obliquely in the choroid. The reflection is switched off in daytime by the melanin of the outer layer of the choroid spreading over and between the plates (FRANTZ, 1942; NICOL, 1963). Those elasmobranchs which lack the reflector-masking device show a rapid pupil-closure in bright light; this reduces the amount of oblique light and so reduces glare and haze due to an exposed tapetum. The yellowing of the human lens may somewhat reduce haze and glare, in a different way, though it is also thought to reduce chromatic aberration (FOX, 1953, p. 145).

It is usually supposed that the carotenoid 'eye-spot', or stigma of *Euglena* and some other green flagellates is a screening or filtering chrome, the photoperceptor proper being a colourless, transparent bulb around one of the roots of the flagellum. However, this bulb proves to be an array of parallel rodlets (WOLKEN, 1960; HOAR, 1966, p. 477), which

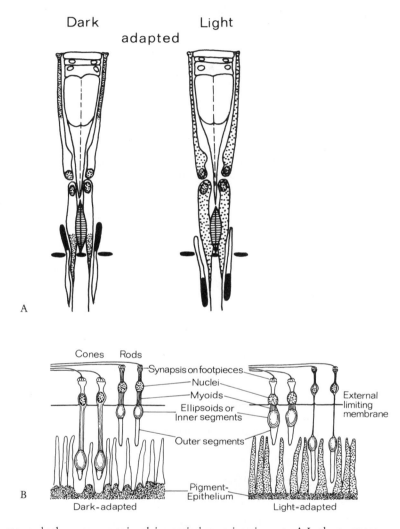

Fig. 7.2. Retinomotor and other movements involving optical screening pigments. A In the crustacean compound eye. B In the retina of the salmon. (After ALI, 1959).
In A the reflecting material is shown as solid black in peripheral cells, proximally. Details in text

are well placed to direct light entering the body by the least impeded route, namely down the gullet, on to the carotenoid chromasomes, which therefore may be the photoperceptor units, as in higher animals. If the carotenoid can act over a distance as great as that of vertebrate retinopsins (p. 165) then it should be able to control the flagellum via its roots or its blepharoplasts. The action-spectrum has a peak at 485nm (BENDIX, 1960), in the rhodopsin range. The directional screening attributed to the carotenoid may be effected by the chloroplasts and other materials, or directionality may depend solely on the orientation of the rodlets. The stigma of *Euglena* has a number of absorption peaks (WOLKEN, 1960), some of which must be due to other chromes.

It seems unlikely that the more hypsochromic retinal pigments, i.e., most pterins, ribitylflavin, and some of the ommochromes, and the carotenoids, can act as good screening chromes, and they are good reflecting materials only when in crystalline form. The presence of some at least of these chromes therefore is still to be explained; this is not a situation where fortuitous accumulation is likely. One possibility is that in an appropriate position some may differentially filter the incident light and so change the effective absorption maximum and the action spectrum of the perceptor chrome proper. If some only of the perceptor elements have their light filtered in this way then a single perceptor chrome could mediate differential sensitivity to more than one modal wavelength and provide the basis for a type of colour vision. The pigeon and some other birds do have filtering chromes in oil droplets in their retinae and these do shift the absorption maximum of the retinopsin by 20 nm (GRANIT, 1942; PROSSER and BROWN, 1961, pp. 363, 375).

Another possibility is that some of these chromes have a direct photoperceptor function. Ribitylflavin and the pterins are particularly photolabile (AUTRUM and LANGER, 1958). WALKER and RADDA (1967) showed that ribitylflavin sustains the photoreactivity of retinal and its derivatives, the usual triplet-triplet energy transitions of flavin photo-reactions being involved. This accessory function is reminiscent of the accessory chromes of photosynthesis (p. 182). Ribitylflavin may even be the main perceptor in some vertebrate retinae (p. 169), since it is present and carotenoids absent in the eye of the teleost, *Beryx* (LEDERER, 1935).

The oxygen-consumption of systems containing pterins is increased by illumination (ZIEGLER-GUNDER, 1956) so that a photochemical effect is indicated. However, the pterin screening pigments of the eye of the mealworm normally depress the sensitivity of the retinulae (YINON, 1972) so that they become more sensitive when the screen is retracted, in the dark-adapted state. Diptera lacking screening pigments, pterins and ommochromes, show a more direct and specific effect on perception (ZIEGLER, 1961; GLASS, 1961, p. 905–7) namely, the loss of two of the three peaks of the normal action spectrum, at 350 and 650 nm. As already suggested (p. 169) the 350 nm peak could be due to a pterin but unfortunately for the total picture no ommochrome absorbs strongly around 630 nm. Nevertheless, in the arthropods the evidence for photoperception by ommochromes is as strong as for any of these accessory chromes. Absorption peaks in the region of 435–440nm in several insects (PROSSER and BROWN, 1961, p. 357; MARAK et al., 1970) are probably due to an ommatin. The eye of the drone bee, but not that of the worker, shows an action-spectrum with a peak here. The phototaxis response of *Drosophila* is proportional in strength to the amount of ommochrome in the eye (KIKKAWA, 1953). YOSHIDA (1968) believes that the xanthommatin in the eye of the anthomedusan, *Spirocodon*, may play a photoperceptor rôle.

There is some evidence that even melanin may act as a photoperceptor in the retina and, more specifically that it may generate the characteristic 'stable' electrical response of the vertebrate eye. Unlike the response of the retinopsins this does not 'adapt' to light, i.e., the rate of discharge in the optic nerve does not fall off rapidly with time of exposure (EBREY and CONE, 1967). The action-spectrum is flat and so must be due to the only dark chrome present in this type of eye. LETTVIN et al. (1965) consider that any tissue with an oriented pigment could generate such a response. The possibility of electron-exchange by melanin (CASSIDY, 1949) is relevant here. When irradiated with visible light melanosomes from the mammalian retina and choroid produce a strong electron spin resonance signal, indicating the formation of free radicals (COPE et al., 1963). These decay rapidly and so minimise adaptation. Some of them are trapped and stabilised inside the macromolecular melanin complexes and this may permit charge-conduction over considerable distances. Zinc is present in the vertebrate choroid and in the presence of extracts of the eye it enhances the light-absorption by melanin (BOWNESS and MORTON, 1953); cobalt, ferrous iron and calcium also have this effect. It has been suggested that the black pigment in the porocytes of *Clathrina* (p. 95) may have a photoperceptor function (HYMAN, 1940, p. 294); its sharp localisation (Fig. 5.1) favours this idea.

In the eye of the duck BENOÎT (1962) found a sensitivity-maximum at 600–750 nm, broad but entirely within the red region of the spectrum. It is quite outside the range of any carotenoid retinopsin but is rather similar to that of the phytochromes (HENDRICKS, 1968); phytochrome is essentially a photoperceptor. There is a bilin in the eye of the water flea, *Polyphemus*, though not in that of the related *Daphnia*.

7.4 Simple Photoperceptors

Many groups of animals are known to have a general photosensitivity in the integument and sometimes in deeper tissues (STEVEN, 1963). Presumably this is mediated by biochromes and usually there is plenty of carotenoid present, as well as various other chromes sensitive to light, but no appropriate structures have been recognised, cellular or extracellular, which would permit the location of relevant chromes. In one of the few cases where a single chrome is known to be near a region of general photosensitivity, the siphons of the ascidian, *Ciona*, the chrome itself does not appear to be photosensitive, and there are other similar instances (STEVEN, 1963). The phenomenon has been demonstrated and analysed mainly by behavioural studies, and in most cases this is the only evidence. It permits the plotting of action-spectra which give clues to the class of chrome concerned, and has yielded considerable general knowledge. Often the general system can be inferred and characterised by eliminating overt photoperceptors; for instance, blinded animals have been found still to have their circadian clocks (p. 139) entrained by external light stimuli (UNDERWOOD and MENAKER, 1970; NEVILLE, 1967b). In the teleost, *Sceleporus*, entrainment still occurs in the absence of eyes, pineal and parietal, and in the locust in the absence of both eyes and ocelli (NEVILLE, 1967b). It is abolished in the latter by covering the general surface with opaque paint.

Taxonomically the phenomenon has been demonstrated in anamniote, but not in amniote, vertebrates. It is most evident in cyclostomes and Amphibia and in a few eyeless species of fish, all relatively sluggish vertebrates. It is shown by members of every major phylum of invertebrates, especially echinoderms and platyhelmia, and may be anticipated also in the phyla not so far studied. It is generally considered rarer in terrestrial arthropods than in the lower and more aquatic metazoa, but NEVILLE (1967b) finds it very important in controlling the circadian clocks of insects. The exoskeleton is no serious light-barrier and indeed pieces of body-wall still respond when implanted in the haemocoele.

Superimposed on the general sensitivity there are often strategical concentrations, for instance round the siphons of bivalves and ascidians. In fact it was on the former that HECHT (1919–27) did his pioneer work showing that the photochemical response to light intensity obeys the Bunsen-Roscoe reciprocity law. Even in such situations no specific perceptor cell has been identified. In many animals the integumental chromatocytes could be the perceptors since they are known to act in this way in the primary colour-change responses. However, Protozoa are usually sensitive all over their cell-surface and yet most of them have no detectable chromes here (CARTHY, 1964); the few ciliates with cortical chromasomes (p. 123) are exceptional. This general sensitivity of the cell surface may have been carried over into all cells of the metazoa. If so, the problem of how these primitive perceptors initiate systemic responses is quite obscure; some of the responses are local, but not all.

In some situations it is certain that nerve cells act as photoperceptors and since the nervous systems of probably all animals evolved from the ectoderm, some kind of connection between the general epidermis and the nervous system may still mediate this photo-response, at least in the lower metazoa. Obscure sub-epidermal nerve-plexuses have been recognised in animals as highly evolved as the arthropods. The best known example of neurons acting as direct light perceptors is in the genital ganglion of *Aplysia* (ARVANITAKI and CHALAZONITIS, 1949, 1957, 1961; KENNEDY, 1960). The cells in question contain a haemoprotein and a carotenoid; when exposed, in the laboratory, they are very sensitive to light in two distinct wavebands, in the carotenoid range and at 579 nm, near the α-peak of the haem. Since insect body-wall is still sensitive when implanted in the body cavity (p. 173) it is possible that the *Aplysia* neuronal carotenoids are stimulated by light even *in situ* and, since longer waves penetrate better, this is *a fortiori* probable for the haem. The nerve cord of *Branchiostoma (Amphioxus)* has direct photosensitivity and this is attributed to the darkly pigmented cells scattered along it; the chrome is probably melanin. The spinal cord of the lamprey also is sensitive to light (YOUNG, 1935) and is not too deep or too obscured to function *in vivo*. Other examples are the supraoesophageal ganglia of *Nephthys* (CLARK, 1956) and the radial nerve of the echinoid, *Diadema* (YOSHIDA and MILLOTT, 1959). In the latter the perceptor chrome may be a naphthoquinone. The nervous system of *Antedon* shows differential sensitivity, with action spectral peaks at 262, 337, and 460 nm, and a shoulder at 520 nm (DIMELOW, 1958); again an echinochrome is implied.

In some sensitive neural tissues no chrome has been recognised. In others there is abundance of a photosensitive chrome but no direct evidence that it functions in a photoperceptor physiological response. Porphyrins in the medulla oblongata and spinal cord of birds and mammals (p. 100) are believed to be capable of mediating photosensory responses *in situ* (RIMINGTON and KENNEDY, 1962). Other chromes in the nervous system (p. 99) therefore also may be relevant in this context. The hypophysis avidly takes up injected porphyrin (KLÜVER, 1944) though curiously enough normally it does not contain any; BENOIT and OTT (1944) showed that the hypophysis is directly sensitive to light, if this reaches it. In a number of the smaller chordates this must happen *in situ* (MENAKER, 1972). Porphyrins are found also in the optic, trigeminal and facial, cranial nerves of birds and mammals but not in the motor cranial nerves, or in any peripheral nerves of the spinal region, and not in the brain itself (KLÜVER, 1944a, b). This sharply contrasting pattern presumably has more than casual significance.

The harderian gland of neonatal rodents is thought to be an extraretinal photoperceptor (WETTERBERG et al., 1970) working in conjunction with the pineal, which has remained a primitive photoperceptor organ in many lower vertebrates (YOUNG, 1950). If the harderian gland is removed from blinded young rats their circadian rhythm is abolished. As already noted (p. 54) the gland contains a special porphyrin, as well as much protoporphyrin. The muscles of the sea-anemone, *Metridium*, also appear to be direct photoperceptors (NORTH and PANTIN, 1958), mediating light and dark adaptations. The tissue shows sensitivity-peaks at 490–520 nm, and 550–560 nm, respectively, in scotopic and photopic ranges, for dark and light phases respectively. The body-wall chromes are not involved since white individuals respond as effectively as the coloured; moreover the colours of the latter are extremely varied between inidividuals (NORTH, 1960).

For the general dermal sensitivity most of the evidence is strongly in favour of carotenoids (STEVEN, 1963), if not specifically of retinals, as the perceptor chemicals. Recorded action-apectra for taxic responses are mainly in the rhodopsin range (PROSSER and BROWN, p. 373).

That of the siphons of the clam, *Mya*, is at 490 nm (HECHT, 1920) and that of a colonial coelenterate at 474 nm (WALD, 1946–47, 1960). The *Metridium* muscle-perceptor for dark adaptation clearly is in this range. The barnacle *Balanus*, however, is most sensitive in the range 530–545 nm, and the hagfish, *Myxine*, shows much the same value, in the porphyropsin range. *Myxine* at least, might have been expected to have a rhodopsin. In most cases the action-spectrum in detail is very much like those of the eye-retinals, with a single maximum in the visible range. However, the circadian rhythm of *Drosophila* is not affected by deprivation of carotenoids although the sensitivity of the visual perceptors of the eye is depressed by three orders of magnitude (ZIMMERMAN and GOLDSMITH, 1971).

There is further evidence that poryphyrins may be the active chromes in some cases, for instance in asteroids (VEVERS, 1963) holothurians (CROZIER, 1914) and insects (VUILLAUME, 1964, 1966). They are the most probable class in earthworms (MERKER, 1926) and some other annelids, and perhaps also in planarians (p. 87) which, like the earthworms, are very sensitive also to u.v. (MERKER and GILBERT, 1932). Porphyrins have been shown to induce oestrus in ovariectomised mice, presumably via hypothalmic stimulation (KLÜVER, 1944). It is relevant that carotenoids, porphyrins and bilins are the most common classes of chrome in the integument (chapter 5).

It has been suggested that ribitylflavin may have a useful photosensitising function in the holothurian integument (RIMINGTON and KENNEDY, 1962) and similarly in the dorsal integument of Crustacea (BEERSTECHER, 1952). It is believed to act as a definite photoperceptor for photokinesis in plants (GOODWIN, 1959). Anthocyanins are another group of very photosensitive biochromes (HARBORNE, 1964) but there is at present no evidence that they function as photoperceptors in the few animals which have exploited them.

Two main groups of response are mediated by these primitive photoperceptions, overt movements and slower responses such as those of the circadian cycle. Muscular movements are usually protective and either locomotor, as in Protozoa, *Branchiostoma* and the lamprey, or local, as in the withdrawal of the siphons of *Mya* and *Ciona*. Orthodox neuro-muscular paths must be involved. Simple kinetic responses of the water-flea *Daphnia* are mediated by this system, as opposed to the optic system which controls oriented movements (HARRIS and MASON, 1956). The circadian chromomotor change (p. 139) is usually optically initiated but where it is a primary response, of the chromatophores acting as perceptor-effector systems, this may be a clue to the mechanism controlling other components of the circadian programme, for instance, the cycle in the mode of deposition of the insect exoskeleton (NEVILLE, 1967, a, b); this activity is independent of the nervous and hormonal systems.

Those perceptors which are actual neurons presumably tap the normal neuro-conduction systems, though the precise mechanism is an intriguing problem. It is thought (ARVANITAKI and CHALAZONITIS, 1961) that the two chromes in the genital ganglion of *Aplysia* promote electron-transfer in the respiratory chain when they are activated by light and that this chemical, scalar transfer operates a vector change in electrical potential across the neuronal membrane leading to a propagated impulse at threshold change. Two separate action potentials can be evoked, corresponding to stimulation by light respectively of the wavelengths of the two sensitivity maxima (p. 174). That due to the haemoprotein is thought to have excitatory effects on the response system and that of the carotenoid inhibitory. Both chromes are in lipid 'granules' which may imply that they could readily become constituents of membranes. The porphyrin of the spinal cord of homoiotherms (p. 100) may be relevant in this context.

On the mode of operation of the dermal photoperceptors the work of Hecht on the rather specialised system in the siphons of *Mya* remains almost unique. His quantitative validation of the applicability of the Bunsen-Roscoe reciprocity law, $E = I \cdot t$, was restricted to a certain range of minimal light intensities (I) and exposure times (t). If the magnitude of the photochemical effect (E) is measured as the reciprocal of the latent period of the response this is related to a wider range of values of the variables I and t by the relation $E = k \cdot t \log.I$, where k is a constant. This effectively is a form of Lambert and Beer's relation. It is also similar to Sumner's relation between integumental chrome-concentration and albedo (p. 136) and to the Weber-Fechner law for difference-thresholds of sense-organs in general. This last resemblance implies that the relation may be a biological adaptation rather than a simple expression of optical physics. For instance, in the production of the muscular response, which is the 'photochemical effect' measured, there is evidence for two separate events, namely a 'sensitisation' followed by a 'latent period'; these may correspond to photochemical and thermal phases recognised in the response in the retina, – if indeed a complete neural conduction phase is not also included. At natural light-intensities E takes 1–20 seconds to reach threshold; this is within the range for smooth muscle responses initiated by other perceptor organs. The threshold intensity is 0.1–1.0 metre-candles, about 10^4–10^6 times that of the human eye.

7.5 Zoochromes of Other Perceptor Organs

A number of biochromes have been implicated in the chemosensory mechanisms of animals. It is logically necessary to distinguish between chromes acting as primary stimuli for the chemosensory cells and those acting as chemo-electrical transducers, the counterpart of the retinals. Quinols stimulate feeding in the bark beetle, *Scolytus*, (NORRIS, 1970), while the corresponding quinones are depressant. Feeding rate is linear in concentration of quinol, implying that the latter has the stoichiometric action of a primary perceptor agent which in this type of organ, at least, is likely to be chemical. Similarly, 1,4-naphthoquinone inhibits feeding in the cockroach, *Periplaneta*, (NORRIS et al., 1971). Here there is evidence that the quinone binds to -SH groups of a perceptor protein, the resulting redox change causing a change in conformation of the protein and allowing a flow of ions to generate an action potential. There is no indication that the perceptor is a chromoprotein, though this is not ruled out.

There is a high concentration of vitamin A and other carotenoids, astaxanthin and other xanthophylls, in the antennae of the males of the amphipods, *Talitrus* and *Orchestia*, which are believed to contain sense organs perceptive to the secretions of the female (BARBIER et al., 1966). The carotenoids are absent from the antennae of females. DINGLE and LUCY (1965) summarise the evidence that carotenoids are concerned with taste and olfactory perception as well as with photoperception.

The mammalian olfactory epithelium (WRIGHT, 1964) may be yellow, brown or black. The yellow is due to vitamin A and other carotenoids (MILES et al., 1939) and the localisation is too precise to be related to the general role of vitamin A in maintaining the mucous state of such membranes (FELL and MELLANBY, 1953). There is also melanin in the olfactory epithelium, and not elsewhere in the nasal mucosa (LIBAN, 1969).

WRIGHT (1964, p. 155) suggests that odoriferous molecules have a frequency of molecular vibration sufficiently near that of the excited state of the olfactory carotenoid to modify it by resonance and accelerate its return to ground state. This would provide the energy for depolarising the membrane of the sense cell. Molecular vibrational energy seems the only basis for the similar odour of molecules with very different structures and *vice-versa*. The major uncertainty is the source of energy for exciting the carotenoid in the first place, molecular vibration being no more than a trigger; conceivably there is a metabolic presetting system, as for the retinals (p. 163). There may be a range of chromes for the different odour-modalities.

Significantly enough, odorifers are mostly aromatic and, as this adjective itself implies, most unsaturated organic ring compounds do have a strong odour. There is a general and basic correlation between odour and colour, therefore and carotenoids in particular tend to have both properties. Most common odorifers are not significantly coloured, however, and so even their most bathochromic absorption bands would not match those of the olfactory chrome; for these the presetting mechanism is unavoidable.

The 'alarm substance' which is released from the skin of injured fish and alerts other individuals to escape is not after all a pterin (HARE, 1971) as initially concluded (HÜTTEL, 1941). Pheromones in general are not coloured. Chromes are not yet known to be common sensory stimuli, therefore, even for chemosensory organs, but they may be the usual transducers for chemosensory stimulants.

7.6 Conclusion

The photochemical transduction step, and possibly also the chemoelectrical step in animal photo-perceptors is mediated mainly by retinals, haplo-carotenaldehydes, in their thermodynamically most improbable isomeric form. This is stabilised by conjugation with a protein, opsin, forming *retinopsin*. Photoexcitation involves dissociation, isomeric changes in the retinal, and a series of regenerative steps. Most vertebrates and invertebrates have the same retinal but some freshwater poikilothermic vertebrates have a second, related form, with absorption maximum at a longer wavelength, befitting the spectrum perceived.

Vertebrates have differentiated scotopsins functioning in dim light from photopsins adapted to bright daylight. These are usually in the rods and the cones of the retina respectively, and have appropriate absorption maxima. In several phyla the photopsins have further radiated into three minor variants with absorption maxima so spaced as to permit colour vision.

It is possible that ommochromes, pterins, melanins, ribitylflavin and other zoochromes may have some photoperceptor functions. These four classes, and also purines, certainly function in screening and reflecting capacities, valuable in bright and dim light, for discrimination and sensitivity respectively. Probably in all animals there is a general dermal light perception, whether or no discrete eyes also are present. Neither perceptor cells nor perceptor chromes have been certainly recognised but action spectra imply carotenoids and possibly porphyrins.

There are chromes in other sensory epithelia, particularly of the chemical sense organs, with the implication that they amplify the chemical signal. Carotenoids are again the predominant chromes.

IV. Biochemical Functions of Zoochromes

Since they concern molecular properties (p. 129) the biochemical functions of these materials are more fundamental than their biophysical functions and they are more numerous. Most of them are directly related to energy-transfer in activating and mediating reactions involving non-ionic bonds, i.e., redox reactions in the broad sense.

Chapter 8

Oxidation-Reduction Functions of Zoochromes

Even if biological redox reactions are defined in the narrowest sense, as reactions which liberate energy from the standard fuels, they are still the most important group of functions of biochromes in all living organisms. It is still more true in the broader sense that all energy-transfer reactions, whether essentially uphill, as in photosynthesis (Fig. 8.1) and other biosyntheses, or downhill as in respiration (Fig. 8.2), come into this category: any movement of an electron and even a charge-redistribution is an oxidation-reduction. Equally clearly, every mode of activation of the molecule of a biochrome, whether by light or by other forms of energy, is basically redox, whatever the nature of the subsequent transductions of the energy. In Bronsted's broadest of all senses every chemical reaction can be considered redox but, as seen in chapter 3, biochromes are not significantly involved in

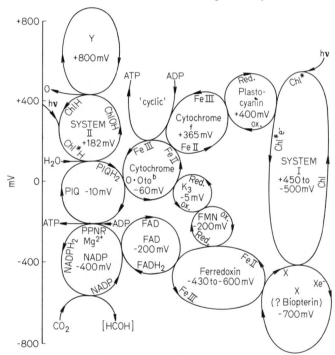

Fig. 8.1. Redox enzyme components of photosynthesis, to show (1) interrelationships of Emerson's two systems and their components, (2) redox potential gradients in the systems (scale on left), (3) resemblances to the respiratory redox system. Chl = chlorophyll; hv = quantum of solar energy; K_3 = Vitamin K_3 (menadione); PlQ = plastoquinone; S, Y = components not yet identified. *indicates activated form

the electrovalent, ionic type of reaction. The coupling of ionic reactions with redox processes still constitutes one of the major problems of biochemistry (MITCHELL, 1966). All coloured materials are potential redox agents and virtually all biochromes which have been adequately studied have been found to undergo redox changes, at least under laboratory conditions.

181

It is an inherent property of such molecules and the leading question about any zoochrome is whether it shows this behaviour under normal conditions *in vivo*. Since living systems have evolved spontaneously (NEEDHAM, 1968c) the probability is that every redox reagent is active in vivo unless there are special devices to prevent this.

A molecule which is excited and then passes on the energy to activate a second molecular species effectively is a catalyst since it reverts to its ground state and can repeat the cycle. The most typical biochromes therefore have become incorporated as prosthetic groups, or co-enzymes, in redox systems. Conceivably some still function as simpler catalysts, in certain contexts.

In photosynthesis (Fig. 8.1) and in those aspects of biosynthesis which occur in all organisms, chromoprotein enzymes catalyse the fixation of energy into various ultimate products; in respiration they catalyse its release from these products as 'substrates', to drive metabolic processes in general. Starting from H_2O and CO_2 photosynthesis establishes a redox potential (E'_o) of up to 1600mV between the ultimate products, oxygen ($E'_o = +820$ mV) and, e.g., glucose ($E'_0 = -800$ mV). In the auto-regulation of biological systems this very large potential cannot be controlled effectively by a single redox enzyme and hence potential-graded series of enzymes are used in both photosynthesis and respiration (Fig. 8.1, 8.2). It is understandable that many of the co-enzyme chromes are common to the two processes and operate roughly in converse order, with appropriate differences in structural detail. Like all catalysts they can speed their reactions in either direction and under appropriate conditions they provide a continuously variable control of the overall rate and of the redox potential gradient.

In their reduced phase a number of zoochromes are colourless (Table 8.1) since the saturation of a single double bond may halve the extent of the conjugated system. A single oxidation-step less frequently 'bleaches' a chrome since this may actually add a double bond in a side chain: $-OH \rightleftharpoons =O$. However, all chromes are bleached by either oxidation or reduction through an adequate number of steps and both methods are used in commercial bleaching. Light is the most potent bleaching agent, as would be anticipated and once more emphasises the availability of photoactivation.

From Table 8.1 it is clear that most single step oxidation-reduction changes in zoochromes involve some change in colour and absorption-spectrum. The change has useful diagnostic value, particularly if the E'_o value of the change is measured. For a number of zoochromes these are virtually the only criteria yet available. In most cases the oxidised form is more bathochromic than the reduced, for the reason given above. A number of biochromes have been used as redox indicators in the laboratory though, as in pH work (p. 77), intensive laboratory research has produced more convenient substitutes.

For a large number of tissue zoochromes, and for some of those in the body fluids, the question of a significant redox function in vivo is still unsettled. It will be considered against the background of those which are already known to function in this way. Here the interest is in their particular contribution rather than in respiratory metabolism as a whole. This is described in detail in most standard biochemical texts and in some more specialised works. It is useful to distinguish two groups of these respiratory chromes, those concerned primarily in energy-release reactions, for the various physiological purposes, and those for which energy-release is more incidental to the particular redox change mediated. The first group work as a closely coupled team (Figs. 8.2) catalysing a universally valid, main series of respiratory reactions and a number of collateral paths, often known as the electron transfer system; the second group are concerned with more parochial sections of metabolism.

Table 8.1. Redox criteria of zoochromes

Zoochrome	Colour and/or absorption peaks		E'_0 and reversibility	Comments	Reference
	Oxidised	Reduced			
Carotenoids					
Astaxanthin	Red	Blue	I	In alkaline medium	Fox, 1953, p. 127
Flavonoids					
Helix: foot secretion	Yellow	Colourless	R	Unstable	Needham (unpublished)
Chromones					
Tocopheroquinones	Red, or Yellow	Colourless (tocopherols)	650–700 mV		Green and McHale, 1965
Quinones	Varied	Varied (simplest are colourless)	R	All redox	Thomson 1957
Ubiquinone	Yellow	Colourless	+ 89 to + 112 mV R		Mitchell and Marrian, 1965; Urban and Klingenberg, 1969
Dopaquinone	Red	Colourless	R		Vuillaume, 1968
1,4 NapththoQs			− 73 to + 71 mV		Mitchell and Marrian, 1965
Echinochrome	Red	Colourless	R		Cannon, 1927
Spinochromes A, C	Yellow	Violet	R		Thompson, 1957; Lederer and Glaser, 1938
Anthraquinones	Yellow	Red	− 266 mV		Morton, 1965
Aphins	Various	Colourless	R		Fox, 1953, p. 311
Zoopurpurin	Blue	Red	? R	Photooxidation	Giese and Grainger, 1970
Lithobius, connective tissue chrome	Violet	Colourless	R	+ 270 mV	Needham, 1960
Arenicochrome	Blue	Yellow	R	Alkaline medium	Van Duijn, 1952
Hallachrome	Red	Colourless	R		Friedheim, 1932, 1933
Pyrroles					
Prodigiosin	Yellow	Red	R	Reduced form is the more bathochromic	Allen, 1967
Helicorubin	Orange-red 534, 572 nm (acid)	Rose pink 532, 563 nm (alkaline)	R	pH change necessary	Keilin, 1957
Cytochrome c	Red-brown 530, 565 nm	Orange-pink 522, 550 nm	R + 260 mV	Not autooxidisable at physiological pH	Keilin, 1925, 1926
Xanthommatin	Yellow 452 nm	Red 485 nm	R	Reduced colour more bathochromic than oxidised	Butenandt et al., 1954
Ichthyopterin	Yellow	Colourless	R	Redox potentals in general low	Polonovski et al., 1946
Other pterins	Yellow to red	Colourless	R		Fox, 1953
Ribitylflavin	Yellow	Colourless	R − 220 mV		Ball, 1939
Unidentified Chromes					
Molpadia, chrome of coelomic corpuscles	Red 569, 612 nm	Red 588 nm	R	Oxygenation -deoxygenation	Crescitelli, 1945

Table 8.1 (continued)

Zoochrome	Colour and/or absorption peaks		E'_0 and reversibility	Comments	Reference
	Oxidised	Reduced			
Calliactis chrome	Purple	Orange	R	Alkaline medium	Fox and Pantin, 1944
Adenochrome	Red 505 nm	Pale yellow	R		Fox and Updegraff, 1944
Vanadochrome	Blue-green	Brown	I	Reversible at low pH, as in vanadocytes	Califano and Boeri, 1950
Bugula, purple chrome	Purple	Yellow	R		Krukenberg, 1880
Chromodoris, tissue and blood chrome	Blue	Yellow	R -102 mV		Pullman and Pullman, 1963, Preisler, 1930
Triops, Cypris, chrome of connective tissue	Green	Colourless	R		Fox, 1955
Ferredoxins	Yellow 390 nm	Paler	R -420 mV		Malkin and Rabinowitz, 1967

R = reversible and I = Irreversible under average physiological conditions.

8.1 The Biochromes of the Standard Respiratory Pathways

The series of enzymes concerned in this pathway are shown in Fig. 8.2 and it is evident that with the possible exception of the pyridinoprotein, all have biochromes in their prosthetic groups. In fact the complete prosthetic group of the pyridino-proteins, nicotine-adenine dinucleotide (NAD), has an absorption-peak in the long-wave u.v. and so is pale yellow; it is in the letter as well as in spirit a biochrome. The prosthetic groups of the series show a bathochromic gradient from this substrate end to the oxygen end so that there is a correlation between bathochromicity and redox potential.

Every dehydrogenation of every respiratory substrate is exploited by this potential-graded series so that at various points the energy is tapped off for storage in ATP. It is worth emphasising that dehydrogenations are effectively the only significant respiratory changes in substrates so that the standard pathway is also a unique pathway. Thus the first step in the oxidation of a fully hydrogenated $-CH_2$-group to the $-CHOH-$ stage involves dehydrogenation followed by simple hydrolysis, as in the substrate sequence succinate $\rightarrow$ fumarate $\rightarrow$ malate: $R-CH_2-CH_2 \xrightarrow[FAD \rightleftarrows FADH_2]{-2H} R-CH=CH- \xrightarrow{+H_2O} R-CH_2-CHOH-$ and this is then oxidised to the $-C=O$ stage by dehydrogenation, as in the conversion of malate to oxaloacetate:

$$R-CH_2-CHOH- \xrightarrow[NAD \rightleftarrows NADH + H]{} R-CH_2-\underset{O}{\overset{\|}{C}}-$$

Further oxidation necessitates removal of any more distal C-atoms, as in the removal of a C_1 unit to form succinate from α-ketoglutarate. A molecule of water is dehydrogenated to provide the $-$ OH group for the now terminal keto group. The removal of the terminal C_1 unit is itself oxidative and constitutes the final reaction of the model series

$$- C{\overset{OH}{\underset{O}{\lessgtr}}} \rightarrow C{\overset{O}{\underset{O}{\lessgtr}}}$$

It is a dehydrogenation of the $-$ OH group donated from H_2O:

$$R - \underset{\underset{O}{\|}}{C} - C{\overset{OH}{\underset{O}{\lessgtr}}} \xrightarrow[FAD \quad FADH_2]{H_2O \quad CO_2} R - \underset{\underset{O}{\|}}{C} - OH$$

As an alternative to the keto structure a terminal aldehyde group $- C{\overset{H}{\underset{O}{\lessgtr}}}$ is at the same level of oxidation. This again is oxidised to $- C{\overset{OH}{\underset{O}{\lessgtr}}}$ by dehydrogenation:

$$R - C{\overset{H}{\underset{O}{\lessgtr}}} \xrightarrow[NAD \quad NADH + H^+]{H_3PO_4} R - \underset{\underset{O}{\|}}{C} - O - \underset{\underset{OH}{\diagdown}}{\overset{\overset{OH}{\diagup}}{P}}{=}O$$

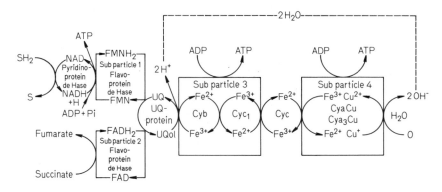

Fig. 8.2. The electron-transport system of enzymes involved in the oxidation of succinate and other substrates (SH_2), showing the usual sequence of action. In the square boxes are the enzymes constituting four sub-particles of the inner wall of the mitochondria; the remainder of the enzymes are free and shuttle between sub-particles. ATP is generated from sub-particles 1, 3, 4

The two oxygen atoms associated with each carbon atom ultimately removed as CO_2 are both supplied in the form of $-$ OH through hydrolysis reactions with H_2O or H_3PO_4 and not through redox reactions proper.

The nitrogen in such substrates as aminoacids is not itself a significant source of respiratory energy while sulphur, the only other relevant element, is quantitatively of minor importance as a fuel substrate notwithstanding the significance of $-$ SH compounds as redox catalysts.

As shown in the reactions above, the initial dehydrogenation of a substrate is by the terminal enzyme of one or other of the two branches of the series, an NAD-protein or an FAD-protein. The hydrogen atoms removed are passed along the series as far as the quinone, UQ, and then released as H^+ ions when the quinol is oxidised back by cytochrome b. At the other end of the series an anionic, reduced form of oxygen is formed and the two ions, H^+ and OH^-, move down their electrochemical gradients to combine and complete the exploitation of the redox potential (Fig. 8.2). All the cytochromes oxidise and reduce by simple electron-transfer between Fe^{2+} and Fe^{3+} states.

Apart from that of UQ, the redox potentials of the series are in logical order for the purpose (EDWARDS and HASSALL, 1971, p. 152), to control the potential:

Component:	Substrate	NAD	FMN	UQ	Cy b	Cy c	Cy a	Cu^{2+}/Cu^+	$\frac{1}{2}O_2/OH_2$
E'_0 (mV):	-800	-320	-210	$+100$	$+40$	$+220$	$+290$	$+350$	$+820$

Apparent anomalies in E'_0 are evident also in the photosynthesis scheme (Fig. 8.1). They imply simply that certain members of the series must be operating some distance from the 50% oxidation level to which the standard E'_0 values refer; more than 50% of UQ must be in the reduced form when the series is in action. In addition the standard values are usually for the free coenzyme *in vitro* and may be considerably different from the values for the holoenzyme, *in vivo*.

The whole system is located in *oxysomes*, the electron-transport particles which form integral parts of the inner membrane of the mitochondria. These particles rather readily break into four subparticles (Fig. 8.2), two containing the alternative forms of the flavoprotein dehydrogenases, the third cytochromes *b* and *c*, and the fourth cytochrome *a*, which is bound to a copper protein (p. 49). Particles 1, 3, and 4 each generate one ATP molecule from ADP for every pair of electrons transported through the system and this is the main source of ATP in respiration. Sub-particle 2 does not itself generate ATP but one is produced at the substrate level, at least in the case of succinic acid (EDWARDS and HASSALL, 1971, p. 173). An important implication is that the steps in the series of reactions and the sites of ATP-generation are spatially organised; in this way living organisms solve the Curie problem that scalar properties such as reaction rates cannot be coupled directly to vector properties such as transport in a specific direction, a coupling mandatory for life. Consequently the traditional mode of representation of reaction-sequences as spatially organised, after all has a functional basis. The NAD, UQ, cytochrome *c* and O_2 itself are mobile components linking the fixed sub-particles and shuttling H or electrons between the latter. Cytochrome *c* is the only water-soluble cytochrome: the apoenzymes of the others are lipoproteins and they are effectively a lipid phase. Hydrophil/hydrophobe segregations play an important part in the whole process.

The system is probably considerably more complex than shown in Fig. 8.2: cytochrome oxidase has two haemoprotein components which have been designated a and a_3 (CAUGHEY, 1967). Their haems may be identical and these may share a single peptide in fact so that they differ only in their immediate ligand fields, for they are within reacting distance of each other. Moreover the associated copper-enzyme also is part of the same molecular complex and one Cu-prosthetic group is fairly closely associated with each haem. Proof that the copper has an active redox cycle is given by the disappearance of an absorption band around 830 nm when cytochrome oxidase is reduced. This peak extends the bathochromic sequence already mentioned (p. 184).

A detailed study of the cytochrome oxidase system implies that in addition to the valency change in its Fe-atom, a_3 may form a complex with O_2 somewhat as the haemoglobins do (p. 228). The indication is that the two haems assume a variety of oxidation-states, electrons flowing through the system by a number of routes. One suggestion (LEMBERG et al., 1968) involves eight distinct reactions:

$$\text{Proximal systems} \rightarrow Cy\ c_1 \rightarrow Cy\ c \begin{matrix} \nearrow a_3 O_2 a \searrow \\ \searrow \begin{Bmatrix} a^{3+} \\ \overset{+}{a}{}^{3+} \end{Bmatrix} \nearrow \end{matrix} \begin{Bmatrix} a^{2+} \\ + \\ \overset{+}{a}_3 O_2 \\ \overset{+2+}{a_3} + a^{2+} \end{Bmatrix} O_2$$

NICHOLLS (1968) suggests an alternative scheme.

Cytochrome b_2, the form of b in bakers'yeast, similarly has hybrid prosthetic groups, again consisting of two pairs, two haems and two flavin nucleotides (MORTON and ARMSTRONG, 1963; HIROMI and STURTEVANT, 1966, 1969). Both types are essential to reduce cytochrome c and they act in the logical sequence, the haem next to cy c. Their mutual spatial organisation is critical and naturally enough this enzyme when dissociated reconstitutes less readily than simpler enzymes. The synergism appears to be relevant to the phosphorylation of ADP at that point. Another hybrid prosthetic group, involving flavin and protohaem IX, is that of L-lactate dehydrogenase: here the flavin may be as the mononucleotide FMN (p. 71) or as FAD (GROUDINSKY et al., 1971). In this type of hybrid flavin and haem interact directly without a UQ intermediary and presumably the value of the association is to permit this simplification.

It is not possible to consider all the details and adaptive radiations in the system. One particular member of the series may exist in multiple forms within the same cell and perhaps even in the same oxysome so that presumably all are functional forms. Cytochrome c is in four such forms in the heart-muscle of the cow (FLATMARK, 1966); they differ in absorption spectra, redox potential, and redox rate and also in structural features. Their functional significance is not yet known. Isozymic and taxonomic variants of course are commonplace.

The general plan of the respiratory pathway also has a number of variants, depending primarily on the physiological and ecological conditions of the organism. The pathway is shortened under anaerobic conditions and cytochrome b or even the flavoprotein may be the terminal enzyme (LASCELLES, 1964; CASTELLO et al., 1966; FLORKIN, 1971). In the last case ribitylflavin must be reoxidised by one of a number of possible intermediary metabolites and the reaction products of the latter usually are squandered. Some microorganisms living at very low O_2-pressures are able to utilise the limited O_2 very effectively with flavoproteins of low redox potential, rather than high-potential haemoproteins, as the terminal enzyme. On the other hand if oxygen is abundant then the high potential gradient between it and ribitylflavin leads to the production of H_2O_2. This is fatal to the normally anaerobic nematode *Ascaris* since the worm has no catalase (p. 193).

The status of the pyridino, flavo and haemo-protein members has been clear for some time and recent interest has centred on the quinones (MORTON, 1965) which indeed may be involved in the coupling with phosphorylation. Ubiquinone (UQ) or Coenzyme Q (CoQ) is the most important, as its name implies, acting in the position indicated in all animals, as well as in gram-negative bacteria, the higher fungi and plants. Naphthoquinones of the vitamin K group replace it in gram-positive bacteria, Actinomycetes and Algae (GAFFRON, 1945). Plants and some bacteria have both: *Escherichia coli* is an example (BRODIE, 1965) and has the vitamin K quinone, MK 9 on the NAD branch and UQ 8 on the direct flavoprotein branch

$$S \longrightarrow \underset{\text{protein}}{\text{NAD}} \longleftrightarrow \underset{\text{protein}_1}{\text{Flavo}} \longrightarrow \text{MK 9} \searrow$$
$$ Cyb_1 \longrightarrow Cya_2 \longrightarrow O_2$$
$$\text{Succinate} \longrightarrow \underset{\text{protein}_2}{\text{Flavo}} \longrightarrow \text{UQ 8} \nearrow$$

The absence of cytochromes c and the reversibility of the NAD $\leftrightarrow$ FP_1 reaction may be noted in passing.

It is possible that both quinones function also in the mammalian system (BRODIE, 1965;

MITCHELL and MARRIAN, 1965; GREEN and BRIERLY, 1965); vitamin K has been found in the mitochondria and rats deficient in the vitamin show defective electron-transport. The commercial analogue synkavit, 2-methyl-1, 4-naphthoquinone, stimulates respiration, while analogues with an antivitamin K action are depressant. They inhibit the oxidation of both succinate and $NADH_2^+$ so that the vitamin is active in both branches. Rats deficient in the vitamin fail to couple phosphorylation normally with the oxidation and this coupling may be the essential role of the NQ; it is itself normally in a phosphorylated state and may be the coenzyme of the phosphorylating enzyme (BRODIE, 1965). It could fill the rôle of the factor postulated in the original phosphorylation hypothesis of LIPMANN (1946). At this point in the system the phosphorylations associated with both sub-particles 1 and 3 might be controlled by the vitamin K-enzyme though in fact it appears to control phosphorylation as a whole (ITO et al., 1970). There is here some parallel to the action of plastoquinone in the photosynthesis system (Fig. 8.1); this quinone effectively controls both cyclic and non-cyclic paths of phosphorylation.

8.1.1 The Molecular Basis of the Action of Respiratory Chromoproteins

Interest in the best known members of the series has reached the molecular level and recent progress on the haemoproteins is recorded in DICKENS and NEIL, 1964; CHANCE et al., 1966; OKUNUKI et al., 1968; KING and KLINGENBERG, 1971. These will be considered later (p. 193). Knowledge of the mode of functioning of the flavoprotein molecule is more complete (SLATER, 1966; PENZER and RADDA, 1967; YAGI, 1969; KAMIN, 1971) and will be illustrated here. The large number of flavoproteins (Table 8.2) justifies this emphasis; they are not all members of the standard series (see p. 184).

Flavoproteins: Ribitylflavin in the reduced form is an outstandingly good electron donor and an equally ready acceptor in the oxidised state (PULLMAN and PULLMAN, 1963). There is an unusually small change in resonance-energy between the two states so that it is a most 'volatile' redox agent. In the redox cycle of this class the isoalloxazine assumes one or other of the three semiquinonoid free radical forms already considered (p. 72). The blue neutral radical is the most common 'active species' (PALMER et al., 1971) but they are readily interconvertible (YAGI, 1971) and the cation reacts with pyridine systems 10^4 times as fast as the neutral radical (PULLMAN and PULLMAN, 1963). The many metal cofactors so characteristic of the flavoproteins bond only to the active free radical, and not to either oxidised or reduced forms, so that no doubt they stabilise the 'comproportionation' of the two to form the quinhydrone (p. 46). It is noteworthy that the anionic red radical $[Fl \cdot R]^-$ does not change colour on forming the metal-bonded cation $[M\ Fl \cdot R]^+$ but this does not imply that the metal has only a stabilising rôle. In the reduction of NO_3 to NH_3 by plants (NICHOLAS, 1957) a different metal is co-factor for each of the series of flavoproteins catalysing successive steps so that the metals probably help to determine the graded redox potentials of that series. In xanthine oxidase, one of the flavoprotein direct oxidases (p. 193), an electron is passed in sequence from the substrate via a molybdenum co-factor, the flavin itself and an iron co-factor to oxygen (BRAY et al., 1966). These transitional metals readily change their oxidation states (HEMMERICH and SPENCE, 1966), here as in so many bio-redox contexts (pp. 50, 228), and it may be noted that ribitylflavin has most affinity for ferrous iron, the most valuable metal of the set.

A long lived triplet excited state, reached as usual via a singlet excited state (p. 12), is the eventual active form (BERENDS et al., 1966; DE KOK et al., 1971). Somewhat paradoxically the lowest singlet excited state of ribitylflavin (RF) is due to $\pi \to \pi^*$ transitions

Table 8.2. Representative flavoprotein enzymes

Enzyme	EC Number	Trivial name	Source	Prosthetic group(s)
Dehydrogenases				
L-Lactate: Cytochrome c oxido-reductase	1.1.2.3.	Lactate deHase	Yeast	2 FMN
D-Lactate: Cytochrome c oxido-reductase	1.1.2.4.	D-Lactate deHase	Yeast	FAD
Pyruvate: Cytochrome b oxido-reductase	1.2.2.2	Pyruvate deHase	Bacteria	4 FAD
Acyl CoA: Cytochrome c oxido-reductase	1.3.2.2	Acyl-CoA deHase	Tissues of mammals	FAD
Succinate: (acceptor) oxido-reductase	1.3.99.1	Succinate deHase	Heart muscle of mammals	FAD
$NADH_2$: Cytochrome c oxido-reductase	1.6.2.1	$NADH_2$: Cytochrome c oxidoreductase	Heart muscle of mammals	FMN
$NADH_2$: Cytochrome b_5 oxido-reductase	1.6.2.2	$NADH_2$: Cytochrome b_5 oxido-reductase	Liver of mammals	FAD
$NAD(P)H_2$: glutathione oxido-reductase	1.6.4.2	Glutathione reductase	Tissues of mammals	FAD
$NADH_2$: lipoamide oxido-reductase	1.6.4.3	Lipoamide deHase	Heart muscle of mammals	2 FAD
$NAD(P)H_2$: 2 methyl-1,4-naphthoquinone oxido-reductase	1.6.5.2	Menadione reductase	Liver of mammals	FAD
$NAD(P)H_2$: ubiquinone oxido-reductase	1.6.5.3	Ubiquinone reductase	Heart muscle	FAD
$NADPH_2$ (acceptor) oxido-reductase	1.6.99.1	$NADPH_2$ diaphorase (old yellow enzyme) {Plants, Yeast}		1 FAD, 2 FMN
$NADH_2$: nitrate oxido-reductase	1.6.6.1	$NADH_2$: nitrate reductase	Plants, bacteria	FAD
$NAD(P)H_2$: nitrite oxido-reductase	1.6.6.4	Nitrite reductase	Fungi	FAD
Ammonia: (acceptor) oxido-reductase	1.7.99.1	NH_2OH reductase	Bacteria	FAD
Oxidases				
L-Lactate: O_2 oxido-reductase (decarboxylating)	1.1.3.2	Lactate oxidase	Bacteria	2 FMN
β-D-glucose: O_2 oxido-reductase	1.1.3.4	Glucose oxidase (notatin)	Fungi	2 FAD
Aldehyde: O_2 oxido-reductase	1.2.3.1	Aldehyde oxidase	Liver (mammal)	FAD
L-amino acid: O_2 oxido-reductase (deaminating)	1.4.3.2	L-amino acid oxidase {Tissues (vertebrates), Venom (snake)}		2 FMN, 1 FAD
D-amino acid: O_2 oxido-reductase	1.4.3.3	D-amino acid oxidase	Kidney (mammal)	FAD
L-4,5-diH-orotate: O_2 oxido-reductase	1.3.3.1	diH-orotate deHase	Bacteria	FAD + FMN
Xanthine: O_2 oxido-reductase	1.2.3.2	Xanthine oxidase	Milk	2 FAD
Pyridoxamine-P: O_2 oxido-reductase (deaminating)	1.4.3.5	Pyridoxamine-P oxidase	Liver (mammal)	FMN

Table 8.2 (continued)

Enzyme	EC Number	Trivial name	Source	Prosthetic group(s)
Mono-oxygenases				
L-Lactate: O_2 oxido-reductase (decarboxylating)	1.1.3.2	Lactate oxidase	*Mycobacterium*	2 FMN
—	—	L-lysine oxygenase	*Pseudomonas*	FAD
—	—	Imidazole decyclase	*Pseudomonas*	FAD
—	—	*p*-HO-benzoate hydroxylase	*Pseudomonas*	FAD
—	—	D-camphor lactonisase	*Pseudomonas*	FMN
Desaturases				
Acyl-CoA: cytochrome *c* oxidoreductases	1.3.2.2	Fatty acid desaturases	{Micro-organisms / Liver-microsomes}	FAD or FMN
—	—	Cyclane desaturase: Aromatising flavoprotein	Liver (guinea pig)	FAD
3-ketosteroid: (acceptor) Δ^4- oxido-reductase	1.3.99.6	3-ketosteroid-Δ^4-deHase	*Pseudomonas*	FMN
Other Redox Enzymes				
$NADPH_2$ ferredoxin oxido-reductase	—	Photosynthetic phosphorylation flavoprotein	Plants	FAD
—	—	Flavodoxin	*Clostridium*	FMN
—	—	Phytoflavin Ferredoxin deputising (flavoproteins)	Algae	—
Other Enzymes				
Mandelonitrile-benzaldehyde lyase	4.1.2.10	Hydroxynitrile lyase	Bitter almonds	FAD
—	—	Acetohydroxyacid synthetase	Tissues (mammals)	FAD

(Based largely on Dixon and Webb, 1964)

and is responsible for the 445 nm absorption-peak (p. 72), requiring less energy than the lowest n → π* transition which produces the 340–371 nm peak (Penzer and Radda, 1967; Song, 1971). The paradox may depend on a 'mixing' of the two transitions (Kok et al., 1971). This is not the only paradox to the chemist; the reduced form although diamagnetic will activate O_2 (Hemmerich et al., 1971). It is analogous in this respect to haemocyanin among the oxygen-carrier chromes (p. 235). Again, although the reduced form 1,5-di H.RF has the conjugated double bonding of the system broken across the middle it is the most stable of the di H-isomers.

Oxidised RF is paramagnetic like the free radical intermediaries. Its ring-system is planar whereas in the reduced form it is somewhat folded about an axis formed by the N_5 and N_{10} atoms (Fig. 3.9) of the pyrazine ring (Kierkegaard et al., 1971). In FAD-enzymes this affects the relation of the isoalloxazine to the adenine ring system of the other nucleotide, in the oxidised state the two being stacked with their planes parallel. This stacking quenches the fluorescence by 90% as compared with FMN and presumably by a smaller but significant amount as compared with reduced FAD. The energy involved probably does useful work in some phase of the redox cycle. Thus although the adenylic acid itself does not change during the cycle it is essential for the normal activity of FAD (VOET and RICH, 1971). There are extensive regions of charge-complementarity between the two ring systems and they form an ideal donor-acceptor pair for charge transfer (PULLMAN and PULLMAN, 1963; PENZER and RADDA, 1967); the transfer has been detected experimentally (SONG, 1971), and the broad absorption band responsible for the colour of the blue radical is due to this (MATHEWS and MASSEY, 1971).

Water also plays an essential part in the formation of this dinucleotide complex (WEBER, 1966) and this is further relevant to the peculiar hydrophil-hydrophobe relations of the whole respiratory chain (p. 186). In aqueous medium FMN will complex also with the pyridine ring and it is noteworthy that in many organisms the flavoprotein of sub-particle 1 (Fig. 8.2) has FMN as coenzyme and not FAD. The NAD coenzyme complex has great geometrical similarity to that of FAD (SZENT-GYORGYI, 1960; VOET and RICH, 1971) again emphasising (p. 184) that in spirit it is one of the redox chromes.

The ribityl side-chain interacts with the isoalloxazine nucleus in the course of the cycle (TOLLIN, 1971) and the apoenzyme is well known to contribute to the process in several respects. Probably every component of the flavoprotein, as well as the metal cofactor, therefore is essential for its normal behaviour; this is already known to be true for haemoglobin (p. 228) and is likely to prove a general phenomenon. It is not surprising that the resultant behaviour sometimes seems anomalous to the chemist and paradoxical to the verge of the miraculous. Free flavin will oxidise free di-H-pyridine nucleotides even in the absence of enzymic adjuvants (METZLER, 1966) so that the rest is all evolutionary sophistry.

The general functions of apoproteins in controlling stability and solubility and in modifying the colour and basic energy-transfer properties of their chrome-conjugants have been considered (p. 29). The active free radicals of RF are specially stabilised by strong H-bridges from the protein to N_5 of the isoalloxazine nucleus and this favours the comproportionation reaction:
$$Fl_{ox} + Fl_{red} \rightarrow 2\ Fl\cdot H$$
(MÜLLER et al., 1971); this gives kinetic stability. The protein carries a positive charge and this facilitates the interconversion between red and blue free radicals (LUDWIG et al., 1971).

Something is known also of the mode of interaction of flavoproteins with such neighbours as the pyridinoproteins and haemoproteins. As already indicated the coenzyme interacts directly with that of the pyridinoproteins. Cytochrome c on the other hand, in horse heart-muscle, has a 'tunnel' in the molecule which could accommodate the isoalloxazine ring system, which actually protrudes from the surface of the molecule of reduced flavoprotein; this is probably because of the fold across N_5-N_{10}, so that it does not protrude from that of the oxidised enzyme (VOET and RICH, 1971). The tunnel is lined by aromatic aminoacids with planar ring-systems which readily form charge-transfer complexes with the isoalloxazine. This would seem to be even more relevant to cytochrome b, in the hybrid enzymes mentioned above (p. 187).

The story is fragmentary compared with that for haemoglobin (p. 231) but already instructive and stimulating.

Cytochromes: Because it is soluble cytochrome c (Cy c) is best known in most respects (DICKERSON, 1972). Its Fe is low spin (p. 228) in both reduced (Fe^{II}) and oxidised (Fe^{III}) states (LUMRY, 1971) and the enthalpy change between the two is 18 kcal compared with 8 for haemoglobin (p. 228). A free radical is formed during the cycle (WILLIAMS, 1966b), as in the ribitylflavin coenzymes (p. 190) and other haemoproteins (p. 194). The absorption-spectrum shows that the haem is not exposed on the surface of the protein and in fact it lies in a cleft, which is closed in the reduced state (DICKERSON, 1972) by a phenylalanine (phe) residue. It is noteworthy that the other three aromatic amino acids all play similarly important roles in the oxygen-transporting chromoproteins (p. 236). In oxidised Cy *c* the phe residue is in an energetically unfavourable environment, another of the many 'improbable' structures in chromoproteins (p. 164) and there must have been strong counter-selection in favour of its role in reduced Cy *c*. The suggestion is that the withdrawal of the phe residue allows the e-transfer group of cytochrome oxidase to enter (DICKERSON, 1972). Conceivably, the reduced isoalloxazine nucleus (p. 191) or whichever enzyme lies reductad to Cy *c* restores the phe residue by forming a charge-transfer complex with it and restoring an electron to the haem.

Opinions about the amount of conformational change in the protein differ greatly (DIKKERSON, 1972; LUMRY, 1971; HARBURY et al., 1966). DICKERSON compares it to a 'two stroke molecular engine' and Lumry emphasises mechanical forces but electrical (electronic) flows. Cytochrome *a* is certainly under molecular strain, judging from its split soret (γ) peak and this promotes a piezoelectric flow of electrons (WILLIAMS, 1966b). The solubility of Cy *c* is reflected in a preponderance of the polar amino acid residues on the outside of the molecule, by no means the most common arrangement in globular proteins. A lipid medium is best for electron transfer (WILLIAMS, 1966b) and the other cytochromes are effectively in such a medium.

A very large number of variants of Cy *c* are now known (KAMEN et al., 1971) both structurally and functionally. Some have more than one haem group and some can be directly oxidised by O_2. The cysteine bonding to side-chains of the haem is not universal but the E'_o of + 250 mV and the absorption-spectrum, 550 (α), 520 (β) and 415 (γ), are consistent.

By acetylating one methionine and one histidine residue of apocytochrome *c* its molecular shape is changed and it acquires the essential properties of haemoglobin (SCHEJTER, 1966). The molecular cleft enlarges sufficiently to take an O_2 molecule, which it binds reversibly; the oxy-complex has the absorption spectrum of oxy-Hb but it is much less stable and the Fe oxidises to Fe^{III} in 10–15 minutes at 20°C. Even in this modified form the Fe of Cy *c* is in a high spin state only below pH 4·7, but cytochrome P 450 normally has high spin Fe^{II} (KELLER et al., 1972) so that there is no universal difference between Hbs and Cys in this respect. Space does not permit further consideration of the other cytochromes.

8.2 Other Redox Reactions Catalysed by Chromoprotein Enzymes

The first criterion (p. 182) distinguishing enzymes of this category, independence of the standard respiratory system, is not easily applied. Most biological materials can serve as fuel and most are oxidised or reduced by enzymes which could have direct or indirect connections with the main electron-transfer system. There is increasing evidence that all may connect in this way so that there is effectively a single intermeshing system. Catalase (p. 193) is a good example of a redox enzyme peripheral to the main system but not independent of it.

The second criterion, that the primary purpose of the reaction is its product(s) rather than energy-release is more easily applied, provided the physiological significance of the product is known. Even so the reaction releases or binds energy and this ultimately feeds into or comes from the main respiratory system. Many of the steps in biosynthesis and in biodegradation reactions are catalysed by pyridino- and flavo-proteins of the main system (KIT, 1960).

In this section consideration will be confined to chromoprotein enzymes of classes already familiar in the main system, flavoproteins, haemoproteins, simple iron-proteins and copper-proteins. Some others are more conveniently dealt with in section 8.3.

Flavoproteins: Of the enzymes listed in Table 8.2 the dehydrogenases must all be considered members of the main respiratory system but nitrate, nitrite and NH_2OH reductases form a biosynthetic series (NICHOLAS, 1957) in plants. Like most of the enzymes of the photosynthesis system (Fig. 8.1) they are then working "in reverse" and it will be recalled that portions of the Krebs' cycle, and other respiratory substrate pathways can be switched to the synthesis-direction when necessary.

The other types of flavoprotein are more relevant here, mostly involving direct oxidation of the enzyme and dealing with substrates which have a known non-respiratory significance. Examples are orotate oxidoreductase, xanthine oxidase, pyridoxamine oxidoreductase, imidazole decyclase, ketosteroid oxidoreductase, and the aromatising flavoprotein. The best known, xanthine oxidase, proves to be involved in pterin biosynthesis (p. 292) as well as in purine excretory metabolism. At the same time it can be coupled with the reduction of cytochrome *c* (BRADY et al., 1971) and so connects with the electron transfer system. Again, a number of these enzymes dehydrogenate in series with a pyridinoprotein (KATAGIRI and TAKAMORI, 1971) and so are simply 'diaphorases', which can be oxidised directly by O_2. It may be noted that this direct oxidation is usually slower than the normal respiratory oxidation via UQ and the main path (MASSEY et al., 1971), and may be an adaptation to special circumstances like the respiratory flavoproteins of some anaerobic organisms (p. 187). As in that situation H_2O_2 is a common product and an $\cdot O_2^-$ free radical is intermediary (MASSEY et al., 1971; BRAY, 1971). It is a 'playing with fire' reaction in many cases.

Haemoproteins: The relevant members are catalase, which destroys the H_2O_2 formed by direct oxidation of flavoproteins and by some other reactions, and the peroxidases, which use H_2O_2 to oxidise organic substrates. Catalase (DEISSEROTH and DOUNCE, 1970) is very specific to its substrate H_2O_2 and, fortunately has the distinction of being one of the most efficient enzymes, catalysing the decomposition of $4 \cdot 2 \times 10^4$ molecules of H_2O_2 per enzyme molecule per second even at $0°C$ (BALDWIN, 1947, p. 168). One special site of action is inside the erythrocytes of vertebrate blood; here it prevents the oxidation of Hb to met-Hb by any H_2O_2 produced. It is stored in special granules called peroxisomes or glyoxysomes. It also helps in protection against ionising radiation since H_2O_2 is a main product of the irradiation of tissues (COLLINSON et al., 1950). It is of course a chemical defence chrome (p. 158).

Peroxidases are widespread in plants and have been found in milk and other animal-materials. The best known are two which are specific respectively to cytochrome *c* and cytochrome oxidase, as substrates, and horse-radish peroxidase on which much of the most analytical work has been done. Cytochrome peroxidase has protohaem IX as prosthetic group and so is quite closely related to the *b* and *c* cytochromes themselves; in fact all haem derivatives have some peroxidase activity (BALDWIN, 1947, p. 168). These peroxidases could be regarded as mediating an alternative terminal respiratory pathway, using H_2O_2 instead of O_2. The substrates cycle between Fe^{II} and Fe^{III} as in the normal pathway.

In lactoperoxidase the haem is mesohaem IX so that this may have originated quite independently of the other peroxidases and the main respiratory path. It may play a part in halogenation-reactions (CAUGHEY, 1967), a rare and specialised biochemical type.

The molecule of the peroxidases differs from that of the cytochromes in two important respects, a high carbohydrate content (MORRISON et al., 1966) and a range of particularly high oxidation states. Fe^{III} is the lowest and there may be up to Fe^{VI} at one phase of the cycle. In horse-radish peroxidase (YONETANI et al., 1966) the inactive Fe^{III} state is brown and in the course of its activity-cycle it passes through a series of coloured complexes, probably free radicals of varied oxidation-state (Fig. 8.3). The first, not detected in cytochrome peroxidase, is a green Fe^{III} complex and is followed by a very stable red Fe^{IV} radical (ES). This is converted to an unstable red Fe^{IV} complex which decomposes spontaneously. The whole phenomenon has considerable parallel to that for the flavoproteins (p. 188).

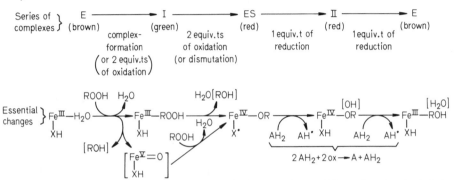

Fig. 8.3. Coloured intermediary complexes formed (above), and essential reactions involved (below) during the course of a complete redox cycle of peroxidases and catalases. Two possible alternative pathways between E and ES are indicated. The oxidation-state of iron is never lower than Fe^{III} and may reach Fe^V. For horseradish peroxidase I and II but not ES have been demonstrated, for cytochrome c-peroxidase only ES and for catalase only I. A = substrate oxidised by the peroxide ROOH; X = operative ligand of the haem-Fe; ˙ indicates a free radical. (After YONETANI et al., 1966)

The interpretation of the intermediaries and of the reaction steps given here (Fig. 8.3) is not universally accepted (EDWARDS and HASSALL, 1971, p. 96) but it has the virtue of generalisation to peroxides (ROOH) other than H_2O_2. It is generally thought that the peroxidase first bonds with the reactant H_2O_2 when its four absorption bands are replaced by two new bands (Keilin's classical method of demonstrating enzyme-substrate complex-formation). The original spectrum returns on adding an O_2-acceptor (A) so that the enzyme is released as the H_2O_2 is decomposed by the second substrate:

Peroxidase – H_2O_2 + A → Peroxidase + AO + H_2O

Peroxidases and catalases show considerable distortion of the molecule (WILLIAMS, R.J.P., 1966a, 1968), the spectrum coming to resemble that of the bilatrienes at one stage; it is recalled that to produce the open chain bilins the Fe must be removed and it is not sufficient merely to open the ring.

Ferredoxins: Ferredoxins play important ancillary roles in the main electron-transfer system (Fig. 8.2), one being associated with each of the sub-particles 1, 2 and 3 in the mitochondria of mammalian heart-muscle (EDWARDS and HASSALL, 1971), i.e., all but the cytochrome oxidase subparticle. One ferredoxin is an enzyme of the photosynthesis system (Fig. 8.1) and another of the nitrate-reduction system in plants. Here an association with flavoproteins is indicated and in fact most of the latter appear to be coupled to a ferredoxin, for instance orotic oxidoreductase, xanthine oxidase, aldehyde oxidase, and the flavoproteins of the main system (MALKIN and RABINOWITZ, 1967).

Adrenodoxin of the adrenal cortex is concerned in hydroxylating the hormone deoxycorticosterone, while another ferredoxin is involved in pyruvate decarboxylation. These are less directly redox in function but that is essentially the function of the class. The iron-sulphur stoichiometry in the molecule is one important indication of this. Indeed they constitute a very important class of redox enzymes (MORTENSEN et al., 1962; SAN PIETRO, 1965; MALKIN and RABINOWITZ, 1967).

Collectively they have a very wide range of redox-potentials:

	mV
Rubredoxin of *Chromatium*	+ 350
Ferredoxin of respiratory subparticle 3	+ 220
Adrenodoxin	+ 150
Rubredoxin of *Clostridium*	− 57
Chloroplast ferredoxin	− 450

so that potentially they are as versatile as the haemoproteins. They have surprisingly restricted aminoacid compositions however which leads to phylogenetic speculations (HALL, CAMMICK and RAO, 1971).

Copper Proteins: The large number of copper proteins now known (Table 3.8) probably all are redox enzymes and most are direct oxidases (PEISACH et al., 1966; WOOD, 1968; GHIRETTI, 1968). Most are peripheral to, or quite independent of, the main respiratory path. Cytochrome oxidase therefore is peculiar in this respect but in any case is a haemoprotein conjugate; there is one copper protein, plastocyanin, also in the photosynthesis system (Fig. 8.1). Some Cu-proteins bind the oxygen so that haemocyanin (p. 235) can be regarded as a specialised member of the class and in fact it retains phenolase, catalase and other activities characteristic of the Cu-protein enzymes (GHIRETTI, 1966). It is noteworthy that these actions overlap those of certain haemoproteins and that haemoglobin also retains some catalase activity. Most Cu-protein enzymes give EPR signals indicating free radical intermediaries like those of the flavoproteins and the peroxidases.

The best known Cu-protein enzyme is o-diphenol: O_2 oxidoreductase, usually called polyphenol oxidase or tyrosinase. It is the key enzyme for the syntheses of the phenolic and indolic melanins (p. 288) as well as the large number of other o-benzoquinone derivatives of arthropods and other taxa. It has four Cu atoms per molecule and two active sites, which may help to explain its remarkable ability to catalyse both the conversion of monophenols to ortho-diphenols and the oxidation of the latter to quinones (BROOKS and DAWSON, 1966). The requirements of the two sites prevent either from complexing with the substrate (KERTESZ, 1966) but the product quinone does bind, and inactivates the enzyme by ejecting the Cu; this inactivation of a terminal enzyme can serve a useful purpose. The enzyme also has the distinction of catalysing the oxidation of substrates such as NADH and ascorbic acid, through coupling with the reversible reduction of the product-quinone back to diphenol. Further it will bind carbon monoxide and presumably also O_2 with the stoichiometry of 2Cu: 1 CO, just as in haemocyanin.

Ascorbic oxidase converts vitamin C to its diketone form, somewhat analogous to the quinone structure, using atmospheric oxygen as oxidant. It has eight Cu atoms per molecule, six of which are in the Cu II state, and only these six actually function in the reaction.

A blue copper protein, ceruloplasmin, in solution in the blood plasma of mammals has become a recent centre of interest since the often fatal Wilson's disease may be due to its deficiency (SCHEINBERG, 1966). Its normal function is still unknown, however: like so many copper proteins it can do many things when tested experimentally. It catalyses the oxidation of Fe^{2+} to Fe^{3+} (ferroxidase) and so may promote Hb-synthesis indirectly through the prior synthesis of transferrin (p. 51) the transport protein for Fe (FRIEDEN et al., 1972). It has long been known that copper is essential for normal Hb synthesis. It will also catalyse the oxidation of polyphenols, polyamines (such as epinephrine), serotonin and ascorbate (SCHEINBERG, 1966; OSAKI et al., 1966). Thus its potentiality overlaps the functions of tyrosinase and other Cu-proteins. It has some resemblance also to Hcy (p. 224) which conceivably evolved from a pre-existing plasma Cu-protein: both have half their Cu atoms in each valency state though the Cu^{II} of ceruloplasmin is paramagnetic and there is no colour-change during its cycle of activity. There are also other differences (OSAKI et al., 1968).

More is known about the molecular physiology of this than of most copper proteins, perhaps not excepting even Hcy itself (p. 235). Four of the eight Cu atoms of the molecule are in close proximity and react in concert, being affected in all-or-none fashion by relevant reagents (OSAKI et al., 1966). They remain in the Cu^+ state while they manipulate the oxidant, O_2 itself, whereas the substrate is handled by any one of the other four Cu atoms, in the Cu^{2+} state; they are widely separated from each other and from the O_2-quartette and yet themselves behave as a second quartette. These and other facts led Blumberg (OSAKI et al., 1966) to propose the model shown shown in Fig. 8.4. The $SH-H_2O$ bridges between outer and inner Cu-atoms are thought to act as signalling devices between quartettes, substrate and oxidant. The spatial separations and provision for competitive interactions between Cu atoms may underlie the paradox that Cu-proteins are good enzymes whereas most copper complexes are much too stable for this.

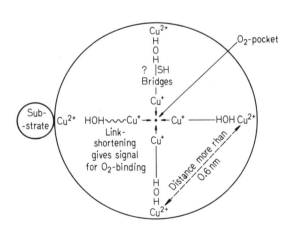

Fig. 8.4. Model to explain structure and behaviour of essential components of the copper-protein ceruloplasmin in oxidising a substrate. The eight Cu atoms behave as two quartettes, one forming a compact group in the Cu^I state, enclosing a pocket capable of taking and binding a molecule of O_2. The other four are more widely separated from each other and from the first quartette and are essentially in the Cu^{II} state. Each is bonded via a ligand H_2O molecule indirectly (? via cysteine residues) to a Cu^I-atom. Substrate binds with one Cu^{II} atom displacing the H_2O ligand which therefore transmits a signal to the Cu^I group to bind O_2. This will now oxidise the substrate. (After BLUMBERG, 1966)

The copper proteins of various mammalian tissues, erythrocuprein, haemocuprein, hepatocuprein and cerebrocuprein do not behave as enzymes and may be stored or scavenged forms of copper (WOOD, 1968) corresponding to haemosiderin; there is some indication that Wilson's disease is due less to deficiency of ceruloplasmin than to excess free copper (SCHEINBERG, 1966). Newborn infants may have as much as 3% of copper in the copper proteins of their liver mitochondria, non-stoichiometric but apparently innocuous and desirable; at that age the liver is still a haemopoietic organ. These tissue Cu-proteins have in common an absorption peak at 280nm, indicative of aromatic aminoacids, no doubt as ligands.

These enzymes vary considerably in their oxidation-state behaviour (Table 3.8). Tyrosinase has Cu^I throughout its activity-cycle so that strictly speaking it is not a chrome at all! Most are blue when oxidised and colourless when reduced, like typical cupric and cuprous salts but, as already noted, ceruloplasmin is blue throughout the cycle. The amine-oxidases cycle between pink and colourless; this unusual colour may be due to their being complexed with a second prosthetic group, flavin or pyridoxal (YARA and YASUNOBU, 1966). Other properties show a similar variety and the class has adaptively radiated very much as the ferredoxins and haemoproteins.

8.3 Redox Functions of Other Zoochromes

Carotenoids: Among carotenoids the retinals (p. 163) undergo oxidation-reduction changes ancillary to their activity cycle (p. 164) and the class as a whole is very sensitive to oxidation. In many of them this tends to be irreversible and PALMER (1922) thought that this ruled out any physiological redox actions for the class as a whole. However, it has since been shown (MOORE, 1945) that a number of carotenoids act as antioxidants (p. 160), protecting other metabolites from oxidation by mopping up the relevant oxidants: here the irreversibility of carotenoid oxidation should be an advantage. In some situations other antioxidants, such as vitamin E, 'spare' vitamin A, this function (HEMMING and PENNOCK, 1965) probably depending on the relative availability of the two. The essential function of the carotenoids of the chloroplasts again may be antioxidant, protecting against any photodynamic effects of the chlorophylls.

In principle the molecular structure of the carotenoids makes them ideal electron donors and acceptors (PULLMAN and PULLMAN, 1963; CHEESMAN et al., 1967) so that, if appropriately coupled, they should be able to operate complete redox cycles, reversibly. The carotenoprotein of spinach chloroplasts will catalyse the photoreduction of cytochrome (KAHN and CHANG, 1965). There is evidence that carotenoids catalyse the dehydrogenation of Δ^5 – 3β – HO steroids, so that they themselves are reduced, and there is the same implication in the catalysis of oxidation reactions by chromodorin (Table 8.1), which may be a carotenoid (PULLMAN and PULLMAN, 1963). However, retinol/retinal remains the only clearly demonstrated regular physiological cycle *in vivo*. A reversible reduction may be suspected for the carotenoid in salmon muscle (p. 103); here it is noteworthy that there is a rather general inverse relation between the amounts of carotenoids and haemoglobin in animals (VERNE, 1926, p. 409).

Fuscins: The fuscin types of chrome may have in common at least an origin from colourless precursors by oxidation; this is indicated by the fact that vitamin E prevents their accumulation in the rat's adrenal (WEGLICKI et al., 1968) and in nerve tissue (BJORKERUD and CUMMINS, 1963). Actually O_2 was found to depress the accumulation of fuscin in nerve cells (VERNE, 1926, p. 410), and CO_2 increased it, but again speculation is pointless until the nature of the material is known. The process of fuscin-accumulation in vitamin E-deficient, and in some normal, conditions often appears to be irreversible but in most cases this is merely a time-trend observation and does not rule out the possibility of a regular redox function which simply expands with age.

At least in the 'brown fat' of the hibernating gland and other regions of the mammalian body the fuscin does seem to be directly concerned with respiration. The O_2-consumption of this tissue is higher than that of white fat and indeed is as high as in kidney tissue (FAWCETT, 1947). Moreover the accumulation of this chrome certainly is reversible; it is most abundant in young animals and often disappears in the adult. The flavoproteins and the terminal cytochrome system also are present in quantity (PRUSINER et al., 1970) and the tissue has a very dense mitochondrial population (AFZELIUS, 1970). There is also lipase, virtually absent from white storage fat, and brown fat is much more rapidly metabolised (NAPOLITANO et al., 1965): its function appears to be simply to generate heat by its oxidation (SMITH et al., 1966; LINDBERG, 1970). This is a purely biochemical mechanism of heat-production, therefore, as contrasted with such physiological methods as shivering, – but probably equally adventitious. It appears to be limited to mammals: presumably birds are adequately insulated by feathers and by the restrictions on heat-loss from legs and other exposed parts.

The evidence for this thermogenic function is very strong. The tissue is best developed in small mammals, particularly those which hibernate, in which it often constitutes a hibernation 'gland' in the shoulder region (SMITH, 1962). In the larger mammals, even including the poorly insulated human species, it is well developed only in the young, which have a large surface/volume ratio and an immature physiological thermoregulatory mechanism, and it atrophies progressively during ontogenesis (AFZELIUS, 1970). That process is accompanied by a decrease in the number of mitochondria per cell. The tissue is particularly well vascularised, i.e., active, in cold environmental conditions and in emaciated individuals (AFZELIUS, 1970). It hypertrophies in individuals with deficient thyroid-activity and consequent low metabolic rate (HULL and HARDMAN, 1970).

LEPKOVSKY et al. (1959) and SMITH (1962) made the important discovery that the fat-oxidation in the tissue could be uncoupled from phosphorylation, making the process purely thermogenic in function and this has been confirmed (HITTELMAN and LINDBERG, 1970). However, if the mitochondria from the tissue are defatted with clean albumin they show *in vitro* a normal oxidative phosphorylation (HOHORST and STRATMANN, 1967; ALDRIDGE and STREET, 1968; GUILLORY and RACKER, 1968). Long-chain fatty acids are known to uncouple phosphorylation from oxidation (PRESSMAN and LARDY, 1956) and so it appears that the lipid itself may turn on the thermogenic uncoupling; the resting gland in fact has normal coupling. Systemic control involves the sympathetic nervous system, and catecholamines have been found to initiate the decoupled metabolism (HULL and HARDMAN, 1970), just as they promote lipolysis in normal fat tissue (REED and FAIN, 1970). They act in the way which appears to be virtually universal for hormones (p. 150), via the cyclic 3′, 5′ – AMP system which here is synthesised by a membrane-bound enzyme and directly activates the lipase. It is interesting that carnitine induces the coupled state (HITTELMAN and LINDBERG, 1970) since this implies that it may have the same significance in muscle, also.

Fuscin granules from some other tissues have a considerable endogenous O_2-uptake (BJORKERAD and CUMMINS, 1963) and as early as 1930 SCHREIBER considered that the fuscin of the nerve cells of *Sipunculus* and some gastropods served a respiratory function. It remains to be demonstrated that the chrome of brown fat plays an essential part in the thermogenic process but this would be very probable if it is a cytochrome or other haemoprotein, as now suggested (PRUSINER et al., 1970). The action could be quite orthodox.

Flavones: While the yellow secretion of the foot and mantle of *Helix* (p. 82) has typical flavonoid redox properties (KUBISTA, 1950) there is as yet no positive evidence that any flavonoid in animals has a physiological function based on this. The rarity of this class of biochrome in animals indicates that even if they have such a function it cannot have any general importance. Conceivably the *Helix* chrome plays a part in the hardening of the epiphragm, a secretion used to seal the mouth of the shell, though it certainly is also secreted when the animal is irritated, and so may be chemically defensive. Both actions could depend on redox changes.

Chromones: For the simpler chromone derivative, α-tocopherol (vitamin E), there is good evidence of a redox type of function, as already seen in relation to vitamin A and the fuscins. It is present in mitochondria in about the same molarity as the other chromes of these organelles and oxidises to tocopheroquinone (p. 37). The semiquinone intermediary in this change also has been detected so that it well may be active in a loop of the main respiratory pathway (GREEN and MCHALE, 1965), as an alternative quinol/quinone system. In chloroplasts, also, the α-tocopheroquinone is present, along with vitamin K, which also assumes a chromane form at one phase of its redox cycle there; α-tocopherol itself may have an active role, therefore. It also may protect intermediaries in the pathway of synthesis of ubiquinone (FOLKERS, 1949); here it is associated with selenium compounds having similar redox properties. Tocopherylphosphate controls the rate of oxidation in muscle, which becomes dystrophic through excessive combustion in states of vitamin E deficiency (HOUCHIN, 1942). There is new evidence on the redox role of α-tocopherol, particularly in relation to membrane structure and probably to transport across membranes (LEWIN, 1971).

Quinones: The innate power, reversibility and versatility of quinones (Qs) as redox agents has been adequately stressed (p. 46) and a number of biochemical instances have been noted already. Collectively the biological Qs cover a wide redox range, since their E_0' falls progressively with the number of condensed rings, from $+534$ mV for the benzoquinones to -226 mV for the anthraquinones (MORTON, 1965). Nevertheless, only the ubiquinones, the vitamin K group and plastoquinone have yet been proved to function as regular redox agents *in vivo*. There is provisional evidence in other cases: thus dopa-quinone, essentially an intermediary in the pathway of synthesis of the indolic melanins (p. 288) will act as a reversible redox agent for the ommochrome biosynthesis pathway, at least in laboratory systems (BUTENANDT et al., 1956; VUILLAUME, 1968). *In vivo* it might in this way control the ratio of melanins to ommochromes synthesised by insects.

From the use of benzoQ derivatives in chemical defence (p. 158), and from laboratory knowledge, it is clear that in general the BQs are too toxic to be used for normal metabolic purposes. The quinols as reducing agents are as deleterious as the oxidant quinones; they inhibit the glycolytic production of lactic acid from glucose (HOCHSTEIN and COHEN, 1963). The phenolic intermediaries in the indolic melanin pathway are strongly cytotoxic through autooxidation and the generation of H_2O_2 which then inhibits essential enzymes with active SH-groups.

Since blood-coagulation normally occurs only on exposure to air, vitamin K might be expected to promote this process also in a redox capacity. Actually its main action seems to be at the earlier stage of prothrombase synthesis (MACFARLANE, 1970) but there is some evidence that the echinochromes promote surface-changes which may be redox. The echinochrome of the coelomic fluid and corpuscles of echinoids may assist in its clotting at wounds. DONNELLON (1938) found that the corpuscles containing the red naphthoquinone explode when the coelomic fluid of the sea urchin is shed, and play a part in the clotting process rather similar to that of the thrombocytes and platelets of vertebrates; potassium salts promote the explosion in both cases (ROTHSCHILD, 1956). In the eggs of echinoids, again, NQ granules collect near the surface, within minutes of fertilisation, and likewise disintegrate explosively (HEILBRUNN, 1934). This also leads to the formation of insoluble material, the fertilisation membrane. The vertebrate fibrinogen → fibrin change involves the oxidative conversion of SH-groups to S-S bridges and it is conceivable that vitamin K is relevant to this change. Cysteine is abundant also in some of the proteins involved in the clotting of echinoderm coelomic fluid (NEEDHAM, 1970c). However, it should be recalled that KUHN et al. (1939) could detect no positive effect of echinochromes on vertebrate blood clotting.

The broader question whether the echinochromes in the various other sites in the echinoderm body have normal redox functions is equally unsettled. The wide distribution of the chromes to virtually all tissues, including gametes, (p. 271) indicates that if they have a function it is likely to be something as general as respiration (FOX, 1953, p. 204). However, the spermatozoa of echinoids contain ubiquinone in their respiratory system (CASERTA and GHIRETTI, 1962) and it is significant that no animals have vitamin K, or any other known NQ in this role. On the other hand echinochrome is in the reduced state in the coelomocytes of *Echinus*, but in the oxidised state in the coelomic fluid, and in the eggs of *Arbacia* (FOX and VEVERS, 1960, p. 140). That of the *Echinus* coelomocytes spontaneously oxidises on exposure to air so that *in vivo* it well may transfer oxidising potential between more and less oxygenated regions, in general from superficial to deeper tissues. CANNAN (1927) and COOK (1928) were unable to measure any O_2-binding by echinochromes of the labile type familiar for Hb, Hcy, etc. (chapter 9) but in echinoderms no doubt any redox role of these chromes is likely to be nearer the level of tissue respiration, operative only at low O_2-pressures. The redox potential of the NQs is low enough to be useful here, and a low O_2-capacity (COOK, 1928) need not be a serious disability provided transport and transfer are rapid enough. It is then not surprising that TYLER (1939), CORNMAN (1941) and BIELIG and DOHRN (1950) could find no evidence that pure echinochromes stimulate the respiration of well oxygenated echinoid eggs and sperm. Some species of echinoid have no echinochromes in their eggs so that NQs certainly are not indispensable there under normal conditions. It is known that 1,4-naphthohydroxyquinone will reduce 'methaemocyanin' (CONANT et al., 1933); that this was not the true analogue of methaemoglobin (p. 223) does not necessarily deprive the NQ action of biological significance.

Since the anthraquinones have a very low oxidation-potential they are unlikely to be able to act as redox agents except at the substrate end of respiratory paths, where the pyridinoprotein and flavoprotein systems appear to be adequate. Hallachrome has reversible redox properties (FRIEDHEIM, 1932). A number of nitrogen, sulphur and oxygen analogues of the AQs, for instance acridines (phenazines), phenthiazines (e.g., methylene blue) and phenoxazines will act in a redox capacity in the laboratory (SEXTON, 1953, p. 210–211). They increase glycolysis and inhibit the pasteur effect so that they do act proximately and not terminally. Pyocyanin:

and iodinin (Fig. 3.9, IV) are two biological phenazines produced by bacteria as antibiotics. They act as structural analogues also of ribitylflavin. Both AQs and NQs competitively inhibit the action of iodinin (SEXTON, 1953), and so they may protect the flavin systems. The phenanthraquinones will catalyse a cyclic redox process in Ehrlich ascites cells (MIT-

CHELL and MARRIAN, 1965): the angular annelation of the extra benzene ring raises the redox potential (MORTON, 1965).

The aphins (p. 48) in general have strong redox properties. Protoaphin, the di-naphthoquinone present in the blood of the black aphis can be reduced to a colourless derivative which spontaneously reoxidises in air. When the blood is shed protoaphin is oxidised with the help of an enzyme, through the series of progressively more bathochromic derivatives already described (p. 44).

Notwithstanding their increased annellation and other changes these remain hydroxyquinones with redox properties. The significance of the series is still obscure but since *in vivo* protoaphin is protected from the enzyme there seems little doubt that the process is an established adaptation, probably defensive. The bathochromic members, and also zoopurpurin (p. 43) the polycyclic quinone of the ciliate, *Blepharisma*, are photodynamic in the presence of oxygen (GIESE, 1946).

Arenicochrome (p. 43) has reversible redox properties (VAN DUIJN, 1952) and so has the violet tissue-chrome of *Lithobius* (Table 8.1). A solution of the latter can be reduced simply by exposing it to a very low O_2-pressure, over pyrogallol and conceivably the chrome is an oxygen-transport agent in the claustral habitat of this animal. The distribution around the tracheae and under the epidermis is consistent with this and so is the coincidence of high concentrations of ribitylflavin. The solid state of the chrome in the cells is no necessary impediment (p. 125) provided there is an O_2-gradient (MORTON and ARMSTRONG, 1963); in any case there appears to be a modicum of this material in the blood and it is incompletely oxidised when freshly shed. There is also some evidence that it is the source of the brown pigment in the exoskeleton of the centipede. Ribitylflavin appears to be incorporated with it in the exoskeleton so that both may continue to function in a redox capacity there.

Porphyrins: In addition to the haemoproteins of established redox-function some other porphyrin derivatives have redox properties which may be physiologically significant. Helicorubin, the haemoprotein of the gastric fluid of pulmonates (p. 241) is so like the cytochromes that a redox function seems highly probable for it and/or for the 'cytochrome h' which is closely related and closely associated in the cells which secrete it (KEILIN, 1957). The redox potential of both is $+200$ mV and cytochrome c will oxidise them. Below pH $6 \cdot 0$ they autoxidise in air but in the alkaline gastric juice of the snail they are held in reduced form. Like the peroxidases (p. 194) but unlike the cytochromes they have polysaccharide associated with the protein.

Although copper porphyrans have poorly reversible redox properties (HILL, 1926; KEILIN and McCOSKER, 1961) laboratory synthesised turacin (p. 88) will inhibit the activity of the simple copper-protein, ascorbic oxidase; this may also be a further indication that the steric relations of the copper have essential similarity in both types of complex. The copper phthalocyanins, in which indole nuclei replace the pyrrole rings of a porphyran, are good redox agents; the manganese phthalocyanins transport O_2 even when they are in the solid state (ELVIDGE and LEVER, 1959).

The corrin coenzymes (p. 209) are concerned mainly with CH_3-transport but the 5'-deoxyadenosyl members are primarily H-transfer agents (STADTMAN, 1971) and promote such reactions as acetate synthesis from CO_2 and H_2. In some of the cobamide systems, the cobalt changes its oxidation state during activity, and therefore it is relevant that C_1-units are transferred only in the fully reduced CH_3 state. The reactions of some of these enzymes require the SH-type of reducing agent as co-factor. The reactions are all redox. Cobamide coenzymes are very readily photo-activable.

Prodigiosin, the independent tripyrryl compound (p. 52) readily changes from a yellow, oxidised to a red reduced form (ALLEN, 1967). Like the ommochromes (p. 65), therefore, it is unusual in having the reduced form more bathochromic than the oxidised.

Bilins: The large number of known bilins (p. 88) is due essentially to the variety of intermediaries in their oxidation-reduction series and the ready reversibility of many of the changes. Among the most important biologically, as well as the most interesting, is the interconvertibility of biliverdin (Bv) and bilirubin (Br). In the bile of the various vertebrates the one or the other predominates. In air Br spontaneously changes to Bv. Apparently a dehydrogenation, as far as the central methine bridge is concerned, it is actually an intramolecular transhydrogenation. The middle methine bridge of Bv is the most electron-poor and so most readily becomes saturated, whether by this method or by external reducing agents. There is some degree of correlation between Br as the bile chrome and a carnivorous habit, but human individuals secrete the verdin when liver glycogen is depleted and the diene when it is abundant, or when glucose is injected. This may depend on glucose providing the source of glucuronic acid to stabilise Br by conjugation: since the reaction is not a true reduction the reducing action of glucose is not relevant; nevertheless it is interesting that lactic dehydrogenase promotes the change from Bv to Br (FOX, 1953, p. 275). In the pericardial cells of *Rhodnius*, again, injected glucose converts Bv to Br (WIGGLESWORTH, 1943). Some orthopterid and other insects may use this change to effect adaptation to background, green in spring and brown in high summer; the change is chromogenic (p. 136) and progressive over the series of nymphal instars (PASSAMA-VUILLAUME, 1965, 1966). It is noteworthy that the normal catabolism of the bilins as a whole is reductive in contrast to that of most metabolites (PULLMAN and PULLMAN, 1963) and consumes rather than releases energy. This is the more surprising in that the bilanes dehydrogenate readily, and often spontaneously, to bilenes. Ferric chloride will oxidise (dehydrogenate) them by steps through bilene and diene to triene (LEMBERG and LEGGE, 1949, p. 137); nitric acid then will oxidatively re-saturate them through seven further stages, as in the Gmelin series, back to colourless bilane (VUILLAUME, 1969). Biliverdin therefore is still quite a strong reducing agent and for instance will reduce OsO_4 to the metal (DIWANY, 1919). It is also an unusually strong electron-acceptor (PULLMAN and PULLMAN, 1963, p. 430), i.e., an oxidising agent, its lowest empty molecular orbital having slight bonding, as opposed to the usual antibonding, properties. This makes even more surprising the conversion to Br intramolecularly rather than by reduction, particularly since Br spontaneously changes to Bv on exposure to air. Collectively the bilins demonstrate the redox versatility of unsaturated carbon compounds particularly clearly, and it is surprising that they are not more important in functions other than integumental coloration.

Indigotins: The coloured indigotins of mammalian urine (p. 59) are due to bacterial action and they are relatively irreversible oxidation-products. This is true also for tyrian purple, otherwise it would not have been so satisfactory as a dye. It is noteworthy that the reaction producing this from the immediate precursor in the secretion of the muricids is not an aerobic oxidation, though it is photoactivated, at least in some species.

Melanins: In addition to its camouflage (p. 136) and thermal functions (p. 160), indolic melanin may have a significant role in respiratory metabolism (HELMY and HACK, 1964, 1965). There is a considerable increase in O_2-consumption by vertebrate neural crest cells at the time of melanin synthesis (FLICKINGER, 1949) and, while this may be largely due to the requirements of synthesis itself, uptake remains permanently above that of the undifferentiated cell, indicating some persisting redox function in the cell, if not specifically of the melanin. MORGAN et al., (1967) suggest a role in anaerobic metabolism, in the lower vertebrates, but most other evidence concerns terminal oxidation. RILEY et al. (1953) detected many of the terminal oxidases and electron-transfer enzymes in the melanosomes of mouse melanoma-cells, and normal melanoproteins do not differ in chemical properties from those of the tumour (HACK and HELMY, 1964). Melanosomes are also capable of oxidative phosphorylation (DORNER and REICH, 1961). These results of course are very relevant to the classical problem (p. 118) of the homology of melanosomes with mitochondria and, by the same token, to the practical problem of isolating preparations of each organelle free of the other. Studies on pure melanins are essential, with controlled additions of what are most likely to be collaborative redox agents *in situ*.

Molecular orbital calculations (PULLMAN and PULLMAN, 1961, 1963) indicate that melanin could act as an electron-transport agent. When irradiated with visible light, melanosomes from the mammalian retina and choroid produce a strong electron spin resonance signal which indicates the formation of free radicals and the potentiality for high redox activity. The free radicals tend to decay rapidly (COPE et al., 1963) and presumably have the redox power of typical semiquinones. Some of the free radicals are trapped and stabilised inside the macromolecular melanin complex (p. 62). This may permit vector conduction over considerable distances (COPE et al., 1963).

Ommochromes: The redox properties of the ommochromes have been stressed already (p. 65). These depend on the nature of the active side-chain groups and not all phenoxazones have reversible redox properties; it therefore seems probable that the biological members, and particularly the ommatins, have been selected for this feature. A surprising number of molecular features contribute to their redox properties and so there may be biological significance in the fact that the reduced form, diH-xanthommatin is more bathochromic than xanthommatin.

It is interesting that the female of *Sympetrum* is yellow and the male red (BECKER, 1942), since the latter sex usually has the higher metabolic rate and therefore possibly the lower O_2 pressure at the tissues. However, if the colour difference were due directly to this it should be possible to turn the female red by exercising her adequately, or the male yellow by respiratory depressants!

The Odonata have ommochromes in their powerful flight muscles (PRYOR, 1962) as well as in the walls of their thoracic air sacs, rather strongly implying a physiological respiratory function. The xanthommatin of the eggs of the echiuroid, *Urechis*, undergoes redox colour changes *in vivo* (LINZEN, 1967), probably in relation to autogenous respiratory activity, since the individual eggs of a batch may be in various states between full oxidation and reduction. However, the eggs of other echiuroids do not have the chrome so that it is not indispensable.

BUTENANDT et al. (1960b) suggest that ommatin D, which is a sulphuric ester, may act as a coenzyme for oxidative trans-sulphuration. The reaction would be analogous to oxidative phosphorylation, which is catalysed (p. 187) by the phosphoric ester of a simpler hydroxyquinone (HARRISON, 1958). It is noteworthy that some carotenoids also have been implicated in trans-sulphurylase activities (CHEESMAN et al., 1967).

Pterins: redox activities are probably the major function of the pterins (VUILLAUME, 1969, p. 20); in this they resemble their structural and biosynthetic relatives, the isoalloxazines (p. 188). With polar side-chains on an aromatic ring-system most of them are effective quinol-quinone agents. They are reductants rather than oxidants (PULLMAN and PULLMAN, 1963) acting at respiratory substrate levels, and at the initial photoactivation stage in photosynthesis (p. 181); the redox potential of biopterin may be lower than -700 mV. Like ribitylflavin pterins are abundant in the superficial tissues and yet many are too pale to affect the visible colour very much. The dorsal scales of the carp and goldfish, which contain both classes of chrome, consume 30% more O_2 than ventral scales, and the pigmented parts of the dorsal scales consume 140% more than the non-pigmented parts of the same scale (Fox, 1953, p. 229). As much as 45% of the consumption of the pigmented region is not sensitive to cyanide and so by-passes the cytochrome system (POLONOVSKI et al., 1946). Experimentally the pterin concerned, ichthyopterin, increases the O_2-consumption of the brain of rats suffering from ribitylflavin-deficiency, so that it may be the coenzyme-constituent of an alternative terminal oxidase to flavoprotein (see below). The oxidation of the tetra-H-pterins has been found to catalyse that of various other biochemical reductants (KAUFMAN, 1961). The O_2-consumption of systems containing pterins is stimulated by light (ZIEGLER-GUNDER, 1956) implying that their redox action can be directly powered by light energy, as in photosynthesis, and possibly in vision (p. 172).

The conjugated pterin, pteroylglutamic acid (p. 66) is essentially a redox coenzyme, existing in di- and tetra-hydrogenated forms, like a number of other pterins, and transferring C_1 units in several states of reduction, – short of CH_3 which is the province of the cobamides (p. 201). Actual transfer is probably always at the formyl stage, and producing this is the rôle of the redox component of its activity. Some Protozoa, like the vertebrates, use the PGA system and demand this vitamin in their diet (KAUFMAN, 1967) but others demand unconjugated pterins, mainly biopterin and neopterin, as coenzymes for a number of effectively redox enzymes (KIDDER, 1967). These function also in mammals but here they are probably endogenously synthesised. A large number of redox reactions are attributed to them (KAUFMAN, 1967). The best characterised are the hydroxylation of phenylalanine to tyrosine (Fig. 8.5) and the further oxidation of tyrosine to dihydroxyphenylalanine,

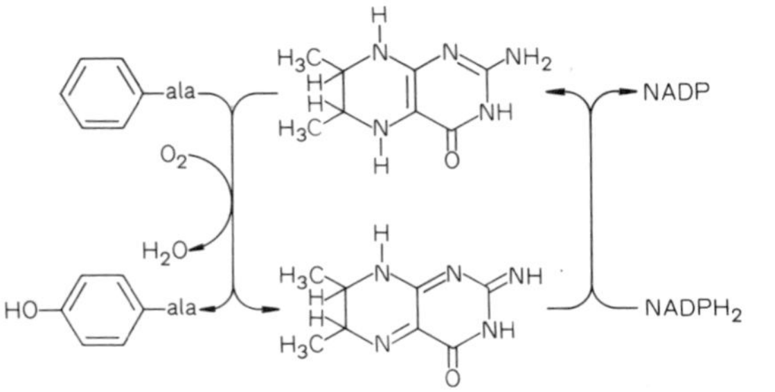

Fig. 8.5. Scheme of action-cycle of a model pterin prosthetic group of phenylalanine hydroxylase. The hydroxilation reaction demontrated in the rat and the regeneration of the reduced coenzyme by NADPH$_2$ in the sheep. In the active phase of the cycle both substrate and coenzyme are oxidised. The model pterin used is 6,7-dimethyl-lumazine (Fig. 13.5) but in vivo the coenzyme probably is di/tetra-H-biopterin. (After KAUFMAN, 1967)

the hydroxylation of tryptophan, the conversion of cinnamic acid to p-coumaric acid, the hydrolysis of fatty acid esters to free fatty acids (TIETZ et al., 1964), the hydroxylation of progesterone at its 17 position and the conversion of 8, 11, 14-eicosatrienoic acid to prostaglandin. The pterin, mostly biopterin, cycles between the tetra- and di-H states and, in the typical case of phe-HO-ase, is co-oxidised with the substrate, by means of atmospheric O_2 (Fig. 8.5). Reduction to complete the cycle depends on NADPH and a suitable fuel substrate, i.e., the pteroprotein plugs into the main respiratory system and acts as an aerobic oxidase! In this case the redox potential must be higher than that of NADP. Tetra-H-biopterin oxidises spontaneously in air, so that *in vivo* this must be controlled by the protein; the view (p. 291) that the hydrogenated pterins are the end-terms of special biosynthetic lines receives further support here.

The role of these enzymes in the whole path from phenylalanine to DOPA explains how pterins are concerned in melanin synthesis (p. 288). Di-H-sepiapterin is a very effective *e*-donor for phe-hydroxylation in the laboratory (GILMOUR, 1965) and may be the active pterin in insects. The parallel HO-ation of tryptophan to kynurenine and 3-HO-kynurenine (VISCONTINI and MATTERN, 1967) helps to explain the presence of pterins in insect eyes and their role in the biosynthesis of the ommochromes. Active radical forms of sepiapterin have been detected.

Ribitylflavin: The redox activities of flavoproteins already considered account for most of the facts about the behaviour and distribution of the chrome, ribitylflavin, except the magnitude of the stores of this material in a number of invertebrates (p. 305). In cirripedes and isopods storage increases under difficult respiratory conditions (FISHER, 1960) so that here, and perhaps more generally, the stores may be mainly for orthodox respiratory purposes. Earthworms, which store large amounts (p. 90), may be expected to suffer difficult respiratory conditions rather frequently in their normal enivornment.

Haemovanadin: The vanadium chrome of tunicates, shows strong redox properties (p. 73). The pale green form, in situ, may be V^{II} (CALIFANO and BOERI, 1950) and will reduce Fe^{III} to Fe^{II}; it will also reduce OsO_4 to the metal (WEBB, 1939). The red-brown solution obtained on haemolysis is probably V^{III} unbound from protein; it still has a strong reducing potential and spontaneously oxidises to the dark blue, V^{IV} state. This is stable *in vitro* and presumably the high acidity inside vanadocytes is necessary to stabilise the more reduced form. At pH 2.4 the value in these cells, haemovanadin will stoichiometrically reduce the cell's cytochrome *c*, as well as extraneously added methylene blue, both aerobically and anaerobically (CALIFANO and BOERI, 1950). The Cy *c* spontaneously reoxidises, even under anaerobic *in vitro* conditions, so that a continuous respiratory cycle may be implied. Hydrogen peroxide or a related metabolite is suggested as the reoxidant, since catalase inhibits the reoxidation. The origin of H_2O_2 under anaerobic conditions *in vitro* presents a problem but perhaps the possibility of photoactivation (p. 247) should be tested, both *in vitro* and *in vivo*. Some tunicates may suffer anoxia when the siphons are closed.

In *Pyura stolonifera* ENDEAN (1955) found a yellow iron compound replacing vanadochrome and this also reduces Cy *c* anaerobically. Both this and the vanadochrome, or niobochrome, of other ascidians therefore may be mediating a respiratory process under rather special conditions, but certainly not transporting O_2 in an orthodox manner, as once supposed. The blood carries no more loosely held O_2 than is found in simple aqueous solution.

Inorganic V-compunds are outstanding as redox agents and will catalyse a number of biochemical reactions. For instance, when conjugated as pyridoxal-V-phosphate they will catalyse cysteine dehydrogenation (BERGEL et al., 1958). They also inhibit the reducing activity of sulphydryl compounds in mammalian liver (SNYDER and CORNATZER, 1958) by oxidising -SH to the -S-S- state. Vanadochrome itself therefore may interact also with this type of biological redox system. The synthesis of carotenoids which involves many reduction-reactions (p. 281) is promoted by $VOSO_4$ (AUSTERN and GAWIENOWSKI, 1969). Paradoxically it does so at high and at low concentrations but not at intermediate concentrations. Could this be related to the paradoxical behaviour of cytochrome *c* in reaction with haemovanadin? It may be relevant to the fact (AZARNOFF et al., 1961) that V-compounds inhibit the synthesis of steroids, in a pathway which competes with the carotenoid path (Fig. 13.1).

Urochrome: This chrome also has redox properties (ECKMAN, 1944), due to SH, phenolic or glucuronic acid groups in the molecule. It is readily converted to a melanin-like form, presumably by oxidation and, also like the melanins, relatively irreversibly. On the other hand urochrome can also be reduced to a colourless product, and this reaction *is* reversible.

Since urochrome is essentially an excretory product probably none of these reactions has any physiological significance but it is noteworthy that vitamin C depresses the amount of the partially oxidised, yellow form normally excreted.

Unidentified Zoochromes: None of the unidentified zoochromes known to have redox properties *in vitro* (Table 8.1) is known with certainty to exercise this property *in vivo*. Without a knowledge of the chemical class and of the active groups of the molecule no very specific information can be expected. However, in view of the strong evidence of functional redox properties for every well known class of zoochrome there must be some with significant redox functions among these more obscure types. The tentative dictum: 'Every biochrome is primarily a physiological redox agent' is probably justified and a secondary diversion of some to the visual-effect type of integumental function must not obscure this. In concluding the sections so far, in this chapter, it is worth emphasising how many classes of biochrome are essentially quinol/quinone for redox purposes and how many have semiquinone active free radicals at a key stage of their redox cycle.

8.4 Photochemical Redox Catalysis by Zoochromes

In view of the logical priority of radiation energy as the means for activating biochromes, this pristine aspect of their redox activities cannot be completely ignored in this chapter. Unfortunately, however, the known biologically useful photoactivation processes are limited to the synthesis of vitamin D_3 from ergosterol in the skin of some higher vertebrates, to photoperception in animals (chapter 7) and plants and to photosynthesis. Various chemical redox reactions have been induced through photoactivated biochromes but most could not have any biological value *in vivo* and are frankly deleterious. These are considered in chapter 11. It seems that except where electromagnetic radiation must be used directly there has been natural selection for the more easily controlled chemical mode of activation.

Photoactivation of a redox reaction through a biochrome is usually called "photosensitisation". It has been extensively studied for ribitylflavin (PENZER and RADDA, 1967; KOK et al., 1971), porphyrins (KRASNOVSKY, 1971; DAVSON, 1970), bilins (SIEGELMAN, CHAPMAN and COLE, 1968), and nicotinamide (KRITSKY, 1971), as the mediators. In the absence of a substrate ribitylflavin is photolysed itself, mainly to lumichrome in the presence of oxygen and to both lumichrome and lumiflavin under anaerobic conditions (BERENDS et al., 1966). In the presence of a suitable substrate, under anaerobic conditions it mediates photoreductions such as the hydrogenation of ergosterol, the reduction of Fe^{III} to Fe^{II} and of NO_3 to NO_2 (PENZER and RADDA, 1967). In the presence of O_2 photooxidations are the rule and all are biologically damaging. The red active radical (p. 194) is involved (KOK et al., 1971) and there is much in common with its chemically activated redox reactions. The photoactivation is strongly inhibited by benzoquinones and by carotenes, in concentrations of 10^{-5} or less (SONG, 1971); this is an instance of the antioxidant action of carotenoids.

Some photobiological processes in fungi have the action-spectrum of ribitylflavin and the effect of inhibitors and competitors of RF supports the view that it is the photosensitiser (KRITSKY, 1971). All told there may be a large number of functional photosensitisation reactions by biochromes *in vivo* but their vital importance would be unquestioned even if photosynthesis and photoperception were the only instances.

8.5 Biochromes and Bioluminescence

Bioluminescence, a phenomenon particularly common in animals (HARVEY, 1960), would be relevant here if only because it is a product of certain aerobic oxidations and uses as enzymes some at least of the standard respiratory enzymes. However, it is even more relevant in the broader biochrome context since it is a chemophotic transduction, the reversal of the fundamental photochemical transduction of the previous section. It might be anticipated therefore that known photochemical biochromes should be able to act in reverse and indeed chlorophyll does bioluminesce: it emits 'delayed light' for several minutes after a period of photosynthesis and the main view is that this is a 'backfiring' of that process (STREHLER and ARNOLD, 1951; WEBER, 1960). The emission is at the same wavelength as the fluorescence of this chrome.

The expectation therefore is that the light-emitter, or luciferin, of bioluminescent systems should be a biochrome and in fact those of the firefly (McELROY and HASTINGS, 1955) and the ostracod *Cypridina* (HARVEY, 1960) are yellow. There is a yellow chrome also in the system of some copepods, while in the luminescent organ of the euphausid *Meganyctiphanes* there is one which is dichroic, rose-purple by transmitted and yellow-green by reflected light (MURRAY, 1885; HARVEY, 1960); it has not been shown that these two are the actual luciferins, however.

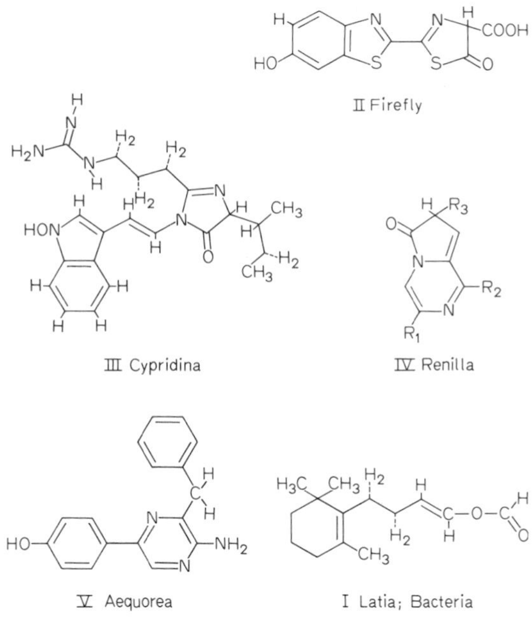

Fig. 8.6. Structural formulae of known luciferin molecules. In that of the Renilla luciferin R_1 is a 3-substituted indole but the nature of R_2 and R_3 is uncertain. All have incomplete conjugate double bonding, i.e. are 'sub-chromatic'. Those of the snail Latia and of luminous bacteria are sub-carotenoids while the others, like most biochromes, have N-heterocyclic ring-systems, but always chain-linked rather than fully condensed

While some luciferins themselves are colourless their oxidation products are coloured. That of the luciferin of the polychaete *Odontosyllis* is yellow (CORMIER, 1969) and that of the sea pansy *Renilla* amber-coloured (KARKHANIS and CORMIER, 1971); the latter has an absorption-peak at 500 nm and a 'shoulder' at 475 nm. That of *Cypridina* is red (GLASS, 1961), i.e., more bathochromic than the unoxidised luciferin. Since the known luciferins are almost aromatic, heterocyclic chain-linked systems (Fig. 8.6), a general bathochromic effect of oxidation is expected, and brings the absorption of colourless members into the visible range. Aequorin (Fig. 8.6, V), the luciferin of the medusa *Aequorea* (SHIMOMURA and JOHNSON, 1972) has effective conjugate double bonding between the directly linked pair of rings and *hyperconjugation* (p. 15) between the pyrazine and the benzene ring so that it approaches closely the necessary structure for visible colour. Its main absorption is at 350 nm but it has a small peak at 470 nm.

Chain-linked and partially unsaturated, rather than fully condensed aromatic, ring-systems appear to be characteristic of luciferins and this may imply that the quanta needed for chemiluminescence are larger than those corresponding to visible light. The aminopyrazine ring occurs in the luciferin of *Cypridina* and probably in that of *Renilla*, as well as in Aequorin (SHIMOMURA and JOHNSON, 1972) so that it may have a specific relevance to chemiluminescence. Most of these molecules are new to biology and no doubt a new field of biochemistry is opening up here.

The luciferin of the fresh-water limpet *Latia* and of luminous bacteria is an aldehyde (Fig. 8.6, I) similar to a retinal (Fig. 7.1) but with a shortened chain (SHIMOMURA and JOHNSON, 1972) so that it again is 'sub-chromatic'. It also has the peculiarity of not being the actual light-emitter and is called luciferin only for consistency: like the luciferins of the other known systems it is the specific substrate of the specific lumigenic enzyme, or luciferase. In the *Latia* and bacterial systems the actual emitter is ribitylflavin, or a derivative, in its capacity of co-luciferase. In biological redox systems and indeed more generally, enzymes are so commonly substrates for a further enzyme that light-emission by co-luciferase rather than by luciferin itself should not be considered a profound difference. There are parallels in other enzymological fields.

Ribitylflavin is rather widely involved in bioluminescent systems though only in the two instances just considered is it known to be a component of the specific reaction. It is present in the system of the earthworm *Eisenia submontana* (WENIG, 1946) and some other annelids (CORMIER, 1969); of fireflies (STREHLER, 1951) and some other animals. However, the wavelength of maximal emission by *Eisenia* is not close to that of the fluorescence of ribitylflavin (VUILLAUME, 1969, p. 30) whereas in all cases tested luciferins do fluoresce at the same wavelength as their chemiluminescent emission (HARVEY, 1960). RF may be functioning in its usual capacity in the main respiratory system, indirectly supplying energy to the lumigenic reaction. Ancillary roles may be suspected for other familiar chromes such as the pterin, 'luciferescein' in the system of the firefly *Photinus* (STREHLER, 1951). Quinol/quinone behaviour has been detected in some systems (HARVEY, 1940, 1952) but of course the luciferin itself might be a quinol.

In vitro many biochromes and commercial chromes chemiluminesce in appropriate systems and this strengthens the view that *in vivo* there has been special selection for sub-chromatic substrates. Ribitylflavin luminesces when oxidised by H_2O_2 (STREHLER and SHOUP, 1953), while chlorophyll and other porphyrin-derivatives respond similarly to such oxidants as tetralin hydroperoxide (ROTHEMUND, 1938; LINSCHITZ and ABRAHAMSON, 1953): this does not demand prior photosynthetic activity by chlorophyll. Acriflavin and some other synthetic chromes emit when oxidised by oxygen (KAUTSKY and MULLER, 1947).

Just as chromes can be both photically and chemically activated, so some luciferins can emit in response to photo-oxidation (FISHER, 1960) and their response has some similarity to the photoinduction of fluorescence and phosphorescence. It differs in still involving the chemical oxidation step and so has a considerable latent period and duration. Photosensitiser-chromes such as ribitylflavin are necessary for the initial step (FISHER, 1960). Of course bioluminescence is so essentially an activity of darkness that photooxidative triggering is not likely to be relevant except in a gregarious luminescent species: even here bioluminescent responses to a flash are usually mediated via the visual system, and not by the luminescent system directly.

The relation between chemiluminescence and fluorescence merits some further consideration. The term phosphorescence would scarcely have acquired its present significance (p. 17) had it been appreciated that this was quite different from the chemiluminescence of phosphorus. All luciferins so far isolated and also a number of their products fluoresce when irradiated at the appropriate wavelength (HARVEY, 1940; MCELROY and GREEN, 1956). In this they resemble all photoactivating biochromes and all photosensitising dyes (DAVSON, 1970, p. 536). Acriflavin and other chemiluminescent laboratory chromes likewise are fluorescent. As already noted the fluorescence is at the same wavelength as the chemiluminescence; that of pure *Cypridina* luciferin is different, but a complex with adenosine monophosphate does fluoresce at the same wavelength as the *in vivo* chemiluminescence (GLASS, 1961) and indeed AMP proves to be a component of the system. The emission by most chemiluminescent animals is in the 480–510 nm range so that their fluorescence must be excited by light of the short-wave visible or the u.v. range; this supports the suggested explanation of the very hypsochromic 'colour' of the luciferins.

Some luciferins may be excited electrically (GILMOUR, 1965, p. 59; NICOL, 1967), by the reverse process to that in a photoelectric transducer. This is to be distinguished from simple, nervous triggering of the chemiluminescent reaction, which is very common. No doubt it is more closely related to the excitation of fluorescence than to chemiluminescence.

Bioluminescence thus is a rather special function of a special group of zoochromes which belong to one or more new chemical classes. The function is rare except in the animal kingdom. In this there is a parallel to the visual display functions in general. As a visual display function in animals bioluminescence shows many exquisite adaptations, rivalling those of colour change (chapter 6). The 'cloud of fire' emitted from the ink-sac of deep sea squids, as the counterpart to the ink-cloud of shallow water forms is an outstanding case (p. 156). Some shallow water members have both ink and photogenic organs and so bridge the evolutionary gap. Light-production is very rare in fresh-water animals and *Latia* is at present unique. This is probably because fresh waters are usually turbid and disturbed. For further details the works of HARVEY (1940, 1952, 1960); GLASS (1961); JOHNSON (1966), and NICOL (1967) may be selected.

8.6 Redox Zoochromes that are Vitamins

It will be evident from this chapter that a considerable percentage of known vitamins, organic microconstituents demanded in the diet, have a redox function, mainly as essential components of coenzymes. These are ribitylflavin (vitamin B_2), retinol (vitamin A), α-tocopherol (vitamin E) and 2-methyl-1,4-naphthoquinone (vitamin K). The only two other vitamins which are strongly coloured, cobalamin (vitamin B_{12}) and pteroyl glutamic acid (vitamin B_{10}) are components of the enzymes which transfer one-carbon units but they transfer these at a particular level of oxidation and so also have a redox function. There is some indication that in the complete dehydrogenase coenzymes nicotinamide (vitamin B_5) has a detectable yellow colour. Again the molecule of thiamine (vitamin B_1) approaches a conjugate double bonded system and is rather easily oxidised to a coloured product, thiochrome; in these respects and in the general form of the molecule it resembles the luciferins (p. 207). It is involved in oxidative decarboxylation and could be considered also a redox agent. Pantothenic acid (vitamin B_3) is a yellow oil (HAWK et al., 1947, p. 1107) and in co-transacetylase (coenzyme A) it again has a major involvement in respiratory processes. Ubiquinone is probably a dietary requirement for most animals and therefore should be classed as a vitamin.

This leaves some colourless vitamins, pyridoxine (B_6), inositol, ascorbic acid (vitamin C) and calciferol (D_3) of which ascorbic acid is probably a normal redox agent. Equally there are coloured redox enzymes which are not demanded in the diet, cytochromes and other haemoproteins, non-haem iron proteins and copper proteins. The correlation between 'vitamin' and 'redox chrome' therefore is not absolute but is rather high. In any case the list of vitamins is not identical in all organisms. A number of blood-sucking parasites demand haem from the diet. Redox zoochromes are indispensable and a number have been offered as hostages to dietary fortune; under natural conditions this is not a serious risk.

8.7 Conclusion

The main respiratory system in the cells of animals and of other organisms is entirely catalysed by enzymes with biochromes as coenzymes. The proximate enzymes of the system dehydrogenate substrates of all kinds and feed the H via various tributaries into a main canal; the redox level of this, between substrates at $-$ 800 mV and oxygen at $+$ 800 mV, is stepped by 'locks' represented by the individual enzymes of the system.

Probably all other redox reactions in the body also are catalysed and controlled by chromoprotein enzymes, mainly of the same chemical classes as in the main system. Most of these reactions therefore probably also feed H and energy into the main stream, though many do not have energy-release as their primary biological function. Some may be quite independent of the main system since some classes of redox chromoprotein, for instance the copper-proteins, are more active in this context than in the main system. However, the ramifications of the latter are very wide and integration is an axiom of metabolic control.

In vitro all zoochromes have redox properties, often readily reversible, and for most of those not yet shown to be co-enzymes there is strong circumstantial evidence of redox functions in normal metabolism.

The pristine mode of activation of chromes, by visible light itself, can be demonstrated in the laboratory for many of the redox zoochromes and all told, there may be still quite a number of normal biochemical processes powered in this way.

The converse transduction, of chemical energy into light, is manifested in bioluminescence, which is a special redox function and involves zoochrome-like substrates of chemical classes as yet unknown elsewhere in living organisms. It is more common in animals than in other organisms.

The importance of the redox zoochromes is emphasised by the fact that a number of them are vitamins for most animals tested.

Chapter 9

Zoochromes that Transport Oxygen

Among zoochromes proper more is known about this functional group than about any other. There are only three distinct types and collectively they are widely, though far from universely, distributed. They occur either dissolved in the blood plasma or sequestered into cells of either the blood or the coelomic fluid and so are largely restricted to the higher Metazoa. All are chromoproteins, the protein having a globular molecule ('globin') suitable for easy transport when free in the plasma and for close packing when inside cells. In fact there is a more fundamental significance in this essentially metabolic protein-geometry (p. 230). In some animals the fluid in question merely surges, or is stirred by sporadic, fortuitous forces and in these animals, as well as in some with a regular circulation, there is uncertainty if the chromoprotein is an effective O_2-carrier in the orthodox manner of vertebrate haemoglobin. In any case it is now clear (PROSSER and BROWN, 1961; JONES, 1963; MANWELL, 1964; GHIRETTI, 1968; WITTENBERG, 1970) that the precise mode of functioning varies considerably between species and sometimes between varying conditions in the same species. Probably some of the modes are still unrecognised.

The most common, best known and efficient type of oxygen-transport chromoprotein is haemoglobin (Hb) of which the protoferrohaemoglobin (PrHb) sub-type is much the most important. It occurs in marine, freshwater and terrestrial animals of many phyla (Table 4.2), in blood cells or plasma or in coelomic corpuscles, but never free in coelomic fluid. In nematodes it is in the pseudocoelic fluid. In cells the molecule is usually an oligomer of an Huffner unit, the unit containing one Fe atom, while the molecule of plasma Hbs may have as many as 192 units (BOELTZ and PARKHURST, 1971). Weight for weight these large molecules cause a lower viscosity than oligomers and so put less load on the circulatory pump; also they have a lower osmotic effect. Sequestration of Hb into cells can lower the viscosity still more.

Chlorohaemoglobin (ChlHb), or chlorocruorin, is closely related to PrHb but has some peculiarities. It is always free in the blood plasma, only of members of the polychaete families Flabelligeridae (Chlorhaemidae), Ampharetidae Serpulidae and Sabellidae and possibly of the small worm *Peobius* of uncertain annelid-echiurid affinities. Since chlorohaem differs from protohaem only in the substitution of $-$ HCHO for $-$ CH $=$ CH$_2$ at position 2 of the porphyrin ring and exists alongside PrHb in some serpulids and sabellids the two have a common biosynthetic origin (p. 286) and only minor differences in properties (Fox and VEVERS, 1960, p. 101).

The chrome in the coelomocytes of the holothurian, *Molpadia*, (CRESCITELLI, 1945) may prove to be a third sub type of Hb.

Haemocyanin (Hcy) the second main type is less widely distributed than Hb (p. 87) and less efficient though in cephalopod molluscs (Table 9.1) it has as much as one third

---

The term *pressure* will be used throughout, rather than substituting 'tension' when applied to fluids, since the tension has the same numerical value as the pressure of the O_2 to which the fluid is exposed. 'Tension' has no conceptual value and gives no better clue than pressure to the actual O_2-content of the fluids. The variation in the content, and its much lower value in water than in air, should be kept in mind. By contrast, CO_2-concentrations in the body-fluids are usually much greater than in the ambient, whether water or air.

of the O_2-capacity of the most efficient of vertebrate Hbs. Most molluscs and arthropods with Hcy are marine but some are limnetic or terrestrial.

The third type of O_2-carrier, haemerythrin (Hery), is a simple Fe-protein and occurs (p. 86) only in cells, in the coelom of sipunculids, priapulids and lingulacean brachiopods, but perhaps also in the blood of the polychaete *Magelona* – all marine invertebrates. Structurally it is probably more closely related to Hcy than to Hb and may have been evolved from the ferredoxins (p. 51), as Hcy from the copper-protein enzymes (p. 196).

Table 9.1. Oxygen-transport characteristics of carrier zoochromes

Taxon	P_{50} (mm Hg)	Bohr effect: change in P_{50} per pH unit (mm Hg)	O_2-capacity: cm³ O_2 per 100 cm³ of		n (Monomer interactions)
			Total fluid	Cells (where relevant)	
Haemoglobin					
Nematodes	0.01–0.20	–	–	–	–
Oligochaetes	0.6–8.0	–1.78	14.0 max	–	1.8–5.4
Polychaetes	0.6–36.0	None	0.18–8.20	–	1.0–6.0
coelomic	7.6	–	–	–	–
Urechis	12.3	None	2.2–6.7	10.2	–
Cryptochiton (radula muscle)	2.5	None	–	–	1.2
Planorbis	1.0–7.0	–	0.9–1.5	–	–
Daphnids	0.8–3.1	Very slight	–	–	–
Triops	6.8	Slight	3.22	–	–
Chironomus l.	0.17–0.60	–	6	–	–
Gastrophilus l.	4.9	None	–	–	–
Cucumaria (coelom)	12.5	None	–	–	1.4
Cyclostomes	2.0–19.0	0.0 to –5.01	–	–	1.0
Elasmobranchs	16.4–26.0	–	4.2–15.7	30.0 (skate)	–
Teleostomes	1.4–1000	–15.85 (mackerel)	6.2–19.75	23.0–41.0	1.8 (eel)
Protopterus (Dipnoi)	11.0	–	–	–	–
Amphibia	4.6–28.0	–1.74 (bullfrog)	2.5–9.8	25.0	2.9 (Rana)
Chelonia	15.0–19.5	–	6.6–10,8	50.0	–
Crocodilia	28.0–38.0	–6.31	6.7–10.0	43.0	–
Lacertilia	19.0–31.0	–	10.0–12.5	–	–
Birds	35.0–58.0	–2.82 to –4.68	12.3–20.0	40.0	3.0 (duck)
Mammalia	22.0–56.0	–2.40 to –9.12	15.0–23.4	30.2–45.5	2.9
Pinnipedia	31.0–40.0	–	19.8–29.3	61.3–68.0	–
Chlorohaemoglobin					
Spirographis	27.0	–4.57	9.1	–	3.4
Haemerythrin					
Sipunculids					
general coelom	2.9–8.0	None	1.6	21.0	1.0–1.4
tentacle coelom	40.0–45.0	–	–	–	–
Lingula	~10.0	–10 app	–	–	–

Table 9.1 (continued)

Taxon	P_{50} (mm Hg)	Bohr effect: change in P_{50} per pH unit (mm Hg)	O_2-capacity: cm³ O_2 per 100 cm³ of		n (Monomer interactions)
			Total fluid	Cells (where relevant)	
Haemocyanin					
Crustacea	8.0–102.0	–ve	0.84–14.0	–	{2.0 (pH 7.0) 4.0 (pH 8.1)
Limulus	11.0–13.0	– (pH 8.3–9.4) + (pH 6.8–8.3)	0.74–2.70	–	{1.0 (pH 5.8) higher at higher pH
Cryptochiton	18.0	± ve (small)	–	–	1.0
Gastropoda	6.0–15.0	{Fusitrition + 131.8 Helix – and + ranges of pH}	1.1–3.3	–	–
Cephalopoda	4.0–36.0	–ve	3.8–5.0	–	–

Notes: P_{50} is the pressure of oxygen which is in equilibrium with the carrier when half saturated (loaded) with oxygen. The 'direct' or 'positive' Bohr effect involves a decrease in P_{50} with increasing pH; hence the negative sign. It is often measured as $\emptyset = (\Delta \log P_{50})/(\Delta \text{pH})$ but ΔP_{50} itself, as here, seems more meaningful. n is measured as the slope of the O_2-equilibrium curve at the point where the O_2-pressure is P_{50}. It is a measure of the interaction between the monomer units of the complete molecule of the OTZ.

An O_2-transport function has not been demonstrated for vanadochrome (p. 73) or for any other chrome in the body-fluids, though this is still possible for echinochromes (p. 200). Since the unfortunate case of the brown manganese-containing 'pinnaglobulin' (GRIFFITHS, 1892) all new possibilities are treated very critically. Plasma chromes such as ceruloplasmin (p. 195), with definite and essential enzyme activities not only indicate how the O_2-transport chromes may have evolved but in some instances themselves may have an O_2-supply function of some kind.

9.1 Physiological Properties of Oxygen-transport Zoochromes

As usual only those properties will be considered which are certainly, or probably, relevant to the function of the chromes themselves. This includes some of the interactions with other constituents of the circulatory system.

9.1.1 Haemoglobins

Vertebrate Hb may be taken as the standard of comparison for all carrier chromes. It is purple when deoxygenated, and bright orange red when oxygenated. OxyHb has its two major absorption-peaks of the visible range, the α, β peaks, at ~ 576–8 and 539–545 nm respectively and in deoxy Hb these are replaced by a single broad peak centred at 556–565 nm. The intense Soret absorption (p. 56) moves from ~ 414 to 425 nm. The precise positions vary with the protein, which is taxonomically specific. In all vertebrates examined Hb acts as an orthodox carrier at normal O₂-pressures (see footnote p. 211), loading with one O_2-molecule per haem-Fe. The total capacity ranges up to 30 volumes of O_2 per volume of blood (Table 9.1) or 46 times the amount in simple solution in distilled water (0.65 cm³% at 20°C). It may be noted that sea water holds only 0.51 cm³% and that the amount in simple solution in vertebrate plasma is only 0.30 cm³% or 0.24 cm³% for whole blood (PROSSER and BROWN, 1961, pp. 153, 198). Other solutes therefore increase the need for a specific carrier. The O_2-association or O_2-*equilibrium curve*, the relation between load, or percentage saturation, and O_2-pressure (Fig. 9.1), is rather steeply sigmoid in most vertebrates so that the Hb loads and unloads rapidly as the local O_2-pressure changes, over a range which is physiologically optimal for that animal. By contrast the curve for some invertebrate Hbs, and other carriers, has a more gentle slope (Fig. 9.1d) and, at the other extreme, the curve for some others has its steep region at such low O_2-pressures that it is nearer hyperboloid than sigmoid (Fig. 9.1e).

Equilibrium curves can be fitted approximately by the relation $C = \dfrac{100(p/p_{50})}{1 + (p/p_{50})^n}$ where $C = \%$ saturation with O_2, p_{50} is the O_2-pressure at which the chrome is 50% saturated, and the index, n, is a measure of the interaction between certain components of the carrier system, in vertebrate Hb mainly intramolecular interactions (p. 230). The index n is also a measure of the steepness of the curve and so of the loading/unloading sensitivity. From the equation it can be seen that $n \log \dot{p}$ is linear in $\log C/(100-C)$ so that the slope of such a plot gives n directly. The three parameters, p_{50}, n and C_{max} physiologically characterise the chrome. In a capacity/pressure curve C_{max} is substituted for the 100 of the percentage saturation curve.

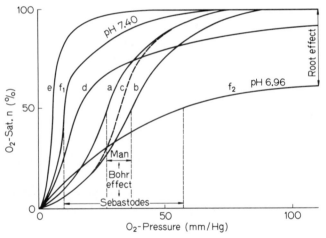

Fig. 9.1. Types of oxygen-equilibrium curves of O_2-transport zoochromes, illustrating Bohr and Root Effects and other features described in text

A very important property of carrier chromes is known as the Bohr Effect (BOHR, 1909), a sensitivity of the O_2-loading to CO_2 and in lesser degree to other acids. The biological rationale of the effect is that the blood acts as a carrier also of CO_2 so that interaction between the two gases is inevitable. In fact the effect over the physiological range works in opposite directions in different animals, but the most usual, 'direct' or 'positive' Bohr effect is that CO_2 promotes the unloading of O_2 (and depresses O_2-loading). Since CO_2-concentration is maximal in the tissues and the gas is rapidly washed out at respiratory surfaces, the effect is adaptive, accelerating the flow of O_2 from blood to tissues without interfering with loading at the periphery. The net effect of CO_2 is to shift the O_2-Eqbm curve to a higher O_2-pressure range (Fig. 9.1, a, b, f) and can be measured by the change in p_{50}. In the circulatory system of course the CO_2-content changes continuously so that for man Fig. 9.1,c is nearer the actual equilibrium curve *in vivo* than a or b. It is steeper than either of these theoretical curves, for pure arterial and pure venous blood respectively, so that the Bohr Effect increases the functional sensitivity of the chrome. Some fish have such a powerful Bohr effect that when out of water they suffocate in the presence of abundant O_2 because they then have no effective means of washing away the CO_2 at the gills.

The Hb of some teleosts also has a Root Effect, i.e., CO_2 depresses the loading capacity (C_{max}) in addition (Fig. 9.1,f). Again this will not normally affect the capacity of arterial blood at the periphery but promotes unloading at the tissues; most importantly it facilitates the special function of O_2-secretion into the swim-bladder, even at high partial pressures of O_2.

A reciprocal effect of oxygenation on the CO_2-load is equally valuable at the periphery. Classically this was explained by saying that oxyHb is a stronger acid than Hb so that a mutual mass action and molecular dissociation mechanism operates between this and H_2CO_3. Now expressed in more sophisticated terms the essence of the adaptation is the same, namely that O_2 and CO_2 each promotes the unloading of the other. A corollary is that deoxyHb is a better buffer against acids than oxyHb.

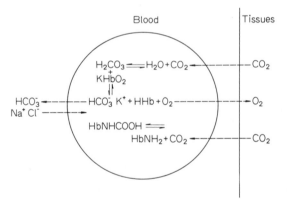

Fig. 9.2. Essential processes in the erythrocyte of a mammal in relation to O_2- and CO_2- exchange between red cell and tissues. Solid arrows denote chemical reactions and footstep-arrows transport movements. All the reactions and movements are reversed at the alveolar surface of the lungs

In the erythrocytes of vertebrates the action of CO_2 on O_2-release at the tissues is further improved by the high concentration of the enzyme carbonic anhydratase, sequestered with the Hb. By promoting the reaction $CO_2 + H_2O \rightarrow H_2CO_3^-$, the product then reacting with the potassium salt of Hb to give free HCO_3^- ion, the enzyme maintains a CO_2 diffusion-gradient into the cell, since the HCO_3^- is free to diffuse out against Cl^- of the plasma (Fig. 9.2). This chloride-shift increases the intracellular acidity and so promotes the release of O_2; it is part of the Bohr effect. A second component is the formation of a carbamino compound of Hb: $R.NH_2 + CO_2 \rightarrow R.NH.COOH$. This carboxylic acid is stronger than H_2CO_3. It accounts for only 6 cm³ in a total of 56 cm³ of CO_2 carried per 100 cm³ of mammalian venous blood but it contributes a substantial fraction of the CO_2 actually eliminated at the respiratory surface. Here both series of reactions are reversed so precisely that there is no need to represent this half of the cycle on Fig.

9.2. The rest of the CO_2 is carried by the buffering systems of the plasma and does not concern the Hb so directly. In many of the invertebrates with an O_2-carrier free in the plasma the chromoprotein itself is virtually the only buffer and CO_2-carrier, in most cases of rather low capacity; however, the amount of CO_2 they actually eliminate at the periphery may not be less than that by vertebrates. Those invertebrates with their carrier chrome in cells have buffering powers almost as good as those of the lower vertebrates (Table 9.2). The general parallel between total buffering power and O_2-capacity is noteworthy; it indicates that the two are closely related, as the above discussion implies, and also that cell-buffering is integrated with that of the plasma in all taxa.

Table 9.2. Buffer value (β) und CO_2 capacity of OTZs compared with oxygen-capacity and concentration of the chrome

Animals(s)		β	CO_2 capacity (vols % of fluid)	O_2 capacity (vols %)	Concentration of OTZ (g/100 cm^3 of fluid)
Man	Hb	30.8	45–53	20.0	16.0
Horse	Hb	25.3	63.0	15.3	10.9
Hen	Hb	–	51.5	10.5	10.9
Alligator	Hb	22.6	38.6–44.0	6.7	6.7
Crocodile	Hb	18.2	–	9.0	–
Amphibia	Hb	–	58.2–70.1	2.5–9.8	6.0–10.2
Teleosts	Hb	–	6.5–47.8	–	9.3–17.6
Skate	Hb	5.5	7.7–10.8	5.0	4.3
Gastrophilus	Hb	–	40.0–131.0	–	– .
Urechis	Hb	4.9	7.0–9.0	4.4	2.5–4.0
Sipunculus	Hery	3.5	–	1.6	0.38
Cephalopods	Hcy	–	11.0–14.0	3.8–5.0	10.0
Gastropods	Hcy	–	45.0–62.0	1.1–3.3	3.7
Crustacea	Hcy	–	25.0–35.0	0.8–14.0	5.6

β is defined as $\Delta BHCO_3/\Delta pH$, the increase in concentration of alkali bicarbonate per unit increase in pH of the respiratory fluid.

The p_{50} of the equilibrium curve is moved to a higher O_2-pressure also by an increase in temperature. For homoiotherms and terrestrial poikilotherms in particular this is adaptive since increased respiratory activity is usually associated with thermogenesis. In small aquatic organisms the heat is probably dissipated too rapidly to have any significance except within the tissues themselves but in fact this is where these metabolic effects operate. Animal O_2-carriers are adapted to their normal temperature range, so that the frog's percentage-saturation curve at 15°C very nearly superimposes on that of a mammal at 37°C; at this temperature the frog's blood cannot load while at 15°C mammalian blood cannot unload!

The equilibrium-curve of the lamprey, which has only one Huffner unit per molecule of Hb, is hyperbolic, indicating that the sigmoid form of the Eqbm curves of gnathostome Hb is indeed due to interactions between the units, if not specifically between the haems. Gnathostome myoglobin (Mb) is a monomer and behaves like lamprey Hb. Their p_{50} is at a very low O_2 pressure, indicating that the Bohr effect also depends on monomer interactions (p. 230); actually lamprey Hb polymerises during its oxygenation-cycle and so it does show a positive Bohr effect. On the other hand some of the highly polymeric invertebrate carriers, for instance the Hb of the lugworm *Arenicola*, nevertheless have hyperboloid curves, with low-pressure loading and rather small Bohr effects (Table 9.1).

In foetal mammals the Hb loads at a lower pressure than that of the mother, so that the O_2-supply to the foetus is assured. Young birds and Amphibia also have Hbs with higher O_2-affinities than that of the adults, perhaps because the morphological and physiological components of their systems are not so efficient as in the adult. The embryo mammal therefore was preadapted to viviparity; heterospecific internal parasites are adapted in the same way, – for instance the nematodes *Ascaris* and *Strongylus*. Even the copepod external parasite, *Lernaeocera*, has Hb with a higher affinity than that of its host, the fish *Urophysis*. *Parascaris* appears to be the only recorded exception (LEE and SMITH, 1965). *Peltogaster* and some other cirripede parasites have an equal advantage by using Hb while their decapod hosts use Hcy. Mb has a greater O_2-affinity than Hb, ensuring a good O_2-flow into muscle tissue.

The Hbs of most vertebrates normally operate only over the steep part of their O_2-Eqbm curves. The system is very well adapted since the normal O_2-pressure range in the body is limited to this central region of the curve and natural selection has learned the law of diminishing returns which operates at the two extremes. In fact the brake on exchange at the two extremes is salutory and sigmoid curves are typical of most controlled biological mechanisms; in consequence they can be pushed out of their sensitive ranges only with great difficulty, under extreme conditions. The Hb is exposed to unloading pressures almost immediately it leaves the ventilation-surface and the whole adaptation is expressed in that simple, steep sigmoid curve. The function of this carrier chrome is the epitome of biochrome utility. The only shortcoming, on this score, is that it is not photoactivated (p. 8); it is a heterotrophe's chrome and also there is an advantage in having the respiratory surface covered.

Like the Hb of gnathostomes, that of earthworms and even that of the larva of the bot-fly, *Gastrophilus*, living in the stomach-wall of the horse, functions at atmospheric O_2-pressure and this is generally true also of the ChlHbs (Fox, 1957). The Eqbm-curves are sigmoid and there are typical Bohr and temperature effects (JONES, 1963). The Hb of *Lumbricus* is fully loaded at 76 mm O_2 pressure, which may be an average ambient value in its underground environment. In the large earthworm, *Glossoscolex*, (JOHANSEN and MARTIN, 1966) the O_2 capacity is 14 volumes percent, comparable to that of the higher vertebrates and much higher than that recorded for most invertebrates. Nevertheless the blood is half loaded even at 7.0 mm of mercury and fully loaded at 14 mm, values in agreement with those of HAUGHTON et al. (1958) for *Lumbricus* and *Allolobophora*, and

Table 9.3. Oxygen transport by the blood of *Lumbricus* in relation to external O_2 – pressure. (From JONES, 1963)

External O_2 – pressure (mm Hg)	152	76	38	19	8
Total O_2 transported (cu mm/g. hr)	8.00	12.30	9.80	3.40	0.00
Amount carried by Hb (cu mm/g. hr)	1.84	4.31	3.92	0.75	0.00
Amount carried in solution (cu mm/g. hr)	6.16	7.99	5.88	2.65	0.00
% carried by Hb	23	35	40	22	—

substantially lower than those given in Table 9.3. However, one result of the low loading pressure is that when oxygenated blood from the skin mixes with deoxygenated blood from other organs the Hb of the latter can take up O_2 from that in simple solution in the oxygenated component. The tissues are supplied by well saturated Hb in consequence and this dissociates sharply where tissue-pressures fall below 10 mm Hg (JOHANSEN and MARTIN, 1966). Both Hb-bound, and dissolved, O_2 components are maximal at 76 mm

Hg (Table 9.3) and the reason for the fall in both at higher pressures remains obscure; a protective significance can be envisaged but the basis is an intriguing problem. High O_2-pressures depress the oxidation of phosphorus and high H_2O_2 concentrations have a reducing effect on Hcy (p. 235) and possibly on vanadochrome (p. 205); the paradox is not rare also in enzyme kinetics.

The Hbs of other invertebrates have varied properties. some still more unusual. A large proportion have the hyperboloid curve (Fig. 9.1, e) with a low p_{50} and high O_2-affinity. There is no positive Bohr effect, which would defeat the purpose of this type of carrier, namely, to ensure loading at the periphery in a relatively anoxic medium. Unloading is ensured by the still lower pressures in the tissues and a number of the invertebrates with this type of Hb, the polychaetes *Arenicola* and *Nephthys*, the snail *Planorbis* and others (PROSSER and BROWN, 1961, p. 221) transport O_2 in an orthodox way at those low pressures. These animals are not necessarily at any disadvantage in O_2-rich media since they are then assured at least the supply which they find adequate at low pressures and are protected againgst any deleterious effects of high O_2 pressures by the limiting capacity of their Hb; they have only to cope with an increased amount of O_2 in simple solution, and a possible difficulty in loading CO_2 at the tissues. *Planorbis* does have a positive Bohr effect and can transfer O_2 to the tissues over a wider O_2 pressure -range; at very low environmental O_2 pressures presumably the animal indeed cannot load O_2 and is inactive, or it disposes of its metabolic CO_2 by other mechanisms, for instance by reaction with the $CaCO_3$ of the shell, or with NH_3 (CAMPBELL and BISHOP, 1970).

A convenient measure of the extent to which the Hb is normally used by an animal is the depression of its O_2-consumption when it is exposed to carbon monoxide. Mammalian Hb has 300 times as much affinity for CO as for O_2 and that of ChlHb is even greater, so that quite moderate concentrations of CO completely inhibit O_2-transport by the Hb. This technique indicates that the oligochaete *Tubifex* transports about one-third of its O_2 as HbO_2, up to an external O_2-concentration of $1.0 cm^3 \%$ (27 mm Hg), which just saturates the Hb (DAUSEND, 1931); above this level, therefore, the percentage falls progressively. Clearly this is another low-pressure-loading Hb, and the medium is difficult for typical aerobes. The Hb of *Tubifex* and some other animals has a much lower affinity for CO than that of mammals (Table 9.4) and continues to transport O_2 substantially in its presence: the technique is valueless in such cases. The larva of the midge *Chironomus* (Table 9.4) is more sensitive than mammals but both it and the larva of the fly, *Tanytarsus*, nevertheless are unaffected by CO until the oxygen pressure is reduced to below half its atmospheric value (WALSHE-MAETZ, 1953). Above 40mm O_2 pressure much of the gas is transported in simple solution and, as in the earthworms (Table 9.3), the Hb makes its major contribution at relatively low pressures, in this case 11–30mm Hg (WALSHE, 1947). In the cladocerans, *Daphnia* and *Simocephalus*, CO has no effect on respiratory rate above an ambient level of 2.5 cm^3 oxygen per litre of water. This is consistent with the fact that these animals do not synthesise Hb until the ambient level falls below this value; evidently the chromogenic mechanism is well geared to the physiological system, as in integumental colour change (p. 136). Above an ambient content of 1 $cm^3 O_2$ per litre, *Daphnia* with Hb consume more O_2 than those without the chrome (GILCHRIST, 1954) so that its useful range is between 1 and 2.5 cm^3 (27–68 mm Hg). Below 27 mm the Hb is fully unloaded and since it also has a positive Bohr effect it is a relatively high pressure Hb; that these Cladocera discard their Hb at high ambient O_2 pressures therefore is surprising.

Chironomus larvae and a number of other animals survive very anoxic conditions in a mild state of anabiosis so that the efficiency of their Hb need not be very critical. In fact they use the chrome particularly during the period of recovery from these conditions and, even so, need the help also of the familiar body-undulations of this 'blood-worm'. Unfortunately these movements preclude the normal filter-feeding movements, which cause a water-current in the opposite direction, and at the same time rapidly use up O_2. Normally

Table 9.4. Relative affinities for CO and O_2 of Hb.s and related haemoproteins
$$\dfrac{HbCO}{pCO} \Big/ \dfrac{HbO_2}{pO_2}$$

Haemoprotein		$\dfrac{HbCO \cdot pO_2}{HbO_2 \cdot pCO}$
Branchiomma	Chl. Hb	570
Chironomus	Pr. Hb	400
Horse	Pr. Hb	280
Man	Pr. Hb	230
Arenicola	Pr. Hb	150
Tubifex	Pr. Hb	40
Planorbis	Pr. Hb	40
Rabbit	Pr. Hb	40
Root nodule	Pr. Hb	37
Myoglobins		28–51
Gastrophilus	Pr. Hb	0.7
Cytochrome oxidase		0.1

(From Prosser and Brown, 1961)

there is an alternation between the two types of activity and the length of the respiratory phase is inversely related to the O_2-supply. One complete 'load' of HbO_2 is a respiratory supply for nine minutes of feeding activity (Jones, 1963) which is adequate under aerobic conditions. The Bohr effect is the strongest known (Glossman et al., 1970).

The echiurid, *Urechis*, and most animals with a typical low-pressure Hb probably have it fully loaded permanently under fully aerobic conditions, since the supply in simple solution is then so much increased. Others, including some of the polychaetes with ChlHb, have an O_2-Eqbm curve like that of Fig. 9.1,d exchanging the gas at relatively slow rates over a wide range of pressures. In fact the slope of the curves even of birds and mammals is much more shallow than that of the low-pressure types.

Barcroft first envisaged the possibility that in the low-pressure type of animal the HbO_2 could act as a store of O_2 against periodic acute anoxia. The feasibility depends on the relative amount of Hb, on its O_2-capacity and on normal respiratory requirements. *Urechis* may have provision for as long as 60 minutes (Jones, 1963) but most animals exhaust their supply much sooner even than the *Chironomus* larva. Even the trematode, *Proctoeces*, can store enough for only 2.5 minutes (Lee and Smith, 1965) and the *Gastrophilus* larva for only 0.5–4.0 minutes (Keilin and Wang, 1946). The latter normally obtains its O_2 from gas bubbles in the contents of the host's stomach.

A somewhat antithetical function envisaged for the Hb of some low pressure types (Jones, 1963; Manwell, 1960a, 1964) is that it acts as a buffer for the tissues against excess O_2 at high pressures. In the simple sense above (p. 218) that the Hb does not increase its O_2 load above a relatively low O_2 ceiling this seems acceptable, but this is viewing from the wrong end a very efficient device for securing peripheral loading at low pressures. Hb itself can do nothing to decrease the O_2-pressure, whether of medium, body-fluids or tissues and under all conditions presumably it can only increase, and never decrease the total O_2-content of the body-fluid, compared with that of an Hb-free fluid. It must also increase rather than decrease the O_2-gradient at the tissues, and therefore the O_2-flow into them. The animal may protect itself by slowing down its fluid circulation and any diffusion processes, particularly at the periphery, but otherwise body fluids and cells must all come rapidly into a maintained steady state with the external medium, the steady state flux being higher with than without Hb. Loss of Hb must decrease the flow and this may be why some antarctic fishes, living in conditions of low temperature and therefore of high dissolved oxygen, yet of low metabolic activity, have no Hb; again this may be why Cladocera eliminate Hb when ambient O_2 is high.

At the respiratory surface Hb might be regarded as the reciprocal of an O_2-buffer, in the sense that buffers keep the intensive property, – H-ion concentration, redox potential, etc., – constant in the face of change in the extensive property, – total acidity, amount of oxidising agent, etc.. Hb in fact increases the O_2-content of the circulatory fluid at a fixed external O_2-pressure. At the tissues, however, the higher O_2-content of the fluid causes a higher O_2-pressure gradient and no simple physical or chemical means can convert this anti-buffer effect into its converse. The earthworm mechanism (p. 217; Table 9.3) seems to be the only one recorded which may protect at the circulatory level against high O_2, controlling both Hb-bound and dissolved O_2; in the present context this instance merits further investigation.

Of course there is the positive danger that an Hb supersaturated with O_2 might not allow sufficiently rapid elimination of CO_2 from the tissues but this could be corrected by an independent plasma CO_2-carrier. Within reason the system is self-adjusting. A further relevant self-regulatory effect is that CO_2-accumulation in the tissues should depress cell respiration and so protect at that level against excess O_2.

Some of the animals adapted to loading at low O_2-pressures are able to extend their p_{50} range by a reversed Bohr effect. This is particularly useful in environments where a low O_2-pressure is due to its depletion by other animals or other heterotrophes since this involves a simultaneous increase in CO_2, particularly in stagnant aquatic environments. The animal's own respiration may contribute to the CO_2 and may even be the major contributor. The reversed Bohr effect makes loading at the periphery possible even at extremely low O_2-pressures; this will also make unloading more difficult but the O_2-pressure at the tissues will always be lower than at the periphery and a steady O_2-flow will always be possible, provided the CO_2 is not much higher in the tissues than in the medium. *Tubifex* Hb has the reversed Bohr mechanism (PALMER and CHAPMAN, 1970) adapted to the high CO_2 + low O_2 conditions produced by bacteria and by itself in its muddy environment. The coelomic Hb of the polychaete, *Nephthys*, also has a reversed Bohr effect.

Nephthys has Hb in both blood and coelomic fluid and the former, unlike the latter, has a normal Bohr mechanism (JONES, 1955). At pH 7.0 the two Hbs have almost identical equilibrium curves, with p_{50} = 6.00 mm Hg, but at pH 7.4 the P_{50} for coelomic Hb is 7.5 and that for the blood Hb is 5.5 mm Hg. This sensitive response may enable the blood to be oxygenated from the coelom at this pH, while under acid conditions the reverse may occur. The mechanism could be adaptive if the coelom is effectively an intermediary both between environment and blood and between blood and tissues; this must further imply an O_2-storage function for the blood Hb. *Nephthys* is a burrowing worm, and has an even lower O_2-pressure in the burrow than *Arenicola*, so that a reversed Bohr effect should be useful. *Nephthys* has storage for only ten minutes respiratory needs yet it does not accumulate lactic or pyruvic acids and must have an efficient means of continuous supply. The above p_{50} values are rather high for a burrowing animal and this is true of some other burrowing polychaetes: the Hb of *Eupolymnia* is only 85% loaded even at atmospheric pressure. A reversed Bohr response could be useful here also.

The Hb of the aquatic pulmonate, *Planorbis*, does not enable it to live in more anoxic media than the Hb-free *Limnaea*. However, it does allow *Planorbis* to exhaust more fully the O_2 in the air it takes down in its lung (JONES, 1963); the O_2 is reduced to 2.8% of the volume of gas in the lung compared with 8.8% for *Limnaea* (21 and 67 mm Hg respectively). *Planorbis* feeds further from the air supply, on the bottom mud, and *Limnaea* up among vegetation.

Anisops, one of the notonectid bugs, has Hb in its tracheal end-cells but uses this for a very different purpose from that of the *Gastrophilus* larva (MILLER, 1964). The HbO_2 supplies O_2 to the boatman's buoyancy bubble, after submersion, so that the animal remains at neutral density. What it loses in respiratory O_2-reserve therefore it gains by not having to swim in order to maintain a constant depth in the water. Oxygen must diffuse down its concentration gradient from the bubble into the tissues and only a special store could replace this. There are some parallels here to the swim-bladder of physoclyst teleosts, which likewise is filled from blood-gases, including O_2, using mainly lactic acid to boost the Bohr and Root effects (ALEXANDER, 1966; DENTON, 1961: HOAR, 1966).

Ascaris has two different Hbs, one in the body-wall muscle and the other in the pseudo-coele. The former is readily deoxygenated but the latter is not, even in an atmosphere of pure N_2. In fact, *Ascaris* has been claimed to survive indefinitely in the complete absence of O_2 (DAVENPORT, 1949), and this was found also for the mud-dwelling oligochaete, *Alma* (BEADLE, 1957). When the nematode, *Strongylus*, is exposed to complete anoxia it does eventually die, but its Hb is still saturated with O_2 (LEE and SMITH, 1965). Presumably these Hb solutions operate an O_2-displacement device which is efficient in transporting O_2 from low concentrations in the ambient medium but must halt when none remains there, to displace that already saturating the carrier. The relation of this to the moving carrier mechanism of WITTENBERG (1959) and SCHOLANDER (1960) is evident, and is discussed below.

These examples illustrate, but do not exhaust, the known variations on the Hb theme among invertebrates. These variations, like the sporadic distribution of the chrome, support the view that it has been acquired many times independently. The many close resemblances in structure and properties (MANWELL, 1964) therefore are due to parallel evolution from a common precursor or group of precursors among the enzymic haemoproteins (p. 191). Since the haem is protohaem in all but the few polychaetes mentioned, it was probably preadapted: Cytochrome *c*, the most common and probably the most primitive haemoprotein, has protohaem and by experimental modification of its protein can be made to bind O_2 (p. 192). Where, as in the annelids, Hb is characteristic of the whole group the function is orthodox apart from adaptation to low ambient O_2 but sporadic evolution, as in nematodes, cladocera, *Chironomus* larva, *Gastrophilus* larva, and *Anisops*, has usually been to meet some special, and rather varied, situation. Among the cases which are still unexplained is that of the nematode, *Tetrameres*, which has no fewer than three Hbs in its pseudocoele alone, with isoelectric points at pH 6.5, 7.0 and 8.0 respectively; this is reminiscent of the range of cytochromes within the individual cell (p. 185).

Myoglobin is so like a monomer of Hb that there is little justification for considering it independently; even its higher O_2-affinity is due largely to the monomer state, and absence of Bohr type interactions. In fact WITTENBERG (1959) and SCHOLANDER (1960) freely used Hb as a laboratory model for investigating the mode of operation of Mb and their very interesting discoveries are relevant to both. The work was extended by WITTENBERG, who has recently reviewed the whole subject (WITTENBERG, 1970), as well as by SCHOLANDER's group. The essential discovery is that Mb normally, and Hb under conditions which may sometimes apply *in vivo*, can transport O_2 by molecular diffusion as opposed to gross hydrodynamic transport. This is not only very efficient through thin layers of a solution of the chromes but its relative efficicency increases with thickness, over a considerable range, so that it is effective through large units of skeletal muscle and could be operative through the tissues of large invertebrates. It is manifested also by packed erythrocytes and could be a factor in their equilibration with each other and with their various environments in the course of their circulation.

If a millipore filter impregnated with distilled water separates air at two different pressures, O_2 passes through at about 50% the rate of N_2; since in air the latter has four times the partial pressure of O_2 this must be about twice as mobile as N_2, in water. If this 'membrane' is now replaced by one impregnated with an Hb solution, O_2-transport rises to 95% of that of N_2. At low pressures, for instance 1/12 of an atmosphere on one side and a vacuum on the other, the figure rises to 400% of that of N_2 – eight times the ratio for simple diffusion. The rate for N_2 remains linear in partial pressure and serves as a reliable marker. The virtue of this mode of transport at low O_2 pressures may be very relevant to the nematode and other situations, and probably to many tissues.

Two main theories about the actual mechanism seemed plausible, either a 'bucket brigade' of Hb molecules, passing O_2 from hand to hand, or an all-the-way bodily diffusion of HbO_2 molecules. Both would satisfy the further observation that a high concentration of the chrome is essential for efficiency. Mathematical treatment by COLLINS (1961), WANG (1961), FATT and LA FORCE (1961) and recently by MURRAY (1968) has decided strongly in favour of the second, moving carrier mechanism. HEMINGSEN and SCHOLANDER (1960) showed that the mechanism was inhibited about 50% by dispersing the Hb in solid gelatin whereas simple diffusion of free N_2 through the fluid capillary system of this typical gel was not affected. The rate of diffusion of MbO_2 is as high as 1/20 of that of free O_2 and since there is 15 times as much Mb as free O_2 in muscle, the mechanism may transport as much O_2 as simple O_2-diffusion, even at atmospheric pressure. With steeper gradients of O_2 concentration, and low absolute values on the side of the active tissue the relative importance of MbO_2 is much greater. Moreover its absolute contribution is invariant from 700 down to 20 torr, whereas simple O_2 diffusion falls progressively. It is most efficient at 10 to 20 torr which is just sufficient to saturate the Hb. Further, the decline in its contribution with increasing thickness of the diffusion layer is much slower than that of simple diffusion so that eventually it accounts for virtually all the transport. An oxygen gradient is essential for net transport and respiring tissues therefore provide the conditions for their own satisfaction. This is in accord with the general principle that evolution has exploited selfregulatory mechanisms wherever feasible (NEEDHAM, 1968c); not all are as direct and simple as this.

Molecular rotary diffusion plays a negligible part in the mechanism of transport, and in any case could scarcely have any advantage over linear or even random movements, in diffusion layers more than one macromolecule thick. Nevertheless the process is in essence a facilitated diffusion already familiar from studies on membrane physiology. Indeed it is one of the few demonstrated examples of this, and has many potentialities in that field of study. As would be expected, it mediates also exchange-diffusion of labelled O_2, in the absence of a pressure gradient (HEMINGSEN, 1962), and this also is faster than simple diffusion. Both this and net transport are depressed by converting the Hb to MetHb, in which the Fe is oxidised to Fe^{III} and no longer reversibly binds O_2. The inefficiency of the Hb in those humans with the sickle-cell trait is due largely to the chrome being in crystalline form in the cell, but it seems doubtful if the Wittenberg effect is important enough in the mammalian circulation to explain this finding; the favoured view is that molecular rotation *is* a critical factor for loading at the ventilation surface and that this is where the sickle state is such a disability.

Eventual dissociation of HbO_2 is an essential part of the Wittenberg effect so that Ascaris Hb, which dissociates 200 times more slowly than that of mammals seems quite unsuitable; it does dissociate, however, and WITTENBERG believes that the moving carrier mechanism explains the function of the anomalous nematode Hbs (p. 221). The loading rate of Ascaris Hb is only twice as slow as that of mammals so that equilibrium is likely to be strongly in favour of the HbO_2 state; earlier experiments therefore may have failed to detect any transport function because of very minute leaks of O_2 into the system. In this event they actually demonstrated the amazing efficiency of the Hbs in transport at very low O_2 pressures, and also their extremely low unloading pressures. It is conceivable that this mechanism contributes to the relative indifference of some Hb-using invertebrates to carbon monoxide, since HbO_2 dissociates so much more readily than CO-Hb. However, in consequence O_2 increases the transport rate of CO by competing with it for Hb on the delivery side. Moreover, CO competes equally favourably with O_2 for cytochrome oxidase so that the net advantage to O_2 may be slight.

The Wittenberg mechanism probably operates in the polychaete, *Thoracophelia*, (SCHOLANDER, 1960) and in the leg-Hb of the root nodules of the leguminosae, where it supplies O_2 to the N_2-fixing bacteria (WITTENBERG, 1970). It seems likely that the haemoprotein of ciliate Protozoa (KEILIN and RYLEY, 1953; SMITH et al., 1962) may function in this way. Haemerythrin also shows the phenomenon (WITTENBERG, 1963) so that it is not restricted to the porphyranoproteins. It is also shown by simple analogues of the prophyrans, such as cobaltodihistidine (WITTENBERG, 1970). WITTENBERG records a number of more distant analogues, transporting materials other than O_2.

9.1.1.1 Chlorohaemoglobins

In all cases examined ChlHb is a high pressure, low affinity, carrier, like vertebrate PrHb (FOX, 1932, 1949, 1951). It is only 90% loaded even at atmospheric pressure (JONES, 1963), and may be expected to have little chromogenic response to a decrease in ambient O_2. The equilibrium curve is shallow so that the p_{50} is as low as 29 mm Hg at pH 7.7 and 20°C. The operative exchange-range is very broad and CO depresses respiration at all values of the external O_2 pressure (EWER and FOX, 1940, 1953). There is a strong Bohr effect (Table 9.1), as for most high pressure PrHbs, and it resembles the latter also in temperature response, and in forming MetHb with strong oxidising agents. As already noted (p. 218) its affinity for CO is even greater than that of the PrHbs (Table 9.4). Where, as in all species of *Sabella*, there is both ChlHb and PrHb in the blood the latter also is of the high pressure type (FOX, 1951); both deoxygenate at the same p_{50} and it is understandable that FOX should consider ChlHb an indifferent or 'neutral' mutant of PrHb. Nevertheless the neo Darwinian school of FISHER and HALDANE no doubt will reserve judgment on this issue.

The ChlHb of *Spirographis spallanzani* has an O_2-capacity of 9.1 cm³ per 100 cm³ of blood (Fox, 1926), higher than that recorded for any invertebrate PrHb except that of *Glossoscolex* (JOHANSEN and MARTIN, 1966). Studies with CO indicate that ChlHb carries only 1/3 to 1/2 of the O_2 consumed by the tissues of serpulimorphs (JONES, 1953) whereas a total capacity of 9.1 cm^3% would imply 94% in the form of HbO_2. The capacity may not be higher than that of most annelid PrHbs, after all. This chrome merits further study at both physiological and biological levels.

9.1.2 Haemerythrins

Except in the sipunculids little is known of the physiology of this class of O_2-carrier. In *Sipunculus* the p_{50} is 8 mm Hg (FLORKIN, 1934) and in *Golfingia* 2.9 mm (KUBO, 1953), indicating a low-pressure exchange. At atmospheric pressure the chrome does indeed make only a minor contribution to the total tissue respiration (JONES, 1963). The coelomic fluid is in equilibrium with O_2 pressures of 32 mm Hg so that the Hery is normally more than fully loaded by diffusion through the body wall. Under these conditions the tissues also probably obtain enough O_2 by simple diffusion from the medium and from solution in the coelomic fluid. As burrowing animals, however, both sipunculids and priapulids must experience periods of low ambient O_2, when Hery-bound gas becomes readily available and easily replaced. The lingulid brachiopods have a good coelomic circulatory system, but sipunculids and priapulids have only incidental movements imposed by the body wall musculature, so that a Wittenberg-Scholander diffusion-carrier mechanism in static conditions may make a major contribution to tissue-supply at low pressures.

Although the O_2-capacity is only 1.6 cm^3 per 100 cm^3 of fluid, about three times the amount in simple solution, it is 21 cm^3 per 100 cm^3 of packed cells; in *Priapulus* the latter tend to adhere to the body wall and other tissues so that the Wittenberg contribution may be considerable. It is established that one molecule of O_2 is carried per two atoms of Fe (MANWELL, 1964). Like Hb, this carrier can be oxidised to MetHery, which no longer binds O_2 reversibly. The implication, that as in Hb the iron remains in the Fe^{II} state throughout the exchange-cycle, is questionable, however (p. 234). A positive Bohr effect has been detected in the lingulid brachiopods but not in the sipunculids (MANWELL, 1960); clearly the former have a more high pressure exchange.

The sipunculid, *Dendrostomum*, has two Herys and, unlike the two Hbs of *Nephthys* (p. 220), both are coelomic. However, one is in the general cavity and the other in the tentacular coelom: this confirms that there is no connection between the two systems (HYMAN, 1959, p. 635). The Hery of the general coelom has a p_{50} of 4.5 mm Hg whereas that of the tentacular system has a p_{50} value of 40–45 mm. The latter therefore is a high pressure type, unloading to a significant extent as soon as the pressure falls below atmospheric (MANWELL, 1960, 1958a).

In *Siphonosoma ingens*, a burrowing form, by contrast, the tentacular fluid has a higher affinity than that of coelom, and receives O_2 via the latter; the two Herys differ also in electrophoretic mobility.

9.1.3 Haemocyanins

The Hcys of Crustacea contain 0.15–0.19% copper and those of cephalopods 0.24–0.26%, compared with the 0.34% Fe in Hb and 0.9–1.0% Fe in Hery (PROSSER and BROWN, 1961). This is equivalent to one atom of Cu per 37,000 and 22,500 daltons of protein respectively, a unit of the same order of magnitude as the Huffner unit of Hb.

One molecule of O_2 is bound per two atoms of Cu, comparable to the binding by Fe in Hcy but only a half of that by haem iron. The O_2-capacity is linear in Cu-content. The Hcys of cephalopods are high pressure exchangers and this is true of some members of the other taxa. None have very low p_{50} values (Table 9.1) and this carrier does not appear to be particularly useful for difficult rspiratory situations. Nevertheless there are a number of interesting variants and peculiarities in the different taxa.

Cephalopod Hcy has a particularly large positive Bohr effect so that the p_{50} of the squid, *Loligo*, one of the most active of cephalopods, is 96 mm in the presence of 6.5 mm pressure of CO_2 but only 36 mm in the absence of this gas. That of the more sluggish cuttlefish, *Sepia*, is only 8 mm in the absence of CO_2, so that it is scarcely a high pressure type except in the presence of at least physiological concentrations of CO_2 (WOLVEKAMP et al., 1942). Another consequence is that, like teleosts, these molluscs die of asphyxia among abundance of O_2 if they cannot eliminate CO_2 promptly at the ventilation surface. None have an O_2-capacity rivalling the most efficient of the Hbs but at the tissues they exchange more than 90% of the O_2 carried, compared with 30% for human Hb. The actual amount of O_2 delivered per unit volume therefore is very nearly as great in the squid as in Man. However, the requirements of this relatively small animal, per unit weight, are three times as great as in Man. For a complete picture, the physiological parameters, blood volume %, heart-rate, stroke volume, etc., all would need consideration. *Sepia* and *Octopus* have progressively higher O_2-affinities and lower Bohr effects, arterial blood being in equilibrium with 55–85 mm pressure of O_2. At higher pressures presumably their more limited requirements are met by O_2 in simple solution.

Among gastropods the pulmonate *Helix* (Fig. 9.1,d) has a p_{50} of 12 mm Hg in the absence of CO_2, at 20°C, and the prosobranch *Busycon*, 15 mm (REDFIELD et al., 1934; REDFIELD and INGALLS, 1932, 1933). That of *Cryptochiton* also is as high as 18 mm (MANWELL, 1958b, 1960b) so that the Hcy of these molluscs appears to overlap the cephalopod range. However, those of *Helix* and *Busycon* have a negative Bohr effect and so are essentially low pressure exchangers. Arterial blood is in equilibrium with an O_2 pressure of 36 mm Hg and 1.7 volumes %; 2/3 of the total capacity is delivered at the tissues. This is ten times the amount carried in simple solution (FLORKIN, 1949) and this Hcy is a continuously functioning exchanger. REDMOND (1968) points out the value of a negative Bohr effect in saving the available O_2 for tissues suffering from a very low O_2 pressure because of the relatively hermetic shell. The shell is permeable to NH_3 (CAMPBELL, 1970) and no doubt to other gases but the transport rate may be inadequate for serious respiratory purposes.

The helicid pulmonates have two types of Hcy molecule, α and β (GRUBER, 1968). Their most important difference is that α has a positive and β a negative Bohr effect so that the whole blood may function effectively as both high and low pressure exchanger, according to conditions. In *Helix pomatia* the ratio of α to β is 3:1 but in *H. aspersa*, from lower latitudes, it is 1:3. A tropical species of the apple snail, *Pila*, has only the β-form. This has a shallow Eqbm-curve so that it is also a wide-range, slow-rate exchanger. The slug, *Agriolimax*, has a strong positive Bohr (MANWELL, 1961) and presumably it has the α-form of Hcy only. *Pleuroploea* also is a positive Bohr type (REDMOND, 1968). The chitons, again, seem to be high pressure types though with the shallow type of curve: this may be characteristic of sluggish animals living in fully aerobic conditions. The Hcy is rarely outside the range 45 to 90% saturated.

Crustacean Hcys have even more curious features, though they are orthodox carriers in some of the larger and more active species. Like the Cu-content the O_2-capacity is lower than in molluscs though it compares favourably with that of all but the cephalopods. P_{50} values are in the same range as those of gastropods, 15 mm Hg in the spider crab *Loxorynchus*, 14 mm in the rock lobster *Panulirus* and the lobster *Homarus*, and 8 mm in the crab *Cancer* (REDFIELD et al., 1929–1933; WOLVEKAMP and VREEDE, 1941; WOLVEKAMP and KRUYT, 1947; REDMOND, 1955). All have a very strong positive Bohr so that the p_{50} of *Homarus* is shifted up to 102 mm by a 38 mm pressure of CO_2; this amount of CO_2 is scarcely physiological but it is interesting to note that the Bohr mechanism continues to operate over such a wide CO_2 range. This type of Hcy certainly is a high pressure carrier but its potentiality is reduced by a rather limited loading and unloading, i.e., a shallow Eqbm-curve. The arterial blood of *Panulirus* is only 54% saturated (JONES, 1963) and there is only 30% unloading at the tissues (REDMOND, 1955). Nevertheless the exchange is still ten times that possible from simple solution in water. *Homarus* also has an outstandingly high O_2-capacity, due to its high Hcy concentration (STEDMAN and STEDMAN, 1925, 1926). The terrestrial crabs also have high pressure loading-ranges (REDMOND, 1968).

By contrast, *Panulirus*, the spider crab *Maia*, and others appear to use their Hcy rather casually. The former delivers about 0.46 cm³ of O_2 per 100 cm³ of blood to the tissues, only two to three times more than could be carried in simple solution (JONES, 1963). *Maia* carries only about half of the oxygen it consumes as oxy Hcy (ZUCKERKANDL, 1956); the total capacity of its Hcy is low and even so it leaves the gills very incompletely saturated (REDMOND, 1968). This implies that loading may be slow in Crustacea and JONES (1963) suggests that even over the respiratory surface of the gills the exoskeleton is relatively impermeable. If this in any way limits exchange then to have Hcy fixing some of the O_2 taken up, and keeping the gradient across the surface steep must be salutary; this will be true even if neither Hcy nor the blood-fluid is completely saturated when it leaves the gills. It is noteworthy that the arterio-venous difference in CO_2 pressure in Crustacea may be as low as 0.4 mm Hg so that this gas may pass into, or across, the exoskeleton more readily than O_2.

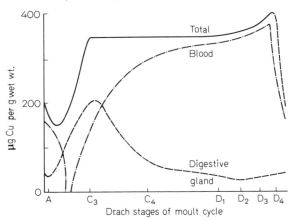

Fig. 9.3. Changes in concentration of copper in blood, digestive gland and whole body of spider crab Maia during one complete moult-cycle. Explanation in text. (After ZUCKERKANDL, 1956, 1957)

In any case it seems that Hcy is not of such vital importance for respiration in *Maia* (ZUCKERKANDL, 1956, 1957), *Crangon* (DJANGMAH, 1970) and some other Crustacea. In *Maia* its concentration in the blood varies greatly over each instar and is virtually zero in stage C_2 (DRACH, 1939), just after the phase of maximal tissue-proliferation (Fig. 9.3). In fact it changes systematically throughout the 'moult-cycle' in a way implying a morphogenetic rather than a respiratory relevance. The concentration is maximal just before the moult, in stages C_4 to D_3, and it falls rapidly at the time of moulting, mainly due to dilution by the water then assimilated to effect the necessary size-increase. Subsequently it continues to fall, however, mainly associated with a rapid uptake of Cu by the 'liver' (Fig. 9.3). The copper content of the latter in fact varies almost inversely with that of the blood, over most of the critical period of rapid change, with a time lag which further supports the view that blood-Cu is transferred there for temporary storage. Blood-Cu begins to increase again before that in the liver begins to fall and so the Cu initially comes mainly from the external medium. Eventually the liver Cu-level reaches a low steady value and blood Hcy a correspondingly high steady level. Much the same changes have been recognised in *Crangon* (DJANGMAH, 1970) so that it may be common in decapods. It would be useful to know if fresh water types also show the phenomenon, since marine Crustacea might be using the protein from the Hcy to produce free amino acids in order to facilitate the osmotic uptake of water at moulting (DUCHATEAU et al., 1959; ROBERTSON, 1960); fresh water members do not have this problem. Alternatively, or in addition, the amino acids may be used for growth, and this would be relevant in both marine and fresh water forms.

It thus appears that Hcy is dispensable from respiratory functions during at least a part of the instar. This is the main morphogenetic phase and it might be expected to make least demands on aerobic respiration (NEEDHAM, 1964); moreover it is a phase of more intense morphogenetic activity than in any large individuals outside this phylum. When the Hcy concentration of the blood first begins to fall, just before moulting, it does saturate more fully at the gills and so over much of the cycle it may be in excess of respiratory needs. Its function as a store of protein may operate throughout the cycle. It is also known to be the main buffer of the blood (PROSSER and BROWN, 1961).

Among the Hcys of chelicerates only that of *Limulus* has been studied in any detail (REDFIELD et al., 1929–1933). Its O_2-capacity is between 0.7 and 2.7 cm^3 per 100 cm^3 of blood and the p_{50} is 11–23 mm (Table 9.1). There is a considerable negative Bohr effect over the physiological range, i.e., below pH 8.3, so that at normal CO_2 pressures it is likely to be a low pressure exchanger. It is fully loaded at 30 mm pressure of O_2, in fact. At more alkaline pH values there is a small positive Bohr which scarcely can have biological significance. The p_{50} is highest at the pH of well oxygenated ocean water. Spending a good deal of its time buried in the sand, Limulus must often operate under conditions of greater anoxia and higher CO_2 content. The spider, *Heterometrus*, has a positive Bohr effect (REDMOND, 1968) so that both Bohr types occur in every major taxon using Hcy. The negative Bohr is most common in animals with an open type of circulation, whether the carrier is Hcy or Hb (REDMOND, 1968), but neither generalisation is exclusive.

Like the vertebrate Hbs, those Hcys with a positive Bohr are better pH-buffers in the deoxy than in the oxy state, and conversely when the effect is negative. The Hcy molecule is too large to have a significant Wittenberg effect (REDMOND, 1968) and presumably this is true of some annelid Hbs. However, that of *Thoracophelia*, which shows the effect (p. 223), is vascular (KENNEDY, 1969) and therefore probably plasmatic, implying a large molecule.

9.2 The Molecular Basis of O_2-Transport

For PrHb there is now considerable knowledge at this level and it gives a picture of adaptation at least as impressive as that of the grosser levels. There is great interest also in the molecular physiology of Hcy and Hery but here major problems are still unsolved, though there is adequate confirmation that the whole molecule is involved, -metal atom, ligand field and protein. The polymeric structure also is significant. The remarkable property of O_2-carriers to be explained is their ability to bind molecular O_2 reversibly, as though it were in simple physical solution, yet stoichiometrically as in a typical chemical reaction.

With a 1:1 stoichiometry between metal atom and O_2 the mechanism in the Hbs must differ in some essentials from that of the non-porphyrin carriers, with their 2:1 stoichiometry. All three main carriers differ in their heats of oxygenation, -8, -13 and -18 Kcal/mole respectively for Hb, Hcy and Hery (KLOTZ and KLOTZ, 1955) and this could be taken as one measure of their relative efficiencies. It may also indicate that O_2 is transported with no significant change by Hbs, but as the perhydroxyl ion, O_2^-, by Hcys and as the peroxo ion, $O_2^=$ by Hery. The ionic forms imply a bridged structure of the general form $M - O_2 - M$, which would explain the 2:1 stoichiometry. Cytochrome oxidase (p. 186) releases O_2 as one or other of these ions (WILLIAMS, 1966a) and it has two copper atoms in the molecule. Since all three types of carrier are related to types of terminal oxidases the study of oxidases and carriers should be complementary.

The essential properties for O_2-transport appear to be inherent to all coordination-complexes of the transitional metals (McGINNETY, PAYNE and IBERS, 1969), the protein being a refinement rather than a *sine qua non:* simple cobalt complexes reversibly bind O_2 with the same correlated changes in magnetic and other properties as the natural carriers (MARTELL and CALVIN, 1952). Of the essential chromatic properties of the transitional metals (pp. 15, 47) those to be emphasised in this context are not so much redox properties as those of facile and labile bonding. For this the extensive overlap of energy levels between 3d, 4s, and 4p electronic orbitals is invaluable: in certain ligand fields even thermal energy is enough to excite electron-transitions (BERNAL, 1954). This second reason to call them the transitional elements of course is not independent of the original reason.

A good index of the versatility of these metals is the wide range of valency and oxidation states in which they have been detected. Iron is usually Fe^{2+} or Fe^{3+} but the ferryl, Fe^{4+} ion has been found in myoglobin derivatives (CAUGHEY, 1967), and occasionally the Cu^{3+} ion has been detected in addition to the usual Cu^+ and Cu^{2+}. The effective oxidation state is even more variable, owing to electron-delocalisation over the whole ligand field: Fe^V and Fe^{VI} are recorded (YAMAZAKI et al., 1966; YONETANI et al., 1966) as well as the more usual Fe^{II}, Fe^{III}, and Fe^{IV}, and the high values are particularly commom in haem compounds. Again although iron normally bonds with up to six ligands occasionally it has seven (WILLIAMS, 1966b), or even eight (CAUGHEY, 1967) in Fe-chromoproteins.

Excitation of the relevant orbital electrons of these metals produces *high-spin* or *spin-free* states with a varying number of orbitals containing only one electron and making the atom paramagnetic. The Fe-complexes of Hb and Hery are very lightly 'poised' near the 'crossover point' between the high-spin and the *low-spin*, or *spin-paired*, ground state (LUMRY, 1971). The deoxy forms have high-spin Fe and so these very readily bond with oxygen and fill orbitals in that way: oxyHb and oxyHery are low-spin. The Cu of Hcy similarly is very lightly poised between two oxidation states (MORPURGO and WILLIAMS, 1968) and oxygenation causes a uniquely anomalous change in its oxidation state (p. 235). The iron of Hb remains ferrous throughout the oxygenation cycle. It might be anticipated that an oxygen-binding chrome should have a redox potential near that of unreduced oxygen, + 800 mV, but in fact that of Hb is lower than that of cytochrome *c* (GEORGE et al., 1966): these features again emphasise that redox properties are of secondary importance.

Another relevant property is that oxygen and nitrogen induce maximal coordination-number in these metals. DeoxyHb has five nitrogen ligands so that these and free oxygen both strongly promote the essential loading of the vacant site with O_2 itself. The ligands of the deoxy forms of the other carriers are unknown but all the evidence points to nitrogen, possibly with oxygen at a limited number of sites. Nitrogen forms the stronger bonds so that the ideal situation is with oxygen only at the carrier site: its smooth and facile unloading is then maximally ensured. Nevertheless oxygenation increases the stability of the O_2-carrier molecules and this effect is proportional to the affinity of the carrier for O_2.

9.2.1 Haemoglobins

In Hb the ligand field is now well known. Apart from the four pyrrole N-atoms of the porphyrin, an imidazole-N of one histidine residue of the globin occupies a fifth site, leaving the sixth vacant for O_2. DeoxyHb is therefore 5-coordinate only, yet surprisingly the empty site does not greatly unstabilise the system; on the contrary if an imidazole N occupies this sixth site, as it does in haemochrome models in oxygenated aqueous solution, the Fe is immediately oxidised to ferric (WANG, 1966). Even MetHb, with an OH ligand at the oxygen site, and the Fe in the ferric form is less stable than Hb (CAUGHEY, 1967). As in almost every respect, the natural chrome is best adapted to its biological purpose. If Hb is lyophilised (freeze-dried) as a thin film on the wall of a container it will oxidise explosively on admitting air (DRABKIN, 1966); this shows that water is a major stabilising component which usually can be ignored because normally it is always present. It is an actual ligand in some of the haem-complexes and was long thought to hold the sixth site in deoxyHb.

A curious further fact is that O_2 in the polar position of the octahedral system typical of 6-coordinate complexes, the position of least strain, in theory should oxidise the Fe to ferric and there is some evidence that in both Hb and Mb it is bonded at an angle much nearer the plane of the porphyrin (CAUGHEY, 1967). The high positive charge on the Fe minimises any tendency for it to reduce oxygen, and itself to be oxidised to ferric, but there are some haemoproteins with a high positive charge on the Fe which do reduce O_2; no doubt much depends on the distribution of orbital electrons in the whole ligand field.

It is very unusual for Fe^{2+} to form paramagnetic complexes as it does in deoxyHb; oxy Hb is diamagnetic, and this is strong proof that its iron is not ferric; ferric iron has an odd number of orbital electrons so that one at least must be unpaired and the atom is always paramagnetic. For ferrous iron to be paramagnetic and high-spin, 2 or 4 of its electrons must be unpaired. In principle the diamagnetic state in oxyHb could be due to spin-paired Fe^{3+} atoms but in Hb the Fe atoms are too far apart for this. The 8 kcal oxygenation-energy (p. 228) is released as O_2-bonding resolves the high-spin state; this is a small difference compared with the 40 to 50 kcal difference between the high and low spin states of some Fe-complexes (LUMRY, 1971) and shows that oxy Hb remains poised near the cross-over point (p. 228). One consequence, of course, is that oxyHb is not much more stable than deoxyHb and unloads its O_2 rather readily, as already stressed

(p. 214). Labelled free O_2 will exchange with O_2 already loaded on Hb (MARICIC, 1966) and this is another index of the relative spontaneity of both phases.

The anomalous stability and other properties of Hb (PERUTZ, 1970) depend to a great extent on the protein moiety and this is probably why, in spite of the fact that taxonomic, and any other, differences between Hbs depend on the protein part, the globins are all relatively similar, notwithstanding also the polyphyletic origin of the PrHbs. The monomer units of all are much the same size (p. 237) and there are many resemblances in gross amino acid composition, primary amino acid sequence and tertiary and quaternary structures. Of course the polyphyletic origins may have been always from a very similar urprotein. Among Hbs, resemblances in tertiary structure are stronger than in amino acid sequence, just as among enzymes (PHILLIPS, 1966); function depends on this more than on primary structure.

As in most conjugates (p. 30) the protein does have a marked stabilising action (SCHEJTER, 1966) and this is further increased by polymerisation. The monomer, myoglobin, is not so stable as gnathostome Hb (CAUGHEY, 1967). In their haemochromes with pyridine and other N-bases, Mb and the monomer Hb of the lamprey show an unexplained twinning of the α- and β-absorption peaks (KEILIN, 1966) but the tetramer Hb of gnathostomes is not affected in this way. Cytochrome c, another monomer, shows the splitting of the α-peak on reduction (ELLIOT, 1966). Among the more specific functions of the Hb-peptide is the formation and control of a 'pocket' to receive the oxygen molecule (p. 232).

These are basic, relatively static, structural properties but the protein also largely controls the events of the oxygenation cycle of Hb; the proteins of other types of haemoprotein are equally specific to their particular cycle (GEORGE et al., 1966). It is now known in some detail (PERUTZ, 1970) how many of the constituent amino acids and tertiary and quaternary structure features of mammalian Hb cooperate to control the oxygenation-cycle. There is a general conformational change, as there is in the reaction-cycle of enzymes (PHILLIPS, 1966). This has been shown both theoretically, by thermodynamic analysis of kinetic and equilibrium data (ELEY, 1943), and by X-ray diffraction studies leading to the construction of steric molecular models. There are changes in crystal form (HAUROWITZ, 1938) and in solubility (MANWELL, 1964). Curiously enough human HbO_2 is much more soluble than its deoxyHb but the converse is true for rat Hb; there must be considerable taxonomic differences in the molecular cycle.

On oxygenation there is increased order, i.e., decreased entropy, of the molecule (WYMAN and ALLEN, 1959) which is relevant to the nature of the increase in stability (p. 229). The magnitude of the entropic change shows that many charged groups are involved (ELEY, 1943; MANWELL, 1964) and is a measure of the number of amino acids which undergo active changes. Since HbO_2 is more resistant to X-rays and proteolytic enzymes than Hb it is clear that the whole set of conformational changes, and not just the behaviour of O_2 (p. 232), is responsible for the stabilisation. In fact the stabilisation of gnathostome Hb by O_2 itself seems to be a property of the particular protein: lamprey deoxyHb is somewhat polymeric but it is destabilised, and dissociates, on oxygenation. This unusual behaviour is shown also by some Hcys (p. 237).

It has always been clear that the very modest degree of polymerisation in vertebrate Hbs, which in any case are intracellular, can have little significance for reducing osmotic and viscosity effects; quite early the interaction between the four was shown to determine the sigmoid form of the Eqbm-curve and the Bohr effect. The index, n (Table 9.1) is a measure of the average number of interacting units. The positive Bohr effect has been correlated with attractive interactions between units and the reversed or negative Bohr with repulsions. The Root effect (p. 215) similarly is due to pH-dependent repulsions (MANWELL, 1964). The interacting units rather naturally were assumed to be the four haems but these are all on the outside of the molecule, in gnathostome Hb (PERUTZ et al., 1960); here they are easily accessible to O_2 but cannot interact directly with each other (PERUTZ, 1970). Proteolytic enzymes depress the Bohr effect and so do protein-reagents such as bromthymol blue. The interactions are between peptide units, rather than specifically between the haems, and indeed the haem itself figures relatively little in present accounts of the O_2-transport process (PERUTZ, 1970).

In the giant Hb molecules of *Arenicola* (Fig. 5.7b) the essential functional unit is probably not the whole molecule but the $1/_{12}$ subunits (WEBER, 1970), of size 250,000 daltons and containing probably 16 monomer units (p. 211). This is still much larger than the vertebrate tetramer and may be a concession to viscosity and osmotic requirements so evident in the whole molecule (p. 211). In fact the value of n for earthworm Hb is 5.4 (BOELTS and PARKHURST, 1971) compared with 1.8–3.0 for vertebrate Hb (Table 9.1) and so the functional unit is likely to have at least twice as many monomers as the latter and may be the whole $1/_{12}$ molecule. At 15,625 daltons the monomer as slightly smaller than that of vertebrate Hb but it would seem that the optimal size of the sphere of influence of one prosthetic group is fairly constant. That in the Herys is smaller and that in the Hcys larger, but of the same general order.

Actually there is also some interaction between neighbouring whole molecules (MANWELL, 1961). Barcroft many years ago demonstrated a shift in the O_2-Eqbm curve of a solution of vertebrate Hb towards a lower O_2-pressure range when the solution was diluted (SCHEJTER, 1966). This implies that these interactions also have some Bohr effect and so the high concentration of vertebrate Hb in the erythrocyte (p. 125) has a further advantage. Interactions throughout the giant molecules of invertebrate plasmatic Hbs also seem likely, therefore, and there is already some evidence for this in Hcy (p. 237).

The emerging picture of the behaviour of the various components of the molecule in the course of the loading-unloading cycle in mammalian Hb (PERUTZ, 1970) is as fascinating as that of any macromolecular microcosm so far explored. So many of the amino acids move and reorientate, and so much of the tertiary and quaternary structures changes (Fig. 9.4) that effectively the whole of the tetramer is actively involved. The deoxy and oxy states of the monomer are both more stable than any intermediate state and their whole conformation clicks over from one to the other once incipient changes reach a certain threshold; it becomes spontaneous both ways. The switch-over property is equally evident in the tetramer as a whole, and moreover has greater lability. Here every component of the transition either way facilitates further components and the situation comes nearer to a smoothly oscillatory phenomenon than to a flip-flop switch. The actual direction of the change at any place and time depends on local concentrations of O_2, CO_2 and other relevant factors. The autogenic facilitation is expressed in the sigmond Eqbm-curve which implies that either transition becomes increasingly rapid with each unit shift in O_2 pressure in the relevant direction. It is expressed also in the finding that an Hb tetramer takes up the second and third O_2 molecules progressively more rapidly than the first whereas taking up the fourth is again more difficult. The same applies to the unloading, so that the molecule is encouraged to remain in its most labile and functional state. It will be noted that for the molecule as a whole this is the intermediate state; intuitively it seems that the combination of this with a flip-flop for the individual monomers must ensure ideal auto-control.

The four Huffner units of the molecule of gnathostome Hb are two of each of two distinct peptides, called α and β, which also behave differently. For instance, the α-peptides are the more important for the Bohr effect (RIGGS and HERNER, 1962). In mammalian Hb this important difference involves a further component of the molecule only recently revealed (BENESCH and BENESCH, 1969), namely, 2,3 diphosphoglyceric acid (DPG). This binds in a pocket of the deoxyHb molecule, stabilises the structure by the equivalent of

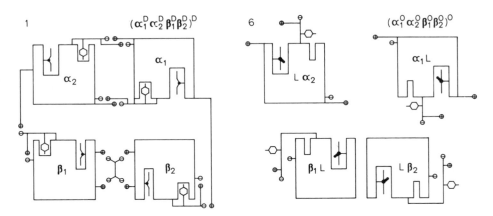

Fig. 9.4. Diagram of suggested conformational differences in a molecule of mammalian haemoglobin between deoxy- and oxy-states. In deoxy-Hb (1) a molecule of diphosphoglycerate (DPG) is bonded between the two β-peptides and all salt-bridges between and within peptides are intact.
In oxy-Hb (6) the operative tyrosine residues of all four peptides have vacated their narrowed pockets, the widened haem pockets are oxygenated, all salt-bridges are broken including the internal bridges of the β-peptides, DPG has been ejected, the β-peptides are closer together and the α-peptides further apart. (From PERUTZ, 1970)

four extra bonds with the two β-peptides and further increases the efficiency of the loading process; incipient oxygenation-changes eject DPG from the molecule and lower the energy required to bind O_2 to further haem units. Like the rest, these component changes are reversible. DPG provides the only direct links between the two β-peptides and its relation specifically to these units implicates it in the Bohr effect. In reptiles and birds inositol hexaphosphate is used instead of DPG and in fishes possibly ATP. For mammalian Hb the latter has an affinity lower by several orders of magnitude than that of DPG; the implication is that DPG is a very efficient recent evolutionary acquisition.

Details of the process of oxygenation are given in PERUTZ's paper. Fig. 9.4 shows diagrammatically the essential differences between the limiting deoxy and oxy states of the tetramer and the main points about the process are as follows. The lability of bonding of O_2 is largely explained by the polar character of the pocket into which it must fit. In the α-unit this pocket is always accessible but those of the β-units are freed of certain blocking amino acids only after oxygenation has begun. Quaternary bonding between the four units increases bond-lengths between Fe and ligands (PERUTZ, 1972) and depresses the O_2-affinity of the haems, particularly of the β-units, but the first and second O_2 molecules loaded, by the two α-units, break these bonds and eventually expose the β-pockets. The important quaternary bonds are between $α_1$ and $β_2$ and between $α_2$ and $β_1$; the gap between the members of each pair widens by 0.7 nm when the bonds between them break. The rupture strains, and so breaks also, the DPG bonds with the β-units. The energy needed finally to clear the O_2 pockets of the β-units is now much less, and O_2 itself is able to complete the process.

Intra-unit movements are equally vital. They depend on the geometry of the haem group, which amplifies the small change in atomic radius of Fe, due to its transition from high to low spin state, into a large movement of the critical histidine ligand, relative to the porphyrin ring. This triggers the tertiary flip-flop within the unit and initiates the inter-unit, quaternary changes. Within the β-units the tertiary changes are further responsible for an essential movement of the 'E'-helix region of the peptide away from the porphyrin ring so as to widen the O_2 pocket enough to take O_2 as a ligand. A movement of the 'H'-helices of these units expels the DPG. It is the binding of an O_2 molecule in the first α-unit which initiates the tertiary changes, while quaternary changes are mainly responsible for transmitting the resultant effects to further units.

The Bohr effect depends on three pairs of weak base groups, two histidine imidazole Ns and the α-NH of a valine residue; these are not ionised in oxyHb but in deoxyHb are brought nearer to COOH groups and so take up a proton from ambient water. These are called the Bohr protons and naturally any stronger acid than water will promote or speed unloading. Equally Hb acts as a buffer by mopping up these H^+ ions.

The great significance of histidine residues, long indicated by studies at grosser levels, is now fairly clear. The histidine which ligates Fe is that conventionally labelled F8. Another histidine residue, E7, lies near the O_2 and contributes to the polar character of the O_2-pocket. Histidine H21 (143) of the β-unit is involved in the electrostatic bonding of DPG, while histidine HC3 (146), the C-terminal amino acid of the $β_1$-unit, is bonded via its α-COOH group to a lysine of $α_2$, and similarly for $β_2$ and $α_1$. Both histidines 146 are also bonded via an imidazole N to an aspartic acid residue of their own units. The two histidine residues, which contribute to the Bohr effect, his. 146 of the β-units once more, and his. 122 of the α-units, do so by the relevant imidazole N becoming a quaternary N-base cation. Tyrosine, valine, arginine and cysteine residues all have specific roles but the emphasis is very much on histidine.

What is known about ChlHb indicates that at the molecular level also it is very similar to PrHb. It seems equally clear, however, that the significant differences in behaviour between the two must have their basis at this level.

9.2.2 Haemerythrins

Knowledge at this level is virtually restricted to the Hery of sipunculids. The nature of the prosthetic group is very uncertain and nothing is known of the tertiary and quaternary structures of the protein. An approach to these problems is now possible since the primary amino acid sequence of the monomer in the Hery of *Golfingia* is now known (KLIPPENSTEIN et al., 1968). It is probable that each Fe atom of the monomer is bonded to an imidazole nitrogen atom of two histidine residues of the peptide (FAN and YORK, 1969) and to one tyrosine residue (FAN and YORK, 1972) so that if the fourth ligand also has an aromatic ring the whole field may have some resemblance to a pseudoporphyran. Electronic orbitals are distorted in the vicinity of the Fe atoms (YORK and BEARDEN, 1969) and this implies an octahedral system rather like that of Hb (p. 229).

The next problems are the states and behaviour of the two Fe atoms of each monomer and the form in which O_2 is carried (OKAMURA et al., 1969; GARBETT et al., 1969; YORK and BEARDEN, 1970; MOSS et al., 1971; DAWSON et al., 1972). DeoxyHery and still more oxyHery present serious technical difficulties so that present views are largely inferences from the behaviour of metHery. However, it seems certain that the two Fe atoms of a unit are sufficiently near each other to cooperate in carrying one O_2 molecule as the 2:1 stoichiometry (p. 228) implies. It is also certain that in deoxyHery the Fe is in a high-spin, paramagnetic Fe^{II} state, (KUBO, 1953; MANWELL, 1964; GARBETT et al., 1969), just as in deoxyHb: there are four unpaired electrons (YORK and BEARDEN, 1970).

OxyHery, like oxyHb, is diamagnetic (OKAMURA et al., 1969; MOSS et al., 1971) and by analogy with Hb the Fe was assumed to be low-spin Fe^{II} (KUBO, 1953; YORK and BEARDEN, 1970). KLOTZ and KLOTZ (1955) found only ferric iron and this has been confirmed by DAWSON et al. (1972) using an ultrasensitive magnetometer. This indicates that the diamagnetic state is due to 'antiferromagnetic' coupling between the two Fe^{III} atoms, which individually must be paramagnetic; this type of coupling exactly opposes the spins of the unpaired electrons of the two atoms as effectively as though they were paired in the same orbital (cf. OKAMURA et al., 1969).

This coupling apparently is effective via an oxygen bridge (p. 228) for which there is considerable evidence (OKAMURA et al., 1969; GARBETT et al., 1969; DAWSON et al., 1972) but a symmetrical $Fe-O_2^=-Fe$ structure is excluded by many indications that the two atoms have different 'environments' in oxyHery (YORK and BEARDEN, 1970; GARBETT et al., 1969; SHUGAR et al., 1972; DAWSON et al., 1972); only one has tyrosine as one ligand (RILL and KLOTZ, 1970). This leads to the conclusion that the bridge consists of one atom only of the O_2, the second being bonded to only one of the Fe nuclei in a complex such as $Fe^{III} - O^{2-} - Fe^{III} - O^- - OH$ (DAWSON et al., 1972). The relatively high energy of oxygenation (p. 228) presumably is required to separate the two O atoms.

There is little change on oxygenation in optical rotatory dispersion and circular dichroism (p. 20) and other optical properties (BOSSA et al., 1970) and this implies no gross conformational changes in the protein. On the other hand crystals of deoxyHery, like those of deoxyHb, spontaneously fragment when allowed to load with O_2. Consequently there must be considerable force generated by movements at some level. There is evidence for inter-monomer interactions, the Fe-atoms helping to bind the monomers and stabilise the octomeric complete molecule (KLAPPER and KLOTZ, 1968; RILL and KLOTZ, 1970). One cysteine residue of each monomer also is involved and if these are blocked the octomer dissociates. The stabilising action of the Fe atoms is related to their O_2-binding function since anions which in the laboratory bind at the O_2 site cause depolymerisation. The free monomer has a higher affinity for O_2 than units in the octomer so that the Fe apparently uses equivalent bonds for O_2 and for neighbouring monomers. The units of vertebrate Hb similarly have a lower O_2-affinity than a free monomer (PERUTZ, 1970), and the fission-products of some Hcy molecules have a greater affinity than the same moiety in the intact molecule (p. 237).

The value of n (p. 214) for sipunculid Herys is usually 1.0 (BATES et al., 1968; FERRELL and KITTO, 1970) or very little more (Table 9.1) and there is virtually no Bohr effect, so that the type of interaction familiar in vertebrate Hb is slight. Lingulid Hery, which does have a positive Bohr, should prove interesting here. The shape of the O_2-equilibrium curve of Herys also implies little interaction between units.

9.2.3 Haemocyanins

More work has been done on the Hcys than on the Herys but the same basic problems remain, the nature of the prosthetic ligand field, the state of the metal in oxyHcy and the way in which O_2 is bound. Details of interactions between sub-units (p. 237) are less immediately important. A more general question, which should be answered first is: why is this copper protein more common as an O_2-carrier than Hery? To suggest as answer that Arthropoda and Mollusca are more abundant than sipunculids, brachiopods and priapulids merely restates the question: why is Hcy more efficient than Hery? For this several reasons may be given: copper has a higher affinity than Fe for O_2. The Cu^{2+} ion in fact forms stronger complexes with most ligands than any other divalent cation. Copper has the most complex and versatile stereochemistry of all the elements, with the possible exception of vanadium (JØRGENSEN, 1966). Moreover copper will reversibly oxidise and reduce under more acid conditions than Fe so that it may be particularly useful under such special conditions as high CO_2-concentrations.

Copper cannot compete with iron in the porphyranoprotein field, however, because of its inadequate coordination-number; for a porphyranoprotein carrier this must be 6 whereas for Cu^{2+} the maximum is 5, and for Cu^+ 4. In fact Cu is unable to bind simultaneously even porphyrin and protein (KEILIN and MCCOSKER, 1961) or porphyrin and oxygen (HILL, 1926) so that protein and oxygen is the only possible combination. Even so two metal atoms are required in Hcy, as in Hery, to bind one O_2-molecule.

With regard to the state of Cu during the oxygenation cycle: it is certainly Cu^+ in deoxyHcy since this is colourless and diamagnetic. OxyHcy is blue, implying the presence of Cu^{2+} which should be paramagnetic owing to its one unpaired electron. In fact, however, OxyHcy is diamagnetic, like the oxy form of the other types of O_2-carrier and gives no electron paramagnetic resonance signal (NAKAMURA and MASON, 1960; MASON, 1964; GHIRETTI, 1966). Chemical tests support the colour in indicating the presence of Cu^{2+}, as they do for the Cu in cytochrome oxidase, which also gives no EPR signal, and neither do a number of the other Cu enzymes with the typical cupric blue colour. Chemical methods should be treated with caution since sensitive biological materials may be changed by the reagents used, but this hazard does not account for the blue colour *in vivo*. Ceruloplasmin does give a strong EPR signal but has its own anomaly in handling O_2 without changing its blue colour. There are reducing ligands which will decolorise ceruloplasmin but the colour can be restored by making the system still more strongly reducing (MORPURGO and WILLIAMS, 1968). The complex anomalies of the redox properties of these systems are reflected also in the paradoxical response of copper proteins to H_2O_2 (FELSENFIELD and PRINTZ, 1959).

Again most Cu^{2+} proteins absorb around 610 nm whereas the peak for Hcy is at 570 nm (WOOD, 1968). Another aspect of the problem is that ageing solutions of Hcy do develop an EPR signal, indicative of Cu^{2+}, but accompanied by a proportional loss of O_2-binding power and of the 570 nm absorption-peak responsible for the blue colour (WITTERS and LONTIE, 1968). As much as 25% of the Hcy of *Homarus* may give an EPR signal *in vivo* (GOULD and EHRENBERG, 1968) probably indicating that ageing occurs even *in vivo*, just as some MetHb is formed in vertebrates. The ageing or denaturation releases simple inorganic ions from organic complexes in which the individual Cu atoms had lost their familiar identity.

Chemical estimations on oxyHcy regularly give only 50% of the copper as Cu^{2+} (SCHULMAN and WALD, 1951; KLOTZ and KLOTZ, 1955) and not 100%, or a variable proportion, as would be expected if the estimate were artefactual. Evidently the two atoms of Cu in each unit behave in some stoichiometric manner, as expected if they must always cooperate in binding an O_2 molecule. It is known that the EPR absorption can be quenched without changing the α^9 configuration of Cu^{2+} (HEMMERICH, 1966).

These facts collectively imply the possibility of a close analogy to the condition in oxyHery (DAWSON et al., 1972), i.e., that in untreated oxyHcy both Cu atoms indeed are Cu^{2+} but they form an oxygen-bridged binucleate complex: $Cu^{2+} - O_2^= - Cu^{2+}$ which is diamagnetic through antiferromagnetic coupling of the unpaired spins of the two atoms. The peroxo, rather than the perhydroxyl, ion (p. 228) is generally favoured now (BANNISTER and WOOD, 1969).

Chemical reagents used to test the state of the Cu presumably disrupt this complex and release one Cu^+, one Cu^{2+} and the perhydroxyl ion O_2^-. This is generally favoured as an intermediate stage in normal loading and unloading and the complete cycle may be represented:

$$
\begin{array}{ccccc}
& \overline{\text{Respiratory Surface}} & & & O_2 \\
Cu^{II} & & Cu^{I} & & Cu^{I} \\
| & & | & & \\
O_2^= & \longleftrightarrow & O_2^- & \longleftrightarrow & O_2 \\
| & & | & & \\
Cu^{II} & & Cu^{II} & & Cu^{I} \\
& \text{Tissues} & & & O_2
\end{array}
$$

The main difference from Hery seems to be that there is little indication of any differences between the ligand environments of the two Cu atoms.

Although there are now many clues, the ligand field is no better known than that of Hery. It resembles that of Hb to the extent that it stabilises the lower oxidation state of the metal in deoxyHcy (LONTIE and WITTERS, 1966) and that oxygenation stabilises the system still more (p. 229). Inorganic cuprous compounds in fact are particularly unstable and it is curious that apoHcy will bond only with Cu^+ and not with Cu^{2+} although this is both stable and much the stronger ligator (JØRGENSEN, 1966). The phenomenon seems to be another manifestation of the controlled instability widely exploited in living systems. It is recalled (p. 49) that in some of the copper-enzymes the metal is bonded in a pseudoporphyrin-field and that Cu^{2+} is rarely found to complex except with ligands in an approximately square conformation (FREEMAN, 1966b).

At one time it was thought that some of the 7–10 peptide cysteine residues per Cu-atom acted as ligands: Cu has a great affinity for SH-groups and it is noteworthy that the number of residues decreases in ageing Hcy in parallel with the loss of O_2-binding power and with the increase in free Cu^{2+} ions (LONTIE and WITTERS, 1966). However, reducing groups near the metal depress oxygen-binding and there are also other reasons for discounting cysteine as a ligand (GHIRETTI, 1966; WOOD et al., 1968).

WOOD and BANNISTER (1967, 1968) showed that at least one ligand per Cu is an imidazole-N of a histidine residue and it is significant what a variety of modes of bonding with the metal this residue shows as the pH of a simple model is varied (IHNAT and BERSOHN, 1970). In *Octopus* Hcy tyrosine residues also appear to be bound by the Cu (NARDI and ZITO, 1968) so that there is a close parallel to Hery (p. 233). Both of these aromatic amino acids also play an important rôle in the Hb cycle (p. 233) though tyrosine is not a direct ligand of the Fe. In addition tryptophan is probably directly bonded to Cu in Hcy since its fluorescence is 40% quenched by Cu in deoxyHcy and 100% by the Cu–O_2 group in oxyHcy (SHAKLAI and DANIEL, 1970; BANNISTER and WOOD, 1971). Moreover the absorption-band of the Cu–O_2 group in the u.v. range, at 320–381 nm, almost precisely corresponds to that of the fluorescence emission-band of tryptophan, 318–375 nm.

The Bohr effect in some Hcys (p. 225) implies inter-unit interactions and in the snail *Helix* GRUBER (1968) finds the functional unit to be twice the size of the monomer, involving four Cu-atoms. The copper-protein enzymes mostly have 2,4 or 8 Cu atoms and much the same size of peptide per Cu-atom as in the Hcys. The molecule of these enzymes behaves as a single unit with strong monomer interactions. The indication is that considerable redundancy of sub-units is built into the molecule (Fig. 8.5) so that it cannot be assumed that the minimal 2 Cu-unit of Hcy is completely autonomous.

The molecule of Hcy very readily and reversibly dissociates (MORPURGO and WILLIAMS, 1968) which would not be expected if the main purpose of the large molecule were solely to ensure low viscosity and a small osmotic effect (p. 211). From the behaviour of lamprey Hb (p. 216), the relationship between O_2-binding and inter-unit binding in Hcry (p. 234) and other clues it seems probable that the dissociation plays some part in the actual oxygenation-cycle. In these very large molecules a small change in binding strength can cause an enormous shift in association-equilibrium (DE PHILLIPS et al., 1970) and it seems inevitable that this should be exploited so as to assist O_2-binding. In the squid *Loligo*, indeed, O_2-binding does cause molecular dissociation as in lamprey Hb (DE PHILLIPS et al., 1969) and the fragments bind O_2 better than the intact molecule. There appears to be the usual safeguard in that high O_2-pressures promote some reassociation. In the snail *Busycon* by contrast *reassociation* is the normal effect of O_2 (DE PHILLIPS et al., 1970) and the whole molecule binds O_2 slightly better than half molecules. The contrast between the two Hcys and between the actions of medium and higher O_2-pressures in *Loligo* is rather reminiscent of the variations in the Bohr effect and this does operate in opposite directions in these two molluscs (p. 225).

Measurements on sedimentation and circular dichroism show that there are conformational changes within the Hcy molecule of *Busycon* during the cycle and it is noteworthy that more is known about events at this level than at lower levels. In fact it is difficult to break down Hcy to the monomer stage though ELLERTON et al. (1970) have done this for the Hcy of the crab *Cancer*. These monomers reassociated to the half-molecule stage as spontaneously as larger moieties and so functional interactions may be anticipated also at these lower levels.

It may be significant also that the molecule of Hcy is built from sub units on a five fold symmetry axis (WOOD, 1968; KONINGS et al., 1969) whereas the macromolecular Hbs have a six-fold symmetry (Fig. 5.7). Conceivably this is ultimately related to the difference in coordination-number between Cu and Fe and so could have functional significance. If so it does not operate at the lowest levels, where a 2×2 polymerisation is evident, particularly in the enzymic copper proteins (p. 195).

9.2.4 Comparisons between the Classes of Oxygen-carriers

Undoubtedly the resemblances between the three classes are numerous and striking (MANWELL, 1964). All are metalloprotein complexes of one of the two most useful transitional metals and the size of peptide per metal atom is very similar, between 6,750 and 37,000 Daltons. All exist as polymers of this Huffner unit, an oligomer when intracellular and a high polymer when free in the relevant body fluid. Interactions between monomers plays a part in O_2-loading and unloading. The aromatic amino acids, and particularly histidine play important structural and functional roles in all and are often direct ligands of the metal.

All stoichiometrically and reversibly bind oxygen, usually without assuming the common oxidation-state of the metal; when this state is produced artefactually the carrier no longer binds O_2 reversibly. Normal unloading of the oxy-form can be induced by simple physical means notwithstanding the chemical implication of stoichiometry. The entropy of oxygenation is low compared with that for the corresponding common chemical bonds. All similarly bind with varying degrees of reversibility such analogues of O_2 as CO, NO, CN and H_2S, molecular structural analogues rather than strict chemical analogues. All have much the same reactions with bound SH-groups and with urea. All have some catalase, peroxidase and other relatively non-specific terminal oxidase activities.

The dynamics and kinetics of O_2-binding also are similar; typically all have hyperboloid to sigmoid O_2-equilibrium-curves with the steep, effective loading/unloading part of the curve corresponding to the normal range of O_2-pressures in the body of each animal. The curve of all can show the adaptation of shifting along the O_2-axis in response to relevant factors. The factors which determine this and the exact form of the curve are much the same for all, – interaction between Huffner units, and structural components sensitive to pH-change (Bohr effect), temperature and some other factors. Among the animals using each type there is a wide range of p_{50} (half-loading pressure) values showing a wide range of adaptability in all. All assist in CO_2-transport which conveniently can be geared to O_2-transport.

Since the three types have evident intrinsic differences the implication is that they have undergone convergent evolution. At the same time probably all animals had terminal oxidases of all three types before there was any need for O_2-transport and therefore in principle all could have exploited the most efficient. On virtually all relevant criteria this is Hb, and since it *is* used sporadically as carrier or as a tissue-Hb in taxa which otherwise have Hcy or Hery it seems certain that there are more positive reasons why the latter also have been selected. *A fortiori* this applies to the evolution of Chl Hb in some polychaetes in the face of already established Pr Hb.

The relative abundance of the three main types as O_2-carriers Hb $\rangle$ Hcy $\rangle$ Hery seems to be proportional to their relative abundance as terminal oxidases and may be a measure of their relative efficiency for both purposes. It is also the order of increasing heat of oxygenation. It is the expected order for the frequency of evolutionary experiment but not of the successful results of such experiments. It is interesting that the colours of the four carriers collectively cover much of the visible spectrum, i.e., Hery: purple-red, Pr Hb: crimson to orange, Chl Hb: yellow-green, and Hcy: blue. This may indicate some functional or ecological basis for the choice between them.

9.3 Other Metalloproteins and Oxygen-transport

It is profitable to speculate on why Fe and Cu alone among the transitional elements of the first series, Ti, V, Cr, Mn, Fe, Co, Ni, Cu, Zn, have produced O_2-carriers and chromoprotein enzymes. The other members have the same essential properties of forming coordination complexes, with several readily inter-convertible oxidation states and, as already noted (p. 223) model O -carriers can be prepared from cobalt complexes. Mo, Mn, Co and Zn all are cofactors in a number of enzyme systems but they are not key atoms in the prosthetic group of an enzyme which might also evolve further into an O_2-carrier. With the exception of Co, in the cobamides (p. 201), Fe and Cu are almost unique also in the redox coenzyme capacity.

One reason is that only Cu and Fe complexes have redox potentials in a physiological range. All the oxidation states of these transitional elements higher than II/III are rare under conditions in the biosphere, and their exploitation by animals therefore is not very feasible. Moreover for some members of the series the redox potential even of the II/III change is much too low for terminal oxidation or for O_2-transport, and so these are completely ruled out: V^{II}/V^{III} has an E_o^1 of -200 mV while that of Cr^{II}/Cr^{III} is as low as -400 mV. Cr^{2+} ions in fact are the strongest known reducing agents in aqueous solution (HESLOP and ROBINSON, 1960, p. 441); they will remove O_2 completely and irreversibly from a mixture of gases and so would not release it to living tissues. At the other extreme values are too high: the Mn^{2+}/Mn^{3+} change has a potential of $+1510$ mV (in 15 N H_2SO_4) and that of cobalt $+1840$ mV (subject to correction for pH), compared with $+160$ mV and $+770$ mV for Cu(I/II) and Fe(II/III) respectively. In general, therefore, the members with the lower atomic numbers are too strongly reducing and those with the higher numbers too strongly oxidising. In the laboratory cobalt forms more coordination complexes, cationic, anionic and neutral, than any other member so that its very high II/III potential is probably the main reason why living organisms have not exploited it more widely. Nickel, the other member lying between Fe and Cu in the series, does not readily oxidise from the II to the III state at all. For the I/II change Cu is uniquely appropriate on the basis of its redox potential, as in so many other respects (p. 235).

Another consideration, not unrelated to the first, is the extent to which the metal forms complex oxy-anions, and in fact behaves as a non-metal. At the one end, V, Cr and Mn and at the other end of the series, Zn, form these complexes most readily. Those of Fe and Cu are most unstable, and this is why they bind O_2 so lightly. For reasons already considered (p. 80) strongly ionic complexes are unsuitable for redox purposes.

Why do no members of the higher series of transitional elements function in these capacities? In principle they are eligible candidates since they still form complexes with coordination-members up to 6, and with a variety of charge-numbers between 0 and 8, though the maximum for both does decline eventually with atomic size. Further, the complexes tend to be increasingly stable with increasing atomic weight of the ligator. Again the affinity for O_2 increases in the order cobalt, rhodium, iridium (McGINNETY et al., 1969) these being the middle members of the central triad of elements respectively of the first three transitional series. The overlap between energy levels of d-orbitals of the nth shell and the s and p orbitals of the $(n + 1)^{th}$ shell increases with the atomic number of the element and this should facilitate transitions to excited states. These heavy atoms have small ionic radii and therefore rarely form uncomplexed cations.

Nevertheless there are over-riding reasons why the higher transitional series are less suitable than the first series both for redox enzymes and for O_2-carriers. The redox potential even of the change from the free metal M to the divalent cation M^{2+} has values as high as $+450$ to $+1200$ mV (HESLOP and ROBINSON, 1960, p. 474) so that the values for the M^{2+} to M^{3+} change must lie above the physiological range for redox enzymes and carriers. The values increase steadily with atomic weight. Because of the increase in number of orbital electrons with atomic weight the nucleus binds progressively less readily with a weak electron-donor such as O_2: the preferred ligands become the more strongly donating halogens. Those oxides which the higher transitional series do form in fact are among their very few *un*complexed, ionic compounds (HESLOP and ROBINSON, 1960, p. 477):

here they donate and do not receive electrons. Fe and Cu are ideally poised between metallic e-donation to, and nonmetallic e-acceptance from, molecules such as O_2. Only with the help of halogens and other strong e-donors as ligands will the higher series bind O_2 (McGinnety et al., 1969). Rhodium then forms reversible O_2-complexes but iridium only irreversible types and so could not act as an O_2-carrier even under the best of conditions. Great stability is not a virtue in O_2-carriers and enzymes.

The complexes of the members of the higher series are more commonly low-spin and diamagnetic than those of the first series, tending to fill up the d-orbitals at the expense of the outer shell. This is one measure of the lower reactivity of heavier atoms, reflected also in the more familiar properties of increased stability of the free metal, insolubility of its compounds, resistance to oxidation, high melting-point, etc. With few exceptions living organisms are found to use only the lightest and most reactive elements of each group (Wald, 1962; Needham, 1965b) and in the present instance these are members of the first transitional series. Finally the members of the higher series are all much rarer than those, particularly Fe, of the first series, so that considerations of fortuity reinforce those of utility. No organisms are known to accumulate or use the rare insoluble compounds of the higher series.

The uniqueness of Fe and Cu therefore is well established and Fe has the additional virtue of abundance in the biosphere. The extent to which copper is used shows that utility is more important than mere abundance. The properties of the Fe and Cu complexes considered in this and the preceding chapter amply bear out this fact.

9.4 Conclusion

To a large extent the preceding two sections, 9.2.4 and 9.3, provide a convenient conclusion for this long chapter, which otherwise is difficult to summarise briefly. It represents the main results in a field of chromatology which has been very intensively studied, mainly because of its biological importance. Consequently it is not surprising that the results are packed with evidence of a high grade of functional perfection at all levels from the atomic upwards. This evidence needs to be perused *in extensio*.

Chapter 10

Zoochromes in Other Aspects of Metabolism

There may be a number of such aspects but at present only two merit detailed consideration, the secretion of chromes into the gut lumen and the role of chromes as end-products of metabolism. Since the gut is an excretory organ for aromatic and other difficult materials these two aspects may not be completely independent.

10.1 Zoochromes and Digestive Fluids

The widespread occurrence of certain chromes in the digestive secretions has been indicated in chapter 5. The significance of this remains very uncertain, even in the vertebrates, where the two bilins, bilirubin and biliverdin are secreted so abundantly and have been so intensively studied (WITH, 1968). In other phyla the relevant chromes are secreted in the main digestive fluid, with the enzymes. Vertebrate bilins arise from the breakdown of haems and, after further changes, much of the material is excreted with the faeces. A certain percentage is resorbed though it is not known to have a specific role in the absorption of the products of digestion, as the bile's salts have. Much of the resorbed bilins is again excreted via the bile or in the urine, so that excretion could be the main significance. Certainly chromes are among the materials excreted via the gut (p. 108) and the bile of vertebrates seems to be the preferred outlet for large aromatic molecules. Difficult materials are excreted also via the general gut-wall, and particularly in the invertebrates these include a large number of biochromes (DIWANY, 1919; WIGGLESWORTH, 1943; PHEAR, 1955). The chromes are sometimes transported across the wall in solid form.

There are reasons for thinking that the chromes of the digestive fluids do also perform a positive function. The elaborate conjugation of vertebrate bilirubin with glucuronic acid, cholesterol and other lipids (GRAY, 1961) could scarcely be necessary for mere excretory purposes, or even for detoxication since that which is resorbed is once more unconjugated (WITH, 1968, p. 139). The invertebrate counterparts also are often protein-conjugated. The horse has a very high content of bilirubin in the plasma, up to 1.57mg per 100cm^3; the level rises when fasting and falls at the onset of feeding (RAMSAY, 1945), presumably correlated with transport to and from the gut-lumen. A considerable circulation and a function in absorption from the gut seems implied. Rats also resorb much of their bilin; after introducing labelled bilirubin into the gut only 10% remained in the lumen after 30 minutes (WITH, 1968, p. 140) and by this time re-secretion was beginning. Man similarly absorbs administered labelled bilin and re-secretion is maximal by 18 hours. Haematin (protoferrihaem) is not resorbed in this way. Bilins are suspected of a transport function also across the mammalian placenta (DIWANY, 1919) and into the 'roots' of Rhizocephala.

It may be significant that all the identified chromes of digestive fluids are haem derivatives. Of course these are the most abundant chromatic products of metabolism in the vertebrates and annelids but the pulmonate gastropods secrete a high concentration of the haemochrome, helicorubin (KEILIN, 1956) in their digestive fluid, notwithstanding their haemocyanin blood-pigment. Malacostracan Crustacea likewise have Hcy in the blood but secrete large amounts of various haem derivatives into the gut (PHEAR, 1955). Some Crustacea were found to secrete bilins (BRADLEY, 1908), and there is a green bilin in the hepatopancreas of the octopus (Fox, 1953, p. 276), and a biliprotein in the gut of the silkworm *Bombyx mori* (Kusuda and Mukai, 1971). Those polychaetes with chlorohaemoglobin in the blood nevertheless have protohaem derivatives in these secretions. It would be instructive to know the nature of the chrome in the digestive fluid of *Priapulus* (p. 108), with haemerythrin in the body-fluid. Green or brown chromes in the digestive secretions are very common.

Like vertebrate bilins, helicorubin appears to recycle and this may be why it is most abundant in the stomach when the animal is fasting; presumably resorption is depressed in the absence of food-products. It remains abundant in the stomach after two months on a diet of filter paper (Fox and Vevers, 1960, p. 107) and so must be conserved very efficiently. There is also provision for keeping it reduced since it promptly oxidises to the ferrichrome on exposure to air. This is reminiscent of the spontaneous oxidation of bilirubin to biliverdin on exposure. In *Helix* it is associated with a related haemochrome having cytochrome properties (Keilin, 1956, 1968) so that a redox function also is highly probable.

It is significant that some of the haemoglobin which *Rhodnius* absorbs from its meal of vertebrate-blood (Wigglesworth, 1943) is passed in the form of a derivative to the salivary glands, and so into the gut with their secretion. There seems no need for the initial absorption if it serves no useful purpose and no need to excrete it via the salivary glands when malpighian tubules and mid-gut diverticula are available. Here and in most animals investigated in any detail there are strong indications of positive and rather specific alimentary functions for these chromes.

10.2 Zoochromes as End-products of Metabolism

There is a long standing and widespread view that the only explanation of the presence of certain zoochromes, at any rate in some animals, is that they are incidental products of metabolism, have no further use, but accumulate because there is some difficulty in excreting them. The direct evidence for this is scanty in virtually every case, and usually equivocal, while the circumstantial evidence is even more ambiguous, and complicated by other issues. It is hoped that this attempt to sift it may stimulate further direct research, particularly of a quantitative nature.

One complication is that in principle a metabolic end-product, at first excreted as having no other significance, eventually could acquire selective value and be retained, e.g., for visual effect purposes; it might do so only in certain taxa. On the other hand if a chrome is synthesised initially for a very specific purpose then *ipso facto* it will be the end product of that particular path and, not knowing the evolutionary history, there is no obvious criterion for distinguishing this from the first situation. It likewise could be taxonomically restricted in distribution; in general it should be more restricted, though not necessarily. For various reasons zoochromes usually will be end-stages of biosynthetic paths, i.e., because they are large, complex molecules, sometimes polymeric, rather highly oxidised and not easily metabolised further.

In laboratory organic reactions a variety of coloured materials are very commonly produced in large amounts as by-products, that is to say, end-products of branch-pathways, (Fox, 1965b) and, unfortunately for the present interest, usually suffer the ignominious fate of such 'nuisance'. This by-product phenomenon can afflict living organisms also, but by comparison they are amazingly free from it. Conditions are milder than in most laboratory techniques but no doubt in addition there has been natural selection for better control of pathways than is possible, except with a great deal of experience, in the laboratory. Various obscure fuscous materials accumulate with age in many tissues (p. 37) but, bearing in mind the time-scale, their rate of production is extremely slow. Uroporphyrin I and coproporphyrin I, and their haems, seem to be formed in incidental side-paths of the main porphyrin III line (p. 286) but the amounts are usually small and moreover they may have been positively selected since a number of animals make integumental use of them. It is difficult to prove the null hypothesis that an apparent by-product has no useful function, however many possibilities have been tested and rejected. In any case, it seems clear that animals are very efficient, by laboratory standards, and do not produce excessive amounts of useless chromes.

An accumulation of a by-product or end-product chrome in the body could be an indication that it is not used further and cannot be eliminated, and the literature on 'kidneys of accumulation' or 'storage excretion' is a confused attempt to express this, and to prove it. Those tissues and organs which store chromes are usually the same as those which store dietary and other products known to be useful, so that rigid proof is needed that any of the former really are useless. The very fact of storage seems to weaken the case for a particular chrome being merely excretory in significance since animals do seem to be able to eliminate what is purely waste, whether in solution in the urine and other fluids or solid, mainly via the gut and epidermis. Uric acid is one of the more insoluble waste nitrogenous materials stored in the appropriate tissue of various animals, including the chloragogen tissue of the earthworm; nevertheless earthworms excrete this stored purine under fasting conditions (HAGGAG and EL DUWEINI, 1959) and so the normal storage must have some positive significance. It is noteworthy that oligochaetes are ammono-, and ureo-, telic so that their production of uric acid is low and there should be no need for it to accumulate; Unlike some groups, the earthworms do not use it as a light reflecting material (p. 67).

A number of chromes are accumulated in the excretory organs themselves, for instance ribitylflavin in the green gland of Crustacea (FISHER, 1960) and the malpighian tubules of insects (FOX and VEVERS, 1960, p. 158); haemochromes in the earthworm nephridium (BAHL, 1947) and several classes in the excretory organs of molluscs (POTTS, 1967). Some of the chromes in these organs show a surprisingly low excretory turnover so that even here they seem to be specifically stored. By contrast many excretory organs have been shown to take up, and promptly excrete, laboratory dyes experimentally administered, so that speaking teleonomically (PITTENDRIGH, 1958) the storage of certain natural chromes would seem to be for a good 'purpose'.

Even so, excretory organs contain fewer chromes than most tissues of the body (Table 5.1). In contrast to these examples, *Daphnia* and related Cladocera, which synthesise haemoglobin when this is needed (p. 309), as promptly destroy it and eliminate the products when the emergency is over. This may be contrasted with the considerable stores of haem and its degradation-products in leeches (NEEDHAM, 1966b) and blood-sucking insects (WIGGLESWORTH, 1943); these animals probably store it for useful purposes. The lower vertebrates store large quantities of purine which they use for countershading. Like earthworms the lower vertebrates are mainly ammono-, and ureo-, telic in their N-excretion and so must synthesise purines specially for integumental use; most birds and many reptiles are uricotelic but do not store purines or use them for countershading! Feeding fishes on a diet high in nucleic acids does not increase the storage of purines (VERNE, 1926, p. 417) so that the process is well controlled. Various other annelids (NEEDHAM, 1970b), most Crustacea (HARTENSTEIN, 1970) and many molluscs (CAMPBELL, 1970) also store purines and again these groups are not primarily uricotelic.

During starvation animals use protein as well as carbohydrates and fats, i.e., they destroy all materials with some functional proportionality; this applies even to integumental zoochromes, such as the ommochromes (BRUNET, 1965), and to the earthworm purines. There are some exceptions, however, for instance the integumental chromes of *Asterias* concentrate (VEVERS, 1949). Again the colour in the condition known as brown atrophy of the heart is due to the haem not being removed as the muscle-tissue atrophies. Similarly in emaciated individuals the carotenoids and other lipid-soluble chromes concentrate. In some cases the increased fuscin-content of ageing tissues conceivably may be a concentration process of this type. On the other hand many species of animals when transferred to darkness, or to a different background, eliminate their integumental chromes without any general atrophy: the two have considerable independence and the concentration of chromes is controlled as necessary.

The very slow rate of accumulation of fuscins really applies to the storage of most chromes. Usually storage rate is only a small fraction of the total turnover of the chrome, most of which is regularly excreted. The leech, *Hirudo*, stores at most a small fraction of its dietary haem; the rest is eliminated at infrequent intervals so that quantitative measurements of the two fractions, and of the leech's own haemoglobin, would be possible. The present impression is that storage is under restrictive control. Another approach to the problem would be to use labelled materials. In a true kidney of accumulation labelled material would disappear only by radioactive decay but in a typical storage organ a continuous metabolic turnover would be expected, as in the fat stores and in most other 'pools' of metabolites. The B-vitamins, including ribitylflavin, have a steady turnover; Man excretes 0.9 mg of RF per day (FOX, 1953, p. 286) and indeed this has been used to estimate his daily requirement.

Supposing that there was a tendency for certain chromes to present difficulty to the excretory mechanisms this might be expected to show a sharp exacerbation in animals after their useful reproductive age, that is the age at which their offspring come into full reproductive capacity. As MEDAWAR (1946) so aptly and humorously puts it, what happens to an individual after this age is not on the agenda of evolution, and there is likely to be no natural selection against genotypes which, for instance, show increasing difficulty in excreting any particular chrome, whether harmless or not. Actually this is much less evident than most of the disabilities of advancing age. For instance the accumulation of fuscins still remains extremely slow, probably because genotypes selected for a slow accumulation earlier in life continue to maintain the same control. Of course in Man and some other animals there has probably been selection for longevity and therefore for physiological adequacy beyond the age defined above. Natural selection has found genotypes which can deal with the rate of production and elimination of chromes as necessary. Coccids have a short life and the species survives concentration of its anthraquinones, which may virtually embalm the adult female.

A phenomenon which might seem to support the idea that some at least of the integumental chromes are casual end-products of metabolism is best illustrated by the experiments of HADDOW et al. (1945). When artificial chromes were fed to dominant white laboratory rodents these were deposited in the hair. Moreover they were deposited in the familiar agouti pattern (p. 326) so that it would seem reasonable to suppose that the natural chromes of other strains are similarly excretory material. The hair is regularly moulted and replaced in these small mammals so that it could be a convenient vehicle for excretion. The same applies to the chromes of birds' feathers; when the ovary is making heavy demands on the body's carotenoids, less than usual is passed into the feathers (PALMER and KEMPSTER, 1919b). Turacin (p. 88) appears to ooze out of the feathers on slight provocation and KEILIN and MCCOSKER (1961) suggest that it is essentially a detoxication-product for both copper and porphyrin. Arthropods cast some pigment in their exuviae though they do also resorb some. The epidermis is known to be a significant accessory excretory organ, as evidenced by the chromes and other materials in sweat. Nevertheless most of the chromes normally in hair, feathers and exoskeletons are there for definite visual display purposes and certainly not because they could not be excreted by any other route. It may be noted that HADDOW et al. used an isoalloxazine chrome and that ribitylflavin itself is usually deposited in the hair, though masked by melanin in the wild type. The experimental isoalloxazine was also rapidly excreted via the urine so that the hair was not the only available outlet. The dominant white does not lack integumental chromes simply because it has none to excrete.

Excessive accumulations of zoochromes are a feature of various pathological conditions (p. 250), and sometimes contribute to the pathological symptoms but this only serves to show how well controlled normally is the metabolism and storage of all chromes. Pathological concentrations do not prove that normal concentrations are at all deleterious.

Most chemical classes of zoochrome do occur in some sites at least in concentrations relevant to the present interest but for none of these is it possible to say that the accumulation is useless or even deleterious or that the animal is unable to excrete it as necessary. The haemochrome of brown fat (p. 197) is useful and excretable and there are other useful fuscins (DIXON, 1970); those of neurons are used up during nervous activity (VERNE, 1926). Naphthoquinones in echinoderms, anthraquinones in female coccids and aphins are among the more difficult cases for the present thesis but useful defensive functions are possible in all cases. In addition the short life-cycle of the insects is relevant. Jaundice and porphyrias show how deleterious bilins and free porphyrins can be but equally how well controlled these chromes are in normal animals. Because of their highly polymeric form, great insolubility and chemical inertness, melanins might be expected to present outstanding excretory problems but in fact they are rarely formed in excess, except in pathological conditions (p. 253), and are very well controlled. They are one of the few classes evolved primarily for integumental purposes and are never stored except in the cephalopod ink-sac.

Ommochromes are another important group in this context because insects, the group which has most exploited them, lack the enzyme-system for catabolising their precursor, tryptophan, in the simple way used by vertebrates. A popular view is that the insects and other spiralian phyla (p. 83) may have lost, or never possessed, the vertebrate system and so must excrete waste tryptophan in the form of ommochromes. There are many arguments against this and in favour of a naturally selected positive function for the class. There are various other, simpler paths from tryptophan, including a number which also start via kynurenine, and some of these lead to other chromes such as the papiliochromes. Ommochromes give evidence of sophisticated molecular structure and properties (p. 65). It is only in the metamorphosis of some insects that accumulations not in chromatophores are found. Actually these are in the meconium (LINZEN, 1967; LINZEN and BÜCKMANN, 1961; LINZEN and HENDRICKS-HERTAL, 1970) and so are the outcome of a successful excretion. Locusts excrete ommochromes when fasting necessitates this (p. 244); in any case they are probably harmless chemically (LINZEN, 1967) and safely could be stored.

The pterins provide some useful comparisons with the ommochromes as primarily integumental and eye-chromes, particularly of insects. Moreover as purine-derivatives they are suspected of an excretory significance and again large amounts are involved in metamorphosis. Fortunately some valuable quantitative measurements have been made of these (ZIEGLER and HARMSEN, 1969). Pierid butterflies, which use them as wing-pigments, each emerge from the pupa with 150–200 µg of simple pterins in the fat body not to mention 1200 µg of xanthine and 1400 µg of uric acid. *Vanessa*, belonging to a family which has no pterins in the wings, synthesises only one-tenth as much and even so excretes 50% of this in the faeces. If one wingbud of a pierid pupa is removed there is an increased excretion and a decrease in synthesis of pterins, indicating a most sensitive feedback. Some insects acquire pterins in their diet but if they do not require these for integumental purposes they promptly excrete all. This is a class readily synthesised by animals, used and stored as appropriate and equally readily excreted when necessary.

The 'brown bodies' of the ectoproct polyzoa provide virtually the only example of undoubtedly waste pigmentary material allowed to remain in the body for considerable periods and evidently giving difficulty in elimination. The Ectoprocta lack specific excretory organs and brown bodies are eventually eliminated by a number of drastic and awkward methods, in the different sub-groups. The material may originate largely from the food since it is stored in the gut-wall but in any case why can the Ectoprocta not re-excrete it into the gut-lumen as so many animals do? Their actual methods amount to gross versions of this, in fact, and all useful material is salvaged. Again, the individual polypides are short-lived.

10.3 Conclusion

Chromes often are the end-products of metabolic paths and some of these accumulate in certain sites, but the accumulation is usually moderate and controlled, for useful purposes and not as a consequence of excretory ineptitude. Some short-lived animals tolerate massive accumulation. Details of the control of storage and elimination are available in a number of cases and all display a labile turnover. Further quantitative measurements are highly desirable. Even insoluble materials are excreted as necessary. The gut is a main excretory route for difficult materials, including a number of chromes.

The gut-secretions of many animals contain green to brown chromes, identified as bilins in some and haemoproteins in others. A proportion of the chrome cycles repeatedly between gut-lumen and body and probably plays some specific role in transport across membranes.

Chapter 11

Pathological States Involving Zoochromes

As natural experiments pathological conditions are always informative about normal function. In the present context of 'chromopathies' it is logical to distinguish between pathological conditions, which affect chromes, and those effected by chromes, but in fact zoochromes rarely have pathological effects unless they are already abnormal, qualitatively or quantitatively. The distinction therefore could be regarded as one between primary and secondary chromopathies.

11.1 The Destruction of Zoochromes by Light

This is a primary chromopathy and it can be regarded as both a quantitative and a qualitative change in the chrome. Carotenoids, and particularly the retinopsins are rapidly bleached by light (Fox and Vevers, 1960 p. 75). Even melanin, for all its stability, is bleached considerably in strong sunlight. Among the polycyclic quinones, zoopurpurin is rapidly destroyed (ib. p. 146). Porphyrins in general are destroyed (Volker, 1939, 1942) and so is the chlorophyll derivative, *bonellin* (Dubois, 1907). Free cobamide is stable but its active coenzyme is rapidly decomposed (Stadtman, 1971). The bilin, *aplysioviolin*, is unstable in the light but the disappearance of biliverdin from the integument of mantids under summer insolation is an indirect response (Passama-Vuillaume, 1965). Haemocyanin suffers photooxidation (Wood and Bannister, 1968) and most of the other cases involve oxidation (Giese, 1946). Perhaps because it is not a component of a terminal oxidase, ribitylflavin is photobleached even in the absence of oxygen (Cairns et al., 1969); the product of photolysis is either lumiflavin or lumichrome, depending on pH (Fox, 1953, p. 284). Pterins also are very sensitive to light (Hama, 1963; Gilmour, 1965, p. 157) and again the action is oxidative (Ziegler-Gunder, 1956). Purines, not being true chromes, are insensitive but it so happens that for biological reasons they are mainly in the ventral integument, least exposed to light! Very few biochromes when free in solution are not sensitive to light, and no chromatologist would leave any chrome exposed to anything stronger than twilight.

11.2 Photodynamic Actions of Biochromes

Chemically photodynamism is defined (Giese, 1968; Davson, 1970) as the non-specific photosensitisation of substrates to oxidation by free oxygen or by such products of its photoactivation as H_2O_2. Biologically it is a traumatic type of photosensitisation and one of the few pathies primarily due to chromes. For the most part animals have protected themselves from such effects of their own chromes by conjugating these with proteins, transitional metals, etc., as well as by segregation and other morphological devices. Protection can be measured by the degree of quenching of the fluorescence of the chrome. Serious photodynamic effects therefore arise mainly from deconjugated chromes such as free porphyrins (p. 53) or from foreign unquenched chromes.

From the basic considerations of chapter 2 it emerges that most photodynamic chromes permit visible light to mimic the direct effects of radiation in the long-wave u.v. range, say 300–400 nm (HOAR, 1966). The action therefore is mainly on proteins (BLUM, 1964) and usually there is no mutagenic effect on nucleic acids, as there is by 200–300 nm radiation. Nevertheless some instances of mutagenesis have been detected (ZETTERBERG, 1964) and certainly dyes (MAYER and MELNICK, 1961) and biochromes (p. 29) do bind to chromosomal nucleic acid; uroporphyrin is an example (NIYAGI et al., 1970). Damage to the aromatic amino acids, including histidine, is most serious but methionine (JORI et al., 1969) and others also suffer, mainly the members which are most demanding in their biosyntheses.

A photosensitised mammal experiences cutaneous irritation in the light, followed by erythrema, oedema, pain and slowly healing lesions (HOAR, 1966). There may be loss of hair, and skin-cancers are common (EPSTEIN et al., 1964); this is the most serious hazard. If deeper tissues are exposed to the trio, – chrome, light and O_2, their enzymes are inactivated. The plasma proteins are denatured and fibrinogen will not clot. Erythrocytes haemolyse: the action is on the membrane itself since the number of quanta required is independent of the concentration of the dynamic chrome. Skeletal muscle twitches and goes into contracture, and heart-muscle beats irregularly; the effect on the latter is completely localised since there is no effect on a second heart perfused in tandem, but kept in the dark (DAVSON, 1970, p. 1540).

The best known instances of photodynamism are due to the polycyclic quinones (p. 43) and the porphyrins (p. 53). Zoopurpurin does not normally harm the ciliate *Blepharisma* itself since it is produced only when the animal is living in dim light but if such an animal is brought suddenly into bright light it is rapidly killed (GIESE, 1953). If the light is increased gradually the ciliate destroys the chrome rapidly enought to avoid damage. In solution it lethally photosensitises ciliates of other species and so may be used offensively by *Blepharisma*. The usual requirement for oxygen has been demonstrated. The related quinone of *Stentor* is photodynamic to other species but not to *Stentor* itself (TARTAR, 1961) and likewise may act as an antibiotic. The protoaphin of the living aphid is not dangerous but its oxidation-products in shed blood are, so that injured individuals may deter predators from attacking further aphids. Hypericin, the polycyclic quinone of St John's Wort, is seriously photodynamic, and often lethal, to sheep; this is a situation which favours the black sheep since the melanin is an efficient light-screen (p. 61)! Buckwheat and other plants also have photodynamic chromes which injure herbivores.

The photodynamic condition 'geeldikkop' in South African sheep is due to the phylloerythrins from dietary chlorophylls (p. 55) so that it is one of the *porphyrias,* due to unquenched porphyrins; even native phaeporphyrans can be photodynamic since Mg, unlike Fe and Cu, does not quench the fluorescence of the porphyrin: plants sometimes suffer photodynamically in bright light, when the ratio of CO_2 to O_2 is low. Sheep do not normally suffer from the derivatives of their dietary chlorophyll, unless the re-excretion of these (p. 88) is prevented by obstruction of the bile duct.

Bilins are non-fluorescent and photodynamically innocuous, while on the biosynthetic side (p. 287) protophorphyrin and even the earlier stages normally are immediately quenched by conjugation with Fe, so that animal porphyrias are rare except in pathological conditions. However, a simple deficiency of iron can result in an accumulation of free porphyrins; other causes include errors in the body's mechanism for handling iron, poisoning by heavy metals such as lead, and disorders of the liver, the main iron-processing organ (RIMINGTON, 1955). In this last instance synthesis goes as far as coproporphyrin but in *porphyria cutanea tarda* it reaches the protoporphyrin stage (GRAY, 1961). Like the chlorophyll porphyrins these in moderation are excreted without harm via both bile and urine. It was suggested

(MERKER, 1926) that the high mortality of earthworms which appear on the surface after rain was due to photodynamism by the protoporphyrins of their body-wall but this would require a prior pathological agency to free the porphyrin from its protein-conjugate *in situ*. It is significant that free porphyrins are usual in the shells of gastropods but in the integument they are protein-bound; in the shells of the eggs of birds again they are free.

If porphyrins accumulate excessively in the spinal cord of birds and mammals (p. 100) there are neuro-pathological consequences (KLÜVER, 1944b). In view of the efficiency of photosensitised neurological responses from deep centres in the brain (MENAKER, 1972) it is conceivable that the neuropathological effects are indeed porphyric. A critical test would be the magnitude of the effect in relation to body-size, to the density of integumental chromes, etc.

The human pathological condition, *xeroderma pigmentosum* may involve photodynamic effects; here small hyperaemic spots tend to enlarge and become necrotic, or even cancerous. This happens only on regions of the skin exposed to sunlight. The condition is genetic and it is interesting that the heterozygote suffers from nothing worse than melanic freckles; there are several possible implications but it seems doubtful if the dynamic effects in the homozygote are due primarily to melanin damaging the Hb. However, it is interesting that there is a photodynamic chrome in the semen of the bull (DAVSON, 1970, p. 1540) and that melanin also in rather a common constituent of mammalian semen (p. 107). This seminal photodynamic agent is inhibited by catalase, indicating that H_2O_2 is the essential photooxidant. The significance of the phenomenon here is obscure, in more senses than one.

Experimentally ribitylflavin (RF) has been found powerfully photodynamic and can be used as a bacteriostatic agent when irradiated by visible light (WAGNER-JAUREGG, 1954). Tryptophan and the vitamin pyridoxine are most affected, presumably by charge-transfer from the isoalloxazine nucleus (PULLMAN and PULLMAN, 1963). The acid photolytic product of RF, lumichrome or 6,7 dimethylalloxazine, is even more potent and acts primarily on DNA (BERENDS et al., 1966). The action is specific to guanine among the nucleotide bases of DNA and this may be related in some indirect way to the fact that guanine is a key intermediary in the biosynthetic paths to RF (p. 290) and the pterins (Fig. 13.2). Lumichrome photodestroys also the tetra-H-pterins and other pyrazine derivatives. Methylene blue, a rather close structural and functional analogue of RF behaves similarly; the acridines and other analogues likewise are mostly photodynamic.

11.3 Qualitative Chromopathies

In the sense of being due to the chemical activities of foreign organisms, the indigoid chromes of mammalian urine, indigotin, indirubin and indiverdin, are abnormal or pathological chromes at any rate when present in freshly voided urine. Here they indicate bacterial infection of the bladder, hepatic carcinoma (BODANSKY, 1938; FOX, 1953, p. 215) or sprue (RIMINGTON et al., 1946). Since the indole and skatole from which the precursors are synthesised by the mammal also in turn are formed by gut-bacteria from tryptophan the phenomenon is doubly pathological. *Urorosein* is a similar bacterial product of the indole acetic acid of dietary plant-material (FOX, 1953, p. 238). The alternative name, *adrectal gland*, for the purple gland of the muricid gastropods (p. 59) raises the possibility that dibromoindigotin evolved as bacterially produced material of this type.

The mildly pathological hereditary human condition known as *alkaptonuria*, due to an error of phenylalanine metabolism (Fig. 11.1) involves the production of an abnormal melanoid pigment in the voided urine. The monomer is now known to be benzoquinone acetic acid (LA DU and ZANNONI, 1969) so that it is a phenolic melanin (p. 60). It is also unusual in being a product of the catabolic pathway of phenylalanine and not of the tyrosine path. No secondary pathological effects are usual but sometimes there is oxidation and polymerisation in the body, a condition called *ochronosis*. There is a selective uptake by connective tissues in the broad sense, *i.e.*, including dermis, sclera and cartilage (LA DU and ZANNONI, 1969). Other chromes such as the bilatrienes (p. 253) also are taken up by these tissues and in most cases no ill-effects have been detected.

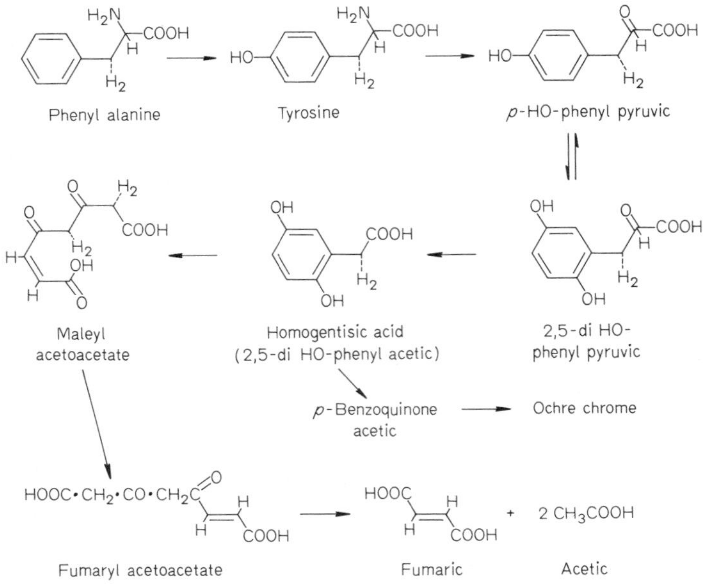

Fig. 11.1. Normal pathway of catabolism of tyrosine to fumaric and acetic acids and the abnormal pathway to the ochre chrome of alkaptonuric patients

Many pathological conditions, including parasitisation, involve the production of abnormal coloured materials (ROBBINS, 1967, pp. 415–429). A number of these still have to be identified but most of the known examples in Man are haem, melanin or lipofuscin derivatives. Most are quantitative rather than qualitative chromatic abnormalities.

11.4 Deficiency and Excess of Dietary Chromes

Natural selection has ensured that the large amounts of some of these chromes consumed have no ill effects. Those admitted from the gut into the body, still in large quantities, the carotenoids and porphyrins in particular, again are relatively harmless there. On the other hand a number of dietary chromes are indispensable vitamins (p. 209) and a deficiency of any of these has a striking pathological syndrome.

Vitamin A (retinol) is not a specific dietary demand for many invertebrates, while some vertebrates are able to synthesise it from carotene so that the latter might be considered a pro-vitamin. Interestingly enough retinol is one of the few vitamins deleterious in excess: the liver of the polar bear is said to be toxic because of its high content of retinol (HEILBRUNN, 1952, p. 193; HOAR, 1966, p. 428). Excessive intake for a considerable period is necessary (HODGES, 1961) and the symptoms have many of the features of a *deficiency* in the vitamin and of some other vitamins. An excessive accumulation of other carotenoids is not uncommon in Man but has not been found to cause any evident pathological symptoms. The tissues become yellow and the condition therefore is called false-jaundice, *carotenaemia* or *xanthaemia* (VERNE, 1926, p. 233; FOX, 1953, p. 174); the high plasma coloration is evident only after the cells have been spun down. The carotenoids may appear also in the sweat. The condition disappears on simply reducing the level of intake. It is rather common in diabetics, who have a 'biochemical lesion' in the endogenous path by which carotene is converted to retinol (WILLIAMS, R. H., 1968). Deficiency of retinol causes night-blindness, rod-function being more affected than that of the cones. There is also excessive keratinisation of the skin, *xeroderma*, which further affects mucous surfaces so that blindness can result from desiccation of the cornea also. There may be neural effects through demyelination of nerves and, in young mammals, through the growing cranial nerves being pinched by a failure of the normal enlargement of the nerve foramina by bone-resorption (FELL and MELLANBY, 1953). A highly deficient foetus may be limbless owing to this effect on spinal nerves, also. In fact most tissues are affected and there is a general stunting of growth. Also there is general debility and liability to infections of many kinds. The general effects are common to deficiencies of other vitamins, including the biochromes ribitylflavin, pteroylglutamic acid, cobalamin (cobamide), and pantothenic acid (NEEDHAM, 1964, p. 304). This is mainly because they are all involved together in the same basic metabolism, mainly the respiratory metabolism of chapter 8. Insects show a general depression of metabolism when restricted to a carotenoid-free diet (GILMOUR, 1965, p. 107) and it would be useful to know if other invertebrates demand some carotenoid. This is the most important type of secondary chromopathy (p. 247).

Deficiency of vitamin E, α-tocopherol, leads to an accumulation of fuscous material in various mammalian tissues, particularly muscle, adrenal and gonad. The effect is probably due to the absence of the normal antioxidant function of this chrome (DEUEL, 1957). In serious cases there is muscular dystrophy and paralysis and also dystrophic effects of bone, liver and other tissues. The gonads particularly are affected and the reproductive activities of both sexes are extensively depressed.

The symptoms of ubiquinone deficiency are not yet known, though some of them could be forecast from those of deficiencies in the other respiratory vitamins and from a knowledge of its specific role in the system. Failure as yet to recognise it as a vitamin is support for the view (p. 304) that it is synthesised endogenously. A deficiency of vitamin K delays blood-clotting but in animals no other deleterious effects are very evident.

A deficiency of ribitylflavin can cause greying of mammalian hair and alopecia. There may be also degenerative changes in nerves and some paralysis of the digits. Cataract of the lens has been observed and a general retardation of growth; it was once called the growth vitamin because of its outstanding potency. Tissues particularly hypotrophic under deficiency are the thymus, testis, and white blood cells. Microcytic anaemia indicates defective synthesis of Hb. Little is yet known of deficiency-symptoms in invertebrates, which all need RF (p. 209), but the 'antennaless' mutant of *Drosophila* will develop normal antennae if given large amounts of RF (GORDON and SANG, 1941).

Deficiency of pteroylglutamic acid (folic acid) also causes greying of the hair and severe alimentary dysfunction a condition known as sprue, which is usually correlated with macrocytic anaemia (below). There is also perosis of the bones and retardation of growth. Cobalamin (p. 209) is an even more crucial antianaemia factor. Deficiency also causes a general depression of growth in Protozoa, birds and mammals and of fertility in hens'eggs (ALMQUIST, 1951). Again perosis is evident and there are degenerative changes in the spinal cord.

Pantothenic acid deficiency causes much the same symptoms, growth retardation, anaemia, greying of the hair, dermatitis, diarrhoea, and muscular and nervous dysfunction. It is very clear that all the coenzyme chromes are essential for the same wide range of metabolic activities, as emphasised at the outset. There could be no more dramatic demonstration of the instructiveness of pathological conditions: the respiratory system is at least as important for maintenance and growth as for the work-functions of the body.

11.5 Deficiency and Excess of Endogenous Chromes

There is no evidence of pathological symptoms due to excess of fuscins and as yet no indication that a deficiency of any but the haemochrome of brown fat even has any meaning. On the other hand a number of pathological conditions are known to cause excessive fuscin-production. In addition to vitamin E deficiency (p. 197) Parkinson's disease causes an accumulation in neurons of the *substantia nigra* and a deficiency of the normal melanoid pigment there (MARSDEN, 1969; ISSIDORIDES, 1971). The so-called lipidoses and some other familiar diseases (BARONE et al., 1969) also are associated with increased lipofuscins (CROCKER, 1969). Intestinal chronic injury also results in fuscous coloration (FISHER, 1969).

The 'haemochromatoses' involve excessive accumulations of haemosiderin (MACDONALD, 1969). They can be primary, due to excessive intake of assimilable iron, or secondary due to excessive destruction of Hb following haemolysis, etc. The spleen (VERNE, 1926, p. 232), liver and skin (RICH, 1924) are most affected and in a lesser degree the pancreas, heart and endocrines (MACDONALD, 1969); the skin becomes very brown. There can be widespread damage to mitochondria and microsomes and this sets up a fatal vicious circle by causing further haemolysis.

Because of the large amounts of haemoproteins, and particularly haemoglobin, synthesised by vertebrates naturally there are a number of pathological conditions connected with this. Excess production of protohaem and its protein-conjugates is not common since there are feedback controls (p. 307), but deficiencies, the *anaemias*, do occur. They can be secondary to dietary deficiencies, whether of iron itself or of the vitamins considered above. There are two distinct main types, microcytic hypochromic and macrocytic hyperchromic or Addison's pernicious, anaemias. In the former, due to iron and ribitylflavin deficiencies, the cells mature normally but they are deficient in haemoglobin. In macrocytic anaemia, due to deficiencies in pteroylglutamic acid and cobalamin, the large erythroblasts do not mature normally into small erythrocytes; each has its full complement of Hb but too few cells are available and Hb-synthesis itself is depressed in proportion. There are also types of anaemia due to innate factors, to poisons which damage the haemopoietic system (Fig. 14.2) and simply to excessive loss of cells through haemorrhage, particularly in women. Also cells, or the Hb within the cells, may be damaged by parasites. Anaemias have certain automatic effects, particularly respiratory and circulatory stresses, general lassitude and weakness; they can also entail a number of more indirect secondary pathies, for instance, chronic degeneration of the spinal cord in pernicious anaemia.

There are some pathies which concern the synthesis of the globin (GOMPERTS, 1969) and these are mainly qualitative abnormalities. They are mostly due to one or a few wrong amino acids at critical sites in the peptides and their genetic basis (p. 327) is evident. The best known, sickle-cell anaemia, results in the Hb crystallizing *in situ* and therefore loading and unloading sluggishly. Being qualitative abnormalities they tend to have secondary consequences. They emphasise once more (p. 230) how much O_2-transport depends on the whole molecule.

Another group of pathies involve destruction of Hb or of the red cells. *Erythroblastosis foetalis*, haemolytic disease of the newborn, is due to an immunological reaction by a mother possessing a Rh-negative genotype against her Rh-positive offspring. This has disastrous secondary effects and is often fatal, if untreated.

In contrast to haem and haemoproteins the by-products of their synthesis and their degradation-products, the bilins, not uncommonly are produced in excess, – while a deficiency would be meaningless. Excess bilins result from excessive destruction of Hb or from difficulty in excreting the bilins themselves. Excess free porphyrins can result from deficiency in iron or inability to ligate it to the porphyrin. The photodynamic secondary effects of the latter are considered above (p. 248).

Pathological states involving excess bilins are called jaundices or *icterus*. In Man bilirubin strongly colours the plasma and most of the tissues, but a green jaundice, due to biliverdin, has been recorded in the pike (Fox and Vevers, 1960, p. 120). In addition to the two main types, due to excess production and faulty elimination respectively, With (1968) recognises a third, due to decreased destruction of bilins, which presumably also implies excessive resorption from the gut. The seven or more clinical types of icterus are variants of these general types, mainly the first two. The resorbed bilin of the obstructive case is already conjugated with glucuronic acid (p. 29) and gives a reaction directly with Van den Bergh's diazo reagent (Table 3.11) whereas in the other types it has not been conjugated and does not react with this reagent until treated with ethanol. This provides a very convenient method for diagnosing the obstructive type.

The elastic fibres of the connective tissue have a great affinity for bilirubin (p. 52) and so have the nucleic acids (With, 1968). It is therefore surprising and fortunate that the secondary symptoms are normally relatively mild, the traditional depression of spirits and a depressed blood-coagulation response being most obvious. The brain appears to have an effective barrier against bilins, as against so many materials which may stray into the circulation. In the newborn, however, which not uncommonly have mild transitory jaundice, the C.N.S. may become coloured (kernicterus). Adults show symptoms at much lower blood-bilin levels than infants, which probably have poorer control but greater tolerance.

There are a large number of pathies causing an excess or a deficiency of melanin. In most cases an excess of melanin in itself does not have secondary pathological effects but a deficiency may. In addition melanomas are so common among tumours that there is some suspicion that irrespective of concentration the chrome may be actively carcinogenic; if so it normally must be very well controlled. A carcinogenic property is consistent with its intense absorption of a wide band of radiation, its high content of free radicals (p. 203) and other properties. A serious deficiency of melanin is pathological in the familiar sense that albinos and other deficient types are more susceptible than normal to disease in general, as well as suffering from lack of the various types of protection given by melanins. In this connection, however, it must be admitted that dominant white genotypes are quite healthy (p. 325); they have been positively selected and vigour has been selected at the same time. They always have adequate retinal melanin.

A further indication that a deficiency of melanin is pathological is the frequency with which abnormalities in melanin-production are associated with neuropathological conditions (LERNER, 1959). Parkinson's *paralysis agitans* is associated with loss of the normal melanin in cells of the relevant brain-stem nuclei (p. 99) and conceivably may be due specifically to this (MARSDEN, 1969; ISSIDORIDES, 1971). A faulty synthesis of the catecholamines may be involved (STEFANIS and ISSIDORIDES, 1970) and the condition is improved by administering L-DOPA. Phenylketonuric patients also have a low melanin content in the substantia nigra, and elsewhere; they are unable to convert phenylalanine to tyrosine.

A genetic deficiency in melanisation with some neural involvement is 'patchy whiteness of the skin' (DEOL, 1970). It is always correlated with patterned abnormalities of the cochlea and saccule of the inner ear, which implies that neural crest components in general are affected (HORSTADIUS, 1949) and is a reminder of this further link between melanocytes and nerve tissue. The patchy pigmentation itself extends to the inner ear meninges and this in particular is correlated with the sensory abnormalities of the genotype. Vitiligo is another type of patchy deficiency in melanin.

Excess melanin is usually due to neoplasia of melanocytes but an exception is the general browning, or 'bronzing' associated with Addisons disease of the adrenal cortex. The pigmentation is due to excessive pituitary output of the MSH hormone (p. 143) and the adrenocorticotropic hormone which is structurally rather closely related to it; the output of these hormones is increased because of the reduced feedback of cortical adrenal hormones in this disease. Not only the skin but also the mucous surfaces and the connective tissue and cartilage all may become melanised. Another non-neoplastic state is *chloasma*, an embarrassing facial mask of melanin associated with uterine troubles during pregnancy. This is not to be confused with the mask due to melanised sweat (FOX and VEVERS, 1960, p. 164), though conceivably this could be regarded as a mild pathy of the same general kind.

Naevi, or birthmarks, are often melanic and mostly benign excesses, but sometimes they become malignant (RILEY and PACK, 1966). Excessive local melanisation can be caused by mere mechanical irritation (VERNE, 1926, p. 231) which strengthens the view that melanocytes are particularly easily stimulated to excess proliferation. A single application of the carcinogen, dimethylbenzanthracene, to the skin of the Syrian hamster can cause multiple, if benign, tumours of melanocytes (RAPPAPORT et al., 1963), this animal being rather prone to tumour-induction. Melanomas are induced to some extent even in white hamsters and they are not uncommon in the albino type of white mammal (MARSDEN, 1969); in such individuals the whole melanogenic system must be normally present but with a block or lesion at some single point. In these albinos the initial oncogenic factor cannot be melanin itself, though conceivably it could be some component of the melanogenic system. Moreover some tumours of melanocytic tissues lack melanin and some lose it after neoplastic activity for some time (BOMIRSKI et al., 1966), so that melanin is not essential for the maintenance of neoplastic proliferation of melanocytes. In fact tumours of the amelanotic melanocytes of albino fishes are more neoplastic than melanotic tissues (GORDON, 1957). These *amelanotic* melanomas still give positive reactions for tyrosine and DOPA and so, like genetic albinos and dominant whites, have only a local lesion in the path.

Melanomas are very rare in tissues which normally never contain melanocytes but on the other hand there is certainly a tendency for pigment-free tumours of other tissues to turn melanotic, and neoplasia does tend to activate melanogenesis in cells with only a very latent power, or perhaps in tissues which have trapped latent melanocytes. Neoplasia is melanogenic in tendency and melanin neoplastic. It is interesting that melanoma should be induced so easily in established colourless tumours, since melanin-synthesis demands a high oxidation potential (p. 202) whereas the respiration of tumours is characteristically glycolytic (GREENSTEIN, 1954), as in other rapidly proliferating tissues (NEEDHAM, 1952, p. 50; 1964, p. 259). Moreover, like the production of other chromes, that of melanin is usually restricted to the differentiating, post-mitotic type of cell, whereas neoplasia demands some degree of dedifferentiation.

Probably the most interesting case of melanotic tumour development in a relatively amelanotic substratum is that common in grey and white horses. Individuals with the relevant genotype show an 80% predisposition to develop melanomas spontaneously later in life. This might seem to support one of the most plausible theories of cancer, that it is the escape from a normal inhibition of proliferation; the melanic system is inhibited in the hair bulbs of grey and white horses and the chances of local tissues escaping from this, by somatic mutation, increase with time, i.e., with age. It may be noted that pale human races are more prone to develop skin melanomas than negroid races. White horses are the result of a rather special genotype, involving the percheron, and the young individual is dark; the grey and white phases therefore may be due indeed to positive inhibition rather than to premature ageing. A further complication is that the albinism is largely restricted to the hair and melanomas tend to be most common where the skin itself is very melanic, for instance in the perineal region. Here there is the further possible complication of a higher content of steroid hormones than in other parts of the skin, steroids being among the most carcinogenic of biological materials. In these horses metastasis of the melanomas also is very common and extensive and is usually fatal. If the melanin is here pathogenic this is indirect, via the pathogenicity of rapid neoplasia.

Melanomas are rather common also in laboratory strains of insects, such as *Drosophila*. Here two mutant genes, on separate chromosomes are necessary for their genesis and there are suppressors for each on the other chromosome, so that the condition is rare in a randomly breeding population. Tryptophan in excess of 1% of the diet inhibits the suppressors (GLASS, 1957; SANG, 1969) and so raises the possibility that this 'melanin' is really ommochrome. Cysteine blocks the action of tryptophan, probably by preventing its oxidative conversion to kynurenine (GLASS, 1957) on the ommochrome path (p. 289). However, the melanin and ommochrome paths intermesh in a number of ways (p. 297) so that the evidence is not conclusive and there is no doubt that insects do produce indolic melanins, as well as the phenolic counterparts (p. 46). It is possible that human alcaptonuric patients synthesise ommochromes from tyrosine (BUTENANDT, 1960) as JOHNSON (1961) found for simple *in vitro* systems, so that very detailed evidence is required in every case. Erythromelanin does not produce tumours, at any rate in fishes (GORDON, 1957).

Insects survive in the laboratory with lesions in the ommochrome path (BRUNET, 1965). An exact measure of their disability would be useful since it might settle the question whether ommochromes have any significance except as integumental chromes and eye-masking pigments.

11.6 Parasitism and Zoochromes

Parasites cause at least a mildly pathological condition in their hosts and they commonly affect pigmentation. VERNE (1926, p. 231) was already aware that parasites sometimes induce excess pigmentation in their hosts and further examples are now known. The effect is often local and appears to have adaptive value to the parasite by making the host conspicuous to a predator capable of acting as a vector. It is an episomatic effect, in fact (p. 131).

Carotenoids appear to be most frequently involved and this is probably because they are so abundant, innocuous and easily transported across membranes. Both cestode and trematode parasites of *Calanus* cause it to become a conspicuous scarlet (MARSHALL, 1934), while dinoflagellate Protozoa cause it to turn orange or coral pink (MARSHALL and ORR, 1955). When parasitised by *Sacculina*, *Carcinus* becomes redder than normal because the parasite absorbs β-carotene differentially relative to the redder oxycarotenoid, astaxanthin (LENEL, 1954). CZECZUGA (1971) has made a comparative study of carotenoids in the crustacean *Argulus* and its host, the stickleback.

In some instances the excess chrome is deposited in the parasite itself, for instance the acanthor larva of the acanthocephalan, *Polymorphus*, which again renders its host much more conspicuous than uninfected *Gammarus* (BARRETT and BUTTERWORTH, 1968). Similarly, Cladocera infected with *Spirobacterium cienkowskii* become red owing to the carotenoid, rhodoviolacein, in the bacterium itself (GREEN, 1959). The ciliate parasites of the copepod, *Idya furcata*, and the hermit crab, *Eupagurus prideauxi*, obtain carotenoids from their hosts and synthetise red, violet, blue and green carotenoproteins from them (CHATTON et al., 1926). The parasites apparently eat the host's eye-pigments so that other classes of chrome also may be involved. Filarian nematodes in the chamaeleon accumulate seven different carotenoids; the host has only five in its fat body so that this may be an indication of the extent to which animals can modify their dietary carotenoids (FRANK and FETZER, 1968). The carotenoids of the *Polymorphus* acanthor differ from those of *Gammarus* and these again from those of the elm leaves on which it feeds (BARRETT and BUTTERWORTH, 1968).

Melanins and porphyrins come next in importance in host-parasite relationships so that there is a general parallel to other pathological situations (p. 252). In the skin of the fresh-water teleost *Oryzia* trematode parasites induce melanomas (IGA, 1965). The melanophores involved remain under the normal control of the pigment-aggregating nerves and this indicates that the melanoma cells do not show the usual dedifferentiation of neoplastic tissues, but continue to function. It would be useful to know if the response differs in any way from that of the melanophores of the normal skin, since this should make the host more conspicuous.

Metacercaria in the cod often become surrounded by a connective tissue capsule on which melanin deposits (HSIAO, 1941) and something similar occurs to parasites in the haemocoele of insects (SALT, 1955, 1957). In the insects the blood cells concerned in the normal blackening reaction in a blood-clot collect round the parasite and form a kind of internal clot. *Malpighamoeba locustae*, a sarcodine parasite of the malpighian tubules of the Saltatoria, induces black cysts both in normal and in albino hosts (KING and TAYLOR, 1936); the melanin system of the blood of arthropods seems to be universal and so independent of any particular integumental system. These appear to be among the few examples of a chromogenic response by the host itself in its own interest. Even here the chrome itself may be rather incidental to the need for a solid capsule. The larva of *Polymorphus* has a somewhat similar effect on *Gammarus* and it is thought that the parasite releases suitable phenolic substrates for the host's tyrosinase (HYNES and NICHOLAS, 1958); the host no doubt keeps its own substrates under control, but Crustacean blood readily responds in this way to added phenolic substrates (NEEDHAM, 1949; DENNELL, 1958). The haemolymph of *Daphnia* is blackened by *Cysticercus mirabilis* (MARSHALL, 1934).

The flies, *Musca autumnalis*, and *Orthellia caesarion*, parasitised by the nematode

Heterotylenchus autumnalis develop large brown to black symmetrical patches in the epidermis and cuticle around the anus (STOFFOLANO, 1971). These are very similar to patches developed in some tumour strains of *Drosophila* and in the epidermis of houseflies fed with carcinogens. In this instance, therefore (p. 250), the parasite does behave like other pathological agents.

The malaria parasite changes much of its host's haem to ferrihaem (p. 241) which is released in solid form and taken up by leucocytes. These are retained by various organs, particularly the spleen, liver and bone marrow, so that the iron may be re-utilised gradually. The organs become very blackened. Sanguisugous animals can be regarded as migrant external parasites; they use and modify the haem in an amazing variety of ways (WIGGLESWORTH, 1943), but this no longer concerns the host as the malarian chrome does. Many internal parasites cause anaemia among their pathological effects, and not always by their direct depletion of Hb; some permanent external parasites also have significant anaemic action.

A number of parasites possibly take Hb from their hosts more as an easy source of haem or of utilisable iron for their own Hb-synthesis than for the globin and other nutrients. Respiratory conditions are often difficult and Hb is commoner in parasites than in their free-living relatives: among copepods and cirripedes it is only parasitic members which do have respiratory Hb (FOX, 1953). The rhizocephalan cirripedes parasitise mainly decapod crustacea and so have to synthesise their Hb *de novo*, without help from the host. The Hb of parasites is adaptive in having a higher affinity for O_2 than that of the host (p. 217).

There is an accumulation of biliverdin in the absorptive roots of some rhizocephala but this scarcely can come from their host's relatively small stock of intracellular enzyme haemoproteins. Nor is it likely to be passed from the parasite to its host. The most plausible suggestion is that already made (p. 241) that it plays some part in transport across the membrane of the roots.

Parasites affect the ommochrome production of their hosts. *Plistophora*, a protozoan parasite of *Locusta*, causes small but intense pink-purple spots in the integument and subcutaneous tissue and these were shown to be ommochrome (GOODWIN and SRISUKH, 1950). The acanthocephalan larvae parasitic in *Asellus* causes an intense darkening of the gill covers (Fig. 11.2), normally rather lightly pigmented. This seems to be due to a vast extension of the normal chromatophore system, the chromes of which are ommochromes (NEEDHAM and BRUNET, 1957). The gill covers are somewhat exposed and also move rhythmically when the animal is breathing so that the dark colour considerably increases their conspicuousness. It seems that various integumental chromes have been used in this way.

If there were any doubt that this serves the parasite the useful purpose of attracting a predatory vector, or definitive host, it would surely be dispelled by the instance of the sporocyst of the digenetic trematode, *Leucochloridium macrostomum* in its intermediate host, the terrestrial gastropod, *Succinea putris* (GAMBLE, 1896; DAWES, 1946). The sporocyst invades the snail's tentacles and prevents their retraction, while introducing a muscular pulsation of its own. It also induces, or itself forms, a sharp pattern of alternating white and green, or white and red annuli round the tentacles, and a red tip, so that they resemble certain caterpillars. The definitive host, an insectivorous bird, snaps at the tentacles, which very easily break off, with their contents of cercaria inside a diverticulum of the sporocyst. Alternatively the tentacles are autotomised, and then the motile sac looks even more like a caterpillar. The tentacles regenerate and repeat the cycle. The parasite does not pulsate in the shade but in any case the snail becomes positively phototactic (GAMBLE, 1896). It would be difficult to imagine a more sophisticated and perfected mechanism for ensuring transfer to the final host.

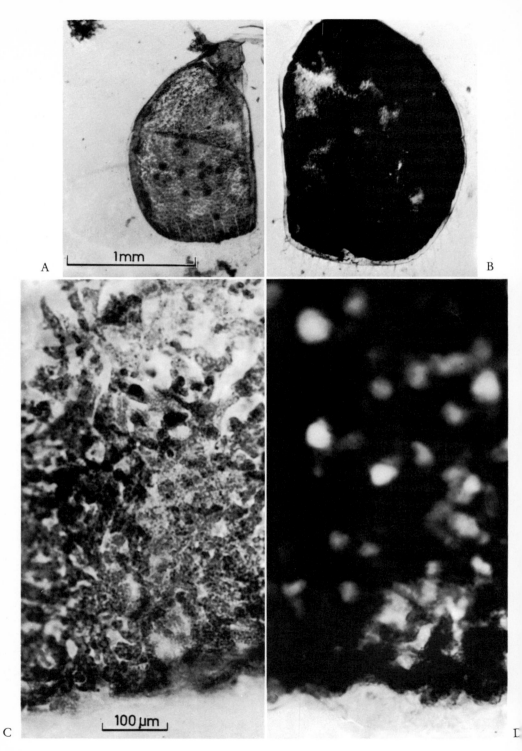

The chromes used by *Leucochloridium* are not yet known, and this is true for that in the amoeba, *Janickina (Paramoeba) pigmentifera*, parasitic in the coelom of the chaetognaths *Sagitta* and *Spadella* (HYMAN, 1959, p. 44). There are many other instances where the chrome has not been identified, for instance the yellow or brown colour of the condition *pityriasis versicolor* caused in Man, by the fungus, *Microsporidium purpur*, and the yellow sometimes caused by the tubercle bacillus.

Since most parasites are scarcely normal when separated from their host it is not so easy to detect any effect by the host on the pigmentation of the parasite as it is the reciprocal effect. Comparison with free-living stages or with free-living relatives could be misleading since, as in the case of the copepod Hb (p. 257), the determining factors may be of a general oecological nature rather than the host-relationship specifically. At the same time the host largely dictates what the parasite receives in the way of dietary chromes. Nevertheless parasites do modify these, as seen above, and they are more or less as independent of their food supply as free living animals. Their respiratory enzymes are adapted to their special requirements (p. 187). More specific examples of host action on parasite chromatology are rare.

11.7 Conclusion

All zoochromes are subject to destruction by the radiation which can excite them. Equally they may photosensitise other biological materials to their detriment. For both effects a strong oxidant such as O_2 also is necessary. These effects may explain why so much of the metabolism of zoochromes today is chemically and not photically excited.

Very few known chromopathies involve chromes of entirely different chemical classes from those normally present, though some minor qualitative abnormalities are known, particularly in the protein component of chromoproteins. Most primary chromopathies involve deficiencies or excesses of normal chromes; the former are the more common and most excesses are surprisingly well tolerated. Deficiencies are best known for the exogenous vitamin type of biochrome, for obvious reasons, but serious deficiencies of endogenous chromes are equally drastic. Even melanin-deficiency is more deleterious than excess but excess is so commonly associated with neoplasia that melanin may appear to be the pathogen; in fact it may be so indirectly, via the general pathogenesis of tumours.

Apart from the episomatic use of integumental chromes by parasites they probably do not affect the host's chromatology more than they do other aspects of its metabolism: the effects are more conspicuous. The parasitic habit dictates properties of the parasite's respiratory chromes. There are few clear-cut examples of the host affecting a parasite's chromatology in the host's interest.

Fig. 11.2. Low power (A, B) and high power (C, D) photomicrographs of ventral surface of gill-operculum of the freshwater isopod, Asellus aquaticus. A, C normal, B, D parasitised by larvae of Acanthocephalus sp., showing the great increase in amount of xanthommatin in parasitised individuals

V. The Significance of Zoochromes for Reproduction and Development

The close relationship, in most animals, between the functions of reproduction by the one generation and the development of the next should not be allowed to obscure the sharp differences between an adult physiological function and a typical developmental process. The significance of zoochromes in reproduction therefore reasonably could have been considered in section III but the significance lies so much in provision for the embryo that the present arrangement seems justified.

Chapter 12

Zoochromes in Reproduction and Development

12.1 Reproduction

The chromes relevant to reproduction are those of the gonads and accessory reproductive organs, those of the integument (in dioecius species) and some elsewhere in the body. On the physiological side it is useful to distinguish chromatic activity related to puberty, and to its seasonal counterpart in perennial species, from activity in the reproductive process itself though in practice the two are not always sharply separated. Another logical distinction is between chromes which are substrates, enzymes and controlling agents respectively, though the first two also can have important controlling actions. In virtually every aspect it is useful to consider sex-differences.

12.1.1 Chromes of the Reproductive Organs

It has been seen (p. 107) that a large number of chromes are associated with these organs, in the male mainly with the accessory organs and tissues and in the female mainly with the ovary. In the male they are usually the chromes characteristic of the integument of the species whereas in the female this is often masked by the quantitative preponderance of carotenoids (GOODWIN, 1950, 1952, 1962), and occasionally other classes, destined for the eggs. There are four times as much carotenoid in the ovary of the cow as in the testis of the bull (Fox, 1953, p. 175) and the differences are even greater in some of the animals producing more and larger eggs. In compensation, perhaps, some of the other organs of the female tend to have less carotenoid than those of the male (CROZIER, 1970). Female crabs at maturity have four times as much fat as the male; this is important in its own right but also as a vehicle for carotenoids. There is usually some carotenoid in the endocrine tissues of the gonads of both sexes (chapter 5). In fact carotenoids tend to accumulate in all organs associated with reproduction, even including such 'peripheral' organs as the vertebrate adrenal.

Little is known about the significance of chromes other than those passed into the eggs from ovaries and yolk glands. There is porphyrin in the shell-secreting region of the oviduct of some birds and this is incorporated into the shell. Coloured shells are common in other animals and again chromes have been detected in the relevant accessory organs. Most of the chromes of the reproductive organs first appear, or strongly increase in amount, at the time of maturity and in perennial animals they wax and wane with the breeding-season. There is a breeding-season cycle even in such a peripheral organ as the hypobranchial gland of *Nucella* (*Purpura*) (LETELLIER, 1889).

12.1.2 Integumental Chromes

Because they so frequently compete with camouflage-requirements integumental colours related to reproduction also tend to appear only at maturity, and at the onset of the breeding season in perennial species. The biology of this competition and of the sex-differences in colour has been considered in chapter 6. Reproductive, courtship and recognition-coloration is common in all those taxa with adequate visual powers to use the signals. Usually only one sex is conspicuous but both sexes in some large birds and in others safe from attack.

In a number of species the relative amounts of specific chromes have been measured in the two sexes. The male of the butterfly, *Argynnis*, has more ommatin D in the wings than the female (BUTENANDT et al., 1960a) and the male pierid contains 66% more uric acid than the female (FOX, 1953, p. 289). In *Fundulus*, the male has 36% more carotenoid in the integument than the female (FOX, 1953, p. 147), giving a conspicuously more yellow colour to the belly and fins. The female is thought to have less because of the demands of the eggs (p. 267) but the male does not deposit his large amounts until maturity and in principle the female could store in advance enough for both purposes. The dorsal skin of the male eel has 85% more carotenoid than that of the female (FONTAINE, 1937).

12.1.3 Pubertal and Seasonal Changes

There are many records of changes in visible coloration at puberty. In the lightly pigmented human races it has again become possible to use skin coloration; the penis and the areola of the nipple darken at this time and the latter still more during pregnancy. The biological significance of these signals presumably antedates the wearing of clothes, but in any case constitutes an interesting problem. The polychaete *Nereis* shows an increase in integumental biliverdin at maturity while haems and carotenoids decrease (DALES and KENNEDY, 1954). At this time locusts exchange astaxanthin for β-carotene (GOODWIN and SRISUKH, 1948); here the effect on coloration must be trivial and some metabolic significance may be suspected. Seasonal changes vary greatly in extent. The male sparrow shows little more than a darkening of the bill (FOX and VEVERS, 1960, p. 43) while the starling shows the reverse change together with a brightening of the plumage. Other birds have an extensive moult before and after the season so that the male is conspicuous in the season and well camouflaged out of season ('eclipse' plumage).

In fishes seasonal changes can be effected by the orthodox morphogenetic type of mechanism (p. 136), but some of these changes develop at shorter notice. In the cichlid fish, *Aequidans latifrons*, a vertical band of chromatocytes in the female becomes very black just a few days before egg-laying, and the male darkens similarly while guarding and fanning the eggs (KRAMER, 1960). It is an easy transition from this relatively rapid chromogenic change to a chromomotor mode of nuptual colour change, as in the stickleback, black sea-bream and cuckoo wrasse (CARLISLE, 1964). *Sepia* and other cephalopods use rapid colour changes for mating purposes (HOLMES, 1955) and the lizard, *Anolis*, likewise shows the chromomotor type in its mating display (HADLEY, 1929). Blushing in human beings also is relevant here. There appears to be a consistent sex-difference in the state of chromomotor dispersion of the chromatophores of the crab, *Uca* (FINGERMAN and FITZPATRICK, 1956), but the significance of this is not clear.

12.1.4 Reproductive Metabolism

Aspects of the general metabolism of biochromes specific to reproduction mainly concern the female and are geared to the function of packing chromes, along with other provisions, into the eggs (Table 12.1) or into the foetus. The blood of female crabs is yellow with carotenoids at the time of ovary-ripening and after oviposition it becomes blue again, due to haemocyanin (VERNE, 1926, p. 508). Carotenoid is mobilised (GILCHRIST and LEE, 1972) from the hepatopancreas to the ovary (ABELOOS and FISCHER, 1926) and is carried as lipoprotein (LWOFF, 1927). In *Simocephalus*, similarly, the carotenoprotein level of the blood is related to the rate of egg-production (GREEN, 1966). Both sexes of the sockeye

Table 12.1. Distribution of biochromes in eggs of animals

Nature of chrome	Animal(s)	Notes	Reference
Carotenoids			
Oxycarotenoid-protein	*Clava* and other hydroids	–	TEISSIER, 1925
Carotenes	Echinoids	⎫ Gonad: probably	FOX, 1953, p. 102
Carotenoproteins	Holothurians	⎭ passed to eggs	⎧ MACMUNN, 1890 ⎫ ⎩ LÖNNBERG and HELLSTRÖM, 1931 ⎭
Pectenoxanthin	*Pecten*	ovary	LEDERER, 1938
Mytiloxanthin and other oxycarotenoids	*Mytilus*	ovary	SCHEER, 1940
Oxycarotenoprotein (ovorubin)	*Pomacea* (Pulmonata)	–	CHEESMAN, 1958
Astaxanthin – protein (ovoverdin)	*Ceratocephale* (Polychaeta)	–	YAMAMOTO, 1935, 1938 KENNEDY, 1969
Astaxanthin – protein (ovoverdin)	Lobster, *Daphnia*	–	GREEN, 1957, 1965
β-Carotene, cryptoxanthin and other oxycarotenoids	*Carcinus*	–	GREEN, 1965
Canthaxanthin	*Artemia*	–	GREEN, 1965
β-Carotene	Locust	–	GOODWIN and SRISUKH, 1949
Lutein and β-carotene	Silkworm	–	MANUNTA, 1933
Red carotenoid	*Styela* (Tunicata)	–	MACMUNN, 1890
Astaxanthin, carotene and ? other oxycarotenoids	Teleosts	–	MANUNTA, 1937b FOX, 1953, p. 146 ff
β-carotene and lutein	Amphibia	–	BRUNNER and STEIN, 1935
Lutein	Chamaeleon	–	MANUNTA, 1937a
Lutein, zeaxanthin	Birds	–	KÜHN et al., 1933
Vitamin A (little carotenes)	Vertebrates	in general	CHEESMAN et al., 1967
Carotenes (little vitamin A)	Invertebrates	in general	CHEESMAN et al., 1967
Flavonoids			
"Bombichlorine"	*Bombyx*	–	⎧ MANUNTA, 1936 ⎫ ⎩ JUCCI, 1932 ⎭
Quinones			
Ubiquinone	Birds	–	CRANE, 1965
Echinochromes	Echinoids	Including jelly round egg	LEDERER and GLASER, 1938
Carminate	Scale insect	–	DIMROTH and KAMMERER, 1920
Simple metalloproteins			
Haemocyanin	*Carcinus, Eriocheir, Macropipus*	First record of intracellular Hcy	BUSSELEN, 1971
?Copper-protein	Hen	Yolk	⎧ J. NEEDHAM, 1942, p. 19
?Iron-protein	Hen	Yolk	⎩
Porphyrin derivatives			
Haemoglobin	Nematodes and trematodes	–	SMITH and LEE, 1963
Haemoglobin	Serpulids	–	KENNEDY, 1969
Haemoglobin	*Daphnia*	Anaerobic conditions	FOX, 1947
Haemoglobin	*Pediculus*	–	WIGGLESWORTH, 1943
Ferrihaemoprotein	*Rhodnius*	–	WIGGLESWORTH, 1943

Table 12.1 (continued)

Nature of chrome	Animal(s)	Notes	Reference
Haemochrome	*Triops*	Shell of egg	Fox, 1955
Protoferrihaem	*Artemia*	Shell of egg	J. and D. M. Needham, 1930
Free porphyrins	Hen	Albumen	Lemberg, 1938
Biliverdin	*Nereis fucata*	–	Green and Dales, 1958
Biliverdin	*Septosaccus*	–	Bloch-Raphael, 1948
Melanins			
Eumelanin	Chondrostei, Choanichthyes, Amphibia	–	–
Indigotins			
Dibromoindigotin	Muricidae	In fluid round eggs, in capsule	Fischer, 1925
Ommochromes			
Xanthommatin	*Urechis*	–	Horowitz, 1940; Linzen, 1959
Pterins			
Various	Insects rather widely	–	Ziegler-Gunder, 1956; Ziegler and Harmsen, 1969
Isoalloxazines			
Ribitylflavin	Polychaetes	–	Kennedy, 1969
Ribitylflavin	Insects	–	Trager and Subbarow, 1938; Bodine and Fitzgerald, 1947
Ribitylflavin	Elasmobranchs	–	Fontaine and Gourevitch, 1936
Ribitylflavin (ovoflavin)	Birds	Albumen and yolk	Kühn and Wagner-Jauregg, 1930
Unidentified			
Vanadochrome	Ascidians		Goldberg et al., 1951
Red, orange and yellow (?Carotenoids)	Ectoprocta		Hyman, 1959, p. 343
Pink (animal pole) Green (vegatative pole)	*Myzostoma*		Pitotti, 1947
Yellow (animal pole) Violet (vegetative pole)	*Perinereis*		Spek, 1930
Pink (MeOH soluble) Yellow (MeOH soluble)	*Lumbriconereis* *Neanthes*		Suzuki, 1968
Green chrome	*Pectinaria*		Tweedell, 1962
Chlorochromin	*Flabelligera*		Krukenberg, 1882
Three green and blue chromes	*Sabella*		Dales, 1962
Yellow chrome (?cytochrome)	*Styela, Ciona*	In presumptive muscle tissue	Conklin, 1924
Slate grey chrome	*Styela, Ciona*	In presumptive gut tissue	
Green 'pigment Y'	*Patella*	Chromoprotein	Bannister et al., 1968

salmon, *Oncorhynchus nerka*, mobilise the carotenoids from their muscle (p. 103) and in the female these are transferred mainly to the eggs but in the male mainly to the skin, to enhance nuptual colouring (CROZIER, 1970). As already noted (p. 245) the transfer by laying hens of their dietary xanthophylls and some other carotenoids to their eggs can compete with the supply to the feathers (PALMER and KEMPSTER, 1919c). On the other hand the amount of carotenoids increases in both ovaries and integument of the female locust at maturity (GOODWIN and SRISUKH, 1919) and of the female of the crab *Emerita* at the onset of the breeding season (GILCHRIST and LEE, 1972).

The cow has a much higher carotenoid metabolism than the bull (Fox, 1953, p. 175); the concentration of carotenoids in the plasma is much higher, especially during pregnancy. In mammals, and particularly in the domestic cow, the reproductive period is extended also at the other end by a prolonged lactation-period so that these sex-differences are very large and persistent.

Some of the chromogen of the hypobranchial gland of *Nucella* is passed into the egg-capsules (FISCHER, 1925). In some of the capsules it develops into the definitive dibromoindigotin but its precise significance here is still uncertain. The gland is present in both sexes so that it is not related solely to egg-formation. LETELLIER (1889) thought it might function in mating, presumably with a chemosensory (Fox and VEVERS, 1960, p. 61) rather than a visual action!

In *Mytilus* carotenoids are used in spermatogenesis as well as in oogenesis (SCHEER, 1940) and the testis is distinctly coloured in some molluscs. On a low lutein diet the fertility of the cock of *Gallus* is reduced (GOODWIN, 1950b) and this may indicate a similar function. The unidentified photodynamic chrome (p. 249) in the semen of the bull (NORMAN and GOLDBERG, 1959) and a melanising activity in the semen of various mammals (BEATTY, 1956; PANT and MUKERJEE, 1971) are relevant here. Both are puzzling since the number of live sperm is inversely proportional to melanising activity. It is important to know the origin of these materials, i.e., whether normal spermatozoa are exposed to them under aerobic and illuminated conditions prior to insemination. Melanising activity is proportional to the amount of integumental melanin in the individual (BEATTY, 1956) and might be simply incidental to this.

There are a number of observations on reproductive chromatology which call for further study. Some insects begin to develop sex-differences in the chromes of the blood as early as pupal and even larval stages. The blood of female larvae of *Xanthia flavago* is yellowgreen but that of the male is colourless; a similar difference holds in the larvae and pupae of *Bombyx* (GEYER, 1913). Of course some of these differences may be related so sex-differentiation in the integumental chromes rather than to differences in reproductive metabolism. Another example at present difficult to interpret is a correlation between the concentrations of bile pigment and haem in the female, but not in the male, of *Chirocephalus* (GILCHRIST and GREEN, 1962). It has been suggested that an increased deposition of purines in the integument of insects, fish, Amphibia, reptiles and probably some molluscs (Fox, 1953, p. 290) may be secondary to reproductive changes in nucleic acid metabolism (VERNE, 1926) but this is questionable (p. 243). Nucleic acid metabolism certainly is much higher in the male than in the female but purine deposition is not. The female of the butterfly, *Sympetrum*, has red ommochrome in the integument and the male yellow (VUILLAUME, 1969, p. 85) but this is not due to a higher respiratory rate in the female (p. 203).

12.1.5 Control of Reproduction

The control of reproduction is mainly hormonal and while carotenoids are associated with the gonadal endocrine tissue (p. 263) so far no functional significance of this has been indicated. The visual-pituitary-gonadal control covers all aspects of reproduction from gonadal maturation to integumental colour-changes. Colour-changes typical of the male are sensitive to androgens but not to oestrogens. It may be significant that melanomas grow faster in female than in male hamsters (ROSENBERG et al., 1963) and that castration abolishes the difference; however, it remains to show that this is relevant to integumental colour rather than to growth-properties in general.

Those chromes known to affect reproduction probably do so as enzymes or at any rate act more directly than the hormones. Vitamin E appears to be particularly necessary (p. 199) for all reproductive processes in both sexes. This has been shown in mammals (ROELS, 1967), the cricket *Achaeta* (MEIKLE and MCFARLANE, 1965), the Crustacea *Daphnia* (VIEHOVER and COHEN, 1938) and *Cyclops* (JACOBI, 1957) and the rotifer *Asplanchna* (GILBERT and THOMPSON, 1968). Flavonoids affect reproductive activities in mammals (HARBORNE, 1965) thought it is not known that they have a normal physiological function. Isoflavones induce the spring flush of milk in cows whereas the flavonoid *genistein* acts as a weak oestrogen and depresses fertility in sheep.

Biopterin promotes sexual activity in the nymphs of aphids (GILMOUR, 1965, p. 153) and it was suggested (BUTENANDT and REMBOLD, 1958) that the high concentration of biopterin in the royal jelly of the honey bee might determine development as a queen rather than as a worker. Later REMBOLD and HANSER (1960) found that biopterin is not metabolised by the larva of *Apis* but this need not prevent it acting as a hormonal type of agent or indeed as a very stable co-enzyme. The alternative suggestion that biopterin accumulates in glands which secrete the royal jelly merely as a selective excretory outlet (ZIEGLER and HARMSEN, 1969) seems improbable in the light both of the discussion in chapter 10 and of the great utility of the hydrogenated bioterins (p. 204). It may be noted that the B-vitamin, pantothenic acid (p. 209) also is highly concentrated in royal jelly (PEARSON and BURGIN, 1941), where there are three times as much as in yeast and liver. There is also a correlation between the pterin-content of the eyes and endocrine organs and the photo-periodic control of reproduction (L'HÉLIAS, 1962).

Protoporphyrin and haematoporphyrin have been found to induce oestrus in mice and to accelerate ovarian development and maturity in younger individuals (KLÜVER, 1944b). The pituitary selectively takes up such injected porphyrins and it is thought that they may sensitise the gland to light. A photoperceptor region mediating reproductive activities has been found in the brain of birds and lizards, but so far not in that of mammals (MENAKER, 1972). BENOÎT and OTT (1944) found that red and yellow light, in particular, stimulated sexual activity in birds and these are the wavebands in which the porphyrins transmit most light. In the female of the nematode *Mermis* stimulation of a red spot, which is due to haemoglobin, induces oviposition (ELLENBY, 1964).

In the ovary, the large amounts of carotenoids serving as substrate-material for the eggs preclude the detection of any other function but where carotenoids are involved in the reproductive activities of the male (p. 267) it is more likely that they are acting as control agents than merely as substrates. HARTMAN et al. (1947) found that astaxanthin acts as a 'fertilisation hormone' in the rainbow trout, *Salmo gairdnerii (irideus)*. A number of carotenoids and some other isopentenyl derivatives at one time were thought to function in the determination of sex and of mating-behaviour in the gametes of phyto-flagellates

(MOEWIS, 1933, 1950) but this proved untrue (PHILLIPS and HALDANE, 1939; HARTMANN, 1955; RAPER, 1957). Flavonols also have been implicated, but again without confirmation.

12.1.6 Conclusion

Of the zoochromes of the reproductive organs only those passed into the eggs have yet a known significance. The pubertal and seasonal changes in integumental chromes are correlated with internal reproductive changes and are under the same external and internal controls. As well as the chromogenic there are some chromomotor colour-changes associated with coutship and mating. Particularly in the female there is much movement of chromes around the body during reproductive activities but a number of chromatological observations are unexplained. Tocopherols, flavonoids, pterins, carotenoids and porphyrins play some part in the control of reproduction.

12.2 Zoochromes in the Ontogenesis of Animals

Ontogenesis is interesting chromatologically for two main reasons: it shows how chromes behave during development and so to some extent, perhaps, how they behaved during evolution; secondly it shows the value of chromes in the control of development, which is the major biological process remaining for consideration. The importance of the study is immediately emphasised by the large amounts and the number of chromes fed into eggs (Table 12.1). As already indicated, carotenoids are outstanding in importance here (GOODWIN, 1950, 1952) and there are very few phyla which do not regularly use them for this purpose. Ribitylflavin (RF) also very frequently is present but virtually every class of biochrome has been recognised in the eggs of at least some taxa. A number of unidentified chromes also have been found in eggs so that the complete tally may be much greater. The chromes of shells, cocoons, capsules, etc., are mainly for camouflage and other forms of protection, and will not be considered in detail here.

It will be noted (Table 12.1) that those animals that supply haems to their eggs are nearly all blood-feeding parasites, and it is known that haem has become a 'demand', or vitamin, for this type of parasite (p. 257), including even the bacterium, *Haemophilus*. Since carotenoids and RF are the most common chromes in eggs it is clear that the general tendency is to supply the eggs with the vitamin type of co-enzyme chrome which the embryo cannot itself synthesise but which it needs as critically as the adult. In other instances, such as *Daphnia*, echinoids, coccids and ascidians, chromes peculiar to that taxon, or to a particular ecological situation, also are supplied in the egg and again it is credible that these are essential to the embryos of those particular animals. In general integumental chromes are among the last to be synthesised by the embryo itself (NEEDHAM, 1964, p. 3); they mostly represent the acme of differentiation. This timing also may have evolutionary significance.

At present it cannot be concluded that all chromes in eggs serve a useful purpose, since artificial dyes such as Sudan III are incorporated into the yolk of the hen's egg (PALMER and KEMPSTER, 1919c). A similar enigma among uncoloured materials is the high content of salts of lead in the eggs of some hens (NEEDHAM, 1942, p. 18). Moreover *Daphnia* seems to pass the most haem into its eggs when it is actually destroying its own Hb, owing to improved aeration of the water (GREEN, 1965). This apparent dustbin-treatment of eggs is most surprising and calls for further study: DRESEL (1948) had found that the amounts of haem in the eggs of *Daphnia* were proportional to the degree of anoxia of the water, and so were adaptive. It is possible that a high concentration of Hb in the egg might inhibit subsequent synthesis by the young *Daphnia* and act as a useful control after all. It seems likely that under normal conditions eggs have some protection from excessive amounts of any dangerous materials. No doubt the large stores of undifferentiated proteins and other materials in the egg-cytoplasm strongly bind stray molecules of many kinds but the binding itself is an effective type of inactivation. Some natural chromes may be bound in this casual way.

At present evidence about the behaviour of oochromes during development is slight and fragmentary, though full of interest. It indicates that a number of them control development in some capacity; in principle any essential constituent, even if only a substrate for some reaction, could exert very critical control. The hatchability of the hen's egg is proportional to the carotenoid content of the yolk (J. NEEDHAM, 1942, p. 23); similarly pigmented eggs of wild type *Ephestia* and *Bombyx* are more viable, and have a more normal development, than the pale eggs of some mutants (KIKKAWA, 1953). Further, there is a general correlation between pigmentation and regions of high developmental activity in embryos (HUXLEY and DEBEER, 1934, p. 381). A number of commercial chromes will activate embryonic processes (J. NEEDHAM, 1942, pp. 189, 210–211); not all act as analogues of biological redox chromes and there is the possibility of fundamental but non-specific activations, which can be met in degree by virtually any chrome. The problem is as broad as causal embryogenesis itself and at present it is possible only to survey the pieces of evidence about each class of chrome: it does not always build a clear or complete picture. It has not always been possible to distinguish between chromes as substrates and those as agents for the process.

12.2.1 Effects of the Classes of Biochrome on Embryogenesis

Carotenoids: The importance of carotenoids, and particularly vitamin A, in the chick embryo is supported by several independent studies, though PALMER and KEMPSTER (1919a) found carotenoids not essential in some breeds of fowl. TEISSIER (1932) came to this same conclusion for *Daphnia* and it is possible that the demand is somewhat sporadic, like that of adult animals of the various taxa for vitamin A. The latter seems to be necessary at least for the development of limbs and eyecups in vertebrates (NEEDHAM, 1942, p. 227). The trout embryo is able to synthesise vitamin A from its carotenoid stocks, as adults of many species do. Vitamin A is often the only carotenoid in the eggs of vertebrates whereas those of many invertebrates have only other carotenoids (KON, 1962).

There is evidence of the usual antioxidant function of carotenoids (GRIFFITHS et al., 1955), which is particularly useful against photo-oxidation. This protection extends to other forms of radiation, and to heat-denaturation (CHEESMAN et al., 1967). In this connection a light-inducible colour-change in the *ovoverdin* of the polychaete, *Ceratocephale,* may be relevant (YAMAMOTO, 1935, 1938); however, this is independent of the presence of oxygen and conceivably may mediate a direct photo-activation. The carotenoprotein of the apple snail, *Pila,* aggregates at the base of the cilia of the velum, as well as in some other locations (FOX, 1953, p. 114). In *Nereis,* similarly, carotenoid accumulates at the base of the prototroche (COSTELLO, 1945), implying a function in relation to ciliary movement. Other chromes also tend to accumulate here, for instance xanthommatin in *Urechis* (FOX, 1953, p. 262). They may act as photoperceptors for local phototaxic responses, or conceivably as respiratory mediators.

The taxonomic distribution of carotenoids in eggs is as complicated and puzzling as that of carotenes and xanthophyllas in adults (p. 303). Moreover it is often significantly different from that of the parental tissues, and perhaps not mainly for the reasons of competition discussed in the previous section.

The pattern of metabolism of carotenoid oochromes also varies greatly. The eggs of Hydrozoa such as *Clava, Hydractinia* and *Sertularia* release carotenoid from the stored carotenoprotein as early as gastrulation (NEEDHAM, 1931, p. 1381) whereas the ovoverdin of the lobster's eggs remains unchanged until just before hatching (FOX, 1953, p. 131) and is not used until then (GOODWIN, 1951). Of course the development of hydroids is more than half completed by the gastrulation stage. In *Artemia* (GILCHRIST and GREEN, 1960) and in the crab, *Emerita* (GILCHRIST and LEE, 1972) there appears to be no decrease in the amount of carotenoid in the egg-embryo system during development and these authors conclude that this class is not used by the embryo. On the other hand MORÈRE (1971) re-emphasises the indispensability of β-carotene for the development of the insect, *Plodia*.

Carotenoids play an important, if still quite obscure, role in the regeneration processes of hydroid Cnidaria (TARDENT, 1963; STEINBERG, 1954, 1955). This was first noticed by DRIESCH (1900, 1901) and MORGAN (1901a). The regenerative powers of the different regions of the stalk of *Corymorpha* show a good proportionality to the amount of red pigment locally (OKADA, 1927). The carotenoid is mainly in the endoderm cells and shows an active aggregation at the sites of the future tentacle-primordia of the regenerate (SCHULZE, 1917; STEINBERG, 1954, 1955; TARDENT, 1963). The aggregation is now thought to be due essentially to a movement of whole cells, which become more closely packed in the critical regions (TARDENT, 1963), but in any case the density of the chrome may be an important factor for regeneration.

In *Corymorpha* the red region induces tentacle-formation in the neighbouring yellow region, more dilute in carotenoid. Reciprocally the latter induces rhizoid formation in the red region (OKADA, 1927) which in isolation tends to produce biaxial hydranths (Janus heads). A morphogenetic system therefore is closely correlated with the chromatic pattern, if not actually due to this. There is an analogy to the coincidence between melanin and morphogenetic gradients in the frog's egg, and similar systems in other eggs. The carotenoid of the regeneration-primordium disappears before visible differentiation (MORGAN, 1901a) so that any action would seem to be related only to the initial induction. The carotenoid is said to be excreted via the enteron (MORGAN, 1901b, p. 268) and the new hydranth then proceeds to accumulate more. There is an accumulation also in relation to gonad-formation, for instance in *Hydra* (SCHULZE, 1917).

It may be significant here that astaxanthin accumulates distally in the zoecia of the ectoproctan, *Bugula* (VILLELA, 1948b). In the polychaete, *Filograna*, opaque pink masses accumulate beside the gut in the segments destined to be involved in the budding process of that worm (BERRILL, 1952); the nature of this chrome has not been studied but it is likely to be a carotenoid.

Quinones: Ubiquinone in the yolk of the bird's egg is passed into the embryo and presumably functions in respiration from an early stage, but from the 11th day onwards the embryo begins to synthesise the coenzyme itself (CRANE, 1965). In fact this is one of the major pieces of evidence that animals can synthesise UQ.

Echinochrome granules in the unfertilised egg of the sea urchin *Arbacia* are distributed throughout the cytoplasm but they move into the cortex within ten minutes of fertilisation (HARVEY, 1910). In the presence of calcium they disintegrate explosively (HEILBRUNN,

1934) and the process is accelerated also by Mg and K salts; it therefore has resemblances to the clotting of coelomic fluid in this group, when coelomocytes supply the granules (p. 199). Following granule-disintegration quinone is detectable in the external medium, and the process may be related to the formation and lifting of the fertilisation-membrane. Not all the granules disintegrate when they reach the cortex. The remainder are bound to the cortex and become crowded in the early cleavage furrows (BELANGER and RUSTAD, 1972) rather like melanin in the frog's egg. They collect here even if cleavage is prevented and their function may be to assist furrow-formation. The views of HARTMAN et al. (1939) that the egg-echinochromes also stimulate respiratory activity and locomotor activity in the spermatozoa have been discussed elsewhere (p. 201).

In *Paracentrotus* a red chrome, initially distributed throughout the egg-cytoplasm and probably again an echinochrome, moves at the time of polar body-formation into the cortex only of the 'vegetative 2' region of the future embryo. It therefore invaginates with the presumptive gut-cells (NEEDHAM, 1942, p. 495) and is thought to protect these cells from infection by microorganisms, – but why those cells more than other regions? Earlier it is thought to protect the surface in general of the egg from organisms which might enter with spermatozoa and this is relevant also to the migration of the echinochrome of *Arbacia*. It also recalls the view of VEVERS (1963) on the function of integumental echinochromes in adult urchins (p. 159).

Non-biological quinones almost invariably have strongly deleterious effects on morphogenetic processes (LEHMANN, 1945, 1956, 1964). In general they are powerful antimitotic agents and they also prevent a normal gelation of the cortex of the *Tubifex* egg, as well as the formation of polar bodies (LEHMANN, 1956). This last effect presumably is due to their action on mitosis (LEHMANN, 1964). Phenanthraquinones appear to act directly on the division-spindle of the cell but naphthoquinones to change the cell cortex. Various hydroxyquinones are carcinostatic (MARXER, 1955; DOMAGK, 1957), presumably for the same reason. Quinones are good bacteriostats (SEXTON, 1963) and fungistats (LITTLE et al., 1949; PETERSEN et al., 1955). Inevitably morphogenetic processes are particularly sensitive to this very potent class of redox agent and it is perhaps surprising that any quinones promote normal development.

Porphyrins: Haemoglobin is often very important in embryos (p. 217), in its normal capacity. Hb passes into the parthenogenetically produced eggs of *Daphnia* a few hours before deposition into the brood pouch (DRESEL, 1948). None enters the fertilised eggs of the late autumn; these suffer no O_2-shortage when they develop at the low temperatures and low population-densities of early spring, and the Hb seems to be actively excluded from these eggs (FOX, 1947). This embryonic Hb of *Daphnia* and that of *Simocephalus* (HOSHI, 1957) has the usual higher O_2-affinity than that of the adult (p. 217), and it deoxygenates at the tissues more readily, so that it would appear to be specially adapted to the respiratory needs of the embryo. The same phenomenon is seen in vertebrates. Experimental conversion of the Hb to COHb depresses the rate of development. The amount of Hb in *Daphnia* embryos decreases during development, perhaps indicating a diminishing need; similarly there is a fall in Hb-content in young mammals after birth and in birds after hatching (NEEDHAM, 1942, p. 648–9). In *Daphnia* some of the surplus is stored in the fat cells and the rest passes as a haemochrome to the gut cells.

During development the Hb in the coelomocytes of *Amphitrite* is progressively replaced by ferrihaem (DALES, 1964); the reason is not yet clear. In serpulids there is a replacement of protoHb by chloroHb (FOX and VEVERS, 1960, p. 101); presumably this is recapitulative. The bilatriene of leeches changes in properties during development (NEEDHAM, 1966b).

These changes all occur at a relatively late stage of development and appear to be producing definitive chromes for post-embryonic functions rather than chromes controlling development itself. Verne (1926, p. 413) emphasised the abundance of porphyrins in developing tissues and believed that they do play the latter rôle.

Melanins: In spite of the heavy melanisation of the animal hemisphere of the egg of many Amphibia more is synthesised as early as the blastula stage (Needham, 1942, p. 655). The synthesis is inhibited by indophenols (Lewis, 1932; Figge, 1938) and this effect is fairly direct, as the chemical structure might imply. Verne (1926, p. 258) believed that new synthesis occurred as early as fertilisation, the amount associated with the sperm-path being greater than that simply displaced from the cortex. It clusters round the nuclei of the blastomeres (Huettner, 1949). In these eggs and embryos the melanin presumably protects against excessive short-wave radiation and absorbs that of the visible and i.r. ranges to accelerate development (Kalmus, 1941, a, b). Heavy melanisation persists throughout larval life in the common frog and camouflage is added to its functions. No more specific morphogenetic action has been detected. In the goldfish and the marine Garibaldi again the young is heavily melanised and this is later replaced by carotenoid, a change striking enough to be considered a metamorphosis (p. 275). Similarly melanin is synthesised by the locust embryo (Goodwin, 1950b) and in the solitary phase disappears in the adult. In all these cases it is likely that the young needs a darker camouflage than the adult but also it may benefit more from the higher capacity for heat-absorption, or the greater protection against radiation.

Ommochromes: Redox agents used experimentally activate the egg of *Urechis* (Brooks, 1947) so that the normal xanthommatin in the cytoplasm may behave similarly. The subsequent movement of this chrome to the base of the locomotor cilia (p. 270) does not necessarily imply a different function: both may be redox in nature.

Pterins: Pterins are known to affect development in several respects. Indeed pteridines in general stimulate cell-proliferation in Protozoa (Dewey and Kidder, 1969) and in higher animals (Jacobson, 1952; Haddow et al., 1972). This is one of the major effects of the vitamin pteroylglutamic acid (PGA) and conceivably other pteridines act only by being converted to PGA. Equally some analogues compete with and block the action of PGA and so inhibit morphogenesis (Lehmann, 1961; Grant, 1965). PGA-enzymes are essential for the biosynthesis of the purine and pyrimidine bases of nucleic acids (Threlfall, 1968; Baserga et al., 1968). It is further noteworthy that xanthopterin stimulates cell-division strongly but inhibits the proliferation of tumours (Lewisohn et al., 1944; Haddow et al., 1972). L'Hélias (1961, 1964) believes that transmissible tumours in some insects are due to an imbalance between pterins and growth-hormones. The further hypothesis is that biopterin acts as a photosensitive and thermosensitive intermediary in the seasonal control of hormone-synthesis (Vuillaume, 1969, p. 22). The possible action of biopterin in caste-determination in bees (p. 268) also is relevant here.

Ribitylflavin: The ribitylflavin content of the eggs of the elasmobranchs falls sharply at fertilisation (Fontaine and Gourevitch, 1936); less than half of the initial amount is later found in the embryo. The stock is subsequently replenished from yolk-sac stores but the decrease is a reliable measure of the wear and tear on an exogenous, heavily used chrome. In the grasshopper, *Melanoplus,* also there is a decrease in RF but a reciprocal increase in pterins (Bodine and Fitzgerald, 1947; Burgess, 1949); the correlation might have a biosynthetic (p. 290) rather than a functional significance. RF has been shown essential for normal development and hatching in fowl (Needham, 1942, p. 23) and deficiency may contribute to a lethal condition known as 'sticky'. This vitamin is essential also for the development of the yellow fever mosquito (TRAGER and SUBBAROW, 1938). Presumably most of these effects depend on its rôle in the FMN and FAD dehydrogenases and oxidases.

Unidentified Chromes: Unfortunately some of the oochromes which have been the subject of interesting observations and experiment are yet unidentified. That of *Perinereis* is a natural pH-indicator and acts as such *in situ:* after the reduction divisions it is yellow (alkaline) at the animal pole and violet (acid) at the vegetative end (SPEK, 1930). The colour-change resembles that of the violet chrome of leeches (p. 62). The egg of *Myzostomum* is green at the vegetative and pink at the animal pole (PITOTTI, 1947) but this may be due to two separate chromes.

A yellow region of the eggs of *Styela* and *Ciona* is probably due to cytochromes; this is consistent with the fact that the colour becomes associated with the presumptive muscular tissue. A slate-grey region becomes the endoderm. In the embryo of *Nereis* three chromes are synthesised at an early stage and become located respectively in cells of the prototroche, hind gut and eyes (COSTELLO, 1945). If the red chrome of *Arbacia* (p. 271) is stratified experimentally, before fertilisation, then the first cleavage is always perpendicular to the layer. However, the orientation of the micromeres has the normal relation to the original egg axis, which therefore is not dependent on the pigment-distribution (NEEDHAM, 1942, p. 660). Similarly the green vegetative pole-plasm of *Myzostomum* normally moves into the 'polar lobe' but the latter can form normally without it. Eventually the green chrome goes to cells $2d$ and $4d$, the two 'somatoblasts'. It is associated with the mitochondria and so with cytochrome oxidase and phenol oxidases (PASTEELS, 1934) and may correspond to the yellow of *Styela*. A green chrome in the cortex of the egg of the ctenophore *Beroe* is later segregated in the presumptive epidermal cells (NEEDHAM, 1942, p. 132). The nephridia of sipunculids develop each from a single yellow cell on either side of the body (GOODRICH, 1945). The primordium of the blood system of phoronids is visible in the larva as two dorsal and two ventral groups of reddish cells (HYMAN, 1959, p. 257) but this is a relatively late stage of development and the colour is probably due to haemoglobin itself.

12.2.2 Pigment Formation by the Embryo

Because pigment synthesis *by* the embryo is usually such a late, differentiative aspect of embryogenesis (p. 315) very little attention has been given to it by embryologists, interested primarily in the major causal relations of the early stages. In that context the following quotation from J. NEEDHAM (1942, p. 417) can be appreciated: "The drawback about all the preceding researches, valuable though they are, is that the end-results of the nuclear inductions are comparatively trivial, consisting as they do of pigmentation changes. The demonstration of nuclear control over morphogenesis requires end-results of a more clearly morphological character." He was referring mainly to the development of body-colour in mammals and insects and of eye-pigments in the latter. These are all rather well known because so extensively controlled by individual genes that produce viable mutants. Unfortunately mutants affecting the major early stages of development are mostly inviable, for evident reasons, and experimentally gross early effects can be produced only by major lesions to the whole genome (NEEDHAM, 1964, p. 392). It seems necessary to accept these limitations with some resignation but also with appreciation of the chromogenic facility.

This is the view of geneticists (chapters 15, 16), and systematists also appreciate any knowledge of chromogenesis: embryo-colour is used as a diagnostic taxonomic character in the Ectoprocta (RYLAND, 1958). For the chromatologist the paucity of embryological evidence is a sad deficiency since it may be expected to throw some light on the evolution of zoochromes (chapter 17). Among the fragments of information that are available there are intriguing features. In most animals the integumental chromes are the last group to develop but many Amphibia and choanichthyan fishes are heavily melanised throughout development: is this an embryonic peculiarity or is it simply that the main chrome of the definitive integument is used also to protect the egg? Many animals laying eggs exposed to the light do have these heavily pigmented but not usually so 'confluently' with the definitive chromes.

In the fish *Brachydanis* the various types of chromatophore develop in sequence and this is 'recapitulated' during regeneration of the integument (GOODRICH and GREENE, 1959): does the sequence have evolutionary significance or does it depend merely on the geometry of development (p. 316)? In the locust embryo melanin develops earlier than ommochromes (GOODWIN, 1950). The insect hormone neotenin when applied to embryos of the locust *Schistocerca* causes abnormalities, the magnitude of which in general is related to the concentration of the hormone (NOVAK, 1969). However, excess pigmentation, which is one of the abnormalities, increases only up to a certain neotenin-level and then decreases again, often to subnormal values. The reason for this difference is an intriguing problem. Young individuals of the leech *Theromyzon* have a mauve chrome, which in the adult is replaced by a yellow one with similar solubility-properties (NEEDHAM, 1966b): this might be due to a change in the conditions governing the biosynthesis of chromes or it might be dictated by changes in camouflage-requirements, a more biological 'cause'. Integumental patterns and hues do change with age and in the relatively late developmental changes of the metamorphic type the actual classes of chrome may be changed.

12.2.3 Chromogenic Changes during Metamorphosis

Most metamorphoses involve a striking change in integumental coloration because they always lead to a radical change in mode of life. Less is known about other types of chrome though in the Amphibia there is known to be a change in the retinal perceptor chrome (p. 166). Among invertebrates many species have colourless planktonic larvae, which metamorphose into benthic, well pigmented adults. Flatfish larvae develop more pigment on the side destined to be uppermost in the adult and lose it from the other side. Some animals have more than one change of metamorphic magnitude during their life.

Little is known even of the integumental changes except in insects, Amphibia and some fish. The transparent leptocephalus larva of the eel becomes a yellow-green adult, camouflaged for fresh water habitats. Before its spawning migration to the Sargasso Sea it metamorphoses again, acquiring the ventral silvering typical of marine fishes. The anadromic salmon makes these metamorphoses in reverse and it is the young parr which shows the change to the marine facies, of the smolt stage. These teleost changes may be under the control of the thyroid (MATTY and SHELTWAY, 1967; OSHIMA et al., 1972), like the more familiar metamorphoses of the Amphibia. In the latter the thyroid has been shown to control the changes among the numerous pterins of the integument (HAMA, 1963; BAGNARA, 1964), as well as that between retinal A_2 and A_1 (WILT, 1959). By contrast, prolactin seems to play the dominant rôle in the control of the metamorphosis of the red eft stage of the newt, when it returns to water to breed (GRANT, 1961): it has been shown to control the changes in skin-chromes and retinal chromes. Probably the whole thyroid-pituitary-gonad axis is implicated.

Insect larvae are usually wingless, and either aquatic or cursorial, so that metamorphosis is extensive. In the endopterygotes, where the whole change is restricted to one or two instars, it is very dramatic. The larvae are usually more cryptically and monotonically coloured than the adults, but in any event nearly always very differently coloured. Consequently much larval chrome is destroyed at metamorphosis (p. 315) and it is usually less amenable to salvage than colourless materials, even including the macromolecular proteins and nucleic acids. Changes in pigment-metabolism are the first signs of metamorphosis in butterflies (LINZEN and BÜCKMANN, 1961). In the hawkmoth this begins about five to six days before pupation (BRUNET, 1965). The large swallowtail, *Cerura vinula*, first synthesises dihydrogen-xanthommatin, in the epidermis, and this soon oxidises to xanthommatin. Then rhodommatin and ommatin D, destined for the imaginal wings, are synthesised in the fat-body and mid-gut, reaching maximal concentration at stage IV (LINZEN and BÜCKMANN, 1965); it is not yet clear if these ommatins are synthesised *de novo* or from xanthommatin. A similar programme has been found in *Phobocampe*. During pupation the ommochrome-concentration of *Cerura* reaches 0.2% of the body weight and from now onwards increasing amounts appear in the gut lumen, often in crystalline form. This elimination, and the fact that the malpighian tubules are one site of synthesis (BUTENANDT et al., 1960a) provide two reasons for the theory (p. 245) that ommochromes are primarily a means of catabolising tryptophan. In addition to providing chromes for the wings a great surplus is eliminated in the meconium of the imago of this insect (p. 350). Ommochrome-production is controlled essentially by the brain-prothoracic gland system (KARLSON and BÜCKMANN, 1956) the analogue of the vertebrate pituitary-thyroid system.

By some Lepidoptera pterins are synthesised almost as extensively at this time (GATES, 1947; ZIEGLER and HARMSEN, 1969). There is less excess than of ommochromes and most of the pterin is stored (p. 246). When the cocoon is made of silk presumably less purine and nucleic acid are squandered during the metamorphosis than amino acids such as tryptophan. The large amounts of chromes handled at this time, both from the larva and for the imago, probably is unavoidable where there is a radical metamorphosis so late in ontogenesis.

When marvelling at the extent of metamorphic colour-changes it should be appreciated that in fact some seasonal and other colour-changes are also very impressive and that integrated over their life-span many animals experience massive changes in integumental chromes. Chromogenic change in response to background, etc., can be almost incessant. The mere maintenance of pigmentation is a large item and again continues throughout life. In this connection it is significant that the hair of many mammals has progressively less melanin in later years. Melanogenesis in the general epidermis also declines with age, in both sexes, the number of melanocytes decreasing by about 11% every ten years in Caucasians (SNELL, 1963; SZODORAY and NAGY-VEZEKÉNYI, 1966). This is in contrast to what tends to be regarded as the invariable tendency for pigmentation to increase with age and shows that neither trend is inevitable. It also emphasises the value of ontogenic studies not merely in embryos but throughout life; the next section, dealing with chromogenesis in the individual, does refer to adult animals as well as to developmental stages.

12.2.4 Conclusion

Among animals collectively virtually all known classes of zoochrome, and some which are still unidentified, occur in the eggs, and in the subsequent embryo. Some may be acquired casually but it is doubtful if this is normally extensive. Those that have metabolic functions in the adult serve the same purpose in the embryo, which is dependent on the stores provided until it can replenish them, either by feeding or by endogenous synthesis. For definitive chromes endogenous synthesis seems to start rather late in embryogenesis and so has been rather neglected by orthodox embryologists. Such chromes cannot be causal agents in the early, main stages of embryogenesis but the neglect is unfortunate because embryonic chromogenesis may have significance for evolution even if not for embryonic differentiation itself. There is considerable evidence that a number of chromes have a general significance in the control of embryogenesis but rarely enough for a coherent account since again attention has not been strongly focussed on the subject. Directed movements and sharp segregations of chromes are a conspicuous feature of early development in many animals.

In later life, and particularly in processes classified as metamorphic, there are great changes in integumental chromes and much unavoidable excretion of material. Little is known of changes in other functional types of chrome at this time though it is known that in Amphibia the type of retinal changes. In most animals there is considerable chromogenic activity and change throughout life; not all chromes accumulate with age.

VI. Evidence from Chromogenesis in the Individual

It is now axiomatic that every useful biological character has become genetically encoded and that its development or 'epigenesis' in the individual provides good evidence of this utility, i.e. of past selection in its favour. Development also should help to explain the origin of the functional properties of the chrome. In the present context development means first the biosynthesis of zoochromes and it may be re-emphasised (p. 137) that this continues throughout life: in fact much of the evidence has been obtained from studies on adults. It is therefore the interest of the biochemist rather than of the embryologist. Other aspects, in particular the neuro-hormonal control of biosynthesis and the transport and manipulation of chromes, including those acquired exogenously, are more the province of the physiologist. For him biosynthesis may be regarded as an algebraic function: an organised decrease in amount of a biochrome is as important as its synthesis, though at the chemical level this aspect has as yet received less attention.

Chapter 13

The Evidence from Chromogenic Pathways

In section 1 the biosynthetic pathways of the known chemical classes of biochrome are outlined. In section 2 their main adaptive features are considered and in the third section their inter-relationships.

13.1 Known Chromogenic Pathways

For completeness the pathways of those zoochromes synthesised exogenously by plants and microorganisms, carotenoids, chromones, flavonoids and isoalloxazines are included. In general the conjugation-steps with protein and other components are not well known.

13.1.1 Carotenoids

The outline of the pathway is shown in Fig. 13.1: further details are given by CHICHESTER and NAKAYAMA (1967) and LIAAEN-JENSEN (1970). It is one branch of an extensive biosynthetic system leading also to terpenes, steroids and other lipids but not including all lipids. Isopentenylpyrophosphate is the characteristic and universal intermediary for all branches, acting as a monomer for serial polymerisation. In the carotenoid line this produces the tetramer, geranylgeraniol; the final C_{40} carotenoid is formed by a higher level head-to-head dimerisation of this intermediary. *Phytoene,* logically if not actually, the first C_{40} product has only three conjugated double bonds, at the site of this final condensation, and so is colourless. True carotenoids therefore must be formed by a systematic removal of pairs of H-atoms (PORTER and LINCOLN, 1950) from the appropriate neighbouring pairs of C-atoms so as to achieve conjugate bonding throughout most of the chain. This is the first and one of the most impressive of 'purposive' or teleonomic (PITTENDRIGH, 1958) manifestations in any chromogenic pathway. It is only fair to point out that in the laboratory the result can be achieved by a simple, though rather special, organic dehydrogenating agent, N-bromosuccinimide (CHICHESTER and NAKAYAMA, 1967); however *in vivo* the problem was more difficult, – to introduce a self-regulating series of oxidations in an otherwise reductive pathway.

There is some evidence that closure of the ends of the chain to form the ionone rings can occur before, and so independently of, maximal desaturation of the chain; their association is the result of natural selection-processes rather than of spontaneous chemical reaction-sequences. Moreover ring-closure reverses the bond-conjugation process by one double bond per ring but no doubt functionally this is more than compensated by the increased planarity of the molecule (p. 33). The formation of oxy-substituents (p. 33) again is independent of ionone ring-formation. With due regard for these labile features the most usual biosynthetic sequence is:

open chain partially desaturated hydrocarbon →
open chain maximally desaturated carotene →
ionone-ringed carotene → oxycarotenoid →

shortened chain carotenoids and other special derivatives. This may have been the evolutionary sequence, also; in fact the sequence has a strong resemblance to the evolution of the anatomy of a phylum of animals in which, for instance, the early members were much more simply metameric than the more specialised later descendents. Probably only a few particular oxycarotenoids readily form carotenoproteins, the keto-groups being used for bonding (Fig. 3.2, VI); however retinal (Fig. 7.1) has no oxo-group on its ring.

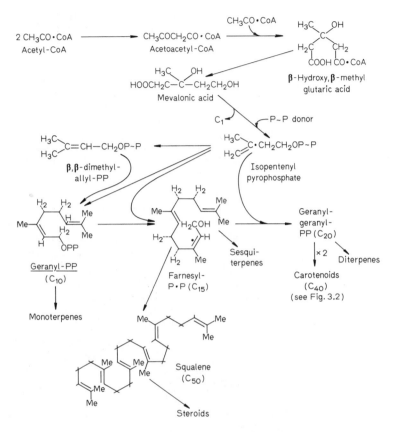

Fig. 13.1. Biosynthetic pathways via isopentenyl pyrophosphate (prenyl pathways). The most important branches lead to carotenoid (C_{40}), steroids (C_{30}), diterpenes (C_{20}), sesquiterpenes (C_{15}), and monoterpenes (C_{10})

13.1.2 Chromones and Flavonoids

The chromone nucleus of the tocopherols (p. 37) is probably synthesised from a benzoquinone-derivative of the ubiquinone type (p. 38) in which the proximal end of the long prenyl side-chain closes to form the pyran ring. Such chromones as the coumarins, with no long side-chain, are formed by a similar ring-closure of the short C_3 side-chain of p-coumaric acid (Fig. 3.3, V), one of the large number of C_6-C_3 derivatives of shikimic acid (Fig. 3.3, IV). The latter is synthesised from the simple precursors, P-enol pyruvic acid and erythrose-4-P, and is the common intermediary also for the phenolic amino acids (Fig. 13.2) among other C_6-C_3 benzenoid compounds.

The flavonoids also (Fig. 3.3) are all derived from this C_6–C_3 group and probably specifically from cinnamic acid (GEISSMAN, 1967) but curiously enough their chromone moiety is not completed by the C_3 chain closing on to the C_6 ring. Instead a chain of three acetate residues condenses on the end of the C_3 chain and themselves ring-close to form a second

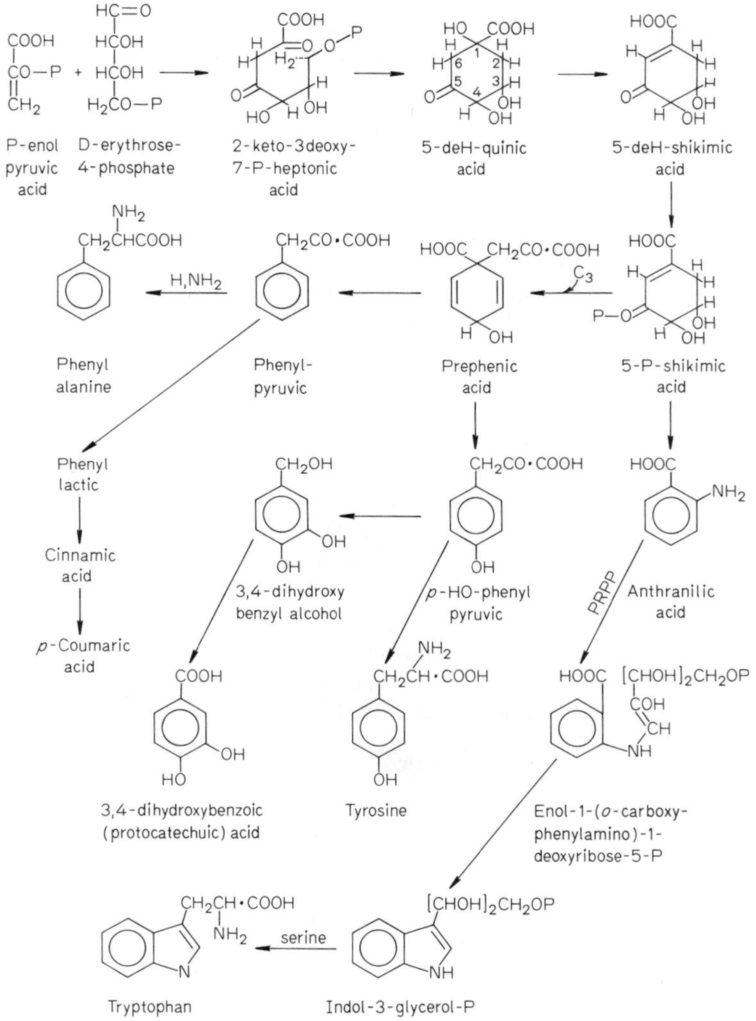

Fig. 13.2. Shikimic-prephenic group of biosynthetic pathways to (1) coumaric acid and flavonoids, (2) phenylalanine (3) tyrosine and protocatechuic acid and (4) tryptophan

benzenoid ring. This then uses the C_3 chain to complete an alternative chromone nucleus, chain-linked to the original benzenoid ring. Although the precise biosynthetic and functional significance of the details of this process are not so clear as in that of the carotenoids it again is clearly a piece of biological 'contriving' rather than a spontaneous chemical sequence: lipid building units are tacked on to an essentially carbohydrate foundation.

The intermediary C_6-C_3-C_6 molecule is a *chalcone* (Fig. 3, VI) which is already a biochrome.

The aurones, which include marginalin (p. 39) may be formed via an alternative path in which an epoxide structure (Fig. 3.2, IX) is formed at an intermediate stage and ensures ring-closure to furan instead of pyran. The chain-linked and rather asymmetrical ring-structure is characteristic of these two classes of biochrome; in other classes with link-chains these connect identical or very similar ring-structures. Linked heterogenous ring-structures are typical of luciferins, however (Fig. 8.6).

13.1.3 Ternary Quinones

The view (p. 40) that because aromatic rings so readily produce the quinonoid structure the ternary quinones probably are a heterogeneous group is supported by a variety in their biosynthetic pathways. Many animal benzoquinones (BQ) are synthesised from phenylalanine or from tyrosine (Fig. 13.3) and so, like the flavonoids, stem from the carbohydrate-shikimic acid pathway (Fig. 13.2). However, some are synthesised via another pathway, simple polymerisation of acetate (BIRCH and DONOVAN, 1953), like the second benzenoid ring of the flavonoids. This is also the way plants and microorganisms synthesise both BQs and anthraquinones (AQ), as well as more highly polycyclic quinones (EHRENSVARD and GATENBECK, 1960) and other polycyclic aromatic compounds (GEISSMAN, 1967). EHRENSVARD and GATENBECK envisage a 'pleiotropic' path of unlimited acetate polymerisation, the polymer then being chopped up and the portions ring-closed to form a great variety of aryl ring-systems. The echinoderms may operate some such system since collectively they have not only many naphthoquinones (NQ) but also some AQs, at least one BQ and the polycyclic quinone (PQ) fringelite (p. 43). However fringelite, and probably PQs in general, are secondary dimers, like the final step in the carotenoid path; by direct polymerisation of acetate no structures are formed larger than those in the fatty acid pathway. It seems likely that longer chains would be in danger of tangling and of wasting material in the tailoring process and this is alien to the economy of biosynthesis. BQs are formed by the condensation of four acetate units, via intermediaries which probably include orsellinic acid (Fig. 3.4, XVI), while AQs use eight units, without waste. *Penicillium* has been shown to form secondary dimers from its AQs. In the BQs all eight of the C-atoms of the four acetate units are retained, and in both sub-classes very useful -OH, -CH_3 and -COOH side-chains are an automatic consequence.

In ubiquinone-synthesis a further, independent step is the conjugation with a polyisopentenyl side-chain. This prenylation of cyclic biochromes is a rather common phenomenon (GEISSMAN, 1967), shown also by vitamin K among NQs, by the tocopherols and by chlorophyll and cytochrome *a* among porphyrin derivatives. The phytyl side-chain of chlorophyll is known to be formed from mevalonic acid via geranylgeraniol and this is probably the path in all cases. Phytyl is bound after the completion of the porphyrin ring-system itself and this again is probably the usual order as well as the evolutionary order. Again it is an addition from a quite independent pathway.

Ubiquinone forms ubichromenols by subsequent ring-closure of the proximal end of the prenyl chain on to the benzene ring (HEMMING and PENNOCK, 1965), a close parallel to the mode of synthesis of the tocopherols (p. 282). In the laboratory, at least, some BQs can be induced to close the side-chain so as to form a NQ instead (MORTON, 1965).

In fact the most usual mode of biosynthesis of NQs is rather uncertain. By free condensation of acetate units it is a less straightforward and economical product than BQ and

AQ. SALAQUE et al. (1967) demonstrated some incorporation of acetate into 6-ethyl-2,3,7-trihydroxy-1,4-NQ by the sea-urchin *Arbacia* but there is some evidence (EHRENBERG and GATENBECK, 1960; GEISSMAN, 1967; ANDERSON et al., 1969) that Wakil's malonyl-C_3 mode of acetate polymerisation, (WAKIL, 1962) as in fatty acid synthesis, may be used (Fig. 3.4, XVII). The synthesis in echinoderms is now thought to be autogenous (GOODWIN,

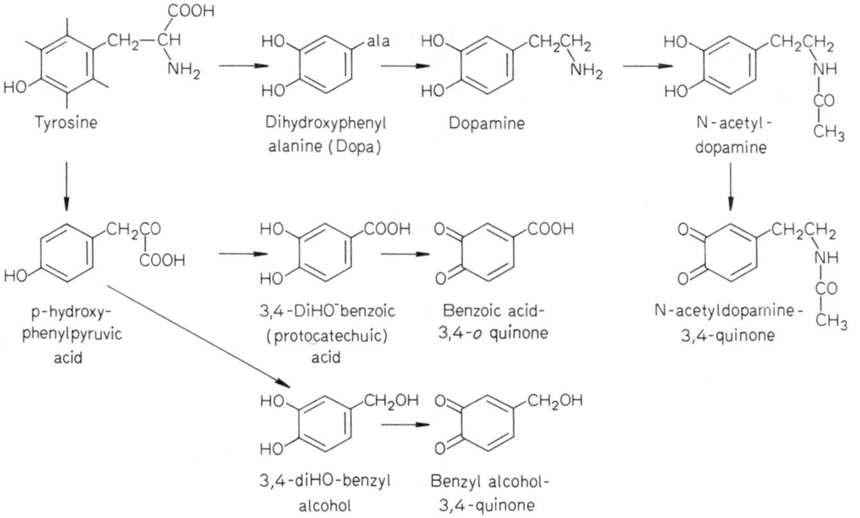

Fig. 13.3. The biosynthetic pathways to the three benzoquinones most used by arthropods for polymerisation to phenolic melanins

1964) and there are more -OH side-chains than in plant NQs (MORTON, 1965); also there are sometimes CH_3O- but never CH_3 side-chains (SINGH et al., 1967). The coccids synthesise their AQs endogenously, since there are none in their food (TIMON-DAVID, 1947), but they probably use the microorganisms of their mycetome for the purpose.

Second stage dimerisation seems to be an inherent property of quinones, rather as is quinhydrone-formation (p. 46). *In situ* the aphins appear to be singly-bonded NQ dimers (p. 43) though the final product of their reactions in shed blood (Fig. 3.4, VI) is effectively a perylene derivative, structurally nearer an AQ dimer. The ciliate PQs and fringelite do appear to be AQ dimers like the plant hypericins and the *Penicillium* product.

Much higher grades of polymerisation, by BQs, probably account for the dark melanins of some insect exoskeletons (p. 46) and for many of the plant melanins, just as for the black oxidation-products of pyrogallol and other quinones in the laboratory. The biosynthetic pathways to the three most common monomers used by insects are shown in Fig. 13.3. Semiquinone free radicals are involved (SWAN, 1963) and yellow dimer intermediaries (Fig. 3.4, XVIII, XIX) have been identified (DAWSON and TARPLEY, 1963). Red and purple intermediaries, no doubt progressively higher grades of polymer, also are evident in the reaction-mixture (LERNER, 1953; PRYOR, 1955, 1962).

13.1.4 Porphyrins

The porphyrin ring is synthesised from monopyrrole units which, from the evidence of the side-chains are initially all identical (Fig. 3.6, XI). This porphobilinogen (PBG) unit is formed from two molecules of δ-aminolaevulinic acid (δ-ALA: Fig. 3.6, X) linked in the curious asymmetrical fashion indicated by the dotted line in XI. Δ-ALA is synthesised from the simple units glycine and acetate, the latter being incorporated into succinate (SHEMIN, 1955; LASCELLES, 1964); again acetate plays a dominant role. Malonate promotes porphyrin-biosynthesis (LASCELLES, 1964), presumably by blocking destruction of succinate in the Krebs cycle. The δ-amino group not incorporated into the pyrrole ring of PBG is essential for linking the unit to its neighbour in the super-ring; bonding occurs at the α-position of the neighbour since this is more nucleophilic than the β-position.

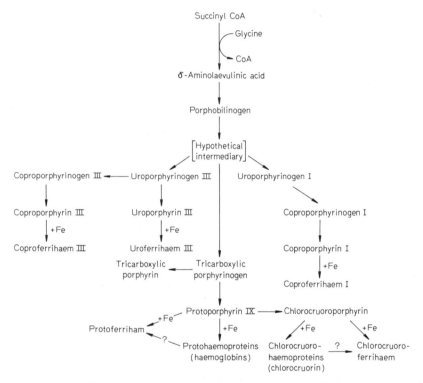

Fig. 13.4. Biosynthetic pathway to proto- and chloro- heams, and collateral pathways. (From NEEDHAM, 1970b, mainly after MANGUM and DALES, 1965)

A simple condensation of four PBG units, with the circular symmetry of alternating short and long side-chains round the porphyrin ring is known as type I. In the biologically more common type III the D-unit is reversed (Fig. 3.6, III) and it is still not clear how and why this difficult manoeuvre is achieved (KARLSON, 1963, p. 175). A special enzyme, uroporphyrinogen III cosynthetase, is involved and the process gives the impression of having been specially selected notwithstanding the synthesis-difficulty. It has not been possible to suppress completely the easier type I path but members of that series are used only for minor purposes. The other two possible arrangements of the two side-chains, types II and IV have been sythesised in the laboratory but chemically are less 'probable' than I and do not contaminate biological systems.

As in the carotenoid path specific paired dehydrogenations are necessary to produce a fully conjugated double bond system, by converting the methylene to methine bridges; the first-formed porphyrinogen thereby becomes a porphyrin. It is possible that the path to protoporphyrin IX passes actually through uro III and copro III, rather than as shown in Fig. 13.4 but in any case there is a general trend towards decarboxylation, chain-shortening and chain-modification (Table 3.9), processes similar to the later steps in the carotenoid pathway. The enzyme for inserting Fe is relatively non-specific so that *Clostridium*, which normally synthesises cobamides but not haems, will insert cobalt into protoporphyrin IX (JOHNSON and JONES, 1964; PORRA and ROSS, 1965).

The biosynthesis of chlorophyll (Fig. 3.6, VII) follows the main path as far as protoporphyrin IX (JONES, 1968); the insertion of Mg is the next step and then a methyl ester-group is bonded at position 6, a 'deliberate' device to permit the closure of the 'isocyclic' cyclopentane ring. Further changes follow and conjugation with protein is virtually the last step.

The cobamides also share the same initial pathway, both δ-ALA and PBG being rapidly incorporated into corrins by relevant biological systems *in vitro*. Uroporphyrinogen III possibly is the point of departure from the main path (GRANICK and SASSA, 1971) but many of the steps are still unknown. The absence of a methine group in the bridge between pyrroles D and A (Fig. 3.6, VIII) once more implies a complication in the closing of the super-ring, and a different problem from that in the haem path, so that the branching point is probably earlier than the closure. Indeed there is evidence (GOODWIN, 1963) that the extra methyl groups on the α and γ methine bridge groups are bound as early as the δ-ALA stage. Most of the other specialisations occur after the insertion of the cobalt atom; like the Fe of the haems this must first be in an organic form. There follows in order the amidation of the side-chain COOH groups, insertion of amino-2-propanol on side-chain 7 and bonding with the appropriate nucleotide (Table 3.10). Guanosine diphosphate may be a common stand-in for the bonding of all of these or else the enzyme concerned has a low specificity (CHELDELIN and BAICH, 1967). Finally adenine becomes the sixth ligand of the Co-atom. The amount of additional molecular engineering involved in this corrin branch seems fantastic and hints at a long history under very specific selective forces.

Prodigiosin seems to be a veritable prodigy among pyrrolic biochromes since it alone is synthesised via a quite independent path. Its pyrrole units are not formed via δ-ALA (BOGORAD, 1963; BOGORAD and TROXLER, 1967) but mainly from proline and ornithine. Further the monomer is synthesised via a path distinct from, though not completely independent of, that of the dimer pair (A, B). This unique peculiar tripyrrole only serves to emphasise how exclusively it is the porphyrin path which has been exploited elsewhere.

In the formation of bilins from haem the latter is first converted to the ferric form and then separated from the globin. It is then oxidised by haem-oxygenase, a complete cytochrome system! Biliverdin is the initial open-chain product and in many vertebrates is converted to bilirubin by a *volte-face* to a peculiar transhydrogenation (p. 58), using an NADP enzyme (GRANICK and SASSA, 1971). It is noteworthy that this is essentially the form of pyridine coenzyme used in synthetic reactions and there are a number of indications that bilins are synthesised for a 'purpose' rather than being mere degradation-products of haems. The variety of their conjugants (GRAY, 1961) is another indication. Four sources have been distinguished in a typical mammal, – senescent Hb, faulty haem from the poietic tissues, senescent haemoprotein-enzymes and free haem. The last normally is a minor component but it is interesting in showing that in most animals free haem is useless; it cannot be conjugated retrospectively but is immediately converted to bilin. By contrast the enzyme haemoproteins have a half life of 12 hours and Hb of 60 days.

13.1.5 Indolic Melanins

Like the phenolic melanins (p. 60) these are biosynthesised from molecules already fairly complex. The classical pathway (RAPER, 1928) from tyrosine to 5,6-dihydroxyindole (Fig. 13.5) is essentially correct. As usual ring-closure resolves double bonds, in this case a quinone =O group and so leucodopachrome must be desaturated again in the interests of chromogenesis. For this purpose I-5,6-Q-2-COOH is the best monomer since including

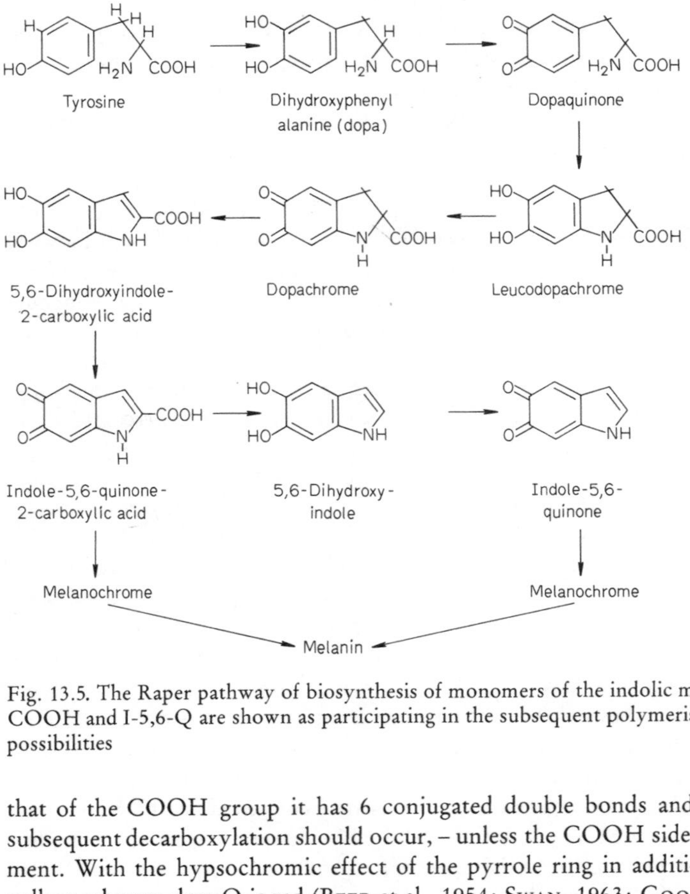

Fig. 13.5. The Raper pathway of biosynthesis of monomers of the indolic melanins. Both I-5, 6-Q-2-COOH and I-5,6-Q are shown as participating in the subsequent polymerisation. See text for further possibilities

that of the COOH group it has 6 conjugated double bonds and it is curious that the subsequent decarboxylation should occur, – unless the COOH side-chain is an embarrassment. With the hypsochromic effect of the pyrrole ring in addition the final I-5,6-Q is yellow whereas dopaQ is red (BEER et al., 1954; SWAN, 1963; GOODWIN, 1964). However the subsequent unlimited polymerisation makes the colour of the monomer of minor importance. As in the pathways of the phenolic melanins (p. 285) a series of melanochrome intermediaries of increasing bathochromicity have been recognised (MASON, 1948). *In vitro* virtually all dihydroxyindole derivatives will undergo some of this series of changes (BEER et al., 1954; CROMARTIE and HARLEY-MASON, 1957) though some stop short of a black, insoluble stage. As long ago as 1893 it was noticed that *in situ* melanin granules are lavender-coloured in their early stages of development (VERNE, 1926, p. 308). Even *in vitro* the process is often too rapid to detect appreciable amounts of intermediaries and rapid recording spectrophotometry is necessary (MASON and PETERSON, 1965). This has also shown that bonding with protein proceeds in step with the polymerisation, -SH and -NH$_2$ groups of amino acids bonding with the quinone groups of the melanin.

Evidence is increasing that the monomers are not all I-5,6-Q, in fact, and indeed may be very varied, both between and within particular products (NICOLAUS, 1968; SWAN, 1970). *In vitro* a variety of precursors, including adrenalin (BU'LOCK, 1961; SWAN, 1963) will form melanic polymers. The melanins obtained from DOPA and 5,6-diHO-I respectively as starting materials differ considerably in properties (ROBSON and SWAN, 1966; HEMPEL and DIENNEL, 1962) whereas according to the classical scheme (Fig. 13.5) DOPA should all pass through 5,6-DHI to a common product. In one natural melanin analysed (HEMPEL, 1966) 16–24% of the residues were dopachrome itself though the remainder were I-5,6-Q. This is in accord with the finding of BU'LOCK and HARLEY-MASON (1951) that 1/6 of the COOH groups of the initial tyrosine molecules persist in the completed melanin. In the melanin of a melanoma Hempel found as much as 50% dopaquinone, 37% I-5,6-Q and 12.5% I-5,6-Q-2-COOH. From his analyses NICOLAUS (1962) concluded that a greater variety of monomers, including even free pyrroles, are incorporated and BLOIS (1965) goes so far as to suggest that all the intermediaries of the Raper pathway may contribute. At the least this is a great contrast to the rigid molecular specificity of most other chromogenic pathways and it implies an unusual type of function, chemically generic rather than specific.

Another interesting feature is the variety and amounts of heavy metal cations associated with the biosynthesis, as with the completed melanin (p. 62). As the essential prosthetic component of tyrosinase copper is the key metal and deficiency of Cu results in achromotrichia, or lack of hair-colour, in mammals (HUNDLEY, 1950). Iron and manganese also promote the oxidation of DOPA though only 1–2% as effectively as Cu (FLESCH and ROTHMAN, 1948). Together with Cu iron has much more effect than either alone, perhaps indicating that cytochrome oxidase (p. 186) or one of the flavoproteins is involved. Ribitylflavin has been implicated on other grounds and becomes trapped in the melanin produced. The metals (p. 62) become trapped owing to the cation-exchange properties of melanin (p. 80). Since Fe, Cu, Zn and Co all promote the biosynthesis also of haems and cobamides it is possible that in completed melanin they function as co-factors of prosthetic groups. Such links between biosynthetic and definitive functional reactions are not uncommon in biochromes (p. 291).

13.1.6 Ommochromes

Whereas indolic melanins have as precursor the phenolic amino acid tyrosine, ommochromes are synthesised from tryptophan via a phenolic intermediary (Fig. 3.8). The first steps were worked out in conjunction with classical genetical studies on eye-colour in several groups of insect (BEADLE and EPHRUSSI, 1937; BECKER, 1938; CASPARI, 1949; BUTENANDT, 1952). An enzyme tryptophan pyrrolase converts tryptophan to the first benzenoid derivative, formylkynurenine and this then spontaneously deformylates to kynurenine (K). Kynurenine hydroxylase then converts K to 3-HO-kynurenine (3-HOK) which is essential for the subsequent formation of the oxazine bridge of the ommochromes; the hydroxylation therefore is one more example of a small, and unexpected yet crucial step in a chromogenic pathway. Two HOK units condense enzymatically, the oxazine bridge constituting a third ring; the condensing enzyme from *Drosophila* is relatively specific to substrate (PHILLIPS and FORREST, 1970) but will also condense the nearly related derivative, 3 hydroxy-anthranilic acid (Fig. 3.8).

In the laboratory condensation of two HOK units is spontaneous and almost instantaneous in the presence of such oxidising agents as $K_3Fe(CN)_6$ (BUTENANDT et al., 1954b) and such biological oxidisers as dopaquinone also promote the condensation (BUTENANDT, 1960), so that the hydroxylation of K may be the only improbable, biologically contrived step. The action of dopaquinone implies the possibility that *in vivo* tyrosinase could catalyse the biosynthesis either of melanin or of ommochrome (GLASSMAN, 1957) according to conditions: the condensation step reduces dopaquinone back to DOPA. *In vitro* other quinones also are effective condensing agents (LINZEN, 1967) so that there is the possibility of considerable intermeshing between melanin and ommochrome pathways. JOHNSON (1961) obtained some ommochrome in addition to indolic melanin by simply irradiating a solution of tyrosine: under this treatment suitable monomers for the condensation must be formed from tyrosine itself, which is fairly closely related to HOK in structure. This is another mode of intermeshing.

The fourth ring of the ommatins is formed by the aspartyl side-chain at position 4 displacing the amino group at 3 by its own α-NH_2, just as in the conversion of 3 HOK to xanthurenic acid (Fig. 3.8, VI) and of K to kynurenic acid (V). Presumably this is a spontaneous step therefore and may precede the oxazine condensation but it is not a prerequisite for this, which occurs without it in ommin A (III). It is also evident that the ommatins are not intermediaries in the ommin path; the pyridine ring of xanthommatin survives even laboratory degradation-reactions (BUTENANDT et al., 1960a) so that it is unlikely to reopen in a biosynthetic path.

13.1.7 Purines, Pterins and Isoalloxazines

Although pteroyl glutamic acid (PGA) is a vitamin for many animals, and biopterin for some Protozoa, in general animals synthesise their own pterins: even Man excretes more than he acquires in his diet (KOSCHARA and HAUG, 1939). The pathway is a continuation of that of the purines which is well known and need not be detailed here. Features of significance include the *ab initio* nature of the process, from glycine, and $-NH_2$ and C_1 units and this also explains why uric acid and guanine are quite commonly the forms in which $-NH_2$ and CO_2 are excreted. Secondly PGA is the coenzyme for the incorporation of two of the C_1 units and so illustrates indirect autocatalysis in biosynthesis. Thirdly the molecule is assembled on an initial foundation of phosphoribose so that purine nucleotides are products prior to the free purines. Finally the guanine nucleotide, G-9-RP, one of the penultimate products is the departure point for pteridine synthesis (Fig. 13.6) after being primed to the triphosphate level. It is already a 2-NH_2, 4-OH precursor on the pteridine notation which unfortunately is different from that of the purines (FORREST, 1960). These are the two positions at which substituents contribute most resonance energy (PULLMAN and PULLMAN, 1963). It is also significant that the resonance energy of the pteridines is greater than that of their purine analogues.

The 9-R-triphosphate chain of GTP is used further in an intriguing piece of molecular engineering, to enlarge the imidazole ring to pyrazine leaving the rest of the ribose as the C_3 side-chain at pteridine position 6. By a happy accident of nomenclature therefore neopterin appears to be the first pterin of the pathway. There is still much uncertainty (HAMA, 1963; KAUFMAN, 1967; SHIOTA, 1971) about the subsequent branching of the path to produce the many other known animal pterins (MATSUMOTO et al., 1971) but a probable scheme is given in Fig. 13.7. The main problem is whether the 7-substituted members arise from a different guanosine precursor, G-7-RP, or (GILMOUR, 1965) by transfer from the 6-position (Table 3.12). A third alternative is a further substitution at 7 followed by removal of the substituent at 6: although the most complicated this seems the most likely. There are a number of 6,7 disubstituted members evidently derived from the 6-series (Fig. 13.7). Moreover, apart from the two xanthopterins the substituents at 7 are never identical with any at 6 so that direct transfer is improbable. The process involves prior shortening of the chain at 6 and it is significant that in the laboratory a C_3 secondary chain can be inserted at 7 in AHP (Table 3.12), which has lost the 6-chain completely (KAUFMAN, 1967; HARMSEN, 1969).

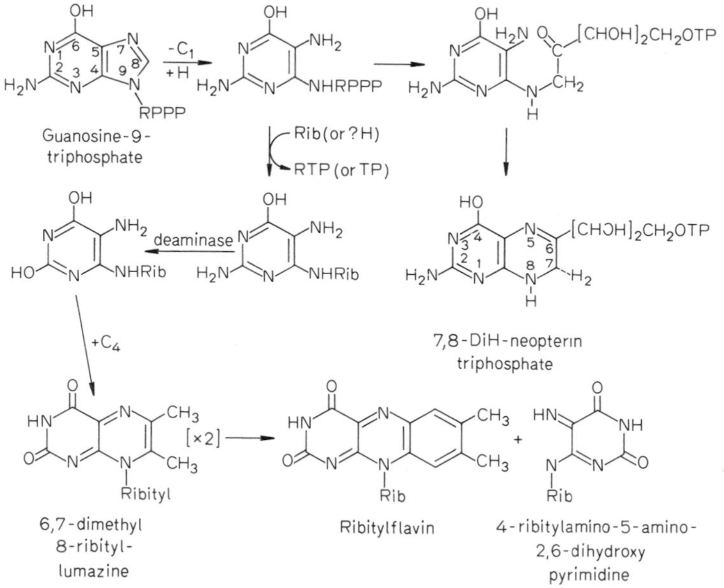

Fig. 13.6. Biosynthetic paths from guanosine-9-triphosphate to pterins and to isoalloxazines

As in other chromogenic pathways progressive oxidative shortening of the side chain appears to be a main feature (HAMA, 1963) and AHP, structurally the parent pterin is actually the ultimate member of the 6-series. Oxidation affects also the rings and pterins with a fully aromatic pyrazine ring (Fig. 13.7, second column from right) are more numerous than the initial di-H-pterin state (third column). This makes the opposite trend, towards tetra-H derivatives (fourth column) all the more conspicuous and again implies biological selection for chemical improbability. Tetra-H-biopterin in fact is very unstable and oxidises spontaneously in air (KAUFMAN, 1967) but it is stabilised *in vivo* and forms the coenzyme of a very efficient reversible redox system (PULLMAN and PULLMAN, 1963), in which its low redox potential is a great asset. The redox change concerns the last step in its biosynthesis and no doubt this is true for other redox biochromes: the logical distinction between genesis and function is biologically artificial, as seen also in the retinopsins (p. 165). It is conceivable that functional destruction of the chrome promotes its further synthesis by simple mass-action.

No pterins are known with a C_2-chain at 6 or 7 so that the first shortening step removes an acetate unit, as in fatty acid catabolism although the chain is carbohydrate in origin.

The PGA-pterin therefore is a fairly direct derivative of neopterin. It seems probable that all the 6-C_1 pterins and the 7-series are derived via the same intermediary (Fig. 13.7). Perhaps it is because animals in general direct all their 6-CH_2OH-7,8-diH-pterin in these two directions that they have come to demand PGA itself ready-made from their diet or from their gut-microorganisms. Two pathways from 7,8-diH-pterin are indicated by the work of GUTENSOHN (1968) and HARMSEN (1969) on biosynthesis in pierid butterflies. Depending on O_2-pressure the two are alternatives and xanthine dehydrogenase catalyses both (as well as the familiar purine oxidations).

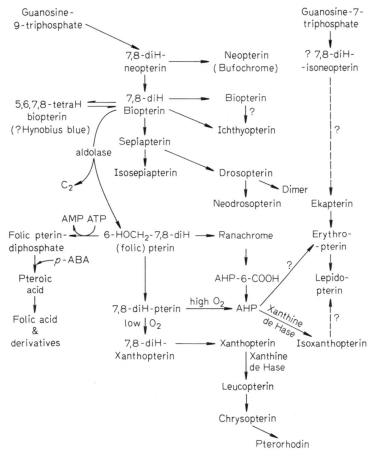

Fig. 13.7. Suggested biosynthetic pathways of the best known pterins. The dotted paths now seem rather improbable

13.1.7.1 Ribitylflavin

Ribitylflavin (RF) is synthesised by microorganisms from the same guanosine triphosphate precursor as the pterins (PLAUT, 1971). The branching-off point is 4-triphosphoribosamine-2,5-diamino-6-HO-pyrimidine (Fig. 13.6) and at the next stage the ribose-TP at 4 is changed to, or exchanged for, the open chain ribityl; the phosphate therefore is lost much earlier than in the pterin path. The next step is deamination to form the 2,4-di HO condition of the eventual RF and then ring-closure leads to the special pteridine intermediary, 6,7-dimethyl-lumazine. The identity of the C_4-donor for this is still unknown but the unbranched, saturated and hydroxylated ribose chain is not an immediate candidate; in any case the ribityl moiety is carried on unchanged to the final RF. The methylated benzene ring of the latter is completed by this same C_4-unit, supplied from a second 6,7-di-Me-lumazine molecule, which thus returns to the pyrimidine stage and perhaps recycles. The synthesis is not an obviously spontaneous sequel to the pyrimidine, purine or pterin paths and again gives the impression of having been selected from among more unusual biosynthetic possibilities solely because the product, RF, has such valuable functional properties (p. 188).

13.2 Adaptive Features of Chromogenic Pathways

In addition to those features stressed in the previous section it is useful to marshall others towards answering the following questions: Has the pathway evolved primarily to produce these chromes or are they the fortuitous products of a prior non-chromogenic pathway? Does the pathway show sophisticated features indicative of strong natural selection? Does it show the economy of energy and material typical of positive biochemical evolution? Has it the same course in all taxa? What evidence is there of correlated evolution, and of other interrelationships, between pathways?

13.2.1 The Primality of Chromogenic Pathways

Although the carotenoid biosynthetic pathway is only one branch of a much more extensive system it is the only branch leading to strongly coloured, highly resonant, elongated yet planar molecules, – which are unique even among biochromes themselves. It must have evolved specifically in the production of these properties which have been enhanced and otherwise exploited progressively in members further along the pathway, by animals as well as by plants. The retinals show the acme of this exploitation.

Flavonoids are such a large group of plant chromes that inevitably their pathway appears to be a major outcome of the shikimic C_6-C_3 line of biosynthesis. In fact it is only one of a number which depend on the specific addition of a C_3 unit to that C_6 ring but the further addition of an acetate-derived benzenoid ring and its development into an alternative chromone structure is peculiar to flavonoids and aurones. The brilliance and range of their colours bears witness to the selective value of the pathway.

Because the quinonoid property is so essentially that of two $=O/-OH$ side-chains on any aromatic ring, quinones appear in many different chromogenic paths and it may seem invidious to talk of *the* quinone pathway. However the $C=O$ group of acetate automatically becomes a quinol/quinone side-chain when acetate polymerises and ring-closes and this provides a natural, coherent group of pathways yielding primarily ternary quinones. It seems likely that biologically the system has been exploited mainly for the production of this functionally versatile class of materials: acetate polymerisation by other pathways produces colourless classes of biochemical compound. The quinone pathway has been extended to produce various polymers, including phenolic melanins.

Like the latter, the indolic melanins are formed by the polymerisation of a variety of quinone monomers and the pathway contrasts rather strongly with the chemical precision of other chromogenic systems. On the other hand melanogenesis appears to have been selected solely to provide dark, protective integumental chromes; such pathological cases as ochronosis (p. 250) show that in fact all quinones have a highly spontaneous tendency to polymerise indefinitely in this way but that normally this is strongly inhibited in all other sites and biochemical systems. Such inhibition of redundant alternative pathways is most conspicuous in this system but is not peculiar to it: selective specificity depends mainly on the phenomenon.

It is particularly clear that the porphyrin pathway evolved specifically to produce that class of biochrome and nothing else. It is an *ab initio* pathway, from C_2-units. The only other examples of this among zoochromes are some of the ternary quinones and the purine 'pseudochromes'. The porphyrins have a very long evolutionary history (p. 336) and this has permitted the many further modifications, particularly in the corrin branch. The bilins also have been positively and not merely neutrally exploited and their biosynthetic pathways are not merely porphyrin-degradation reactions. Bilins are frequently conjugated with other components.

The ommochrome pathway has been discussed already from this aspect (p. 245); it is a pathway from tryptophan selected specifically to produce this class of chrome. Papiliochromes and possibly some other classes represent separate, minor exploitations of tryptophan pathways.

If in the senses of both chapters 2 and 6 the purines are accepted as biochromes then the whole purine-pterin-flavin pathway can be regarded as purely chromogenic *ab initio*. Thus virtually every chromogenic pathway produces primarily or solely that class of chrome. Of course biochromes necessarily are relatively complex and extensively dehydrogenated molecules so that their pathways cannot conveniently produce other materials. More significantly each chromogenic path is of considerable length and involves some unusual reactions and enzymes. None of the chromes in living animals appear to be merely fortuitous products of non-chromogenic pathways.

13.2.2 Sophistication of Chromogenic Pathways

This is the aspect mainly emphasised in section 1. Wherever a pathway is long and/or involves steps reckoned by the chemist to be improbable reactions under the circumstances then the implication is that it must be the result of strong selective forces proportional to the adaptive value of its products. The porphyrin path, and particularly the corrin branch, is the best example of this but most others show it in degree. In the few cases where the structure of the protein part of a chromoprotein is known it is clear that its biosynthesis is equally sophisticated and adapted to the function of the entity.

If space permitted the sophistication could be illustrated in almost every detail that is well understood. Succinate and glycine not only unite to form units capable of dimerising in a specific asymmetrical manner to form pyrrole rings but they also provide the side-chains necessary for subsequent linkage between rings and other side-chains for specific adaptive radiation in the completed porphyrin. In the light of our after-sight they are seen to show great ingenuity and fore-sight.

Probably most sophisticated of all is the chromogenic cycle of the retinopsin molecule (Fig. 7.1) no doubt because it is also the reiterative functional cycle of an exquisitely perfected physiological process. In every revolution of this cycle a most improbable and thermodynamically unstable *cis*-form is built, stabilised, used physiologically and returned to the stable all-*trans* state. In fact the whole carotenoid pathway is an outstanding illustration in the present context. Dehydrogenation and, ionone ring-formation to produce strong chromatic properties require special devices and these must be segregated so as not to imperil the synthesis of other prenyl polymers.

13.2.3 Economy in Chromogenic Pathways

This is usefully considered against the background of a prodigal use of some exogenous chromes, particularly the carotenoids (p. 302). Endogenously synthesised zoochromes are rarely stored in large quantities, their production being well geared to current requirements. For O_2-carrier chromes these requirements are rather high and so the chromes sometimes are stored, at least by annelids (Fig. 5.7). Integumental chromes are stored only in special circumstances such as those of metamorphosis in insects. Excess production of haemoglobin is prevented by feedback-controls (p. 307) and these probably operate also in other paths. Economy is shown again by the suppression of certain endogenous pathways when the relevant chrome becomes available in the diet; haemophilic parasites tend to lack even the haem path because they obtain what they need from their hosts (p. 257). Another manifestation of economy is the relatively limited number of chromogenic paths used; in animals the only known chromogenic paths are those of the ternary quinones, the porphyrins, those starting from tyrosine and tryptophan and the purine-pterin-flavin system. Chromogenesis therefore obeys the general biochemical laws of economy in amount, and austerity in variety (LWOFF, 1944; FLORKIN, 1960).

Austerity is evident also within classes of chrome: all carotenoids are derived from the same C_{40} precursor and all porphyrins via porphobilinogen. A common prochromogen is probable also for the flavonoids and for the pterins. Of course in all these classes there has been wide radiation from that precursor and the radiation probably has been always adaptive. By analogy with morphological evolution it seems likely that austerity of chemical classes is due to the fact that only one is optimal for any particular situation but once a class, like a morphological type, has been selected for that reason there is added virtue in detailed radiation. Another major manifestation of economy is the absence from living systems of the variety and amount of coloured by-products which characterise organic reactions in the laboratory (pp. 242, 338).

13.2.4 Chromogenic Pathways in the Different Taxa

The taxonomic distribution of zoochromes shows that some pathways may have been independently exploited by the various taxa so that probably there has been parallelism and convergence. It therefore might be expected that a particular chromogenic path varies in details in the different taxa although the general outcome has been similar in all. Unfortunately as yet very few paths have been traced in more than one phylum and only for the haem path is there any significant comparative information. This appears to be very similar if not identical in vertebrates, insects (PASSAMA-VUILLAUME and BARBIER, 1966), nemertines (VUILLAUME, 1969) and annelids (NEEDHAM, 1970b). However, it is probably identical also in plants and microorganisms and had a unique origin in the common ancestor of all (p. 307) so that it is no test of parallelism. On the other hand haemoglobins have independently evolved in a number of animal taxa from haemoproteins of the enzymic type and the globins probably do illustrate parallel and convergent evolution. (PADLAN and LOVE, 1968; HUBER et al., 1971); they show considerable general resemblance but differ in amino acid sequence and other details. An identical $retinal_1$ appears to have been evolved independently in a number of phyla of animals, probably by an identical path, albeit a short one.

There is some indication (MATTY and SHELTWAY, 1967) that fishes may synthesise purines for their integument via a different pathway from that known in other organisms; however here there is also a contextual difference, the latter pathway referring to synthesis for nucleic acids or for excretion. It would be useful to know the pathway for integumental purines in other taxa. The pterin pathway must have considerable resemblance between arthropods and lower vertebrates but there may be differences in detail.

13.3 Interrelationships between Chromogenic Pathways

The last question on page 293 is composite and merits further analysis. Relevant interrelationships include (1) similarities between pathways due to the common properties of all chromes, (2) paths originating from a common precursor, (3) connecting paths between members of different main pathways, (4) production of a class of chrome via more than one pathway, (5) interaction between members of different pathways as chemical reagents or as biological agents, (6) a member of one path acting as coenzyme for steps in other pathways. Interactions at the higher level of the control of pathways will be considered in chapter 14.

13.3.1 General Similarities between Chromogenic Pathways

These are rather numerous (section 1) and many of them are related to the properties common to chromes: because of the emphasis on conjugate double bonding all paths involve desaturation, dehydrogenation steps and are typically oxidative. All involve a good deal of polymerisation because this is the easiest way to achieve a large molecular size and bathochromicity; both arithmetic and geometric polymerisation are used. Prenylation is rather common and likewise ring-closure, using the base of this or of other side-chains. Acetate is a very common building unit, mainly because it has already one double bond per two C atoms. Ribose closes the pyrazine ring of pterins and the pyrrole ring of tryptophan (Fig. 13.2) though not that of the melanin indoles.

13.3.2 Paths Beginning in a Common Precursor

Where this occurs it again is largely because of properties common to all biochromes; the carotenoids are a clear demonstration that the converse is not equally true for they share precursors and path with no other class of biochrome. The prenyl side-chains of some other classes are not conjugately double-bonded and do not contribute to chromacy. Since the other classes are all condensed aromatic ring-systems it is not surprising that some have common precursors. Acetate polymerisation produces most of the sub-classes of ternary quinone and part of the flavonoid molecule. The C_6-C_3 unit which provide the rest of the flavonoid (GEISSMAN, 1967) is also the source of phenylalanine and tyrosine and therefore of the phenolic and indolic melanins. Tryptophan and therefore the ommochromes and papiliochromes also originate from this source. This leaves only the tetrapyrrole class and the purine-pterin-flavin group each with its own distinct path. In all, therefore, there are no more than seven main chromogenic pathways: carotenoid, flavonoid, quinone, tyrosine, tryptophan, porphyrin and purine and three of these have a common precursor.

In the shikimic acid set of pathways there are mechanisms controlling the differential synthesis of phenolic and indolic products, and of other alternatives, from a common precursor (MARGOLIN, 1971). The term 'metabolic interlock' has been used for such mechanisms (JENSEN, 1969).

13.3.3 Connections between Pathways

It is noteworthy that VERNE (1926, p. 198–9) believed that there was very extensive interconversion between the chemical classes of chrome: "... il est permis de se demander s'il n'existe pas entre eux (biochromes) des relations, et si même tous no sont pas plus ou moins reducibles les uns aux autres." It is now clear that this was a major over-speculation, no doubt motivated by the observation that in the integument the various chromes wax and wane in close correlation, direct or inverse. The correlation is biologically mandatory and is ensured by common control-mechanisms (chapter 14) but for the most part it does not involve direct interconversion, except within classes. VERNE was probably influenced also by the fact that most integumental classes can produce much the same wide spectrum of colours.

Some degree of interconversion between classes has been detected *in vitro* and may operate also *in vivo*. Benzoquinones such as ubiquinone are interconvertible with chromones (GLOVER, 1965). Some flavonoids can be converted to anthraquinones and others degraded to benzoquinones. The production of indolic melanins from tyrosine and of phenolic-based ommochromes from tryptophan might be regarded as relevant examples. Tyrosinase and tryptophanase have broad specificities to a wide range of both phenolic and indolic quinones and amines (GLASSMAN, 1957). These depend on a considerable degree of structural relationship and could be considered more relevant to the previous sub-section.

13.3.4 Convergence between Pathways

Benzoquinones can be synthesised via at least four pathways, by polymerisation of acetate (p. 284), from shikimic acid (p. 283) and as noted above from chromones and from flavonoids. The first two at least operate *in vivo* in some animals. There is also some indication that naphthoquinones and anthraquinones may be formed via more than one pathway but this seems to be peculiar to quinones among biochromes. The implication is that for each other class there is only one most spontaneous and economical pathway and that this alone has been exploited however frequently the exploitation has occurred. Being entirely carbocylic the ternary quinones are less specific products than heterocyclic molecules.

13.3.5 Interaction between Members of Different Chromogenic Paths

The most interesting example under this head is the action *in vitro* of dopaquinone in oxidising 3-hydroxykynurenine (3 HOK) to ommochrome (p. 289) provided tyrosinase is present to reoxidise the DOPA formed (INAGAMI, 1954; BUTENANDT et al., 1956). If more dopaquinone is formed than is necessary to react with the supply of 3-HOK then melanin also is synthesised. The colour of the product and the amount of melanin precipitate is quantitatively related to the ratio of DOPA to 3-HOK:

Relative amount of DOPA	Relative amount of 3-HOK	Colour of product	Amount of precipitate
∞	0	Black	+++
2	1	Brown-black	++
1	1	Red-brown	+
1	2	Red	(±)
1	4	Red	−
0	∞	Yellow	−

In vitro various intermediaries also are produced so that the system has many possibilities. The oxygen-consumption is much higher than for the same total amount of DOPA or of 3-HOK alone, proving that there is an interactive component of the consumption. The system probably does also operate *in vivo*, – in the blood of the silkworm *Bombyx*; its *rb* mutant when homozygous shows reddening instead of the usual blackening of shed blood.

Dopaquinone here is acting as a catalyst, enzymatically regenerated and itself differing from a coenzyme only in lacking an apo-protein. Since the DOPA remains freely available as a substrate for the melanin pathway this is a type of interaction distinct from the more familiar enzymic type of the next sub-section. It is a type which merits more attention since in the present case it is potentially a powerful mechanism for controlling the ratio of melanins to ommochromes.

13.3.6 Biochromes Acting as Coenzymes in Chromogenic Paths

Biochromes that act as coenzymes are definitive products of their own path and therefore the latter need not be affected by any catalytic action they may have on other paths, except to the extent that depletion of the coenzyme through functional use may have become adaptively geared to promoting its further synthesis. In any case the relationship is an important one chromogenically and many examples of chromogenic biochrome coenzymes could be given, since most chromogeneses are redox processes. Each step may use the whole battery of electron-transport enzymes (p. 184) quite apart from any more specific enzymes. Only a few of the more intriguing examples will be considered.

In the purine-pterin-flavin path the three subclasses are not only related directly, as products of the path: ribitylflavin (RF) is also the essential component of the coenzyme of xanthine dehydrogenase, the enzyme acting on a number of both purines and pterins (p. 292). In turn the tetra-H pterins are coenzymes of the hydroxylases of both phenylalanine and tryptophan (KAUFMAN, 1967), as well as for some other steps in the shikimic acid complex of paths. This may be one reason why pterins are so commonly associated with the darker chromes of the integument, in both Amphibia (FOX and VEVERS, 1960, p. 157; BALINSKY, 1970) and arthropods (ZIEGLER-GUNDER, 1956, 1961; HARTENSTEIN, 1970). The same applies to integumental RF (FOX, 1953, pp. 285, 287; BALINSKY, 1970).

The position is complicated by the facts that in the light RF inactivates tyrosinase (GALSTON and BAKER, 1949) presumably photodynamically (p. 247) and that isoxanthopterin inhibits melanin-synthesis in the elytra of the polychaete *Harmonia*, at some stage between dihydroxyindole and the final polymer (ISAKA, 1952; OSHIMA et al., 1956). These actions could be useful in controlling the integumental colour. Both RF and pterins are coenzymes for reactions in the conversion of tryptophan to 3-HOK so that they may favour the differential synthesis of ommochromes relative to melanin. This may be the relevance of the fact that the RF and ommochrome contents of the silkworm are controlled by a common gene (KIKKAWA, 1953).

Pteroylglutamic acid (PGA) is coenzyme for essential reactions in the porphyrin biosynthesis pathway (JACOBSON and SIMPSON, 1946; DOAN et al., 1946); this concerns the transfer of C_1-units to and from the porphyrin intermediaries and so is particularly important (CHELDELIN and BAICH, 1967) for the corrin section of the path (p. 287). Where the C_1-unit is transferred in the CH_3-state vitamin B_{12} is the actual cotransferase so that the corrins are essential for their own biosynthesis. In addition various substituted purines are components of the completed cobamide coenzymes (Table 3.10) so that in all this is a good example of the complexity of biochrome interactions in chromogenesis. Since PGA-enzymes catalyse C_1-additions in the biosynthesis of the purines they promote the synthesis of the pterins themselves. There is a C_1-removal step in the actual pterin path (Fig. 13.6) and in chromogenic pathways collectively the C_1-transferases are an important group of enzymes. Acetate transfer is equally widespread so that pantothenic acid (p. 209) in coenzyme A is another important chrome in this context.

13.4 Conclusion

Chromogenic pathways show the characteristic features of naturally selected biochemical processes. Each leads specifically to a class of biologically important chromes and none of these are by-products of paths leading primarily to other materials. Some steps in the pathways would be difficult under the same conditions in the laboratory and could have become organised only biologically, – under strong selective forces. The paths show the same economy of energy and materials as most biochemical processes; for each chemical class all phyla probably use the same pathway, which is optimal in these and other respects. Because the similarities in basic properties between all chromes dictate that many of them are rather closely related, their pathways have various types of interrelationships. Members of different paths interact also in ways relevant to the control of chromogenesis. One type of interaction, the widespread use of redox coenzymes in all pathways is not peculiar to chromogenesis, but may be more pronounced here than in other aspects of metabolism because the syntheses are so extensively oxidative.

Chapter 14

Evidence from the Control of the Supply of Biochromes

'Supply of Biochromes' covers the uptake and subsequent handling of exogenous chromes and the biosynthesis and further handling of those synthesised endogenously. The mechanisms controlling these processes have been accessible to study longer than the actual chromogenic pathways and so provide much additional avidence, mainly at the systemic level.

14.1 Uptake and Handling of Exogenous Chromes

It must be emphasised that the distinction between exogenous and endogenous groups of zoochrome is not rigid. A number of endogenous chromes are supplemented from dietary sources; once absorbed the latter join a common pool with any synthesised endogenously. On the other hand there is some evidence that animals may synthesise a certain amount even of carotenoids, pteroylglutamic acid and ribitylflavin.

14.1.1 Carotenoids

This is the most abundant and varied class of exogenous biochrome, even in carnivores which receive them at second or third hand. Some appear to accept passively any carotenoids in their food but others control their uptake, quantitatively and qualitatively. Animals also modify what they absorb and they conjugate some carotenoids. Since some phyla at least synthesise steroids and other materials of the isopentenyl pathway (Fig. 13.1) it would seem that if necessary they should be able to synthesise carotenoids also but at present there is little evidence of any of them producing significant amounts.

MORGAN (1901b, p. 268) thought that the local increase in concentration of carotenoids in regions of regenerative activity in hydroid Cnidaria (p. 271), confirmed photometrically by GOLDMAN (1953), was due to endogenous synthesis but another explanation is now offered (p. 271). LWOFF (1927) detected endogenous synthesis in the copepod *Idya* but later decided that the evidence was not conclusive (FOX, 1953, p. 120). There is an 8% increase in the carotenoid content of the brine-shrimp *Artemia* between egg and nauplius stages (NEEDHAM and NEEDHAM, 1930), when no external source is available. GILCHRIST and GREEN (1960) found a similar increase in other Crustacea though they were reluctant to believe that there could have been endogenous synthesis. Enzymes of the carotenoid path are present in mammalian liver (CHICHESTER and NAKAYAMA, 1967), even in that of the pig which avoids absorbing carotenoids from the gut (PALMER, 1916). In vitro AUSTERN and GAWIENOWSKY (1969) obtained good evidence for carotenoid biosynthesis by the corpus luteum of the cow. Significantly they had best results in an atmosphere of nitrogen, the inference being that O_2 is necessary for the other isopentenyl path, to the steroids (Fig. 13.1); it is known that anaerobic bacteria cannot synthesise steroids for this reason (RAFF and MAHLER, 1972). Here are two possible reasons why animals usually do not synthesise carotenoids, – the low redox potential required and the competition with the steroid path.

Another possible reason why endogenous synthesis of the class by animals is at most minimal is that it is not essential in large amounts, notwithstanding the high concentrations found in so many animals. Cavernicolous gammarids seem to receive little or none in their diet, have no detectable amounts in the body and yet suffer no apparent ill-effects (BEATTY, 1949). This has been confirmed experimentally in the insects *Tribolium, Tineola* and *Blatella* (FOX, 1953, p. 136) and in the Crustacea *Carcinus* and *Panulirus*, which normally contain very much (FISCHER, 1927). Leghorn hens survive depletion of all detectable amounts and lay fertile eggs which give chicks capable of growing to maturity still without external supplies (PALMER and KEMPSTER, 1919a). At least retinal-deficiency would be expected in all these animals and also vitamin A-deficiency in birds; the retina in particular should be examined very carefully in such animals.

The converse situation of heavy and variable concentrations equally may be taken to indicate that they are not of critical importance. At the same time it also shows that they are relatively innocuous, as already seen for Man (p. 251). *Carcinus* normally has large stores in the labyrinthine excretory organ and hepatopancreas; it also moves large amounts to and from the exoskeleton and stores high concentrations in the eggs. Nevertheless it rapidly excretes its stores when fasting or on a carotenoid-free diet (FISCHER, 1927). Other animals found to excrete their stores even when none is being ingested are the sea-anemone, *Actinia* (M. and R. ABELOOS-PARISE, 1926), *Hydra* (FOX, 1953, p. 85), the tadpole of *Pelobates* (VERNE, 1926, p. 414), *Octopus* (NIXON, 1968), the green-fish *Girella* (FOX, 1953, pp. 147–8; 150–1)and the flamingo (FOX and VEVERS, 1960, p. 72). Some fish, such as the atlantic minnow *Fundulus*, do maintain a steady level under fasting and other dietary regimes (FOX, 1953) and in the integument of the starfish *Asterias* the concentration of carotenoids actually increases when fasting (VEVERS, 1949), owing to the differential decrease in other constituents. It therefore appears also that there is no absolute need to excrete the class when fasting. Animals simply vary greatly in their handling of carotenoids and the manner is not critical.

VERNE (1926, p. 414) thought that the varied handling might be related to two distinct types of carotenoid, a 'variable' and a 'constant' component, such as Terroine had found for the simpler lipids. The latter component becomes part of the fabric of the body while the former does not; the idea is discouraged by the extent to which all carotenoids, including those of the integument, can be depleted without ill effect.

Some animals not only can manage on a very low uptake of carotenoids but actually avoid absorbing them from the gut (Table 14.1). This is common among carnivorous mammals and birds and is shown also by the pig, guinea-pig, rabbit, goat and some cephalopods; it may be true also for the gephyrean group of worms which rarely contain much carotenoid yet must be exposed to it in their diet. The pig absorbs less than 10^{-7} of the carotenoid in its diet and its lard is quite white. The goat has a destructive enzyme, a carotenase, in its gut. An occasional mutant of the rabbit does absorb these chromes so that again the normal individual must have some absorption-barrier. Cell-membranes are rather permeable to lipids (DAVSON and DANIELLI, 1943) and carotenoids are among the lipids found to be regular components of the membranes. Vitamin A protects cell-membranes and promotes the absorption of lipids in general (KENNEDY, 1969); carotenoid-uptake therefore may be essential in most animals to facilitate the uptake of other lipids.

This may be why some animals do not exclude them from the body yet oxidise them to colourless forms after absorption (FOX, 1960). Others while not excluding them completely do exercise qualitative selection (GOODWIN, 1952, p. 285ff.). Some taxa absorb mainly carotenes and others the oxycarotenoids (Table 14.1); at one time it seemed that this might be a distinction between herbivores and carnivores respectively but the actual position is less simple. Further, some taxa absorb both types and then differentially excrete the one but the horse and some invertebrates definitely select at the absorption stage (Fox, 1953, p. 188ff.). Mammals absorb mainly β-carotene because it is more resistant than the oxycarotenoids to the digestive enzymes (PALMER and ECKLES, 1914). When feeding on the tomato the locust *Schistocerca* is more narrowly selective Fox, 1953, p. 134): it takes up carotene but excludes the much more abundant lycopene. The surf-perch *Cymatogaster* hydrolyses certain carotenoid esters to xanthophylls in the gut and absorbs some but not all of these compounds (Fox, 1953, p. 152). In all these cases absorption is quite specific.

Table 14.1. A. Forms in which carotenoids are absorbed from the gut

Carotenes by	Oxycarotenoids by	Both by	None by
Porifera	Some invertebrates	Echinoidea	Some invertebrates
Thoracophelia and other annelids	Most fishes	Some insects	Decapod Cephalopoda
Gastropoda	Many birds	*Octopus*	Some birds (Carnivorous)
		Rana	Some mammals: rabbit, goat, guinea pig, pig, Carnivora
Sacculina		Primates	
Insecta			
Mammals (e. g., cow, horse)			

B. Predominant carotenoids in tissues of animals

Carotenes in	Oxycarotenoids in	None or rare, in
Porifera	Actiniaria	Priapula
Echinoidea	Asteroidea	Sipuncula
Holothuroidea	Ophiuroidea	Echiura
Polychaeta	Most Arthropoda	Decapod Cephalopoda
Gastropoda	Bivalvia	Non-absorbing mammals (above)
Daphnia	Integument of vertebrates	
Eggs of *Carcinus*		
Insecta	Most tissues of fishes and birds	
Fat of vertebrates	Mystacoceti	
Tissues of Amphibia, some Mammals		

By contrast the primates, the frog, echinoids, some insects and the octopus show indiscriminate absorption (Table 14.1) and the absorption varies freely with the diet. Again, whale oil has a high content of astaxanthin the main carotenoid of the euphausids on which many of the Mystacoceti feed. The carotenoids of the lady-bird beetle are related at second hand to those of the plants on which its aphid prey feed. Pugmoth caterpillars store carotenoids when feeding on plants containing them and flavonoids when on certain other plants; it is noteworthy that flavonoids also are absorbed non-specifically by this animal (Fox and VEVERS, 1960, p. 168).

The various changes made by animals to carotenoids before or after absorption from the gut are nearly all oxidative; in aerobic organisms the oxidation is relatively spontaneous and this helps to protect other materials from oxidation (p. 197). The formation of retinol from half a carotene molecule is oxidative and astaxanthin, probably the most common animal carotenoid, is a diketonic form, very rare in plants. Its formation from dietary carotenes has been confirmed (Goodwin and Srisukh, 1949). By feeding adult *Artemia*, previously depleted, with a range of pure carotenoids Hsu et al. (1970) found that they converted β-carotene to echinenone and further to canthaxanthin and these are the only two carotenoids it passed to the eggs. On the other hand zeaxanthin, 3,3'-di HO-carotene, one of the few oxycarotenoids of plants is absorbed but neither used nor further oxidised. Yet the mussel *Mytilus* does store zeaxanthin which it synthesises from dietary precursor-carotenoids.

In summary the following main methods of dealing with dietary carotenoids may be recognised:
1. Assimilate indiscriminately and store unchanged.
2. Assimilate indiscriminately and selectively excrete.
3. Assimilate indiscriminately but subsequently convert some to other forms.
4. Convert some in the gut and then absorb indiscriminately.
5. Assimilate selectively.
6. Assimilate and destroy, or excrete.
7. Exclude from absorption.

Usually carotenes are transported as carotenoproteins and xanthophylls free in solution in the blood-lipid (Fox, 1953, p. 175). This might seem paradoxical but the blood-lipids are not completely apolar, as shown by their solubility in the vertebrate bile. This could be relevant to differential absorption from the gut. The ova have first call on the relevant carotenoids of female animals (p. 263) and this has been shown not only for birds (Heiman and Wilhelm, 1937) but also for the chamaeleon (Needham, 1942, p. 652) and for the crab *Carcinus* (Fox, 1953, p. 121). After a period of deficiency little carotenoid accumulates in the blood of *Carcinus* until the sixth day of refeeding with carotenoids and in the hepatopancreas until the 15th. day (Fox, 1953, p. 121) the ovary by then being well stocked. The handling of dietary carotenoids thus is a major and complex section of metabolism demanding considerable provision in the genome. It also presents many puzzles.

14.1.2 Other Exogenous Zoochromes

The uptake of dietary chromones and flavonoids (p. 82) is probably unrestricted; the flavone of the larva of the Marbled White butterfly, *Melanargia galatea*, like that of the pugmoth is identical with that of the plant on which it feeds. It is passed on unchanged to the adult butterfly but this does not indicate that animals cannot modify this class; some animals at least destroy them metabolically (Harborne, 1965). Anthocyanins are absorbed and used unchanged by weevils feeding on mullein. Conceivably sertularid Hydrozoa obtain their flavones from phytoflagellates, direct or at second hand: otherwise endogenous synthesis must be suspected.

It is generally believed that animals obtain ubiquinone from dietary sources though there is some evidence for endogenous synthesis (p. 251); the prenyl side-chain shows considerable taxonomic variation but this could be due to variation in the dietary source. The sea-otter certainly derives from dietary sources the echinochromes it deposits in its bones and coccids probably depend on microsymbionts for their massive supplies of anthraquinones.

Cobamides (Cobalamins:): There is little vitamin B_{12} in plant materials (West and Todd, 1957, p. 765) other than that derived from soil-microorganisms (Mervyn, 1971) and animals depend mainly on gut-microorganisms for their supply. Carnivores receive considerable amounts also from the liver and other tissues of their prey. Since gut-microorganisms are most active in the posterior region of the gut the absorption of cobamides occurs mainly in the distal part of the ileum (BOOTH, 1963). A special receptor, restricted to this part of the gut-wall, is essential and the uptake is an active transport. Because of the long journey there is no uptake of ingested B_{12} for about an hour.

Pteroylglutamic Acid: Strictly speaking PGA is of exogenous origin but in many animals it again is synthesised by microorganisms of the gut. The unconjugated pterins are extensively synthesised by animals (p. 290) but these also are often present in food and it may be significant that there is some storage of pterins in mammalian liver (FOX, 1953, p. 298). Dietary sources of these pterins have a significant anti-anaemia effect in mammals and fishes (JACOBSON and SIMPSON, 1946) indicating that the gut-microorganisms are able to use them to synthesise PGA.

Ribitylflavin: While there is little doubt that most animals acquire their RF from dietary sources there is some evidence for endogenous synthesis in insects such as *Periplaneta* and *Tineola* which are able to thrive on diets qualitatively very deficient. They manage even to store RF in amounts greater than they could possibly acquire from the food (BUSNEL et al., 1943; METCALF and PATTON, 1942; GOODWIN, 1963).

Relatively little is known about the movements of dietary RF in the animal body partly because the amounts involved are usually small and the turnover-rate high. In Man the excretory output is about equal to the average intake (FOX, 1953, p. 286) and in rats it falls sharply when dietary sources are curtailed. Most animals do store some, and a number have massive stores (p. 243): these fluctuate with dietary supply again indicating a continuous turnover. In general phytophagous insects have much more than carnivores (FOX and VEVERS, 1960, p. 159) and this has the same implication, supposing that both absorb all they ingest. On the other hand marine isopods and cirripedes contain about 1–2µg per gram body-weight compared with 3–14µg in amphibious and terrestrial types (FONTAINE et al., 1943) and this is thought to indicate that uptake is related to respiratory needs, though it is not clear that in general terrestrial and fresh-water animals need larger quantities of flavoproteins than their marine relatives.

The RF content per unit weight remains fairly constant throughout growth in a number of animals investigated, indicating that storage also is controlled. The total content of the body therefore increases greatly with size, by a factor of 900 in the eel *Anguilla* (FONTAINE, 1937) and a comparable increase is recorded for insects (FONTAINE, 1938; BODINE and FITZGERALD, 1947). Turnover shows much the same relationship: the excretion per 24 hours by an adult man is up to 900 µg compared with values as low as 22µg for infants (FOX, 1953, p. 286).

RF is absorbed without chemical change and apparently without any special active transport-mechanism (BOOTH, 1963). After absorption no doubt there are controls between immediate conjugation for use and storage. In vertebrates the main storage organs are the liver and kidneys, in insects the malpighian tubules and in annelids the chromogogen tissue. Animal tissues often have much higher concentrations than in plants (FOX, 1953); that in the malpighian tubules of *Attacus pernyi* reaches 1 mg per gram. Bivalve molluscs appear to store locally so that the content of each tissue is related to its metabolic needs (BROOKS and PAULAIS, 1939). In two snakes, *Bothrops jararaca* and *Eudryas bifossatus* high concentrations, 200µg per 100ml, have been found in the blood (VILLELA and PRADO, 1945) but the significance of this is not clear.

14.1.3 Chromes Acquired Both Exogenously and Endogenously

Haems: Some of the dietary haem is absorbed by the leech *Hirudo* (DIWANY, 1919) and by the bug *Rhodnius* though not by all sanguisugous insects (WIGGLESWORTH, 1943). The rest is evacuated as protoferrihaem by both bug and leech (WIGGLESWORTH, 1943; NEEDHAM, 1966 b). In *Rhodnius* the absorbed fraction has a surprisingly varied fate. Some at least is absorbed as intact Hb or as a haemochrome of the denatured protein; the latter is cherry red and is stored in the maturing oocytes and the salivary glands; from the ovum it passes to the salivary glands of the larva, implying a regular digestive function (p. 242). The remainder of the absorbed haem is degraded to biliverdin and stored in the pericardial cells or is excreted once more via the gut as protoferrihaem, verdohaem or biliverdin. Iron as well as biliverdin accumulates in the gut-wall and persists throughout life. *Hirudo* also stores protoferrihaem and a bilatriene, mainly for camouflage purposes (NEEDHAM, 1966 b). Non-sanguisugous leeches also store both and the rhynchobdellids at least must acquire this entirely from their food. Most animals which receive dietary haem appear to absorb some to supplement their own haemogenesis.

Bilins: In contrast to the bug and leech, orthopteran insects appear to obtain their integumental bilins from their own haemenzymes (GRANICK and SASSA, 1971) since they have no Hb and are herbivorous yet do not degrade chlorophyll to bilin (VUILLAUME, 1969). They incorporate labelled glycine eventually into bilin (VUILLAUME and BARBIER, 1966; RUDIGER, 1970) and so probably synthesise it via the orthodox porphyrin path. Some animals absorb plant-bilins (p. 81) and possibly some do degrade chlorophyll as far as this. The ostracod *Heterocypris* obtains and stores in the gut-wall the bilins of the blue-green algae and they are depleted when the crustacean feeds on green algae which lack bilins in their photosynthetic system (FOX and VEVERS, 1960, p. 122). This animal does not degrade chlorophyll to bilin. The sea-hare *Aplysia* likewise obtains several bilins (p. 81) from its algal food (RUDIGER, 1970) and uses them, conjugated with protein, as the three chromes of its ink. The squash-bug *Anasa* has in its fatbody and pericardial tissue bilin which may be derived from dietary chlorophyll (METCALF, 1945); this insect makes more use of the latter than most animals, – having crystalline phaeophorbides in various tissues and free Mg in the blood.

Purines: Perhaps mainly because the metabolism of the purines has been so much more studied than that of most chromes the relationships between endogenous and exogenous supplies seem particularly complex. It is known that vertebrates digest dietary nucleic acids and absorb their purines as free bases and these supplement endogenously synthesised adenine and guanine in the synthesis of nucleic acids, but some may be excreted without ever entering the fabric of the animal. The purines excreted by Man are believed to come from both exogenous and endogenous sources. Animals which are more fully ammonotelic or ureotelic break down purines of both sources before excretion; at the same time these animals often have purines in their integuments and in deeper stores. Indeed the lower vertebrates have much more here than the purinotelic reptiles and birds: they have excretory ureotely or ammonotely but 'retention purinotely'. This is a possible reason why their method of purinogenesis is unorthodox (p. 295). Excretory purinotelic types, building up even their protein waste nitrogen into uric acid probably minimise uptake and storage, and maximise excretion of purines. Unlike most insects the cockroach is ammonotelic and not uricotelic (MULLINS and COCHRAN, 1972) and it absorbs and stores any uric acid it ingests.

Purine stores are depleted by fasting and this has been demonstrated not only for the earthworm (p. 243) but also for the frog, *Microhyla ornata* (VERNE, 1926, p. 414). The latter has integumental purinocytes and even their purine is depleted. On the other hand there is an increase in purines in some fasting Amphibia (ib., p. 340); like the carotenoids of *Asterias* (p. 244) they may concentrate because other constituents, including nucleic acids, are depleted more rapidly. Also as for the carotenoids there is a complete range from high tenacity to great prodigality by animals. As already indicated (p. 243) MILLOT could detect no increase in integumental purine when feeding Amphibia a diet rich in nucleic acids but this is not surprising if camouflage was already adequate since the Amphibia have a complete series of purinolytic enzymes. If any purine was retained it probably was deposited in store. Exogenous and endogenous purines are widely used by animals but like carotenoids they are handled in a variety of ways, still not fully understood. In degree this is probably valid for the other chromes, also.

14.1.4 Conclusion

Among animals collectively most of the known classes of zoochrome are acquired to some extent from the food but only carotenoids, chromones, flavonoids, ribitylflavin, cobamides and pteroylglutamic acid are totally exogenous in source for virtually all animals. Indeed even carotenoids and ribitylflavin possibly are synthesised endogenously by some animals, at least in indispensably mimimal amounts. Ribitylflavin, cobamides and pteroylglutamic acid are not modified during or after uptake into animals whereas carotenoids and chlorophylls are changed in a variety of ways, as also is haem by sanguisugous parasites. Dietary chromes are stored more abundantly than those synthesised endogenously; they may become more concentrated than in the tissues of the food-source. By most animals they are also used more prodigally than endogenous chromes though some animals control uptake and output very precisely. Dietary chromes of classes which are also synthesised endogenously join the common pool. For those classes which have been studied extensively, the carotenoids and purines in particular, the handling shows great variation among animals and still presents a number of problems.

14.2 The Control of Chromogenesis

Knowledge in this field is still rather patchy and it seems preferable to consider only the better known, instructive examples rather than attempt a complete survey. The genetic basis of control will be considered in chapter 16 and the present interest is in the modulation of this basic mechanism by relevant environmental factors, both external and internal, in relation to contextual contingency. Examples related to the control of integumental chromes have been considered in chapter 6; the neurohormonal pathways of mediation of modulation-responses are probably common to most externally triggered responses. The control of haemoglobin-synthesis by oxygen-level is an example where the modulating agent is both external and internal. Internal modulating agents are most relevant to controls at the level of the actual chromogenic reactions; here the feedback type of control will be considered, again best known for the porphyrin pathways (GRANICK and SASSA, 1971). All these modulations show that chromogenesis is a very sophisticated product of evolution.

14.2.1 Feedback Control of Chromogenesis

Facultative anaerobic microorganisms such as *Rhodopseudomonas* synthesise haems, chlorophylls and cobamides (LASCELLES, 1968) and control each of these branches of the porphyrin system very efficiently. Vertebrates and other animals have only the haem path and therefore a simpler control-mechanism (GRANICK and SASSA, 1971). All enzymes of the haem-pathway except δ-ALA synthetase (p. 286) are normally present and active in unlimiting amounts, and control depends primarily on completed haem repressing further synthesis of this one enzyme, at the beginning of the whole pathway. *Rhodopseudomonas* both represses the synthesis and inhibits the action of any of the enzyme already formed.

Haem also depresses the rate of entry of δ-ALA synthetase into the mitochondria of the haemopoietic cells and the rate of entry of Fe into the cell (NEUWIRT et al., 1969). These also are feedback devices in the sense that they are inhibitory actions by the product of the path on processes essential for earlier steps in the system. Repression of the synthesis of δ-ALA synthetase is at the usual level of transliteration (transcription) of genic DNA into *m*-RNA but there is also repression at the translation level of *m*-RNA into peptide (GRANICK and SASSA, 1971). For this purpose haem is the prosthetic group of a protein 'holorepressor'. Destruction of haem through use, etc., lifts the repression and effectively induces synthesis of the enzyme. In addition the bilins, products of haem-destruction, promote haem-synthesis (LEMBERG, 1937; WITH, 1968); for the bilins themselves this is a positive feedback but for haem itself it is another aspect of negative feedback. WITH (1968) suggests that the mode of action of bilins here is to provide pyrrole rings or even complete tetrapyrrole chains for porphyrin-resynthesis although he is very conscious that bilins are salvaged in such a way only very exceptionally. Like most biologists, perhaps under the spell of WILLIAM of Occam, he feels bound to advocate a direct type of action; this is relevant to the general theme below.

Haem promotes the synthesis of globin, in accordance with the general principle that key metabolites promote as substrates for, and depress as products of, any particular pathway. In this instance the haem is thought to act by facilitating the essential tertiary folding of the globin chains (GRANICK and SASSA, 1971). Actually the haem unites with a globin dimer and the prior dimerisation stage is controlled by the microsomes without its help (MACLEAN and JURD, 1972) so that the action of the haem may be at the quaternary stage. Another view is that it controls at the translational level. There is some evidence that it promotes the synthesis also of cytochrome apo-proteins and this may be relevant to the differential control of the various haemoproteins of the body. The further synthesis of both haem and chlorophyll depends on a supply of their respective proteins so that the promotion is mutual (LASCELLES, 1968). It has mass-action properties.

In organisms such as *Rhodopseudomonas* haem (NEUWIRT et al., 1969), chlorophyll (LASCELLES, 1968) and corrin (BYKHOVSKII et al., 1969) all control their own synthesis by negative feedback and so ensure a fair supply of the commom precursor to the other branches of the pathway. Corrins actively stimulate haem synthesis (BYKHOVSKII et al., 1969) and inhibit only their own branch, no doubt as usual at the first convenient step; corrin-synthesis therefore is not increased simply by adding δ-ALA. Since haem inhibits and represses at the δ-ALA step this automatically throttles back synthesis in all three branches. The situation implies that the haem path antedated the other two in evolution. Chlorophyll-synthesis is restricted by the limiting concentrations of the enzymes in its branch as well as by the negative feedback-mechanism but there is a balancing control since free Fe stimulates the synthesis of both chlorophyll and corrin (GOODWIN, 1963).

Under aerobic conditions still a third feedback control at the δ-ALA stage operates in *Rhodopseudomonas* which possesses an activation-mechanism for the precursor zymogen of δ-ALA synthetase; this activator is inhibited in proportion to the O_2-pressure. The chlorophyll path seems to be the one most affected: the transliteration and translation of the whole group of chlorophyll genes is blocked in parallel. In its aerobic, non-photosynthesising state therefore this organism synthesises only its haemoprotein respiratory enzymes. The complexity of these porphyrin controls, collectively, is very impressive; controls of haem-synthesis at more systemic levels are equally so.

14.2.2 The Systemic Control of Haem-Synthesis

Knowledge mainly concerns haemoglobin (Hb), with O_2-pressure as modulating factor (GORDON, 1959; FISHER, 1969). As already implied, O_2 is normally considered an external factor but probably always acts after intake to the body. Apart from annelids (FOX, 1955) most animals that have Hb, mammals, turtles, goldfish, the snail *Planorbis*, the larva of the midge *Chironomus* and cladoceran Crustacea (PROSSER and BROWN, 1961, p. 223) all show a similar haemogenic response to an increase in O_2-need, whether due to increased activity or to decreased O_2-supply. The mechanism is reversible. In the cladocera *Daphnia* and *Simocephalus* (FOX, 1953; 1955) the response is all-or-none but in the others it is a typical modulation.

Under the O_2-deficit of high altitudes Man increases his Hb by 20% which is only 1/10 of the *pro rata* synthesis by *Daphnia* though the time for maximal response, 10 days, is very similar and compares with that by the camouflage chromes of the integument (p. 136). Increase in the number of red blood cells is the main vertebrate response but there is also some increase in the amount of Hb per cell. CO_2 has no effect on the response so that it is not so closely geared to the functional response (p. 140) as the chromogenic response of the integumental chromes. This is biologically rational since one of the physiological responses to a decrease in O_2 is increased ventilation which washes out CO_2 making the latter an uncertain index of O_2-level.

It is not self-evident that a deficiency of O_2 can be compensated by increasing the amount of Hb in circulation, unless the loading at the acceptor-surface is near 100%. In fact this is true for vertebrates and most other groups investigated, but possibly it does not apply in those annelids that lack the chromogenic response to O_2-level. The response must impose an increased mechanical load on the circulation which must not outweigh the advantage of the increased Hb per unit volume. In Man there is an increased blood-volume so that viscosity need not be increased and there is an increased heart-rate and cardiac output which fully exploits the increased Hb and blood-volume.

Other haemoproteins, myoglobin and cytochrome *c*, respond chromogenically in the same direction as Hb to a decrease in ambient O_2 (PROSSER and BROWN, 1961, p. 223) but there are complications. In some animals, at least, and in some microorganisms there is a marked increase in the amount of the cytochromes in *direct* relation to O_2-level (KING, 1971). The rationale for this may be that the cytochromes must be adequate to handle the O_2 presented to the tissues by the Hb and this is maximal when ambient O_2 is maximal; certainly an increase in amount of cytochromes alone under anoxic conditions would not solve the problem. In microorganisms there is also the fact that some cytochromes are formed only under aerobic and others only under anaerobic conditions. In yeasts the aerobic cytochromes, as well as the catalases and peroxidases, increase 30 fold at the transition from anaerobic to aerobic conditions whereas the anaerobic cytochrome b_1 does vary inversely as O_2-level.

On land O_2-pressure shows a simple decrease with altitude but in water the O_2-content varies with depth in a more complex fashion, depending on temperature and pressure, the nature and density of living organisms, water circulation, etc.. It also decreases with increase in salinity (GILCHRIST, 1954) and this induces the anostracan *Artemia* to synthesise

Hb. Similarly *Daphnia* synthesises more at high than at low temperature (Fox and Phear, 1953); here there is also a speeding of metabolism and so an increased demand for O_2 as well as a decreased supply.

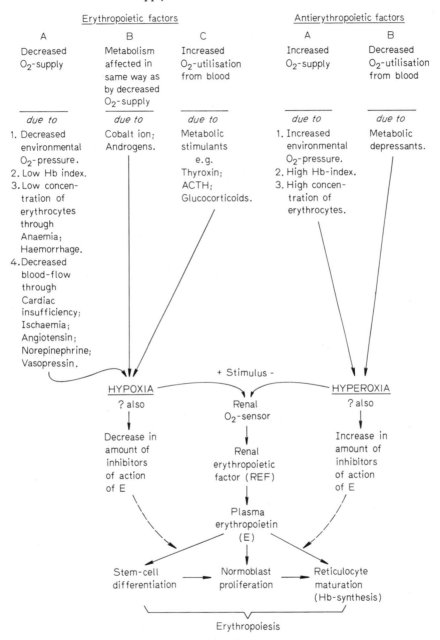

Fig. 14.1. Erythropoiesis in mammals. Summary of the factors which affect the process via the O_2-level in the blood and the control-mechanism which this modulates. Note that erythropoietin acts at at least three points in the development of the erythrocytes. (Based mainly of Fisher, 1969)

There is now considerable knowledge of the mechanism by which O_2-supply and demand control haemogenesis in mammals (GORDON, 1959; FISHER, 1969; GOMPERTS, 1969). The kidney is the main perceptor of the effective O_2-level and produces a renal erythropoietic factor in proportion to the oxygen-deficit in the blood leaving the organ. This reacts with plasma-globulin to produce a glycoprotein, *erythropoietin*, and this is the agent which directly stimulates the production and maturation of erythrocytes in the haemopoietic tissues. Hyperoxic conditions depress the production of this hormone and in addition the activity of erythropoietin already formed may be depressed by a circulating inhibitor the level of which is directly dependent on the O_2-level (Fig. 14.1). All the known factors affecting the O_2-level of the blood actuate this mechanism; they include factors that change the ambient O_2-supply to the blood and factors that affect the rate of O_2-withdrawal from the blood by the tissues. Thus in contrast to the modulation of integumental chromogenesis there is a single and internal triggering agent, a single internal perceptor organ and a purely hormonal effector limb to the response.

14.2.3 The Control of Phase-coloration in Insects

The control of this further aspect of integumental chromatology merits some consideration here particularly because in most cases the significance of phase-coloration is still uncertain. Triggering agents for phase-changes include both external and internal, and biotic and abiotic types. The best studied instance is the *solitaria/gregaria* phase-change in locusts; the solitaria phase has good cryptic colouring but the gregaria has bold yellows, reds and blacks, due to carotenoids, ommochromes and melanins. The pattern probably is biologically useful in promoting and maintaining gregariousness by making individuals conspicuous, and secondly by accelerating the early morning absorption of the heat necessary to initiate locomotion, particularly flight by the imago. There may be other functions also.

Gregaria-type darkening of a nymph is induced by the presence of other individuals and if such a nymph is isolated it returns to the solitaria colouring over the subsequent instars. Even one companion constitutes a significant stimulus towards gregaria-facies and the effect increases in proportion to the number per unit space. It appears to be mediated mainly through visual stimuli though olfactory, tactile and other types of mechanical information also contribute, at least in some species. The response to visual stimuli is mediated through the neuro-hormonal system and probably has some interrelationships with other integumental chromogenic mechanisms; the efferent hormones are those common to other aspects of morphogenesis and chromogenesis (p. 144), the tropic hormones from the brain, *neotenin* (juvenile hormone), *ecdysone* (growth and differentiation hormone) and the corpus cardiacum hormone (KENNEDY, 1956). Blood from a gregaria-phase nymph will induce bold coloration in a solitary individual (NICKERSON, 1956). It is neotenin, the corpus allatum hormone, which is believed to mediate the promotion of gregaria colouring (JOLY, 1951) and it causes excess pigmentation if applied to embryos of *Schistocerca* (NOVAK, 1969). Ecdysone as usual acts antagonistically to neotenin, inhibiting melanin-synthesis in the cricket (JONES, 1956). Probably the other types of stimulus feed into the same central and efferent pathways.

Increased atmospheric CO_2-content also has been found to induce gregaria-colouring (FUSEAU-BRAESCH, 1967); presumably crowding causes an appreciable local increase in CO_2-content. The same applies to atmospheric humidity which again has been found a gregaria-stimulus, in *Locusta* (ALBRECHT, 1964, 1965). The gregaria is more active than the solitaria phase and movement in itself seems to stimulate the production of bold chromes;

conceivably this is related to CO_2-production and a raised internal level. It was originally suggested that the production of gregaria-chromes was the direct effect of the increased metabolic rate associated with increased activity but a direct effect seems as improbable here as in almost every other case. The carotenoids are only transported, not synthesised *de novo;* moreover the main carotenoid in *Locusta* is β-carotene (Fox, 1953, p. 136) whereas a more oxidised derivative might be expected if high metabolic rate were the direct cause. If this and all the other stimuli act indirectly all may relay through a common central and efferent pathway, as already suggested. The large number of effective stimuli, mainly biotic in origin make this a particularly interesting modulation.

A large number of other orthopterids show the same type of response to crowding. However some, including the cricket *Gryllus*, are dark when solitary and pale when crowded (Fuseau-Braesch, 1960); unlike the acridids *Gryllus* does not migrate and this may be relevant to the converse type of colour-change. The larvae of a number of Lepidoptera darken if crowded (Fox and Vevers, 1960, p. 57) but the biological significance of this is not yet clear.

14.2.4 The Control of Exogenous Integumental Chromes

This merits further amplification because of the indication above that these chromes are subject to the same controls as endogenous chromes of the same site and function. When exposed to light the integument of Crustacea increases its carotenoids in parallel with the other chromes (Verne, 1926, p. 329) and this is true for other taxa. In *Daphnia* there is an increase in carotenoid also in the gut, fat-body and eggs (Green, 1957). The implication is that the primary response is an increased uptake from the gut, storage in integument, fat-body and eggs then following automatically. The primary response clearly is indirect and its mechanism is not yet known. Deposition of carotenoid only in the illuminated part of the integument will probably prove no more a simple direct effect of light than that of the endogenous chromes (p. 313).

It would be interesting to know how far the changes which animals make in the carotenoid molecule occur at the site of deposition, since endogenous chromes *are* synthesised *in situ*. It is known that vitamin A is formed from carotene in the gutwall and travels to the retina via the liver; this is a rather special case but it emphasises again the general systemic nature of the controls. At the same time it is noteworthy that retinal subsequently cycles repeatedly *in situ* (p. 164) and that it provides one of the few examples in animals of a biologically useful direct effect of light on a biochrome.

Carotenoid-transport to the integument is increased also by low temperature, just as the synthesis of the synergistic endogenous chromes, melanins, ommochromes, and red pterins. Some other aspects of the control of exogenous chromes are indicated in section 1 of the chapter.

14.2.5 The Indirectness of Chromogenic Controls

Although this has been stressed in chapter 6 and elsewhere it seems to call for special emphasis here. It is not peculiar to chromogenic responses and indeed should be axiomatic for all biological controls since the direct effect of virtually every environmental factor acting on living organisms necessarily is deleterious to their steady-state dynamics. Life is precisely the ability to circumvent by pathways of indirect intermediation the disruptive direct effects of these factors. The axiom perhaps is questioned particularly in relation to integumental chromogenesis because logically light and colour should be closely interrelated. Particularly in the primary colour-change responses the relationship can appear so manifestly direct.

In fact, however, the sun-tan response of the human skin is as indirect as any secondary response (p. 142), although it corresponds spatially with the prior erythraemic response which *is* a direct photodynamic effect, of short-wave radiation (p. 247). The indirect response of melanin-synthesis is induced maximally by the short-wave visible rays not only in Man but also in the integuments of tadpoles (VERNE, 1926, p. 332) and pierid and vanessid caterpillars (BRECHER, 1938). By contrast melanogenesis for the ink-sac store of the cuttle-fish *Sepia* is not sensitive to induction or acceleration by light and this has biological rationale.

The orange integumental pigments of salamanders are synthesised most abundantly in response to red light and the yellow of the frog *Rana* in response to the yellow range (MILLOT, 1923; VERNE, 1926). These are the biologically relevant incident and reflected colours but they are the hues least absorbed by the relevant chromes if already present in the skin. The latter therefore cannot be the actual perceptors for the response and in any case may be virtually absent if the previous light-conditions were very different.

The response of integumental chromes to temperature appears to be equally indirect: the direct effect should be a rate of chromogenesis directly proportional to temperature but in poikilotherms and in phenocopies of the Himalayan rabbit dark pigments show the converse response. In the Himalayan rabbit itself the pattern is purely genetically controlled and is a hint that the phenocopy depends on external factors triggering an innate mechanism rather than affecting chromogenesis directly. In poikilotherms there is most melanin at low temperatures because the rate of synthesis is slowed down less than the rate of destruction (FOX and VEVERS, 1960, p. 42) and this is therefore a particularly simple example of the way circumvention can work.

In the height of the mechanistic era of biological thought MELL (1931) suggested that the butterfly *Hestina* has pale wings in the dry season because melanogenesis is directly inhibited by desiccation but the likelihood of any appreciable seasonal difference in humidity inside the very impervious pupacase of these always rather desiccated animals seems small. On the other hand insects have hygroperceptors sensitive to a 5% change in relative humidity (PROSSER and BROWN, 1961, p. 39) or less (WILLIAMS, 1971) so that an indirect response to relative humidity is feasible.

14.2.6 The Chain of Intermediation in Chromogenic Controls

It is known (p. 142) that secondary chromogenic responses of integumental chromes are evoked ultimately by motor nerve transmitter agents or by hormones, both acting at the cell-membrane of the chromatophore. In internally triggered responses such as haemopoiesis (p. 309) again hormones constitute the effector limb of the response. In primary integumental responses it seems likely that an external signal causes the local production of a specific hormone of the same functional type. The best known hormone of a secondary response, the MSH factor of the vertebrate pituitary (p. 143) acts via the universal hormonal intermediator, cyclic AMP (BITENSKY and DEMOPOULOS, 1970) and it seems possible that this will prove the common link in all chromogenic responses, including primary integumental mechanisms. MSH ultimately causes an increase in tyrosinase activity in melanocytes (KIKUYAMA et al., 1969) such as occurs in the sun-tan primary response in Man.

It does this by inducing the production of the *m*-RNA for tyrosinase-synthesis (p. 195) and nucleoprotein from a normally pigmented mouse will initiate melanin-synthesis in an albino (OTTOLENGHI-NIGHTINGALE, 1969). Supposing that cyclic AMP derepresses the operator gene for the tyrosinase locus then a virtually complete causal chain may be constructed for a vertebrate secondary response and perhaps also for the analogous primary response. The scheme would be:

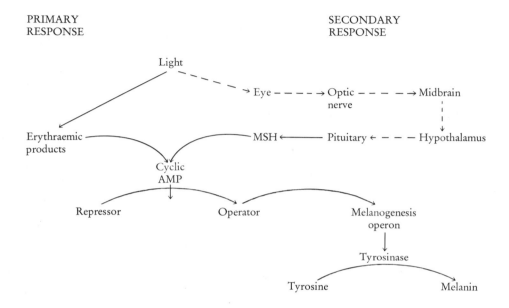

In all chromogenic responses the chain of intermediation must be somewhat of this length, though varying in its details.

14.2.7 Conclusion

At the level of the chromogenic reactions efficient controls by negative feedback operate as in other biochemical pathways. Little is yet known of other controls at this level. At the higher levels external signals trigger either a primary response by integumental chromatophores or secondary responses via the neuro-hormonal system. The best known example of control by an internal signal, that of haemoglobin synthesis by oxygen-pressure, operates entirely through blood-borne chemical factors. Phase-coloration in locusts is a special response of integumental chromes to a variety of mainly biotic signals. The same controls as those which promote the synthesis or degradation of an endogenous chrome at a particular site also promote the uptake of exogenous chromes and their transport to and from the site. No controls, even of primary responses, operate simply as direct physical or chemical effects but always as indirect biological responses, with a relatively long chain of intermediation. Direct effects of the relevant external factors are usually both deleterious and uncontrollable and adaptive responses must circumvent this. Many long-standing puzzles are due to misconceptions on these points.

Chapter 15

The Development of Integumental Colour-Patterns

"We have had so much to do with the study of Form that pattern has been wellnigh left out of the account although it is part of the same story."
D'Arcy W. Thompson, p. 1090 (1942)

This is a further important aspect of chromogenesis. Although embryologists have largely neglected the more basic aspects of chromogenesis in the early embryo (p. 274) some have taken a lively interest in the eventual development of colour-patterns since these constitute good morphological features. Further these patterns are among the more convenient and useful subjects for genetical studies (chapter 16) and for the study of certain aspects of chromatic evolution.

It should be emphasised that animals with characteristic developmental patterns often do not have the ability to change them subsequently except in a metamorphosis or in cettain other post-embryonic changes. These static developmental patterns, stand in contrast to the labile patterns used by other animals for chromogenic and for the still more rapid chromomotor colour-changes (chapter 6).

Colour-patterns usually develop before the young animal meets any of the relevant external factors so that they are ontogenetically predetermined, though phylogenetically determined only through the usual after-sight of natural selection. Their sensitivity, in some animals, to dynamic modulation in the light of the experience of the individual is just one example of the genetic basis also of physiological mechanisms. Metamorphic colour-changes likewise are genetically determined and it is noteworthy that there is a metamorphic change in colour-scheme in many animals which have no major morphological change, for instance between juvenile and adult in birds and mammals. The colour-change does not depend on the experience of the juvenile; as in the more dramatic metamorphoses (Wigglesworth, 1959) the genome has two separate codes which are translated at different stages of ontogenesis.

Seasonal colour-patterns also are genetically based though these again are adapted to be modulated by relevant environmental variables such as light-period and temperature (p. 137). If these and the background-responses are entirely genetically determined and merely triggered by environmental signals then there can be no question of any direct vector relationship (p. 155) between the particular environmental colour-scheme and that of the animal. In the present context the more important corollary is that the types of pattern-development described below in principle are relevant also to these changes in pattern.

It is also important here to reconsider the scale of colour-pattern in animals (p. 132): what may appear at the macro-level to be a dull, monotone coloration may be a complex pattern of brilliant colours at the micro-level. Moreover this could undergo striking changes during what at the higher level might appear a mere unpatterned change of hue or indeed no gross change whatever. In fact, however, the basic physical and structural principles governing photoperceptor organs dictate that acuity cannot be much better than ours, within our focal range. For this, pattern-discriminating eyes and the animals which possess them all are relatively large and the types of pattern considered below have general validity. While in principle micro-patterns might be resolved by eyes of very short focal length, diffraction patterns and other aberrations set a very severe limit to this and no effective examples are known.

The morphogenesis of colour-pattern has been studied in a number of instances though still in very few taxa. Interest has centred on some of the more striking patterns which sometimes prove to have a rather complex developmental basis. An intensive study of some of the simpler systems should be very rewarding. Most patterns have as foundation a simple dorso-ventral countershading gradient. Vertebrates have a very appropriate provision for this since all chromatocytes originate from the dorsally situated neural crest; they therefore soon form a dorsal cape and only the hypsochromic types in quantity make the long journey to the ventral side. These in consequence differentiate later than the bathochromic group (BAGNARA, 1966). A few of the latter do reach the ventral side, of course, or the reversal of countershading (p. 137) could not be induced. Most other taxa have a similar countershading-pattern, although invertebrates develop from a ventral and not a dorsal blastoderm so that most of their chromatocytes must make a long journey. Sometimes vertebrate chromatocytes show also an anteroposterior pattern, corresponding to the distribution of the segmental nerves, probably because of the common origin of the two tissues.

One of the simplest patterns superimposed on the foundation, at least from the present aspect, is that of the Himalayan rabbit and Siamese cat (p. 138), a melanisation of the tips of all extremities. It can be phenocopied in a number of other breeds and species simply by keeping the individual at a low temperature during the period of development of the pelage; melanin then develops wherever local temperature is below a critical value (p. 137). In the Himalayan rabbit and Siamese cat however this pattern has become genetically determined and appears whatever the ambient temperature during development; has a pattern originally environmentally determined here become genetically fixed or do environmental factors induce the phenocopy not directly but simply by triggering a latent (? recessive) genetic trait?

If the first alternative is correct then it is an example of the biological exploitation of a situation which originally arose fortuitously. Another example is the colour-pattern of isopod Crustacea, depending on the exclusion of integumental pigment from areas where muscles are inserted on to the exoskeleton (VANDEL, 1938; NEEDHAM, 1942); the consequent broken pattern, as seen under the microscope (Fig. 15.1) resembles larger-scale camouflage-patterns and so may be an actual instance of a pattern which is more cryptic in the eyes of smaller animals than ourselves. Curiously enough ALLEN (1888) found that in some mammals, also, the underlying muscles determine features of the integumental colour-pattern. Here they do not insert into the skin and the pattern probably is another manifestation of the metameric segmentation of vertebrates.

A pattern to which a number of students have attributed a simpler developmental basis than the pattern itself seems to imply is the striping of the tiger and zebra. HEDGES (1932) thought it might be determined by the type of phenomenon responsible for Liesegang's Rings (NEEDHAM, 1964, pp. 59–61) and TURING (1952) developed a mathematical theory of the production of such periodicity through spontaneous metabolic activities. No analytical evidence is yet available and there is little to add to D'ARCY THOMPSON's entertaining hints of the possible mechanism (THOMPSON, 1942, p. 1090). In this case there appears to be no relation to metameric segmentation. Most theories to explain this and other periodic structures are based on the idea of the interaction of two or more temporally oscillating factors of different phases and perhaps also different periodicities.

A valuable clue to the proximate mechanism of banded patterns is provided by the experiments of HADDOW et al. (1945) already mentioned (p. 244); the isoalloxazine derivative fed to albino rats is deposited in those parts of the pelage actively growing at the time.

The collective results on individuals fed at a sequence of times shows the occurrence of a spatio-temporal cyclic pattern of hair-growth which in normally pigmented strains could give a striped pattern in normal ontogenesis if chromogenesis also had appropriate periodicity. Actually the small rodents normally do not produce gross colour-patterns of this kind (SEARLE, 1968) but the mechanism may apply also to larger mammals which usefully could be subjected to similar experiments. One normal feature of hair-coloration which was mimicked in these experiments is the agouti banding along the individual hairs (p. 326).

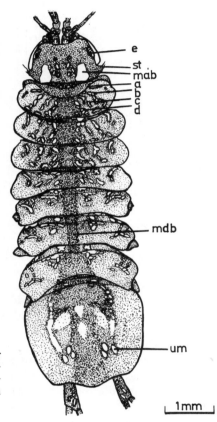

Fig. 15.1. Sketch of dorsal surface of body of freshwater isopod, Asellus aquaticus, to show pigment-free patches, e.g. a, b, c, d, mab, um, due to insertions of muscles through epidermis on to exoskeleton. (From NEEDHAM, 1942)

In the Plymouth Rock breed of fowl there is a gross pattern of stripes which has a very sophisticated basis (COHEN, 1966), because feathers are relatively large anatomical structures, grow to different lengths and at different rates each from a localised growing point and come to lie tangentially to the surface. In spite of all these complications the stripes of neighbouring feathers often are in register and a gross pattern covers the whole plumage as though it had been painted on; it would seem to be an excellent example of the natural selection of simplicity out of complexity. Even within the individual feather pattern-development shows this same phenomenon. A regular series of transverse bands of pigment, as in the Plymouth Rock, is rather common and these represent 'isochrones' within that feather-germ, i.e. regions of simultaneous development. However they cross the barbs of the feather at an angle ('ESPINASSE, 1939; NEEDHAM, 1964, p. 49) because of the curious intrinsic basis for the shredding of the initially 'entire' cylinder, to produce the rachis and barbs. All the barbs are formed simultaneously whereas pigment-deposition is progressive and periodic, each band representing from one to twenty days of activity (COHEN, 1966). Spotted patterns are similarly produced as transverse rows of spots, a spatial periodicity in chromogenesis being added to the temporal discontinuity (Fig. 15.2, A).

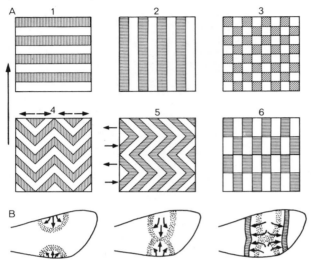

Fig. 15.2. A Diagrams showing mode of production of six common simple colour-patterns in animals produced by a linear generative zone starting along the lower edge of each area and proceeding in the direction of the large arrow. Details in text. B Generation of a class of colour-pattern in the wings of butterflies: three successive stages each of which is the terminal stage in certain species. (Based on HENKE, 1947, 1948)

The pattern is actually simplified, and not complicated as might have been anticipated, by the further phenomenon of several independent gradients of developmental activity along the transverse axis of the feather. The general developmental rate increases from the rachis to each edge but there is a gradient of decreasing sensitivity to relevant hormones in the same direction as well as one of decreasing latent period of response (LILLIE and JUHN, 1932; JUHN and FRAPS, 1936; WADDINGTON, 1939, p. 198). The resultant of all three gradients is that pigment-deposition is simultaneous across the whole vane. Experimental interference, as by augmentation of a particular hormone, causes asynchrony and so more complex patterns. Early transplantation of feather-buds shows that the pattern becomes determined early, before pigmentation itself begins so that local differences in pattern between feathers must be based on already developed spatial differential sensitivities to hormones and other chromogenic factors.

The development of the wing-patterns of butterflies has been studied by GOLDSCHMIDT (1927), KUHN (1932) and others and reviewed by WADDINGTON (1939, p. 192ff.; 1956, p. 418ff.). There is strong evidence that the variety and complexity of these patterns depends on a common and basically simple process with which other factors interact. General and chromatic morphogenetic processes all tend to start from a few consistent points, in particular the middle of the leading and trailing edges of the wing (Fig. 15.2, B). In *Ephestia* the final pattern is roughly symmetrical about the line joining these two points (KÜHN, 1932, 1955), the process spreading along this line from either end until confluence occurs and then longitudinally towards each end of the wing. SCHWARTZ (1953) recognises three minor patterns which are superimposed on this and modify it to some extent. The simplest condition in a mature wing is a pigmented patch around each point of origin and the next a complete bar across the wing. In more advanced species further transverse bars are added, roughly symmetrically on proximal and distal sides of this; a periodic discontinuity thus is again evident. The edges of the transverse bars vary between simple and very serrate, depending on rate-differences which have a periodicity along the transverse axis of the wing also. As in the bird's feather the production of the series of transverse bars is due to discontinuous pigment-deposition interacting with a basic growth-process which also is not simultaneous throughout. Chromogenesis is maximal where cell-proliferation is most actively producing cells ready for differentiation and so most responsive to relevant chromogenic stimuli. Spatiotemporal waves of cell-proliferation have been demonstrated in the wing of *Drosophila* (WADDINGTON, 1950). These patterns, developing gradually with interaction between component processes, have been termed 'diachronic condition-generated forms' (WADDINGTON, 1962) and they may be peculiar to colour-patterns. A further clue to the mechanism is that a wound in the path of flow of chromogenesis will be by-passed if made early enough but if it is healing at the time of contact the chromogenic wave is halted permanently. Having 'missed the tide' through a competing act of morphogenesis there is no second flow.

The colour-pattern of gastropod shells is even more diachronic, continuing from early juvenile stage to maturity or later. It has considerable resemblance to that of feathers, and likewise develops on the perimeter of the base of an exoskeletal cone, which in this case usually is spirally coiled. Pigment, like the other materials of the two outer layers of the shell, is secreted by specialised tracts of cells in the mantle edge. In some species a background chrome is secreted by an unbroken tract which also is continuously active (COMFORT, 1951). Discontinuous activity, giving bands parallel to the mantle edge (Fig. 15.2, A) are less common than transverse bands in feathers whereas longitudinal stripes, due to the continuous activity of spatially discontinuous chromogenic centres are rather common. The activity may be temporally periodic also, giving the usual regular array of dots or blotches. In the freshwater prosobranch, *Theodoxus fluviatilis*, there may be either longitudinal stripes, or transverse bands, or blotches (Fig. 15.3) and an individual may start producing bands and then change to blotches or *vice versa*, depending on pH, salinity, temperature, etc. (NEUMANN, 1959). The genetic basis is more flexible than is implied in the other examples so far considered, and would repay further analysis.

In some gastropods chromogenic activity moves from its initial series of foci along the mantle edge progressively to other cells, in one direction along the edge, so that oblique pigment stripes are formed; presumably a new centre periodically starts at the departure end of the mantle edge. An alternative mode is for the activity to oscillate to and fro along the edge so that longitudinal zig-zag stripes result (Fig. 15.2, A). A third is for the activity to proceed both ways from the initial points until each meets its counterpart from the neighbouring centres when the activities mutually extinguish; a new cycle then starts at the original points along the edge and the resulting pattern is a series of zig-zag bands parallel to the lip of the shell (Fig. 15.2, A). It is interesting that in this one biological class there could be selection for the same essential pattern-motif by two or more quite distinct and sophisticated methods. Other patterns also are found and it will be interesting to know if they involve still further methods or merely variants of those already described.

As in birds, mammals and insects chromogenesis is not rigidly tied to other components of morphogenesis though normally coordinated with these. Albino shells and those with only the background colour show slight morphological specialisations where pigment normally would be deposited and these may represent a process normally correlated with chromogenesis. Alternatively they may be due to the chromogenic cells going through the motions of deposition under conditions of inhibition or deprivation.

Conclusion: The genetic basis of all colour-patterns is to be emphasised: even where chromomotor and chromogenic modulation is extensive and persists throughout life the whole physiological mechanism has a genetic basis. A pattern of the static type may be replaced once or more often during the life-history, each replacement having its genetic basis. Some of these also are subject to chromogenic modulation by environmental factors, but not reversibly and indefinitely as in the typical dynamic patterns. There are probably few significant patterns not visible to the human eye.

Fig. 15.3. Shells of the freshwater prosobranch, Theoduxus fluviatilis, to show the variety among individuals in the colour-pattern. Patterns 1, 2, 6 of Fig. 15.2, A are evident and some individuals switch from one to another during the course of their growth. (From WADDINGTON, 1962; after NEUMANN, 1959)

The developmental basis of a number of patterns in several different taxa has been analysed. Most have been geometrically regular, i.e. periodic, patterns and therefore the developmental basis shows some similarities between taxa. It differs in details, however, in some taxa being as simple as the pattern produced, in others much more complex. Within particular taxa the mechanism has 'radiated' to produce a variety of final patterns which scarcely could have been interrelated without this knowledge of the processes of development. In one species at least the pattern produced can have three distinctive variants and an individual may oscillate between these during its growth. The precise mode of control of the spatial and/or temporal discontinuities in chromogenesis which give these patterns is still obscure but as yet none are manifestly the consequence of an interaction between two simple oscillatory processes such as those which produce Liesegang periodicities. The pattern is often superimposed on a self-coloured foundation or on a countershading pattern.

Chromogenesis interacts with cell-division activities and other general morphogenic processes but there is the usual flexibility in the coupling. Some colour-patterns actually appear to be the fortuitous result of general morphogenesis and one at least of these has become genetically fixed in some races. Analysis usefully could be extended to other taxa and systems, to simpler patterns and to more complex patterns.

Chapter 16

The Genetic Basis of Chromogenesis

The genetic basis of some aspects of chromogenesis has been noted already. In this chapter the aim is to survey an adequate representative sample of genetic mechanisms for their evidence on biochromatic evolution and significance. At present knowledge is restricted mainly to those taxa most suitable for genetical work and even here the genetic basis has been related to the actual process of chromogenesis in relatively few instances. In this respect the now classical works of BEADLE and EPHRUSSI on the eye-chromes of insects (p. 330) and of SCOTT-MONCRIEFF on anthocyanins (p. 329) are still outstanding.

The analysis of the genetic control of chromogenesis in any particular case is greatly facilited once stocks of breeding individuals bearing more than one allele at the relevant locus are available. If only one colour-morph, the 'wild type' is known then the action of its gene is not easily defined; additional allelomorphs at that locus constitute natural experiments in colour-polymorphism which analytically have much the same value as pathological states (p. 247). Fortunately natural spontaneous mutations are not excessively rare; still more fortunately colour mutants are among the most easily recognised phenotypically and have been invaluable to geneticists. Reciprocally chromatology owes a great debt to the geneticist and particularly to those interested also in embryogenesis (WADDINGTON, 1939, 1956, 1962; HADORN, 1961).

Naturally the integumental group of chromes have been studied most and again the blood-chromes, mainly haemoglobin (Hb), rank second. As many as 150 mutations affecting Hb are known in Man alone (PERUTZ, 1971). There is also some genetic information on the metabolic group of chromes, notwithstanding the fact that these are too indispensable for many of their mutants to be viable in the wild. As usual, therefore, knowledge is most abundant for chromes the biological significance of which is already well established on other grounds. There is an urgent need for genetic research directed to chromes at present of uncertain significance, particularly among those of the deeper organs of the body. There is already a little evidence here which shows that the project is feasible: in the rabbit white fat is dominant to yellow, as shown by a mutant which deposits orange-yellow xanthophylls (PEASE, 1928). The mutant lacks an oxidase system which normally destroys dietary xanthophylls absorbed from the gut (p. 302).

This example is also a reminder that there is good genetic control of the handling of exogenous chromes and that as in the preceding chapters no sharp distinction need be made between exogenous and endogenous groups. Other examples involving the former have been studied in silkworm silk; some strains of *Bombyx mori* secrete flavone in their silk while others do not; some have carotenes and all strains have xanthophylls (MANUNTA, 1933, 1935, 1937c, d). Gene C controls the uptake of carotenoids from the gut and gene Y its transport from the blood to the silk-gland (UDA, 1919), perhaps depending on the permeability of the gland. VERNE (1926, p. 528) concluded that heredity works mainly with endogenous chromes and he was probably correct to the extent that more genes are required for them, collectively and individually, than for the exogenous group. Nevertheless he was aware of the work of GEROULD (1921) showing that the ability of caterpillars to use exogenous chromes is hereditary: the blue mutant of *Colias* lacks the ability to assimilate carotenoids.

A number of genes controlling the individual chromogenic pathways (p. 281) also are already known. CHICHESTER and NAKAYAMA (1967) cite a large number controlling steps in the carotenoid pathway of plants and BIRCH (1960) in the flavonoid path. Some controlling endogenous zoochromes are considered below, though in many cases the exact site of action is not yet known.

16.1 Genetic Colour-Polymorphism

Considerable information has come from natural instances of genetic types of polymorphism and potentially much more is available in such animals as the sea-anemone, *Metridium* (Fox and PANTIN, 1941, 1944) and the juvenile stages of the crab *Carcinus*. Some isopods show a modest degree of colour- polymorphism (HOWARD, 1952; 1953; BOCQUET, 1953): in *Armadillidium* there are two different though allelomorphic genes causing a red body-colour in place of the normal slaty hue. Phenotypically the two are indistinguishable but the one is dominant and the other recessive to the wild type (HOWARD, 1952). The term polymorph is restricted to alternative phenotypes not maintained solely by recurrent mutation, so that all are 'wild type'. FORD (1953, 1971) distinguishes transient polymorphism, as in the early stages of the response of *Biston betularia* to the industrial environment (p. 133) from balanced polymorphism which in a constant environment persists because all morphs have selective advantages.

The proportions of the various morphs are maintained by the simplest feasible mechanism, the mutant(s) being dominant to the wild type but lethal in homozygous state. Consequently only the two genotypes *Mm* and *mm* and the corresponding distinct phenotypes survive to reproduce in each generation. True genetic polymorphism concerns only individuals of an interbreeding population and it is only in this situation that the genetic basis can be checked. Where geographical variants can be crossed experimentally they also provide useful information. The seasonal type of polymorphism, as in the butterflies *Colias* (p. 160) and *Hestina* (p. 313), modulations of a single genotype, are not relevant here.

In the snail, *Cepaea nemoralis*, three alleles at the same locus determine most of the familiar colour-morphs (SHEPPARD, 1958, p. 142). *Brown* is dominant to *pink* and this to *yellow*. The Swallow-tail butterfly, *Papilio polytes*, has three colour morphs but the form *polytes* differs from *cyrus* at one locus and *romulus* from *polytes* at a different locus. *Romulus* is dominant to *polytes* and *polytes* to *cyrus*. The two dominants each mimic a different species of butterfly.

Occasionally genetic polymorphism arises in a single individual as a result of a somatic mutation. A particularly striking example, described by ATZ (1939) in the lobster (Fox, 1953, p. 131) involved a sharp sigittal demarcation between green and orange carotenoproteins. The eyes by contrast had the same dark ommochrome on both sides so that this organ or this chrome is controlled by a locus not affected by the mutation. Such Pied Piper colour-mosaics have been found more frequently in insects (GOLDSCHMIDT, 1927, 1938), probably because of their absolute abundance, and occasionally in birds (DURKEN, 1932, p. 233.). Bond's pheasant, a sagittal dimorph like the lobster, had the further complication that the two colour-patterns were male and female respectively and there was the appropriate gonad on the two sides. In birds at least hormonal controls might have been expected to erase the effects on local feather-colour of a genetic difference in this respect between the two sides, but in fact this must have survived at least the juvenile to adult colour change. The dimorphism extended even to the two sides of the individual feathers of the tail so that there is no doubt that control depends on local factors. It is difficult to see how any simple somatic mutation could produce the condition in the tail feathers and some kind of transverse morphogenetic gradient may be implied. This is independent of the structural transverse gradient in these feathers which showed bilateral symmetry about the bird's sagittal plane.

16.2 General Properties of Chromogenic Mutants

Most of the mutations from wild-type chromogenic genes are deficiency-changes. In most phyla the extreme mutation can cause complete albinism but there are also various intermediate grades. In mammals mutations at the *extension* coatcolour locus reduce only the amount of eumelanin but those at other loci affect both eu- and phaeo-melanins, usually differentially. Those at the *pink-eyed* locus affect eumelanin the more strongly whereas the converse is true for mutations at the main *albino* locus.

Also like mutations of other types of gene, most colour-mutants are recessive to the wild type since the latter has become dominant through natural selection (HALDANE, 1927; FISHER, 1931). Mutations either way, to deficiency or excess relative to the wild type, tend to be recessive to it. In fact chromogenic mutations to excess-production are rare, indicating that selection has favoured high production already as the wild type. One of the few known 'excess' mutants is *extension of pigment* in birds (HADORN, 1961, p. 292) but even this is due to deficient cell-proliferation, i.e. excess per cell rather than to an absolute excess.

Some dominant chromogenic mutations are known and in fact there are more than predicted by the Haldane-Fisher theory of the evolution of wild type dominance (SEARLE, 1968, p. 250). In part this may be due to polymorphism (p. 324). Of 32 dominant mutations known in the mouse as many as nine are chromogenic (HADORN, 1961, p. 132). This is a very high percentage of the 40 coat-colour genes known in the mouse and implies that dominant mutations are rather characteristic of colour genes. The *dilution* gene of the palomino horse is dominant and there is a dominant white in many mammals so that dominance/recessivity is not necessarily positively correlated with excess/deficiency. The dominant white is not so deficient in pigment as the true albino and in particular has the normal eye-pigmentation, so that as in Atz' lobster the pigmentation of this organ can be controlled independently of that of the integument. It would seem that the retention of essential pigment by the dominant white has been naturally selected and that presumably its whole phenotype has been selected. Arctic white races and species no doubt have this type of allele. True recessive albinos lack tyrosinase (ONSLOW, 1917; FOSTER, 1952; BARNICOTT, 1957; FITZPATRICK et al., 1958) or have the enzyme but no free tyrosine (VERNE, 1926, p. 535; FOX and VEVERS, 1960, p. 36) but in the dominant white the enzyme is present in the skin, under inhibition.

Complete albinos not only suffer the inconvenience and disability of retinal exposure to light but are more susceptible to diseases and in various ways are less healthy than well-pigmented individuals. In general white flowers also are due to recessive alleles and are less healthy than the coloured strains. A high percentage of colour-mutants show pathological complications (HADORN, 1961; SEARLE, 1968) and the general susceptibility to diseases is one aspect of this. Again it is not peculiar to chromogenic mutants but it re-emphasises that chromes are at least as important as any other group of biochemical materials. This is consistent with knowledge of the intrinsic properties of chromes in general (chapter

2) and with the fact that virtually all vitamin-deficiencies result in pigment-deficiencies and increased susceptibility to disease (p. 250).

Anaemia is commonly correlated with mutations at the loci of integumental chromes: this might be taken to indicate a specific relationship between the two types of chrome but it could be just one more manifestation of the general debility of mutants. An essential implication is that most genes are highly pleiotropic so that deficiency can affect the whole genome. Certainly there is no specific relationship between integumental effect and anaemia in the *dominant spotting* mutant of the mouse since the ratio of the two effects differs between heterozygote and homozygote (HADORN, 1961, p. 49). It may be relevant that anaemia is significantly more common in human blondes than in brunettes; the general correlation therefore must be very strong not to have been completely selected out of the blond genome, which evidently has had selective value in the appropriate latitudes.

Pleiotropy is very evident also in the albinos of insects (BELLAMY, 1958) and Crustacea, which have so many different chromes in the integument, – ommochromes, melanins, carotenoids, pterins and others. The *albino* shows that its mutant-locus controls the pathways of all these chromes, possibly in the way indicated below (p. 329) for the *white* mutant of *Drosophila*; in any case there is a 'master control'. The *lemon* gene of *Drosophila* controls the paths of the pterins, purines and melanins, and probably also of the ommochromes since it affects tryptophan metabolism (HADORN, 1961, p. 268). In Man there is a positive correlation between colour-blindness and *xeroderma pigmentosum*, i.e. between an abnormality possibly in the retinal chromes and one affecting the integumental chromes. However this is merely an instance of linkage between two separate gene-loci. Both are on that part of the X-chromosome which is absent from the partner Y-chromosome in the male so that when the mutant allele of both loci is on the X-chromosome of the male both abnormalities show in the phenotype.

16.3 The Alleles at Individual Loci

There are a large number of alleles at some of the loci of the mammalian hair-chromes, as many as 13 at the *agouti* locus of myomorphs (SEARLE, 1968). At the *albino* locus of mice there are six and these can be arranged in the typical quantitative series (p. 325) controlling a progressive depression of the two melanins in order. The chinchilla stage of the series has no phaeomelanin but still the full complement of eumelanin; the rest of the series depress eumelanogenesis in the same graded way. In some mammals, but not in myomorphs, similar quantitatively graded series are known for the *extension* locus: they depress only eumelanogenesis (p. 325). There are seven graded alleles at the *pink eyed* locus of myomorphs, four at *brown* and five at *dilute*. The last two constitute a proof (p. 328) that mutants at vector-type loci and others not directly concerned in the biosynthetic reactions nevertheless are essentially quantitative in their phenotypic effect.

The *agouti* locus in myomorphs is not only outstanding for its large number of alleles but also unique in the precisely converse effects of two sub-groups on the widths of the bands of the two kinds of melanin which in the wild type alternate regularly along the individual hairs. The allele at one end of the series produces self-coloured black and that at the other end self-blond so that the wild type *agouti* is in the middle of the series and not at one end, as is more usual. Chromogenically the locus has a number of other interesting features (CLEFFMAN, 1963). *In vitro* studies on material from individuals of this series indicate that the same line of chromatocytes produces both types of melanin, in conformity with the view that both types are on the same pathway. Reducing agents such as – SH compounds will induce the cells to produce phaeo-, instead of eu-, melanin and this is interesting in view of the presence of a cysteinyl-DOPA complex (Fig. 3.7, VII) both in phaeo- (PROTA, 1970) and in erythro-melanin (FLESCH, 1970). When transferred to simple *in vitro* conditions in the absence of reducing agents all the cells previously producing phaeo- start forming eu-melanin so that the latter is the standard product. As in flower-colour (p. 329) the alleles for the more reduced chrome are recessive to those for the more oxidised. This is relevant to the fact (p. 296) that chromogenic pathways are usually oxidative.

16.4 Alleles at Chromoprotein Loci

Since a single gene codes for a complete peptide a chromoprotein needs little more genetic information than a free chrome; the surprise is that so much more information is required to synthesise the relatively small molecule of the chrome than the large peptide molecule. It is a general biochemical phenomenon of some importance. Even so there is considerable scope for mutation at each of the mutons (codons) in the protein-coding gene and since many of the substitutions of a single amino acid do not seriously impair the function of the peptide many mutants of the globin locus of mammals have been viable enough to permit analysis. This is the main source of the 150 Hb-mutants (p. 323).

The classical example, *sickle-cell* in Man, involves merely the substitution of valine for glutamic acid in one (β) of the two types of peptide in globin. In this case the effect on the function of the Hb is profound, even in the heterozygote which has only 50% of this HbS; it crystallizes in the deoxy state and so loads less readily with O_2; the solid state also constrains the whole cell into a characteristic saucer-shape (a sickle in profile). It is not lethal in the heterozygous state and since it confers more resistance to malaria than normal Hb (HbA) a balanced polymorphism between the two genotypes has evolved in malarious countries.

HbC has lysine in place of glutamic acid at the same locus but it also has another substituent amino acid on the same chain (MACKUSICK, 1969). Altogether 44 mutations are known in the β-chain; they are much rarer in the conservative α-chain, Hb-Hopkins[2] being one of these. The difference between β and α in this respect is reflected in the fact that all human Hbs have two α-chains whereas the two β-peptides represent one of a series of variants, including γ, δ and ε. The early foetal Hb has two α and two ε peptides and is designated $α^A ε^F$. This is replaced quite early in development by HbF ($α^A γ^F$) and soon after birth by the adult HbA ($α^A β^A$). The δ-peptide is found only in a second form of postnatal Hb, HbA_2 ($α^A δ^A$). Comparison of amino acid sequences indicates that the various peptides have evolved serially from α (INGRAM, 1962) in the order α → γ → β → δ. The fact that γ is now prenatal and β postnatal raises intriguing problems. Unfortunately it has not been possible as yet to obtain enough ε-peptide to work out its amino acid sequence. From comparative studies on the number of amino acid differences between the globins of vertebrate groups of known evolutionary dichotomies it appears that successful amino acid substitutions occur at a relatively steady rate of one every 1.4×10^7 years and this indicates that γ, β, and δ arose respectively about 600, 260 and 44 million years ago (ZUCKERKANDL and PAULING, 1962; FLORKIN, 1966, p. 21) in the late Precambrian, Permian-Triassic and Eocene-Oligocene therefore. The evolution of these peptides represents a large number of mutations within the locus of that one original globin peptide of 146 amino acid residues.

A single amino acid replacement at the locus controlling the protein which is conjugated with the chrome of the pupa-case of Diptera and Lepidoptera produces a black in place of the usual brown case. It has become the normal wild type is some species. It is due merely to the absence of β-alanine from the protein (FUKUSHI and SEKI, 1965). The example is interesting because of this unusual amino acid in a protein and because it shows that the protein moiety can affect the colour even of a melanin.

16.5 The Number and Types of Chromogenic Loci

In all the higher Metazoa a large number of loci are concerned in the control of zoochromes. The range of eye-colour in Man shows that it must be controlled by a number of factors and more than one locus is involved (ROSTAND and TÉTRY, 1965). Blue eyes lack a brown chrome in the anterior or stromal layers of the iris and behave as recessive to all other genotypes. However an eye homozygously recessive for only the main eye-colour gene is hazel in colour and the double recessive state at other loci also is necessary for a really blue iris. Of course the blueness is a structural effect (p. 4) and not a true chrome: it depends on reflection from the dense black melanin of the most posterior of the retinal layers of the iris, with Tyndall scattering of the longer waves by the stromal proteins.

Hair-colour in Man also is multigenically controlled (PENROSE, 1959, p. 121) and there may be as many as ten loci concerned with human skin-colour (ib., p. 47). There are known to be at least 40 controlling coat-colour in the mouse (p. 325), although eu- and phaeo-melanins are the only chromes involved. In *Drosophila* as many as 30 genes controlling eye-colour were known already 30 years ago (WADDINGTON, 1939, p. 159). Here there are two main classes of chrome, ommochromes and pterins; in the bee 14 genes, about half of the total, are concerned with the ommochrome path (DUSTMANN, 1968). The chromogenic genes of the mouse are scattered throughout the genome, showing to what extent they must have relationships with non-chromogenic loci.

VERNE (1926, p. 531) made a useful classification of the eleven gene-loci for mammalian hair-colour then known; he recognised five functional types, a general chromogenic gene the homozygous recessive of which causes complete albinism, three genes each specific to one of the three forms of melanin, two intensity genes determining the amount of chrome formed, two controlling topographical distribution or pattern and the remaining three regulating the uniformity of the most common pattern. Further discoveries have added many new members of these groups (CASTLE, 1940, 1954; WRIGHT, 1941; LITTLE, 1957; SEARLE 1968) as well as some new types such as the agouti (p. 326) controlling pattern along the individual hairs. VERNE's five types may be further grouped into two main categories; the first three affect simply the amount of chrome and the last two pattern i.e. local variations in the activity of the first group. In a sense they are scalar and vector groups respectively. There are likely to be fewer of the second type in the control of insect eye-pigments so that most of the 14 ommochrome genes of the bee may be concerned with the chromogenic pathway itself (p. 289). However there are much fewer than 14 enzymic steps in the pathway so that some must perform other auxillary functions.

For mammalian coat-colour the mode of action of some of these auxillary genes is known. Some control the supply of substrate so that mutants of these loci cause one type of albinism; the relevant enzymes are quite normal and the lesion is in the transport of substrate into the cell (Fox and VEVERS, 1960, p. 35). The *brown* locus in mammals controls the binding of formed melanin to the matrix-protein of the premelanosomes (SEARLE, 1968).

The *white* mutant of *Drosophila* lacks an enzyme which normally binds both ommochrome and pterin to their matrix-protein (ZIEGLER, 1961) so that the locus controls the pathways of both classes at this final step of chromogenesis, in diametric contrast to the typical feedback type of control (p. 307). The synthesis of this matrix protein is controlled by several other loci (MOYER, 1963), presumably not all concerned with the amino acid sequence itself. The *dilute* locus of mammals determines the diameter of the branches of the chromatocyte and some mutants produce branches too narrow to admit the melanosomes (RUSSELL, 1948; MARKERT and SILVER, 1956).

16.6 Interactions between Chromogenic Loci

Between the various loci controlling a particular functional type of chrome synergistic interrelationships are very sophisticated. This is particularly evident for mammalian hair-colour which involves only the one class of chrome so that for melanogenesis Verne's classification is an outline of the complete mechanism. For comparison it is interesting to consider also the interactions between the genes controlling the flavonoid class in flowers (LAWRENCE and PRICE, 1940; BIRCH, 1960). Here a number of different loci each appears to control one branch of the flavonoid pathway and to compete for a common precursor. Each gene has two or more allelomorphs and the phenotypic set of chromes actually produced in the flower depends on the possible combinations of alleles at the four loci. In the dahlia gene I by itself causes the production of apogenin, Y causes that of butein while A and B both produce anthocyanin, but at different rates. Moreover A and B produce either a pelargonidin or a cyanidin according to conditions. The dominant allele at a particular locus determines the most oxidised product in its path so that the dominance-sequence for production is delphinidin > cyanidin > pelargonidin. The steps of biosynthesis are in the reverse order so that longer pathway is dominant to shorter pathway: in this sense also (p. 325) recessive mutants are deficiency mutants. The interaction of loci to produce such mammalian patterns as that of the tortoiseshell cat is fairly completely known and is an equally fascinating system (SEARLE, 1968); the great advantage analytically of the plant flavonoid system is the large number of branch-paths producing series of distinctively coloured monomer chromes.

16.7 Chromogenes and Taxonomy

On the evolution of characters which have left little fossil record (p. 336) and show little embryological recapitulation (p. 269) comparative genetics eventually should give invaluable information. Although considerable knowledge of neighbouring and synergistic genes is necessary to prove the genetic homology of loci controlling the synthesis of one of the relatively small molecules of a chrome monomer it is a reasonable article of faith that amino acid sequences in a chromoprotein can decide this question for chromoproteins, even without direct genetic research. For this reason the comparative molecular biochemistry of cytochrome c and the vertebrate Hbs (p. 229) has been very rewarding, not only for chromatology but for biochemical evolution in general. A wider consideration of all Hbs introduces the further theme of parallel and convergent evolution (p. 238). At present genetics has nothing to add to this.

Indeed as yet there is very little taxonomic chromogenetics. This is largely restricted to integumental and eye chromes of insects and mammals. Here chromogenes and their alleles are very similar in the various orders of the class, or at least in the higher orders. This implies a common origin in the exploitation of the chromes, probably not very early in the evolution of the taxon. The six mammalian hair-colour loci with the largest number of viable alleles, namely *agouti* (A), *brown* (B), *albino* (C = colourless), *dilute* (D), *extension* (E) and *pink-eyed* (P) are common to all the higher orders of the class (SEARLE, 1968). They are absent from the Edentata. It must be recalled that hair originated in the mammalia so that hair-colour genes are relatively neoform. It so happens that the higher orders of insects and the angiosperms originated at much the same general era of evolution and indeed also the birds, teleosts, coleoid cephalopods and other brightly coloured or patterned taxa.

Among insect chromogenic systems the best known instance of homology is the classically analysed ommochrome path of the compound eyes (BEADLE and EPHRUSSI, 1937; BUTENANDT, 1952; LINZEN, 1967; DUSTMANN, 1968). The *v* gene of *Drosophila*, *a* of *Ephestia* and *s* of *Apis* (KERR and IAIDLAW, 1956) all control the synthesis of tryptophan pyrrolase, the enzyme for the first step in the pathway, and the same gene has been found in other members of the three orders (GREEN, 1955). The *cn* gene for the synthesis of kynurenine hydroxylase in *Drosophila* similarly is homologous with *i* in the bee. There is less comparative knowledge about the pterin path in this eye though a number of relevant genes are known in *Drosophila*.

Homologies between more distant taxa are much less certain and in any case information is very sparse. The *sepia (se+)* gene of *Drosophila* determines the conversion of a yellow to a red pterin in the eye, and in the course of its life-history *Rana temporaria* normally converts the same yellow pterin to a red product in its skin pterinocytes. This may be a spontaneous or otherwise highly probable step and there is no doubt that pterin radiation occurred independently in arthropods and vertebrates (p. 89).

In groups other than the insects and mammals there is little genetical knowledge on any chromes. From the skins of some species of the fishes *Salmo, Squalus, Chondrostoma* and *Helmichthydes* iridocytes are sometimes absent (VERNE, 1926, p. 465), indicating that their purines are under similar genetic controls. Genetic polymorphism has been detected in the Hbs of the cod and whiting (SICK, 1961). GORDON (1957) studied in detail the genetics of certain integumental colour variants, mainly involving melanin, in suitable small rapidly breeding fishes such as the platyfish and swordtail. This confirmed a number of the general genetical features in mammals and insects. In the Crustacea albino and ommochrome mutants have been mentioned. In the amphipod *Gammarus* a rapid-darkening red eye-mutant produces normal dark ommochromes at high temperature whereas low temperature depresses synthesis and produces a red eye (FORD and HUXLEY, 1927), in contrast to the usual response in insect eyes and integument. Albino and colourless eye-mutants of *Gammarus* lack the necessary morphological differentiation of the eyes rather than the actual chromogenic mechanisms (FORD, 1929; HUXLEY and WOLSKY, 1932). The green variety of *Allolobophora chlorotica*, due to a bilatriene, is probably recessive to the pink form which owes its colour only to the blood-Hb showing through (KALMUS et al., 1955).

16.8 Conclusion

Biochromes are ideal material for genetic studies and many chromogenes are known, as usual mainly from their mutants; they concern coat-colour in mammals, haemoglobin in mammals, insect eye-pigments and to some extent other chromes. Natural genetic polymorphism is a further source of information. Although chromoprotein-synthesis demands few more genes than that of the chrome prosthetic group alone the protein introduces much additional scope for mutation. It is useful to distinguish between scalar genes controlling the actual steps of chromogenesis, vector genes which determine local patterning and auxiliary genes, but effectively all have a quantitative type of action. Virtually all known genes control the synthesis rather than the subsequent functioning of chromes.

Mutations of chromogenes resemble those of other genes in mostly causing deficiencies and in usually being recessive to the wild type. They tend to cause a general debility and susceptibility to disease, probably more pronounced than in mutants of most other types of gene. Master genes pleiotropically controlling all other genes of a functional group are evident in various taxa; mutations at these loci give pleiotropic deficiencies of varying grade down to complete albinism. The effects on the different chromes are not necessarily strictly proportionate. A higher percentage of mutations of chromogenes are dominant to the wild type than predicted on the Haldane-Fisher theory of the evolution of dominance and more than among other types of gene. For a number of loci many alleles are known; those which determine a longer and more oxidative path are usually dominant to those producing less oxidised products by shorter paths.

Within the only well known taxa, mammals and insects, there is much homology of genes for a particular functional group of chromes. These taxa and/or their integumental chromatic patterns are of relatively recent evolutionary origin. There is not yet enough information for comparative studies between phyla and little within most of these.

VII. Evolutionary Evidence and General Assessment

Assessments of biological significance using functional, developmental and genetic evidence inevitably converge on evolution and this is true for zoochromes. A general consideration of their evolution therefore is the subject of the next chapter. It will then be possible to make a final assessment.

Chapter 17

Zoochromes and Evolution

Just as to ontogenesis (p. 269) there are two distinct aspects of the relation of zoochromes to phylogenesis, namely their own particular evolution and their contribution to the general evolution of animals. Of course the two aspects are closely interrelated, particularly at this 'final' level of Aristotelian causation and no attempt will be made to keep the two sharply separated.

If we had as much knowledge about the past history of zoochromes as about that of gross anatomy this could contribute much to the picture provided by living animals, which represent only end-on views of particular branches of evolution. Unfortunately there is little direct knowledge of the chromes of extinct animals, – though perhaps more than about most biochemical constituents, so that this chapter must be based mainly on inferences from extant organisms and from general evolutionary knowledge and theory. The functional significance of biochromes is used to deduce their evolution rather than vice versa.

In the study of comparative anatomy the structure of living members of phyla at the various levels of organisation appears to give a reliable picture of general anatomical evolution and usually is validated by the actual fossil types discovered. It would be reasonable therefore to assume that in chromatology likewise comparative studies of living taxa should be very helpful but in fact they are almost useless (chapter 4). The best known zoochromes of the Protozoa are the polycyclic quinones (p. 43), the most complex of known monomer biochromes. The Protozoa may be primitive in having relatively few biochromes but then these might be expected to be simple in structure. In most other phyla all classes of zoochrome occur and no evolutionary trends are evident. Either the major chromatological events occurred before any significant amount of anatomical evolution or they have been experienced very much in parallel in all phylogenetic lines. Ommochromes, echinochromes and some other quinones seem to be exceptional and to have a more restricted distribution. The ommochromes are particularly instructive since they are common to all phyla of the spiralian line and virtually peculiar to this line. Echinochromes are peculiar to echinoderms and common to all living classes of this phylum so that they also were acquired at a relatively early stage of evolution. On the other hand some quinones occur sporadically in rather small taxa and so represent rather recent and trivial evolutionary history. In fact such variation in integumental chromes between the smaller taxa is rather common (p. 93) and can be used for evolutionary studies at that level; information is most scanty at the highest levels where it is most needed.

There are indications that much more will be achieved when the subject is approached in the right ways. ELIAS (1942) has shown that comparative-evolutionary studies are possible when full use is made of the anatomical relationships of the chromes. Also working within the Class Amphibia BAGNARA and OBIKA (1965) have given a most interesting picture of the evolution of their pterins. A somewhat similar study of echinochromes in echinoderms was made by SINGH et al. (1967). These are at intermediate levels but could be extended.

Also by analogy with comparative anatomy it would be expected that embryonic development might throw light on evolution, through recapitulation, but unfortunately this has been obscured by two adventitious trends, (1) to store many maternal chromes in the yolk or in the cytoplasm of the egg and (2) to delay *de novo* synthesis of chromes until a late stage of development. Nevertheless there is evidence in favour of the principle; this comes from the second-stage recapitulation, – of embryogenesis by regeneration. As already noted (p. 275) the fish *Brachydanis* during regeneration recapitulates the embryonic sequence of appearance of the various kinds of chromatophores. Again, tetrahydrobiopterin is present in the skin of the larva of the newt *Triturus*, disappearing at metamorphosis, but it reappears in regenerating limbs (KOKOLIS and ZIEGLER, 1968). The leaf-insect *Phyllium* shows a similar colour-recapitulation during the regeneration of its legs (NEEDHAM, 1965c). Of course if the main classes of chrome were acquired very early in animal evolution then ontogenetic evidence scarcely could be more informative than that from adults of the lower phyla.

Needless to say geographical distribution and other useful sources of evidence on general evolution cannot help here. Finally there is the difficulty common to all biochemical constituents (FLORKIN, 1963, 1966, 1971) that isology is no proof of homology; precisely the same chemical substance may have been exploited many times independently. This puts a premium on any direct knowledge, however slight, of biochromes in fossil material.

17.1 Fossil Zoochromes

In fossils, as in living animals, chromes are effectively 'labelled' by their conspicuousness and so have attracted more attention than most other materials. Even allowing for this, however, it does seem that they are commoner than most other organic materials, because they are more durable and better preserved in fossiliferous deposits. Melanin was found in the ink-sacs of the fossilised Liassic squid *Geoteuthis* and BUCKLAND, with his usual sense of the dramatic, illustrated his account of these animals using their own ink (FOX and VEVERS, 1960, p. 23). Crinoid quinones have been isolated from their jurassic fossils (BLUMER, 1965) while porphyrins have survived in fossil material as old as the Silurian (TRIEBS, 1935). Indeed even Precambrian porphyrins have been recognised (MEINSCHEIN et al., 1964) and this probably is the most stable class of biochrome. Colours are very common in such ancient fossils as trilobites (WELLS, 1940) and brachiopods (FOX and VEVERS, 1960, p. 165–6); no doubt much further useful chromatological work could be done on this and other fossil material.

Unfortunately chromes in this material rather commonly undergo secondary changes after deposition. Vanadium and nickel porphyrans are found in coal and other fossil material (BLUMER, 1965) but it is very unlikely that these metals were present *in vivo;* a further implication is that *in vivo* there must be devices preventing these metals from forming porphyrans whether directly or by displacing Fe and Mg. Also fossil porphyrins lose all their -COOH side-chains and become aetioporphyrins. The polycyclic quinones of Jurassic crinoids, known to geologists as fringelite (Fig. 3.4., XIII) are most probably dimers of the anthraquinones found in living crinoids (p. 42), so that the dimerisation may have occurred post-mortem. Certainly there have been extensive changes in the fringelite molecules throughout their fossil history (BLUMER, 1965). These conform to the general tendency in buried fossil material towards chemical reduction by anaerobic bacterial action; this is particularly familiar in the high hydrocarbon content of coal, petroleum and other fossil materials. In passing it is noteworthy that this particular change determined the whole outlook of organic chemists, so that they take the hydrocarbons as the origin of all organic materials; by contrast biologists take as their starting point the same simple and fully oxidised materials with which autotrophes themselves start, CO_2 and H_2O (NEEDHAM, 1965b).

Recent carotenoid-containing deposits already show a considerable enrichment in the carotene to xanthophyll ratio in the lower layers of the sediment (Fox, 1953, p. 184–187). In the upper layers in contact with air the change is in the opposite direction and this may be why the quinones of Jurassic crinoids are still only 0.5% pure hydrocarbon.

17.2 Prebiological Evolution of Organic Chromes

The relative stability of biochromes is consistent with laboratory knowledge about aromatic ring-compounds in general. Because of their stability it seems likely that chromes, and porphyrins in particular, must have accumulated even in the earliest pre-organismal chemical system (GAFFRON, 1965; BLOIS, 1965). They would function spontaneously as photocatalysts and in other activating capacities by virtue of their intrinsic properties; in available labile materials they could mediate cyclic series of changes, forerunners of the familiar metabolic cycles of the discrete organisms which later evolved. Further it seems that there would be an inevitable tendency for the more stable materials to increase progressively in amount relative to the more labile and so chromes may have been always too abundant rather than the converse. Since the total of material available for organic combination in the biosphere is strictly limited there would be great advantage from any reactions and processes which could reverse the trend and return pigment and other stable materials into circulation. Even in preorganismal systems this would be favoured by mass-action and by general Le Chatellien tendencies but it seems that the present point of balance, or steady state, is much more in favour of the labile constituents than is explicable by these forces alone. It depends on the advantage of discrete organisms on which natural selection can act, though the precise method calls for explanation.

The problem probably was not solved until discrete organisms did appear. By having a selectively permeable membrane they could sequester and separate materials and so institute increasing differences and gradients in potential between 'organism' and 'environment' which today we regard as an innate contrast. Thenceforward under the action of natural selection the behaviour of all organic materials could become increasingly controlled to promote the welfare of the organisms: no doubt the faculty of self-perpetuation was a prerequisite for the permanent selection of all such controls. Equally certainly the control of chromes would be one of the most urgent of these. Along with the ability to use them as a source of labile material no doubt the ability to use a modicum for chromatic purposes, particularly as redox agents, was selected. Living organisms therefore may have acquired their functional chromes from the medium before they acquired the ability to synthesise them endogenously, though initially endogenous chromogenesis may have been as spontaneous as in the medium and its inhibition would be an essential part of the control in question. There would next be selection for the organisation of the energy-manipulation by the redox chromes.

The rate of synthesis of these relatively stable materials in the pre-organismal and later environments need not have been very high to result in a differential accumulation which could bring organic cycling to a standstill in a cosmically short time. A low rate of synthesis might be anticipated because of the relative complexity, large size, and relatively high state of oxidation of their molecules but in fact the rate may have been very high (Fox, 1965a). Laboratory experiment has shown that in spite of these considerations some familiar biochromes are synthesised in quantity from simple precursors under feasible primitive tellurian conditions. Examples are the porphins themselves (SZUTKA, 1965a),), pteridines (ORÓ, 1965) and possibly anthocyanins among a number of other, unidentified chromes (SZUTKA, 1965b); in fact coloured products are very common (FOX, 1965b) though they rarely attract positive attention and few have been identified. FOX (1965b) found such products along with the proteinoid formed by thermal polymerisation of amino acids and ORÓ (1965) obtained pink and yellow chromes in his 'prebiological' synthesis of pyrimidines and purines from simple mixtures of HCN, NH_3 and H_2O. Thus chromes may have been invariable by-products of the synthesis of the two most important classes of biological materials, proteins and nucleic acids. The laboratory chromes are mostly dark, indicating high polymerisation (SZUTKA, 1965a; BLOIS, 1965) and great stability. It is conceivable therefore that the whole prebiological system was essentially a 'rich chromatic mess' and that the many wonderfully transparent organisms and cells we can now see were evolved later when the control mechanisms postulated above had become established. It is now possible to see more meaning in those two mystic lines from SHELLEY's ADONAIS:

> "Life, like a dome of many coloured glass
> Stains the white radiance of Eternity."

Organic chromes are not rare also beyond the Earth. Traces of porphyrins have been detected in some of the lunar fines of the 'Apollo' missions (HODGSON et al., 1970) and certainly chromes fluorescing visibly when irradiated at 310 and 350 nm (RHO et al., 1970). The optical spectrum of certain regions of interstellar space indicates benzoporphyrins, possibly chelated with Mg and with pyridine derivatives (JOHNSON, 1970). Jupiter has an enormous red mass in its atmosphere, varying in size, hue and position. It also has a yellow cloud-layer due to the simple organic, or potentially organic, materials NH_4HS and $(NH_4)_2S$ (LEWIS and PRINN, 1971). There are also blue cloud-bands (RASOOL, 1972). The seasonal colour-changes on Mars are less spectacular but also symptomatic of the wide distribution of chromes in the universe and of their responses to environmental cycles; there is a possibility that these colours again are due to organic materials (BURGESS, 1972).

17.3 Chromes and the First Discrete Organisms

The extent of the prebiological accumulation of chromes would be affected by the time-interval before discrete organisms appeared and acquired the ability to metabolise them. This may have been a relatively short interval since in the modern laboratory, using generous concentrations of proteins and other macromolecular electrolytes the first step, the production of small cell-like bodies, occurs spontaneously and instantaneously: they are stable-state forms therefore. A two-component type has been called a coacervate by DE JONG (1935) and a simple protein type a microspherule by FOX (1959, 1960). There are natural forces which would tend to produce the necessary local concentrations and enrichments for the formation of such bodies repeatedly in many appropriate places on the eobiological Earth. These bodies assimilate selectively, grow and reproduce by budding so that the only very time-consuming stage may have been the evolution of genetic veracity, precise self-perpetuation as distinct from mere perpetuation. The abilities to absorb, synthesise and use chromes in the ways indicated would be among the essential properties demanding accurate perpetuation. A stable cell-membrane would be another, minimising the 'death'-rate of the bodies as well as providing the whole basis for the segregation from the environment (p. 337).

If this picture of an early plethora of spontaneously synthesised chromes is correct then the number of classes at present exploited by living organisms may seem very small. It is also a very small fraction of the number synthesised by the organic chemist from the products of coal and by other means. The law of austerity in biological evolution (LWOFF, 1944; FLORKIN, 1960) therefore seems to have operated here as elsewhere. Not only are there few chemical classes of biochrome but often few biological members of a class, though certain particularly successful members have 'radiated' into a number of minor variants; examples of this are the pterins and echinochromes, and in plants the carotenoids, flavonoids and quinones.

This picture of austerity must be taken with the reservation that as yet very few animals or other organisms have been systematically examined for all their biochromes. In those that have a number of chromes have been found that are not recognised members of known chemical classes (p. 73) so that altogether there could be quite a number of further classes, even if not very widely distributed. On the other hand a large percentage of the zoochromes newly examined regularly prove to belong to known classes, which undoubtedly are the most significant in quantity and breadth of distribution. There are more classes widely distributed among animals than there are among plants and this is consistent with the greater variety of activities and requirements in animals; also relevant is the fact that chromogenic pathways in general are oxidative and that heterotrophes use such pathways much more than autotrophes.

As already implied, it seems certain that the first chromes to be harnessed by discrete organisms would be those which catalyse or otherwise control basic metabolism. In effect this means redox reactions involving energy-manipulation for synthesis or for work so that the porphyrans, isoalloxazines and quinones are likely to have been the best candidates, with simple metalloproteins and carotenoids next on the list. HALL et al. (1971) believe that the simple iron-proteins, the ferredoxins (p. 51), may have been the first chromoproteins since they are among the simplest of all proteins, containing only a restricted number of amino acids, all of the 'non-essential' group; moreover as chromoproteins they have a prosthetic group derived from the protein itself. However some proteins with a restricted variety of amino acids are very specialised and again the prosthetic group of the ferredoxins in fact is rather sophisticated (Fig. 3.5, A, D) and depends entirely on the protein whereas porphyrans, ribitylflavin and other chromes would have considerable redox activity even as free chromes. Today the ferredoxins certainly have a varied redox function but they appear always to be ancillary to enzymes with aromatic prosthetic groups. GAFFRON (1965) believes that porphyrans antedated discrete organisms and the high degree of homology in the amino acid sequence of the protein moiety of cytochrome c, even between bacteria and mammals, is strong evidence that they were protein-conjugated from a very early stage of the organismal era.

A question of some importance here is which type of porphyran was exploited first, the respiratory haems or the photosynthesising chlorophylls? Until fairly recently the unquestioned answer would have been: chlorophyll because all life depends on autotrophy. However it is now envisaged that the early autotrophes could have used photoactivators other than chlorophyll (OPARIN, 1957), possibly not porphyrans at all. The biosynthetic path (p. 286) also implies that chlorophyll arose by a modification of the haem path; it is also a longer and more demanding branch of the path. Again in organisms which synthesise both, haem controls the synthesis of chlorophyll just as it does that of the other specialised porphyrans (p. 308). Nevertheless there remains the problem that the haems themselves have very poor photosynthetic potential (CALVIN, 1962) and so are much poorer precursors of chlorophyll biologically than chemically.

In principle there are plenty of alternatives to the porphyrans as the photoactivating chromes of the first photosynthesising organisms: any fluorescent chrome that is sufficiently stable could suffice (GAFFRON, 1965). There is also a specific reason for entertaining the possibility that the first autotrophes did not use chlorophyll. In most modern autotrophes chlorophyll releases free oxygen whereas some estimates give no more than 8×10^8 years ago for the first appearance of free O_2 on Earth (OPARIN, 1957, 1960; GRANICK, 1965; BRINKMANN, 1969). However this is questioned by others (RUBEY, 1955; ABELSON, 1966) and an origin of free O_2 as far back as 2.7×10^9 years is possible (MARGULIS, 1972). Moreover in any case some of the most primitive chlorophyll-based autotrophes today use reductants other than H_2O and so do not release free O_2 (GAFFRON, 1965). There is no compelling reason to suppose that the key chrome of photosynthesis was ever anything but chlorophyll.

If the first photoautotrophes had a chlorophyll system which did not release free O_2 then the whole cytochrome oxygenase system of respiration, including catalases and peroxidases, would have been unnecessary but this does not imply that all respiratory haemoproteins would have been absent: modern anaerobic bacteria use cytochromes in conjunction with oxidants other than O_2 (RAFF and MAHLER, 1972). Consequently there could have been both photosynthesising and respiratory haemoproteins geared to terminal oxidants other than O_2. What does seem inevitable is that the terminal oxidant of respiration will always have been also the terminal product of photosynthesis. Equally the terminal products of respiration must always have been the substrates for photosynthesis; at present these are CO_2 and H_2O, both volatile and virtually ubiquitous in the biosphere. They are virtually irreplaceable for these and other invaluable properties and if both must always have been used for the purpose then there is no alternative to the equally volatile O_2 as the product of their photosynthesis.

It is the purple bacteria among living autotrophes that fix CO_2 without releasing O_2 and they are obligate anaerobes. In fact they are entirely peculiar autotrophes since they use the light-energy only catalytically (GAFFRON, 1965) the actual energy coming from the special reductants used instead of water, $- H_2S$, H_2 and Fe^{2+}. Since these reductants must arise mainly from the biosynthetic activities of other organisms the purple bacteria and other 'chemoautotrophes' are really heterotrophes or partial heterotrophes. They do at least show that the photosynthetic system is very versatile and that there are many intermediate states between typical autotrophy and typical heterotrophy, but they are modern specialists with all the accessory chromes and redox enzymes of the typical photosynthesis system. At the same time it should be emphasised that these components are likely to be as ancient as the chlorophylls themselves.

Autotrophes always must have been able to use some of the materials they synthesised to provide energy for reproduction, repair and other physiological activities so that heterotrophes could have evolved smoothly by the hypertrophy of this catabolic function, eventually supplemented by powers of capturing macro-amounts of material. This classical view is in contradistinction to the revolutionary thesis (OPARIN, 1957, 1960) that the first discrete organisms were heterotrophes living on abiotically synthesised material in the environment. The further thesis is that some of these became the first autotrophes when the abiotic materials became exhausted. The feasibility of this thesis depends very much on such parameters as the number of organisms involved and their general rate of metabolism: under present conditions on Earth heterotrophes would consume all available organic material in the biosphere within a few years at most (NEEDHAM, 1959, 1965b, 1968c) in the absence of autotrophic activity. Under any conditions it would probably be too rapid to permit evolutionary 'avoiding action'. Under any conditions indeed it seems likely that neither type of organism could have originated significantly ahead of the other. Together they constitute a steady-state biochemical system and operate a metabolic cycle which can rotate efficiently only provided each increases the speed of its section of the cycle in close correlation with the other.

It seems likely (Figs. 8,1, 8.2) that autotrophes have always used much the same coenzyme chromes in their anabolic and catabolic systems though both initially may have been much simpler than at present. If so it is very surprising how little difference there is between the respiratory systems of modern heterotrophes and autotrophes. The simpler conclusion is that neither has changed much from the common aboriginal system though it is possible that they have experienced parallel evolution, aided by the fact that autotrophes form the diet of heterotrophes which today obtain all their proximate respiratory coenzymes from their food. The dependence stands in sharp contrast to the independence shown by animals in their integumental and other chromes: it may be opportunist rather than obligatory, but in any case the respiratory chromes are very similar in all organisms.

17.4 The Evolution of Chromes in Animals

An inventory of the biochromes probably already in use in the earliest unicellular organisms indicates that the only chemical classes acquired later, specifically within the animal kingdom, were the melanins and ommochromes. Both are integumental and retina-screening chromes and both *in situ* are dark chromes absorbing radiation over a wide range. Plants make some use of melanins but mostly of the phenolic N-free group (p. 60). Indolic melanins and ommochromes like most animal chromes are more prodigal with nitrogen.

It does not necessarily follow that melanins and ommochromes were first exploited rather late in the evolution of animals or of integumental zoochromes. They may have been among the first integumental chromes if protection against irradiation was the earliest function of chromes in this site (PROSSER and BROWN, 1961, p. 530). It is clear that all visual-effect functions would be pointless until the evolution of photoperceptors with relatively good powers of pattern-discrimination whereas anti-radiation screening would have been valuable much earlier.

A mixture of a variety of monochromes may screen as effectively as melanin and this may be the reason why so many different chromes are found in animal integuments. Significantly enough all but melanins and ommochromes are classes used elsewhere in the body for other purposes: only these two were evolved primarily for integumental purposes. There is some indication that they may be supplanting the others in this site not only for screening purposes but even for visual display. The Tortoiseshell cat is an illustration of the effective use of melanin for this purpose.

It seems likely that the oxygen-transport chromes were the next type to be evolved and that the myoglobin (Mb) type of tissue-transporter antedated those of the body fluids, which can only benefit from such chromes in relatively large animals with some effective movement of the fluid. Mb is found in particular critical tissues even of animals with no circulating fluid or with a chrome other than haemoglobin (Hb) in such a fluid (Table 4.2) so that it would seem to have evolved from the tissue haem-oxygenases to cope with local tissue-problems in all organisms. In some of these Hb evolved from it as the circulatory O_2-carrier and the α-chain of vertebrate Hb is still very similar to the single peptide of Mb. Other taxa exploited simple Fe-proteins or Cu-proteins although they may already have had Mb in certain tissues or exploited it subsequently. Clearly Hb like Mb has been evolved many times independently and probably at many different stages of evolution.

Since primitive photoperceptors and simple eyes must have evolved before visual-effect coloration the retinopsins presumably were derived from protective carotenoids of the general integument; these are intrinsically photosensitive. Directional and pattern discrimination depended on anatomical and physiological adaptations and on help from melanin or ommochrome as screening pigments. The eyespots of leeches and some other animals strongly imply that localised eyes evolved from a general integumental pigmentation so that photoprotection facilitated the evolution of photoperception and this eventually provided the rationale for visual-effect coloration of all kinds.

The selection of carotenoids and specifically of the retinals as photoperceptor chromes (p. 163) either was so very early in animal evolution or they are so uniquely appropriate, or both, that no major alternative has been exploited for the purpose. It seems possible that the unique molecular and chemo-electrical transduction properties of the retinals may be shared in degree by the carotenoids of the photosynthetic system, the only other undoubted photo-chemical system in living organisms. In this context it is interesting that the carotenoid eyespot of phytoflagellates shows some evidence of having been derived from chloroplasts. One view is that the carotenoids of chloroplasts normally absorb light of appropriate wavelength and transduce it into quanta more auitable for absorption by the chlorophylls. In the retina of course the ultimate transduction is electrical but this may prove relevant also to the photosynthesis system.

It is evident that once any predator had evolved eyes with pattern-discrimination there would be selective value to any potential prey in evolving camouflage, whatever the efficiency of its own eyes. This applies equally to some sematic colour-schemes and may explain the bright colours of such animals as the Porifera and Cnidaria. Some echinoderms and leeches are capable of chromogenic and even chromomotor colour-change in spite of their own primitive visual powers which however are adequate to trigger these responses; indeed some animals change their colour-scheme by primary responses alone (p. 141).

With only one perceptor chrome, vision at first must have been purely monochrome, 'black-and-white', as it is still in many animals. Nevertheless colours in the visual field would contribute to discrimination, as they do in a black-and-white photograph, and this should have provided the selective force for a gradual evolution of colour-vision. There would also be selection for the discrimination of specific wavebands, for instance of green by herbivores. If colour-vision did evolve from monochrome in this way without discontinuity of any kind it is not surprising that it should have been based entirely on a further differentiation of retinopsins and that no new chromes should have been exploited (p. 168). Although it may seem that one shortcoming of this is that the absorption-peaks of the relevant chromes are rather crowded towards the centre of the visible range they do gain from the relatively high solar output in that region. Moreover the information they transmit approximates to that expected of evenly spaced peaks; a prepared filter passing the whole of the 400–500 nm range appears indigo-blue, a similar 500–600 nm filter appears to transmit green and a 600–700 nm filter red, equally spaced on the subjective colour-circle (Fig. 2.1). The absorption-bands of the retinopsins are much narrower than the pass-bands of these filters, as well as being more bunched. Nevertheless, assisted by central analysis and synthesis the human visual information builds a surprisingly rich colour-picture. It is even possible to synthesise a complete polychrome image from the information transmitted through two colour filters only (LAND, 1959).

Although it is quite true that visual-effect patterning and coloration could scarcely antedate eyes of the appropriate efficiency it is equally evident that some, for instance those of camouflage could not afford to postdate such efficiency. On the evolutionary time-scale both must have been in step. Moreover in all cases any improvement in the visual effect would immediately select for improved vision in the relevant eyes.

In general the integuments of animals today are heavily pigmented from a combination of the protective and visual-effect causes considered here and elsewhere (chapters 6,11) and therefore it is particularly intriguing that so many planktonic animals are almost to completely transparent, and almost crystal clear in some cases (HARDY, 1956). The condition is very common in Protozoa and so might seem pristine but the main reason probably is that the Protozoa are too small to be relevant to even the finest-grained retina so that no kind of visual effect coloration has had selective value. In addition many Protozoa are planktonic and by being transparent in that environment may escape even the simpler type of eye, sensitive merely to shade and movement. Much the same factors may apply to many of the transparent larvae of the plankton and they again might be taken to imply that transparency is pristine and is recapitulated in the ontogenesis of such animals; their adults usually are pigmented.

A large proportion of the transparent planktonic animals are adults and many of these are large enough to benefit from camouflage and other visual-effect devices. In fact transparency has cryptic value in this environment whether viewed from above or from below. Indispensable pigmented eyes and other minor patches of pigment need not vitiate this effect provided they confer no geometrical regularity and do not pick out characteristic outlines. The largest of these animals, such as the large medusae have considerable areas of delicate coloration which conforms to these requirements and probably also serves two positive purposes, to tone down the slight opalescence of large masses of macromolecular material and to break up the outline. Such large relatively placid masses might be rather conspicuous in the turbulent water if completely transparent and colourless.

Unless they are near the surface throughout the day aquatic animals do not need u.v.-screening chromes since the water absorbs the short waves so effectively (p. 7) and in fact many of the transparent planktonic animals move down in relation to the altitude of the sun (HARDY, 1956). Being pigment-free they are also free of the risk of photodynamic damage by light of the visible range. Presumably their respiratory chromes are minimal in amount and well quenched by conjugation.

By contrast animals with a heavy integumental pigmentation can safely store large amounts of chromes in the deeper organs and this would have been the last major step in chromatic evolution. In general the amount of such deep chromes is related to the amount of integumental pigment and incidentially to body-size since the screening value is proportional to the absolute thickness of the screen and also because the storage-requirement is roughly proportional to body-size. It is useful to store exogenous chromes because the supply may be erratic (p. 301).

This is as far as the general evolutionary sequence of zoochromes can be traced with any confidence since the exercise is dependent on a knowledge of function (p. 335). The sequence suggested seems reasonable and has not led to any obvious inconsistencies with present knowledge and intuition. In some instances it may even strengthen present tentative views on function. However, undoubtedly for some time to come progress will depend primarily on the direct investigation of function.

17.5 Conclusion

At present there is little hope that evolutionary studies will make a major contribution to knowledge of the significance of zoochromes. A number of chromes have been found in fossil material and this proves their stability. Comparative taxonomic studies, embryogenesis and other approaches which are so informative of morphological and general evolution for various reasons have little value in chromatology. The reciprocal exercise of using available knowledge of functions and innate properties to trace probable evolution is more profitable and gives a feasible and reasonably complete picture of the sequence of events.

It seems likely that coloured materials always were abundantly available on Earth, were exploited by biotic systems from the outset and have become controlled by these systems. Photosynthetic and respiratory energy-manipulations were the first major functions of biochromes. Screening against radiation may have been second and photoperception third though this may have been followed closely by the visual-effect functions. Oxygen-transport by chromes of specific tissues preceded that by chromes of circulatory fluids and both may have evolved many times independently, the first occasions probably before visual-effect coloration. Storage of chromes in the viscera carried a photodynamic risk until integumental pigmentation provided an effective screen.

Chapter 18

General Assessment

The instrinsic properties of all coloured molecules (chapter 2) show that they are ideal for exciting reactions of the type characteristic of biochemical systems and there is a strong *a priori* case that in general biochromes must play a significant role in physiology. This is reinforced by a study of the specific properties of the individual classes of chrome that have been exploited by living organisms (chapter 3) and by the formation of chromoproteins and other conjugates. The relatively small number of main chemical classes of biochrome indicates that selection has favoured only those with optimal properties for the various biological purposes.

The differences between the biochromes of animals and plants (chapter 4) is in keeping with their very different modes of life. By contrast most phyla of animals have most of the classes of zoochrome, whether by homology or through convergent evolution; in any case evolution has tended towards the same class of chrome for the same purpose in all taxa. Nevertheless there are significant taxonomic differences among animals in the complement and distribution of zoochromes and these may be due mainly to the fortuitous element in evolution. In addition, at the lowest taxonomic levels there may have been positive selection for divergence, at any rate among the integumental chromes, for which chemical constitution may be of minor importance. At this level chromatic evolution manifestly is very active today and this may be why such a high percentage of dominant mutants are found (p. 325).

The distribution of chromes within the body (chapter 5) is consistent with the view that most of them have functional significance in their respective sites: distribution is rarely indiscriminate notwithstanding the fact that in animals collectively, and often individually, many tissues house many classes of zoochrome. The state of the chromes in the body-fluids and cells, and particularly in the chromasomes (p. 118) of integumental cells, shows great structural sophistication indicative of functional specialisation. This is even more evident in the structure of mitochondria and chloroplasts.

The adaptive value of integumental chromes for 'visual effect', concealing or advertising, and for protection against irradiation, mechanical protection, chemical defence and thermal regulation is shown in detail in chapter 6. Compromise responses often are necessary and may account for some which are difficult to interpret (see below). It is clear that the retinals and their protein-conjugates have become uniquely adapted to the photoperceptor function (chapter 7).

Adaptation in the redox chromoproteins (chapter 8) is universally recognised and needs no belabouring. Much the same set of biochromes mediate photosynthesis, the other half of the redox cycle of life. There is strong circumstantial evidence that members of almost every other class of zoochrome may have significant redox functions *in vivo*. This shows very clearly what was implied in chapter 2, that the redox property is innate to all chromes, whatever biological function they perform. They may be photo-excited more commonly than at present realised. Bioluminescence is chemo-photic, i.e. an interesting reversal of the usual transduction by chromes. The importance of the function of the oxygen-transport zoochromes also is proved by their amazing efficiency and sophistication at all levels from the molecular upwards (chapter 9). In some Crustacea, molluscs and worms the carrier is a facultative aid rather than a necessity but this does not diminish its general significance: the carriers meet a variety of needs.

There is evidence that chromes in digestive secretions have a positive function (Chapter 10) though as yet there is no firm proof in any instance. Except perhaps in the Ectoprocta there is no evidence that any chrome accumulates in any animal merely because the latter cannot excrete it; all stores of chromes in the deeper tissues probably have some useful significance. Qualitatively abnormal chromes are very rare (chapter 11) and probably there has been selection against the synthesis of such materials. Excessive amounts of normal zoochromes are rarely very deleterious but deficiencies always are (see below): the value of chromes is nearly always positive and rarely negative. There is some danger to and from biochromes, through photoactivation, but normally there is good protection against this risk.

Virtually every known class of zoochrome has been found in the eggs of animals and a considerable number in the reproductive system; chromes characteristic of the particular taxon are always present here. In reproduction and development the respiratory and other metabolic chromes perform their usual functions and in addition there may be some chromatic functions specific to these processes (chapter 12) but they are less well understood than those of other physiological fields.

Chromogenic pathways give much evidence of having been exploited and evolved to produce those particular classes of biochrome and for no other purpose (chapter 13). They have all the characteristics of naturally selected biochemical pathways. There are interrelationships between chromogenic pathways just as the various chromes are interrelated functionally and genetically. In the control of chromogenesis (chapter 14) there is even stronger evidence that chromes evolved in the performance of important biological functions. At the morphological level the development of colour-patterns in the integument (chapter 15) is as sophisticated as any developmental process and demonstrates the importance of the patterns. These biogenic processes are all encoded in the genome (chapter 16) along with those of the proteins eventually conjugated with the chromes. Chromogenes have the properties common to all genes and possibly also some specific qualities. Mutations to the deficiency of a chrome are usually both recessive and deleterious: abundance of pigment has been strongly selected as the wild type.

In general therefore both direct and indirect evidence indicate a functional value for all known zoochromes in their respective contexts. Where the function has been known for some time and investigated in detail the chrome has been shown to be adapted in almost every detail. Some outstanding examples are summarised below.

18.1 Outstanding Adaptations of Zoochromes

Among the adaptations shown by integumental chromes the wonderful precision of many cryptic patterns and mimetic patterns (COTT, 1940) can never be overemphasised. Many advertisement colour-schemes, for instance that of the parasite *Leucochloridium* (p. 257), are equally striking. At its most sophisticated, in the cephalopod molluscs, the chromomotor mechanism of colour-change is an adaptation scarcely surpassed in perfection by that of any other physiological mechanism. Among defensive functions of integumental chromes the ingenuity of the mechanism of the bombardier beetle may be singled out (p. 159).

The photoperceptor function of the retinals (p. 163) is an amazing evolutionary achievement at all levels from the molecular upwards. Particularly striking is the use of a sterically hindered and thermodynamically unstable *cis*-form as a basis for a rapid triggered response.

For redox functions the ideal suitability of the flavin-adenine dinucleotide (SZENT-

GYORGYI, 1960; B. and A. PULLMAN, 1963) and of the flavoproteins in general (p. 188) is outstanding even in a system where in fact every member shows very efficient adaptation. An ingenious division of labour between two variants of the FAD coenzyme in L-amino acid oxidase (WELLNER, 1967) may be cited. This permits the complete oxidation of the amino acid molecule without the formation of awkward amino acid free radicals as intermediaries. These would be unavoidable if only a single type of prosthetic group were involved. The two Fe and two Cu atoms of cytochrome oxidase again enable it to reduce a complete molecule of O_2 at one step.

The regular gradation of redox potential along the series constituting the electron-transport system (p. 186) is another outstanding manifestation. The photosynthetic system of autotrophes shows a similar property (Fig. 8.1) and is a most sophisticated adaptation in almost every respect. It shows adaptational plasticity for instance in the ability of the Cyanophyceae to adjust the ratio of phycocyanin to phycoerythrin so as to suit the spectral composition of the light they receive (LEMBERG and LEGGE, 1949, p. 597). Biolominescence also shows some impressive adaptations, but especially in its physiological and anatomical aspects not considered here.

The exquisite adaptation of the haemoglobin molecule to O_2-transport (PERUTZ, 1970) crowns previous knowledge of its outstanding efficiency at the physiological level. Its biogenetic and functional adaptations to variations in O_2-supply, in metabolic conditions and in the ecological circumstances of the animal all are very impressive. Among these ecological conditions are the relations of mother to foetus and host to parasite. In the biosynthesis of haem and other porphyrin derivatives the feedback control system is as sophisticated as in any metabolic pathway (p. 307). The other O_2-transport zoochromes are less efficient but nevertheless also highly adapted in their molecular structure and behaviour. The O_2-transport chromes show that iron and copper are ideal elements for the purpose. They also emphasise the extent to which chromatic properties may depend on a protein moiety.

It is probably too early to recognise the full extent of biological adaptation in the actual chromogenic pathways which have evolved but already some impressive features are evident for instance the systematic desaturation of a polyisopentenyl molecule to produce a carotenoid. No doubt it is not fortuitous that the ring-closed chromes all have useful side-chains in addition: these are built into the particular precursors used and only occasionally do special side-chains have to be added later. In the control of chromogenesis and of chromomotor responses adaptation is abundantly evident, particularly in the ways that the deleterious direct action of external factors is circumvented by indirect routes.

Thus beyond question most of the known zoochromes have been naturally exploited to serve essential biological functions. Nevertheless there remain some problems and uncertainties which also should be summarised. Do these indicate incomplete knowledge or have some zoochromes in certain sites no useful significance?

18.2 Chromatological Problems

Most of the problems concern integumental chromes and one of the most outstanding of these is the reason for the bright and varied coloration of so many species of Cnidaria. The most obvious possibility is warning coloration, their nematocysts making the Cnidaria highly dangerous and deterrent to potential predators. However for this purpose a common pattern of sharply contrasting colour-patches is usual whereas in many Cnidaria there is also great colour-polymorphism within the species and certainly between species (FOX and PANTIN, 1944). The intraspecific variety holds even within small local populations so that exploring animals would gain minimal and not maximal advantage from their encounter with any one pattern. This prompts a second suggestion, that the colours function as an attraction to potential prey; in this event there would be advantage to the cnidarian in a maximal variety of bright colour-schemes. Most Cnidaria are macro-feeders and such prey as Crustacea have good colour-vision so that the theory could explain the bright colours of reef-corals, sea anemones and some others. Unfortunately *Metridium*, a sea-anemone outstanding for its colour polymorphism (FOX and PANTIN, 1941), is a filter feeder so that here at least some other explanation is needed. In this and some other anemones a clonal method of reproduction helps to maintain the variety but this does not account for it.

Many Cnidaria look absurdly like the flowers of some terrestrial plants both in shape and in being brightly coloured; this is particularly intriguing since their purposes must be as different as their clientelle, which never have any chance of learning from experience of the other type of lure. In any case it would be to their disadvantage if they could! The explanation is simply that common principles have governed both: concentric and radial patterns round a focal point of mouth or calyx is the ideal lure (THOMPSON et al., 1972). The radial spread of tentacles in the Cnidaria may have arisen to facilitate searching a wide area in an active method of predation and so will have been preadapted for this more passive method. The bright colours of motile animals frequenting coral reefs probably is mainly mimetic camouflage, whether offensive or defensive.

Another problem, is the pallor of many cavernicolous animals and the contrasting blackness of some insects of the same environment (PRYOR, 1962). The explanation may be that in the dark (virtually no fresh water animals are bioluminescent (p. 208)) it is immaterial whether there is much pigment or none in the integument. The majority of animals here may reduce pigmentation of necessity or for economy whereas insects need the melanin for hardening the exoskeleton. Exceptionally some insects produce colourless hard exoskeletons but if the first cavernicolous insects lacked this property there would be no subsequent selection in its favour. Some of the normal surface-dwelling animals that have a dynamic colour-change become pallid at night while others become dark or intermediate in shade (p. 139); an albedo of 0/0 effectively registers as any value from zero to infinity depending on the error or asymmetry in the response-system of the species. Here again there appears to be no selection against variety. In caves there is little possibility that black animals can use the chrome for heat-absorption or for protection against radiation which in any case would be equally relevant to all cavernicolous species.

Also rather enigmatic is the prevalence of self-coloured reds and blacks at intermediate depths in the oceans. It would be more understandable if all were black. However, at these depths where at most only a little green light penetrates from the surface and any bioluminescence is mostly in that same general range deep reds will be virtually indistinguishable from blacks and both will be less conspicuous than any other colours, and than white. Only at greater depths do light conditions approximate to those in caves and are pallid animals at all common. It is conceivable, if not very probable, that the blacks and reds also have some heat-exchange function but antiradiation protection seems quite out of question here.

It is surprising to find bold blacks and reds again (p. 136) characteristic of animals of the lakes of high mountains and high latitudes (BREHM, 1938):

> "My love was clad in the black velvit
> And I myself in Cramasie."
> Anon: 'O Waly, Waly'

In Tibetian lakes the waters are a turquoise blue so that both red and black are particularly conspicuous. The colour is too prevalent to be fortuitous and a positive advantage of considerable power is implied, to outweigh the anti-camouflage disadvantage. Of BREHM's 11 possible advantages the two most probable are heat-absorption and protection against radiation of short wavelength. Cold rather than heavy u. v. irradiation may be the more important factor common to high altitude and high latitude and it is relevant that terrestrial poikilotherms also are darkest at high altitudes and latitudes (RAPOPORT, 1969). However the clarity of the air at high latitudes may compensate to some extent for the longer path of sunlight through the atmosphere and reflection from snow also increases the shortwave exposure, so that neither function can be ruled out. In East African lakes the brilliant red planktonic fauna collect in the surface waters in the early morning; after mid-morning like marine plankton (p. 343) they move downwards (Dr. M. J. COE; personal information) presumably because either the light or the heat then becomes too great.

It is interesting that *Euglena* and other normally green Protista are often red in these environments. However they are red also in hot springs; here conceivably the carotenoids concerned may have an anti-oxidant function (p. 197) the high temperature no doubt accelerating oxidation processes. It may be recalled that short-wave radiation also stimulates oxidation in the presence of oxygen so that the carotenoids may function similarly in both types of environment. It is further intriguing that hydracharines are mostly red or black even in temperate fresh waters and so are many terrestrial mites. Most of them have massive stores of carotenoids.

18.2.1 Problems Concerning Specific Classes of Integumental Chrome

Carotenoids: Certainly there remain more problems concerning carotenoids than any other class of chrome and probably this is not due solely to the fact that they have been so much more studied than most other classes. Their abundance and variety is one reason for both the interest and the number of problems. One problem is that although abundance is usual quite a number of animals rigidly exclude carotenoids (p. 302), the pig is one example yet it needs vitamin A (HALE, 1933), has retinopsins in its retina and carotenogenic enzymes in its liver. Another enigma concerns the distribution of carotenes relative to oxycarotenoids (p. 82). A detailed knowledge of the sites and redox conditions of each location will probably help: in general carotenes are most abundant in deeper tissues and oxycarotenoids in the integument. Further the significance of the more unusual animal carotenoids is almost entirely unknown yet animals modify carotenoids more than plants do and more than animals modify other exogenous chromes.

Melanins: Here again the number of problems is related to interest and importance. The familiar and rather striking range of skin-colour in the human species is not explicable simply by the varying intensity of short-wave radiation in the local insolation (BLUM, 1961). The total insolation is as high at latitude 40° as at the equator and in any case the human epidermis absorbs much of the u. v. around 300nm without the help of melanin, which is situated below it as a second line of defence (TREGEAR, 1966). In response to insolation the skin particularly of fair, vitiliginous and albino individuals improves its protection against the 300nm range by thickening the stratum corneum (BLUM, 1961; VAN DER LEUN, 1964): both defences therefore are improved. In the visible and i.r. ranges white reflects more and absorbs less than black skin so that it might seem the more suitable colour in low latitudes. However, melanin probably is of critical importance in absorbing the short-wave radiation which does penetrate the epidermis, the most dangerous range, below 300 nm. Moreover in the near i.r. which accounts for much of the heating radiation from the sun melanin reflects as well as white (BLUM, 1961; TREGEAR, 1966). The genetic gradient from equator to poles therefore is rational and so is the suntan response.

The free radicals trapped in melanin during its own biosynthesis (BLOIS, 1965) indicate that it should function in mopping up not only ionising radiation, which otherwise causes erythrema and widespread molecular damage, but also any free radicals liberated as a result of such damage (TREGEAR, 1966). The energy of the u.v. radiation which penetrates the corneum, around 4eV, is adequate to break C–C and C–N bonds and to cause extensive damage particularly in proteins and nucleic acids (COLLINSON et al., 1950). Damage by ionising radiation leads to local greying of the hair and this might be one demonstration that melanin indeed does 'take the rap'; however the greying response is rather common and non-specific and often results from simple chemical and cellular disturbances (p. 251) so that it may not be critical in the present context.

There are some anomalies in the geographical distribtuion of human skin colour and these may be due to a number of causes. Relatively recent human migrations and interbreeding have probably contributed and it is also possible that the factor which led to the evolution of black panthers and of the dark skin of forest anthropoids has played some part. Altitude and other oecological variables inevitably complicate simple geographical trends. Even so it is possible to recognise a general downgradient in human melanisation from equator to poles, with adaptive value for a naked homoiotherm.

Dibromoindigotin: The function of this product of muricid gastropods (p. 59) remains quite obscure and also the reason why it is secreted as the precursor. It is one of the few known biological bromine compounds, all very sparsely distributed since Br is the least outstanding halogen in any respect (NEEDHAM, 1965b). It is only marginally responsible for the brilliant colour of tyrian purple. A possible clue is that the bromine derivatives of a number of aromatic polycyclic molecules have a particularly high electrical conductivity (AKAMATU et al., 1954); however this applies only to the crystalline state which is scarcely relevant to an aquatic gastropod.

Ommochromes: Those butterflies that have considerable amounts of these chromes in the wings appear to synthesise much more than necessary at the pupa stage and then excrete massive amounts in the imaginal meconium (BUTENANDT et al., 1954a; LINZEN, 1967). LINZEN's conclusion (p. 245) that in general ommochromes are essentially excretory products only partly tapped as integumental chromes therefore cannot be ruled out. It is noteworthy that the Robin Moth *Cerura* needs 120 mg of tryptophan-free protein for its cocoon-silk while its total production of ommochromes is around 1 mg; this corresponds very closely with the amount which could be produced from the waste tryptophan from 120 mg of mixed larval proteins broken down to provide amino acids for the silk. Insects do not catabolise tryptophan completely but some animals of the spiralian phyla do and the insect condition may be a consequence rather than the cause of ommochrome synthesis. This remains one of the problems of chromatology.

Purines and Pterins: The main problems concerning purines were discussed in section 11.5. Some Lepidoptera synthesise purines and pterins in large amounts at pupation, again as wing-chromes but unlike ommochromes these are not squandered (p. 246). They do present some problems, however (HARMSEN, 1969; ZIEGLER and HARMSEN, 1969). The particular pterins synthesised (p. 292) vary with the O_2-supply and the significance of this is unknown; those formed at high O_2-pressures are extensively excreted, in fact. Another enigma is why so many different pterins are formed in individual species, both of insects and of Amphibia (p. 89); at present there is no knowledge of a useful division of labour between them. This kind of multiplicity is not peculiar to pterins: there may be as many as seven ommochromes in the cricket *Gryllus* (LINZEN, 1966); there is a plethora of echinochromes in some echinoderms (p. 42) and of carotenoids in some taxa.

18.2.2 Problems Concerning Chromes Other than those of the Integument

The role of the retinals in light-perception (p. 163) and spectral evidence indicate that carotenoids probably are also responsible for the general dermal light-sensitivity but if so where are they sited? Again there are indications that they may mediate chemical perception but the evidence is still circumstantial. In both cases other chromes also may be involved.

Although there is much evidence that almost all classes of zoochrome may have members acting as redox agents *in vivo* more direct or more complete proof is needed in many cases (p. 197). Even concerning the O_2-transport zoochromes there are many minor problems such as the amazing tenacity with which O_2 is bound by the haemoglobins of the oligochaete *Alma* and the nematodes *Ascaris* and *Strongylus* (p. 221). These Hbs may pass on O_2 so long as more is available to displace it (WITTENBERG, 1970) so that only the last load is denied to the tissues, an extreme situation as rare as it is irremediable in nature. There are also the minor enigmas of the great variation, even among mammalian Hbs, in their crystal-form, solubility, etc. (p. 230). Some of these already, and no doubt others in due course, can be explained in terms of the structure of the molecule but it is still curious that molecular structure varies so much and in those precise ways within a recent and probably monophyletic group of animals. Not surprisingly (p. 232) oxyHb and deoxyHb differ in crystal form but each might be expected to be similar in all mammals. It is not clear also why in Man oxyHb is more soluble than deoxyHb and conversely in the rat (MANWELL, 1964) or why in the cow and sheep foetal Hb is 6–20 times more soluble than that of the adult but conversely in Man. Can all these differences be functional?

The reason for the occurrence of chlorohaemoglobin (ChlHb) in only a few groups of polychaete worms is particularly obscure. ChlHb is so similar to the common Hb(PrHb) that it cannot have more than marginal advantage over the latter which moreover is known to be highly efficient and versatile. If it has an advantage then this should have given it a wider distribution, at least among annelids. The special formyl side chain does occur also in other haems, of the cytochrome group, so that it may be a fairly common mutation but must be selected very rarely. In view of the extent to which haemerythrin and haemocyanin have converged on PrHb in their properties (p. 238) it seems unlikely that there could be selection for the converse process of wide dichotomy in the properties of Hb. The most conspicuous difference between PrHb and ChlHb, the higher CO-affinity of the latter can only be a disadvantage. ChlHb also has a high O_2-capacity and low O_2-affinity so that it should be almost as efficient as PrHb for active fully aerobic animals but in fact most of the polychaetes which have ChlHb are sedentary and sluggish. Many of them have PrHb also and among serpulimorphs collectively there is almost every conceivable arrangement in the distribution of the two Hbs in blood and tissues, with no evident division of labour between them (FOX and VEVERS, 1960, p. 101; FOX, 1964). Fox therefore was driven to suggest that ChlHb arises sporadically as a chance mutation from PrHb and persists in those polychaete groups because there is inadequate counter-selection. This still leaves the question why it does not persist in other taxa. The high O_2-capacity could make it an efficient carrier by the Wittenberg mechanism, for which its low affinity need not be a serious drawback and this remains the only tentative clue to a positive virtue.

For both haemerythrin and haemocyanin the precise nature of the ligand field and the way in which the metal binds O_2 are still uncertain (p. 235). The curious behaviour of Hcy in *Maia* and *Crangon*, in relation to the moult-cycle, is difficult to explain except on the lines suggested (p. 226). No doubt many animals do operate with little reserve of protein (in contrast to their large stores of some chromes!) and consequently when Crustacea are fasting and at the same time very inactive except morphogenetically it may be feasible to sacrifice Hcy to the need for free amino acids, the latter being used both for osmotic purposes and for building structural proteins. Recent studies indicate that the transport of O_2 in simple solution is adequate for the basal respiration of some Crustacea. The role of haemovanadin remains a tantalising problem. What is the significance of aphins in the blood of aphids, and of their chemical changes in shed blood? There are other relatively permanent blood-chromes of obscure function.

There would seem to be a risk of porphyria from the presence of free porphyrins in the spinal cord of birds and mammals (p. 100) and in some other animal tissues. However it may prove that they are never in positions exposed to dangerous levels of illumination. In the skin porphyrins are usually conjugated with an effective quencher or screened by other chromes. More obscure is the reason why bilins accumulate in the bones of some teleosts, but not of others, and quinones in the bones of some mammals (p. 102).

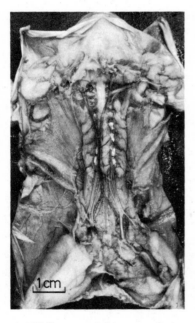

Fig. 18.1. Dissection of frog, Rana temporaria, to show melanin veil around systemic arches, dorsal aorta, and iliac and other arteries as well as around the sympathetic nervecords lying laterally to the dorsal aorta and systemic arches. The pigment picks out the entry of these cords and of blood vessels into the posterior end of the skull

While none of the accumulations of chromes in the internal organs of animals is demonstrably due to a difficulty in excreting the material the problem of the significance of some of the stores remains unsolved. It might seem relevant that plants also tend to deposit large amounts of chromes in various parts of the body, particularly in the wood of roots and in the bark. Among these are quinones, which also are responsible for the greatest accumulations in animals (p. 245). The main sites in plants imply a protective function against predators and the heaviness of the deposits may be the result of mutations to increased tolerance by predators. Some stores of chromes in animals may serve a similar function of passive chemical defence, as contrasted with the active secretion of most defensive quinones of animals (p. 159); at the same time it is noteworthy that when irritated the earthworm *Eisenia* extrudes chromes from the coelomic tissues.

Among the most obscure problems concerning deep chromes is the rationale for the veil of melanin round blood vessels and nerves in the abdominal cavity of the frog and other Amphibia (Fig. 18.1) and for the complete mantle of melanin round the abdominal cavity of the lizard (Fig. 18.2); this ends sharply at the abdominal-thoracic boundary. Fishes often have a mantle of purine, or purine mixed with melanin, round their abdominal cavity. It is conceivable that the melanophores of the frog's nerves and blood vessels may have chromomotor properties capable of forming a complete screen. A possible explanation

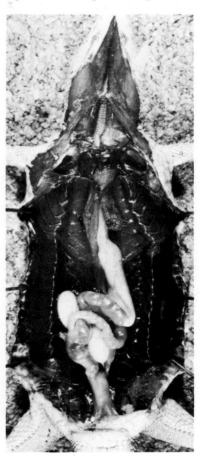

Fig. 18.2. Dissection of common lizard, Lacerta viridis, to show the complete cloak of melanin around the abdominal cavity and ending sharply at the junction between abdomen and thorax. The thoracic wall is red with haemoglobin so that the contrast in a fresh specimen is very striking

of these screens is that they protect particular structures or even the body-wall musculature from irradiation emanating from the viscera. Radioactive or other dangerous materials may be acquired in the food or generated during digestion. Also there is a high rate of cell-division in the gut-mucosa and the testes and this gives a small output of u. v. radiation (BARENBOIM et al., 1969), a type of chemiluminescence called 'mitogenetic radiation' (GURVICH, 1959). Alternatively the melanin may speed the radiation of muscular and nervous heat so as to cool these organs and to warm the viscera.

Altogether there are many functional and evolutionary questions still to be resolved but most of them could be answered by methods already available. Biochromes have fundamental biochemical and physiological importance and not simply an aesthetic interest for the naturalist. The solution of chromatological problems will be the solution of major problems in biochemical evolution. An evident preliminary problem is the complete identification of all zoochromes.

18.3 Unidentified Zoochromes

There is a conspicuous absence of any significant reference to fuscins (p. 35) in this chapter and indeed in any but chapters 3,5,8 and 10; this is an emphatic hint that little can be discovered about the function of any but an integumental chrome until its chemical nature is known. *A fortiori* no complete and balanced picture of the role of zoochromes can be expected until all have been identified. It is recalled that a considerable number of zoochromes so far studied still have not been identified (Table 3.14).

No doubt when appropriate methods are used a high proportion of these will prove to belong to known chemical classes of chrome but it seems likely that some at least are members of classes not yet known in animals or even in any organism. When all the pigments of a fully representative selection of animals have been analysed the number of classes is likely to be greater still. There is every reason to go ahead as rapidly as possible with this essential groundwork.

18.4 Conclusion

The known zoochromes perform useful functions in all instances that have been adequately studied. Most are very well, and many extremely well, adapted to their functions. There remain a number of problems which emphasise the need for more research at all levels, particularly in connection with the chromes of the internal organs. The need to identify chemically all relevant chromes cannot be overemphasised: except in the integument there is little chance of establishing the function of an unidentified substance.

Advances have been made since Poulton, studying only the integumental chromes was led to conclude that many of these are useless and irrelevant. Few would now agree that "... colour as such is not necessarily of any value to an organism ..." (POULTON, 1890, p. 12) though some may wonder if 'as such' was intended to have an implication. Still fewer would accept "... all colour was originally non-significant" (ib., p. 13). Most zoologists would feel uneasy about a chapter headed 'Non-Significant Colours', for reasons more positive than the difficulty of proving a null-hypothesis.

Had I the heaven's embroidered cloths
Enwrought with golden and silver light,
The blue and the dim and the dark cloths
Of night and light and the half light,
I would spread the cloths under your feet:
But I, being poor have only my dreams;
I have spread my dreams under your feet;
Tread softly because you tread on my dreams.
W. B. YEATS: "Aedh Wishes for the Cloths of Heaven"

Bibliography

ABE, K., BUTCHER, R.W., NICHOLSON, W.E., BAIRD, C.E., LIDDLE, R.A., LIDDLE, G.W.: Adenosine-3',5'-monophosphate(cyclic AMP)as the mediator of the actions of the melanophore-stimulating hormone (MSH) and nor-epinephrine on the frog skin. Endocrinology 84, 362–368 (1950).
ABELOOS, M., ABELOOS, R.: Sur les pigments hépatiques de *Doris tuberculata* Cuv. (Molluscque Nudibranche) et leurs relations avec les pigments de l'éponge, *Halichondria panicea* (Pall.). C.R. Soc. Biol. 109, 1238–1240 (1932).
ABELOOS, M., FISCHER, E.: Sur l'origine et les migrations des pigments carotenoides chez les crustacés. C. R. Soc. Biol. 95, 383–384 (1926).
ABELOOS, M., TEISSIER, G.: Notes sur les pigments animaux. Bull. Soc. Zool. France, 51, 145–151 (1926).
ABELOOS-PARIZE, M., ABELOOS-PARIZE, R.: Sur l'origine alimentaire du pigment carotenoide d' *Actinia equina* L. C. R. Soc. Biol. 94, 560–562 (1926).
ABELSON, P. H.: Chemical events on the primitive Earth. Proc. nat. Acad. Sci. U.S. 55, 1365–1372 (1966).
ABRAMOWITZ, A.A.: Color-changes in cancroid crabs of Bermuda. Proc. nat. Acad. Sci. U.S. 21, 677–681 (1935).
ABRAMOWITZ, A.A.: The double innervation of the caudal melanophores in *Fundulus*. Proc. nat. Acad. Sci. U.S. 22, 233–238 (1936).
AFZELIUS, B.A.: Its gross anatomy, histology and cytology. In: Brown Adipose Tissue, pp. 1–32. Ed. by O. LINDBERG. New York: American Elsevier Publ. Co. Inc. 1970.
AKAMATU, H., INOKUCHI, H., MATSUNAGA, Y.: Electrical conductivity of the perylene-Br_2 complex. Nature (London), 173, 168–169 (1954).
ALBRECHT, F.O.: État hygrometrique, coloration et resistance chez l'imago de *Locusta migratoria migratorioides* (R.F.), Experientia 20, 97–98 (1964).
ALBRECHT, F.O.: Groupement, état hygrometrique et photoperiode en resistance au jeune de *Locusta*. Bull. Biol. France Belg. 99, 287–339 (1965).
ALDRIDGE, W.N., STREET, B.W.: Mitochondria from brown adipose tissue. Biochem. J. 107, 315–317 (1968).
ALEXANDER, N.J.: Differentiation of the melanophores, iridophores and xanthophores from a common stem-cell. J. Invest. Dermatol. 54, 82 (1970).
ALEXANDER, P., BARTON, D.H.R.: The excretion of ethylquinone by the flour beetle. Biochem. J. 37, 463–465 (1943).
ALEXANDER, R. McN.: Physical aspects of swim-bladder function. Biol. Rev. Cambridge Phil. Soc. 41, 141–176 (1966).
ALI, M. A.: The ocular structure, retinomotor and photobehavioral responses of juvenile Pacific salmon. Can J. Zool. 37, 965–996 (1959).
ALLEN, E.G.: Conditions of the colour-change of prodigiosin. Nature (London) 216, 929–931 (1967).
ALLEN, H.: The distribution of colour marks in the Mammalia. Proc. nat. Acad. Sci. U.S. (1888), 84–105 (1888).
ALLEN, R.L.M.: Colour Chemistry. London: Nelson 1971.
ALMQUIST, H.J.: Nutrition. Ann. Rev. Biochem. 20, 305–342 (1951).
ANDERSON, H.A., MATHIESON, J.W., THOMSON, R.H.: Distribution of spinochrome pigments in echinoids. Comp. Biochem. Physiol. 28, 333–345 (1969).
ANDERSON, R.E.: Is retinal-phosphatidyl-ethanolamine the chromophor of rhodopsin? Nature (London), 227, 954–955 (1970).
ARCANGELI, A.: 1926: quoted from FOX and VEVERS, 1960.
ARVANITAKI, A., CHALAZONITIS, N.: Inhibition ou excitation des potentiels neuroniques à photoactivation distincte de deux chromoproteides (carotenoide et chlorophyllien). Arch. Sci. Physiol. 3, 45–60 (1949).
ARVANITAKI, A., CHALAZONITIS, N.: Activation par la lumiere des neurones pigmentés. Arch. Sci. Physiol. 12, 73–106 (1957).
ATZ, J.W.: Half-boiled lobster. Bull. N.Y. Zool. Soc. 42, 128 (1939).

AUSTERN, B.M., GAWIENOWSKI, A.M.: In vitro biosynthesis of β-carotene by bovine corpus luteum. Lipids, **4**, 227–229 (1969).
AUTRUM, H., LANGER, H.: Photolabile pterine im auge von *Calliphora erythrocephala*. Biol Zentralbl. **77**, 196–201 (1958).
AZARNOFF, D.L., BROCK, F.E., CURRAN, G.L.: A specific site of vanadium inhibition of cholesterol biosynthesis. Biochim. Biophys. Acta **51**, 397–398 (1961).
BÁBÀK, E.: Über den Einfluß des Lichtes auf die Vermehrung der Hautchromatophoren. Pflüger's Arch. Gesamte Physiol. Menschen Tiere **149**, 462–470 (1913).
BAGNARA, J.T.: Hypophyseal control of guanophores in Anuran larvae. J. Exp. Zool **137**, 265–283 (1958).
BAGNARA, J.T.: Pineal regulation of the body lightening reaction in Amphibian larvae. Science **132**, 1481–1483 (1960).
BAGNARA, J.T.: The pineal and the body lightening reaction of larval Amphibians. Gen. Comp. Endocrinol. **3**, 86–100 (1963).
BAGNARA, J.T.: Analyse des transformations des pteridines de la peau au cours de la vie larvaire et à la métamorphose chez le triton *Pleurodeles waltlii* Michah. C.R. Acad. Sci. **258**, 5969–5971 (1964).
BAGNARA, J.T.: Interrelationships of melanophores, iridophores and xanthophores in lower vertebrates. J. Invest. Dermatol **54**, 82 (1970).
BAGNARA, J.T., OBIKA, M.: Comparative aspects of integumental pteridine distribution among Amphibia. Comp. Biochem. Physiol. **15**, 33–49 (1965).
BAHL, K.N.: Excretion in the Oligochaeta. Biol. Rev. Cambridge Phil. Soc. **22**, 109–147 (1947).
BALDWIN, E.: Dynamic Aspects of Biochemistry. 2nd. edn. Cambridge: University Press 1947.
BALINSKY, J.B.: Nitrogen metabolism in Amphibia. In Comparative Biochemistry of Nitrogen Metabolism. Vol. II. pp. 519–637. Ed. by J.W. CAMPBELL. London-New York: Academic Press 1970.
BALL, E.G.: The role of flavoproteins in biological oxidations. Cold Springs Harbor Symp. Quant. Biol. **7**, 100–110 (1939).
BANNISTER, W.H., NEEDHAM, A.E.: Connective tissue pigment of the centipede, *Lithobius forficatus* (L.). Naturwissenschaften **58**, 58–59 (1971).
BANNISTER, W.H., WOOD, E.J.: Free electron model for the absorption of oxygen-bridged binuclear complexes in the near u.v. Nature (London) **225**, 53–55 1969).
BANNISTER, W.H., WOOD, E.J.: U.v. Fluorescence of *Murex trunculus* haemocyanin in relation to its binding of copper and oxygen. Comp. Biochem. Physiol. **40B**, 7–18 (1971).
BANNISTER, W.H., BANNISTER, J.V., MICALLEF, H.: The green pigment in the foot of *Monodonta* (Mollusca) species. Comp. Biochem. Physiol. **24**, 839–846 (1968a).
BANNISTER, W.H., BANNISTER, J.V., MICALLEF, H.: Biliprotein from the ova of a keyhole limpet. Nature (London) **217**, 754–755 (1968b)
BANNISTER, W.H., BANNISTER, J.V., MICALLEF, H.: Soluble pigment-protein complex from the shell of *Monodonta turbinata* (Archaeogastropoda). Comp. Biochem. Physiol. **35**, 237–243 (1970).
BANTOCK, C.R.: Natural Selection in experimental populations of *Cepaea hortensis*. J. Biol. Educ. **5**, 25–34 (1971).
BARBIER, M., LEDERER, E.: 1957: cited from VUILLAUME, 1969.
BARBIER, M., FAURÉ-FREMIET, E., LEDERER, E.: Sur les pigments du cilié *Stentor niger*. C.R. Acad. Sci. **242**, 2182–2184 (1956).
BARBIER, M., CHARNIAUX-COTTON, H., FRIED-MONTAUFIER, M.-C.: Presence dans les secondes antennes des mâles de *Talitrus saltator* et *Orchestia gammarella* (Crustacés Amphipodes d'un taux relativement élevé d'astaxanthine. C.R.H. Acad. Sci. **263**D, 1508–1510 (1966).
BARDEN, H., MARTIN, E.: Histochemical and elemental studies of neuromelanin and lipofuscin. J. Invest. Dermatol. **54**, 83 (1970).
BARENBOIM, G.M., DOMANSKII, A.N., TUROVEROV, K.K.: Luminescence of Biopolymers and Cells. New York: Plenum Press 1969.
BARNICOTT, N.A.: Human pigmentation. Man **57**, 114–120 (1957).
BARONE, P., INFERRERA, C., CARROZZI, G.: Pigments in the Dubin-Johnson syndrome. In: Pigments in Pathology, pp. 307–328. Ed. by M. WOLMAN. New York-London: Academic Press 1969.
BARRETT, J., BUTTERWORTH, P.E.: Carotenoids of *Polymorphus minutus* (Acanthocephala) and its intermediate host *Gammarus pulex*. Comp. Biochem. Physiol. **27**, 575–581 (1968).

BARTLEY, J. A.: A histological and hormonal analysis of physiological and morphological chromatophore responses in the soft-shelled turtle, *Trionyx* sp. J. Zool. 163, 125–144 (1971).
BASERGA, R., THATCHER, D., MARZI, D.: Cell-proliferation in mouse kidney after a single injection of folic acid. Lab. Invest. 19, 92–96 (1968).
BATES, G., BRUNORI, M., ANTONINI, E., AMICONI, G., WYMAN, J.: Studies on haemerythrin. I. Thermodynamic and kinetic aspects of oxygen-binding. Biochemistry 7, 3016–3020 (1968).
BATTERSBY, K. A.: Applications of colour in everyday life. In: Colour and Life, pp. 125–132. Ed. by W. B. BROUGHTON. London: Institute of Biology 1964.
BAUER, V.: Zur Hypothese der physikalischen Wärmeregulierung durch Chromatophoren. Z. Allg. Physiol. 16, 191–212 (1914).
BAUMANN, C.: Regeneration of porphyropsin *in situ*. Nature (London) 233, 484–485 (1971).
BEADLE, G. W., EPHRUSSI, B.: Development of eye-colours in *Drosophila:* Diffusible substances and their interrelations. Genetics, 22, 76–86 (1937).
BEADLE, L. C.: Respiration in the African swampworm, *Alma emini* Mich. J. Exp. Biol. 34, 1–10 (1957).
BEATTY, R. A.: The pigmentation of cavernicolous animals. III. The carotenoid pigments of some Amphipod Crustacea. J. Exp. Biol. 26, 125–130 (1949).
BEATTY, R. A.: Melanising activity of semen from rabbit males of different genotype. Proc. Roy. Phys. Soc. Edinburgh 25, 39–44 (1956).
BECKER, E.: Über das Pterinpigment bei Insekten und die Färbung und Zeichnung von *Vespa* im besonderen. Z. Morphol. Ökol. Tiere 32, 672–751 (1937).
BECKER, E.: Die Gen-Wirkstoffe-Systeme der Augenausfärbung bei Insekten. Naturwissenschaften 26, 433–441 (1938).
BECKER, E.: Die Pigmente der Ommin und Ommatin Gruppe, eine neue Klasse von Naturfarbstoffen. Naturwissenschaften 29, 237–238 (1941).
BECKER, E.: Über Eigenschaften, Verbreitung und die Genetischentwicklungsphysiologische Bedeutung der Pigmente der Ommatin- und Ommin-gruppe (Ommochrome) bei den Arthropoden. Z. Indukt. Abstam. Vererbungsl. 80, 157–204 (1942).
BEER, R. J. S., BROADHURST, T., ROBERTSON, A.: The chemistry of melanins. V. Autoxidation of 5,6-dihydroxyindoles. J. Chem. Soc. pp. 1947–1953 (1954).
BEERSTECHER, E.: The nutrition of the Crustacea. Vitam. Horm. 10, 69–77 (1952).
BELANGER, A. M., RUSTAD, R. C.: Movements of echinochrome granules during early development of sea-urchin eggs. Nature, New Biol. (London) 239, 81–83 (1972).
BELLAMY, D.: The structure and metabolic properties of tissue-preparations from *Schistocerca gregaria* (Desert Locust). Biochem. J. 70, 580–589 (1958).
BENAZZI-LENTATI, G.: Ricerche istochimiche sulla branchia die *Mytilus edulis* con particolari riguardo al pigmento. Arch. Zool. Torino, 27, 63–91 (1939).
BENDIX, S.: Pigments in phototaxis. In: Comparative Biochemistry of Photoreactive Systems, pp. 107–127. Ed. by M. B. ALLEN. New York: Academic Press 1960.
BENESCH, R., BENESCH, R. E.: Intracellular organic phosphates as regulators of oxygen-release by haemoglobin. Nature (London), 221, 618–622 (1969).
BENOIT, J.: Hypothalamo-hypophyseal control of sexual activity in birds. Gen. Comp. Endocrinol. Suppl. 1, 254–274 (1962).
BENOIT, J., OTT, L.: External and Internal factors in sexual activity: Effect of irradiation with different wavelengths on the mechanisms of photostimulation of the hypophysis and on testicular growth in the immature duck. Yale J. Biol. Med. 17, 27–46 (1944).
BENTLEY, K. W.: The Natural Pigments. New York: Interscience 1960.
BERENDS, W., POSTHUMA, J., SUSSENBACH, J. S., MAGER, H. I. X.: On the mechanism of some flavin-photosensitised reactions. In: Flavins and Flavoproteins, pp. 22–36. Ed. by E. C. SLATER. Amsterdam-London-New York: Elsevier Publ. Co. 1966.
BERGEL, F., BRAY, R. C., HARRAP, K. R.: A model system for cysteine desulphydrase action: pyridoxal phosphate-vannadium. Nature (London) 181, 1654–1655 (1958).
BERGMANN, W., BURKE, D. C.: Contributions to the study of marine products XL: The nucleosides of sponges: IV: Spongosine. J. Org. Chem. 21, 226–228 (1956).
BERGMANN, W., STEMPIEN, M. F. Contributions to the study of marine products XLIII: The nucleosides of sponges, V: Synthesis of spongosine. J. Org. Chem. 22, 1575–1577 (1957).
BERNAL, J. D.: The origin of Life. New Biology 16, 28–40 (1954).
BERRILL, N. J.: Regeneration and budding in worms. Biol. Rev. Cambridge Phil. Soc. 27, 401–438 (1952).

BERSIN, T.: Exchange adsorption in Man. In: Ion Exchangers in Organic and Biochemistry, pp. 486–501. Ed. by C. CALMON and T.R.E. KRESSMAN. New York: Interscience 1957.
BIELIG, H.-J., BAYER, E., DELL, H.-D., ROHNS, G., MOLLINGER, H., RÜDIGER, W.: Chemistry of haemovanadin. In: Protides of the Biological Fluids. Vol. 14, Ed. by H. PEETERS. pp. 197–204. Amsterdam-London-New York: Elsevier Publ. Co. 1967.
BIELIG, H.-J., DOHRN, P.: Zur Frage der Wirkung von Echinochrom A und Gallerthüllen Substanz auf die Spermatozoen des Seeigels *Arbacia lixula (A. pustulosa)*. Z. Naturforsch. 5B, 316–338 (1950).
BIELIG, H.-J., WOHLER, F.: Bauplan des Hämosiderins. Naturwissenschaften 45, 488–489 (1958).
BIRCH, A. J.: Biosynthesis of flavonoids and anthocyanins. In: Symposia on Biochemistry, B. 1: Natural Pigments and Their Biogenesis. 17th. Int. Congr. Pure Appl. Chem. Munich, 1959: Main Lectures, Vol. II, pp. 73–84 (1960).
BIRCH, A. J., DONOVAN, F. W.: Studies in relation to biosynthesis I: Some possible routes to derivatives of orcinol and phloroglucinol. Aust. J. Chem. 6, 360–368 (1953).
BISONETTE, T. H., BAILEY, E. E.: Experimental modification and control of molts and changes of coat-color in weasels by controlled lighting. Ann. A.Y. Acad. Sci. 45, 221–260 (1944).
BITENSKY, M. W., DEMOPOULOS, H. B.: Activation of melanoma adenyl cyclase by melanocyte stimulating hormone. J. Invest. Dermatol. 54, 83 (1970).
BJÖRKERUD, S., CUMMINS, J. T.: Selected enzymic studies of lipofuscin granules isolated from bovine cardiac muscle. Exp. Cell Res. 32, 510–520 (1963).
BLOCH-RAPHAEL, C.: Localisation, formation et destruction de l'hemoglobine chez les Annélides Polychetes. Ann. Inst. Océanog. (Paris) 19, 1–78 (1939).
BLOCH-RAPHAEL, C.: Evolution de l'hemoglobine et de ses dérivés au cours du developpement de *Septosaccus cuenoti* (Duboscq). C.R. Soc. Biol. 142, 67–68 (1948).
BLOIS, M. S.: Random polymers as a matrix for chemical evolution. In: The Origins of Prebiological Systems, pp. 19–38. Ed. by S.W. Fox. New York: Academic Press 1965.
BLOIS, M.S. Recent developments in the physics and chemistry of the melanins. In: The Biological Effects of Ultraviolet Radiation, pp. 299–304. Ed. by F. URBACH. Oxford: Pergamon Press 1969
BLUM, H.F.: Does the melanin pigment of human skin have adaptive value? Quart. Rev. Biol. 36, 50–63 (1961).
BLUM, H.F.: Photodynamic Action and Diseases caused by Light. New York: Hafner Publ. Co. 1964.
BLUMBERG, W. E.: Discussion. In: The Biochemistry of Copper, pp. 578–579. Ed. by T. PEISACH, P. AISEN, and W.E. BLUMBERG. New York: Academic Press 1966.
BLUMER, M.: Organic pigments: their long-term fate. Science 149, 722–726 (1965).
BOCQUET, C.: Recherches sur le polymorphisme naturel de *Jaera marina* (Fabr.) (Isopodes Asellotes). Arch. Zool. Exp. Gen. 90, 187–450 (1953).
BODANSKY, M.: Introduction to Physiological Chemistry. New York: Wiley 1938.
BODINE, J.H., FITZGERALD, L.R.: Riboflavin and other fluorescent compounds in a developing egg (Orthoptera). Physiol. Zool. 20, 146–160 (1947a).
BODINE, J.H., FITZGERALD, L.R.: A spectrophotometric study of a developing egg (Orthoptera) with especial reference to riboflavin and its derivatives. J. Exp. Zool. 104, 353–363 (1947b).
BOELTS, K.J., PARKHURST, L.J.: Ligand-binding kinetics of haemoglobin from the earthworm. Biochem. Biophys. Res. Comm. 43, 637–643 (1971).
BOGORAD, L.: The biogenesis of heme, chlorophyll and bile pigments. In: The Biogenesis of Natural Compounds, pp. 183–231. Ed. by P. BERNFELD. 1st Edn. Oxford: Pergamon Press 1963.
BOGORAD, L., TROXLER, R. F.: Biogenesis of heme, chlorophyll and bile pigments. In: The Biogenesis of Natural Compounds, pp. 247–313. Ed. by P. BERNFELD. 2nd. Edn. Oxford: Pergamon Press 1967.
BOHR, C.: Blutgase und respiratorischer Gaswechsel. In: Handbuch der Physiologie des Menschen. Band I, pp. 54–222. Ed. by W. NAGEL. Braunschweig: F. Viewig u. Sohn 1909.
BOHREN, B.B., CONRAD, R.M., WARREN, D.C.: A chemical and histological study of the feather pigments of the domestic fowl. Amer. Nat. 77, 481–518 (1943).
BOLKER, H.I., BRENNER, H.S.: Polymeric structure of spruce lignin. Science 170, 173–176 (1970).
BOMIRSKI, A., NOWINSKA, L., PAUTSCH, F.: The tyrosine positive amelanotic melanoma in the golden hamster. In: Structure and Control of the Melanocyte, pp. 252–259. Ed. by G. Della PORTA and O. MÜHLBOCK. Berlin-Heidelberg-New York: Springer 1966.
BOOTH, C.C.: Absorption from the small intestine. Sci. Basis Med. (1963) 171–196.

Bossa, F., Brunori, M., Bates, G.W., Antonini, E., Fasella, P.: Studies on haemerythrin II. circular dichroism and optical rotary dispersion of haemerythrin from *Sipunculus nudus.* Biochim. Biophys. Acta **207**, 41–48 (1970).

Bowman, T. E.: Chromatophorotropins in the central nervous organs of the crab *Hemigrapsus oregonensis.* Biol. Bull. **96**, 238–245 (1949).

Bowness, J.M., Morton, R. A.: The association of zinc and other metals with melanin and a melanin-protein complex. Biochem J. **53**, 620–626 (1953).

Bowness, J.M., Morton, R.A., Shakir, M.H., Stubbs, A.L.: Distribution of copper and zinc in mammalian eyes: Occurrence of metals in melanin fractions from eye-tissues. Biochem. J. **51**, 521–530 (1952).

Bozler, E.: Über die Tätigkeit der einzelnen Glattmuskelfaser bei der Kontraktion II: Mitteilung: Die chromatophoren Muskeln der Cephalopoden. Z. Vergl. Physiol. **7**, 379–406 (1928).

Bozler, E.: Weitere Untersuchungen zur Frage des Tonussubstrates. Z. Vergl. Physiol. **8**, 371–390 (1929).

Bradley, H.C.: A green bilin in the digestive fluid of *Cambarus.* J. Biol. Chem. **4**, 36–37 (1908).

Brady, F. O., Rajagopalan, K. V., Handler, P. Preparation and properties of flavin-free flavoproteins. In: Flavins and Flavoproteins, pp. 425–446. Ed. by H. Kamin. Baltimore: University Park Press and London: Butterworth 1971.

Braterman, P.S., Davies, R.C., Williams, R.J.P.: The properties of metal porphyrins and similar complexes. In: Structure and Properties of Biomolecules and Biological Systems, pp. 359–407. Ed. by J. Duchesne. London: Interscience 1964.

Bray, R.C.: The role of molybdenum in milk xanthine oxidase. In: Flavins and Flavoproteins, pp. 385–398. Ed. by H. Kamin. Baltimore: University Park Press and London: Butterworth 1971.

Bray, R.C., Chisholm, A.J., Hart, L.I., Meriwether, L.S., Watts, D.C.: Studies on the composition and mechanism of action of milk xanthine oxidase. In: Flavins and Flavoproteins, pp. 117–132. Ed. by E.C. Slater. Amsterdam-London-New York: Elsevier 1966.

Breathnach, A. S.: Normal and abnormal melanin pigmentation of the skin. In: Pigments in Pathology, pp. 353–394. Ed. by M. Wolman. New York: Academic Press 1969.

Brecher, L.: Der Weg der Farbanpassung bei Schmetterlingspuppen vom Rezeptor bis zum Effektor. Die Puppenfärbungen der Vanessiden (*V.io* L., *V.urticae* L.) IV und V. Biol. Gen. **14**, 212–227 (1938).

Breder, C.M., Rasquin, P.: Further notes on the pigmentary behaviour of *Chaetodipterus* in reference to background and water transparency. Zoologica (New York) **40**, 85–90 (1955).

Brehm, V.: Rotfärbung von Hochgebirgsseeorganismen. Biol. Rev. Cambridge Phil. Soc. **13**, 307–318 (1939).

Breton-Gorius, J.: Etude au microscope electronique des cellules chlorogogenes d'*Arenicola marina* L. Ann. Sci. Nat. Zool. Biol. Animale, (12), **5**, 211–272 (1963).

Breuer, M.M.: Binding of aromatic compounds to hair. Nature (London), Phys. Sci. **229**, 185–186 (1971).

Brindley, M.D.H.: On the occurrence of oxyhaemoglobin in *Macrocorixa geoffreyi* Leach. Trans. Entomol. Soc. London (1929) 5–6.

Brinkmann, R.T.: Distribution of water vapour in the stratosphere as derived from setting sun absorption data. J. Geophys. Res. **74**, 5355–5368 (1969).

Brodie, A.F.: The role of naphthoquinones in oxidative metabolism. In: Biochemistry of Quinones, pp. 355–404. Ed. by R. A. Morton. London – New York: Academic Press 1965.

Brooks, D.W., Dawson, C.R.: Aspects of tyrosinase chemistry. In: The Biochemistry of Copper, pp. 343–357. Ed. by J. Peisach, P. Aisen, and W.E. Blumberg. New York: Academic Press 1966.

Brooks, G., Paulais, R.: Répartition et localisation des carotenoides, des flavins et de l'acide-L-ascorbique chez les Mollusques Lamellibranches: cas des huitres et des gryphées. C.R.H. Acad. Sci. **208**, 833–835 (1939).

Brooks, M.M.: Activation of eggs by oxidation-reduction indicators. Science **106**, 320. (1947).

Brown, C.H.: Quinone tanning in the animal kingdom. Nature (London) **165**, 275. (1950).

Brown, F.A.: Color changes in *Palaemonetes.* J. Morphol. **57**, 317–333 (1935).

Brown, F.A.: Background selection in crayfishes. Ecology **20**, 507–516 (1939).

Brown, F. A.: The source and activity of *Crago*-darkening hormone (CDH). Physiol. Zool. **19**, 215–223 (1946).

Brown, F.A.: Action of Hormones in Plants and Invertebrates. New York: Academic Press 1952.

Brown, F. A., Cunningham, O.: Upon the presence and distribution of a chromatophorotropic principle in the central nervous system of *Limulus*. Biol. Bull. **81**, 80–95 (1941).
Brown, F. A., Ederstrom, M. E.: Dual control of certain black chromatophores of *Crago*. J. Exp. Zool. **85**, 53–69 (1940).
Brown, F. A., Klotz, I. M.: Separation of two mutually antagonistic chromatophorotropins from the tritocerebral commissure of *Crago*. Proc. Soc. Exp. Biol. Med. **64**, 310–313 (1947).
Brown, F. A., Meglitsch, A.: Comparison of the chromatophorotropic activity of insect corpora cardiaca with that of Crustacean sinus glands. Biol. Bull. **79**, 409–418 (1940).
Brown, F. A., Saigh, L. M.: The comparative distribution of the two chromatophorotropic hormones (CDH and CBLH) in Crustacean nervous systems. Biol. Bull. **91**, 170–180 (1946).
Brown, F. A., Sandeen, M. I.: Responses of the chromatophores of the fiddler crab *Uca* to light and temperature. Physiol. Zool. **21**, 361–371 (1948).
Brown, F. A., Thompson, D. H.: Melanin dispersion and choice of background in fishes with special reference to *Erycymba buccata*. Copeia, (1937), 172–238.
Brown, F. A., Webb, H. M.: Studies on the daily rhythmicity of the fiddler crab *Uca*: Modificatins by light. Physiol. Zool. **22**, 136–148 (1949).
Brown, F. A., Bennett, N. F., Ralph, C. L.: Apparent reversible influence of cosmic ray induced showers upon a biological system. Proc. Soc. Exp. Biol. Med. **89**, 332–337 (1955).
Brown, F. A., Hines, M. N., Fingerman, M.: Hormonal regulation of the distal retinal pigment of *Palaemonetes*. Biol. Bull. **102**, 212–225 (1952).
Brown, F. A., Webb, H. M., Sandeen, M. I.: The action of two hormones regulating the red chromatophores of *Palaemonetes*. J. Exp. Zool. **120**, 391–420 (1952).
Brown, P. K., Wals, G.: Visual pigments in sigle rods and cones of the human retina. Science **144**, 45–52 (1964).
Brunet, P. C. J.: The metabolism of aromatic compounds. In: Aspects of Insect Biochemistry, pp. 49–77. Ed. by T. W. Goodwin (Brit. Biochem. Soc. Sympos. No. 25). London – New York: Academic Press 1965.
Brunet, P. C. J.: Sclerotins. Endeavour **26**, 68–74 (1967).
Brunet, P. C. J., Kent, P. W.: Observations on the mechanism of a tanning reaction in *Periplaneta* and *Blatta*. Proc. Roy. Soc. London, B. **144**, 259–274 (1955).
Brunner, O., Stein, R.: Über die Carotenoide von *Rana esculenta*. Biochem. Z. **282**, 47–50 (1935).
Bu'Lock, J. D.: Intermediates in melanin formation. Arch. Biochem. Biophys. **91**, 189–193 (1960).
Bu'Lock, J. D.: The formation of melanin from adrenochrome. J. Chem. Soc. (**1961**), 52–58 (1961).
Bu'Lock, J. D., Harley-Mason, J.: Melanin and its precursors. Part II: Model experiments on the reactions between quinones and indoles, and consideration of a possible structure for the melanin polymer. J. Chem. Soc. (**1951**), 703–712.
Burgeff, H., Fetz, H.: Strahlenmessungen an Falten der Gattung *Zygaena* Fab. zur Erklärung des Littoralmelanismus. Biol. Zentralbl. **87**, 689–703 (1968).
Burgess, E.: Mars since the dust settled. New Sci. **53**, 420–423 (1972).
Burgess, L. E.: A preliminary quantitative study of pterin pigment in the developing egg of the grasshopper, *Melanoplus differentialis* Arch. Biochem. **20**, 347–355 (1949).
Burkhardt, D.: Spectral activity and other response characteristics of single visual cells in the Arthropod eye. In: Biological Receptor Mechanisms (Symp. Soc. Exp. Biol. No. 16), pp. 86–109. Ed. by J. W. L. Beament. Cambridge: University Press 1962.
Burtt, E.: Exudate from millipedes with particular reference to its injurious effects. Trop. Dis. Bull. **44**, 7–12 (1947).
Burwood, R., Reed, G., Schofield, K., Wright, D. E.: The pigments of the stick lac II: The structure of laccaic acid A_1. J. Chem. Soc. C, (1967), 842–851.
Bush, C. A.: Optical rotary dispersion and circular dichroism. In: Physical Techniques in Biological Research, Vol. I, A, pp. 347–408. 2nd. Edn., Ed. by G. Oster and A. W. Pollister. New York: Academic Press 1971.
Busnel, R.-G., Drilhon, A.: Presence de riboflavine (Vitamine B_2) chez un insecte, *Tineola bisselliella* Hum. alimenté avec un régime privé de cette substance. C. R. H. Acad. Sci. **216**, 213–214 (1943).
Busselen, P.: The presence of haemocyanin and of serum proteins in the eggs of *Carcinus maenas*, *Eriocheir sinensis* and *Portunus holsatus*. Comp. Biochem. Physiol. **38 A**, 317–328 (1971).
Butenandt, A.: The mode of action of hereditary factors. Endeavour **11**, 188–192 (1952).
Butenandt, A.: Über Ommochrome, eine Klasse natürlichen Phenoxazonefarbstoffe. Angew. Chem. **69**, 16–23 (1957).

Butenandt, A.: Über neue Farbstoffe, ihre Biogenese und physiologische Bedeutung. In: Symposia on Biochemistry, Opening Lecture. 17th. Int. Congr. Pure Appl. Chem. Munich, 1959. Main Lectures, Vol. II, pp. 11–31 (1960).

Butenandt, A., Rembold, H.: Über den Weiselzellenfuttersaft der Honigbiene II: Isolierung von 2-amino-4-hydroxy-6-(-1,2-dihydroxy propyl)-pteridin. Hoppe Seyler's Z. Physiol. Chem. 311, 79–83 (1958).

Butenandt, A., Schiedt, U., Biekert, E., Kornmann, P.: Über Ommochrome I: Isolierung von Xanthommatin, Rhodommatin und Ommatin C aus den Schlupfsekreten von *Vanessa urticae*. Liebig's Ann. 586, 217–228 (1954a).

Butenandt, A., Schiedt, U., Biekert, E.: Über Ommochrome III: Synthese des Xanthommatins. Liebig's Ann. 588, 106–116 (1954b).

Butenandt, A., Biekert, E., Linzen, B.: Über Ommochrome VII: Modelversuche für Bildung des Xanthommatins *in vivo*. Hoppe Seyler's Z. Physiol. Chem. 305, 284–289 (1956).

Butenandt, A. Biekert, E., Kübler, H., Linzen, B.: Über Ommochrome XX: Zur Verbreitung der Ommatine im Tierreich. Hoppe Seyler's Z. Physiol. Chem. 319, 238–256 (1960a).

Butenandt, A., Biekert, E., Koga, N., Traub, P.: Über Ommochrome XXI: Konstitution und Synthese des Ommatins D. Hoppe Seyler's Z. Physiol. Chem. 321, 258–275 (1960b).

Bykhovskii, V. Y., Zaitseva, N. I., Bukin, V. N.: Some aspects of the regulation of the biosynthesis of vitamin B_{12} and porphyrins in *Propionibacterium shermanii*. Dokl. Akad. Nauk. 185, 459–461 (1969).

Cain, A. J., Sheppard, P. M.: Selection in the polymorphic land snail, *Cepaea nemoralis*. Heredity 4, 274–294 (1950).

Cairns, W. L., Treadwell, G. E., Metzler, D. E.: Products of anaerobic photolysis of ribitylflavin. In: Flavins and Flavoproteins, pp. 189–191. Ed. by K. Yagi. Tokyo: University Press and Baltimore: University Park Press 1969.

Califano, L., Boeri, E.: Studies on haemovanadin III: Some physiological properties of haemovanadin, the vanadium compound of the blood of *Phallusia*. J. Exp. Biol. 27, 253–256 (1950).

Califano, L., Caselli, P.: Ricerche sulla emovanadina I: Dimostrazione di una proteine. Pubbl. Sta. Zool. Napoli 21, 261–271 (1947).

Califano, L., Caselli, P.: Ricerche sulla emovanadina II: Analisi spetrofotometrica. Pubbl. Sta. Zool. Napoli 22, 138–145 (1950).

Calvin, M.: Evolution of photosynthetic mechanisms. Perspect. Biol. Med. 5, 147–172 (1962).

Cameron, D. W., Cromartie, R. I. T., Kingston, D. G. I., Scott, P. M., Hamied, Y. K., Joshi, B. S., Calderbank, A., Haslam, E., Sheppard, N., Watkins, J. C., Lord Todd: Colouring matters of the Aphidae XVII–XXII. J. Chem. Soc. (1964), 48–104.

Campbell, J. W., Bishop, S. H.: Nitrogen metabolism in Molluscs. In: Comparative Biochemistry of Nitrogen Metabolism, pp. 103–206. Ed. by J. W. Campbell. London – New York: Academic Press 1970.

Candeloro, S., Grdenić, D., Taylor, N., Thompson, B., Viswamitra, M., Hodgkin, D. M. C.: Structure of ferroverdin. Nature (London), 224, 589–591 (1969).

Cannan, R. K.: Echinochrome. Biochem. J. 21, 184–189 (1927).

Carleton, H. M., Leach, E. H.: Histological Technique. 2nd. Edn. London: Oxford University Press 1938.

Carlisle, D. B.: Niobium in ascidians. Nature (London) 181, 933. (1958).

Carlisle, D. B.: Colour change in animals. In: Colour in Life, pp. 61–67. Ed. by W. B. Broughton. London: Institute of Biology 1964.

Carlisle, D. B., Knowles, Sir F. G. W.: Endocrine Control in Crustacea. Cambridge: University Press 1959.

Carpenter, G. D. H.: Experiments on the relative edibility of insects with special reference to coloration. Trans. Entomol. Soc. London (1921), 1–105.

Carthy, J. D.: The physiology of colour perception in invertebrates. In: Colour in Life, pp. 69–78. Ed. by W. B. Broughton. London: Institute of Biology 1964.

Caserta, G., Ghiretti, F.: Crystalline ubiquinone (Coenzyme Q_{10}) from sea-urchin sperm. Nature (London) 193, 1079–1080 (1962).

Casnati, A., Nencini, G., Quilico, A., Pavan, M., Ricca, A., Salvatori, T.: The secretion of the myriapod, *Polydesmus collaris collaris* (Koch). Experientia 19, 409–411 (1963).

Caspari, E.: Physiological action of eye colour mutants in the moths *Ephestia kuhniella* and *Ptychopoda seriata*. Quart. Rev. Biol. 24, 185–199 (1949).

CASPARI, E.: Discussion. In: Pigment Cell Growth, p. 41. Ed. by M. GORDON. New York: Academic Press 1953.
CASSIDY, H. G.: Electron exchange polymers I. J. Amer. Chem. Soc. 71, 402–406 (1949).
CASTLE, W. E.: Mammalian Genetics. Cambridge, Mass.: Harvard University Press 1940.
CASTLE, W. E.: Coat colour inheritance in horses and other mammals. Genetics 39, 35–44 (1954).
CAUGHEY, W. S.: Porphyrin-proteins and enzymes. Ann. Rev. Biochem. 36, 611–644 (1967).
CHALAZONITIS, N., ARVANITAKI, A.: Identification et localisation de quelques catalyseurs respiratoires dans le neurone d'*Aplysia*. Bull. Inst. Oceanog. No. 996, 1–20 (1950).
CHANCE, B., SCHOENER, B., SCHINDLER, F.: The intracellular oxidation-reduction state. In: Oxygen in the Animal Organism, pp. 357–392. Ed. by F. DICKENS and E. NEIL. Oxford: Pergamon Press 1964.
CHANCE, B., ESTABROOK, R. W., YONETANI, T.: Hemes and Hemoproteins. New York-London: Academic Press 1966.
CHARLTON, H. M.: The pineal gland and colour change in *Xenopus laevis* Daudin. Gen. Comp. Endocrinol. 7, 384–397 (1966).
CHASSARD-BOUCHARD, C.: L'adaptation chromatique chez les Natantia (Crustacés Decapodes). Cah. biol. Mar. 6, 469–576 (1964).
CHATTON, E., LWOFF, A., PARAT, M.: L'origine, la nature et l'évolution du pigment des Spirophyra, des Polyspira et des Gymnodinioides. Presence de carotinalbumins dans la mue des Crustacés Décapodes. C.R. Soc. Biol. 94, 567–570 (1926).
CHAUVIN, R.: Sur le rougissement du criquet pèlerin. C.R.H. Acad. Sci. 207, 1018–1020 (1938).
CHAVIN, W.: Pituitary-adrenal control of melanisation in xanthic goldfish, *Carassius auratus* L. J. Exp. Zool. 133, 1–36 (1956).
CHEESMAN, D. F.: Ovorubin, a chromoprotein from the eggs of the gastropod Mollusc *Pomacea canaliculata*. Proc. Roy. Soc. London B, 149, 571–587 (1958).
CHEESMAN, D. F., LEE, W. L., ZAGALSKY, P. F.: Carotenoproteins in invertebrates. Biol. Rev. Cambridge Phil. Soc. 42, 131–160 (1967).
CHELDELIN, V. H., BAICH, A.: Biosynthesis of the water-soluble vitamins. In: Biogenesis of Natural Compounds, pp. 679–742. 2nd. Edn. Ed. by P. BERNFELD. Oxford: Pergamon Press 1967.
CHICHESTER, C. O., NAKAYAMA, T. O. M.: The biosynthesis of carotenoids and vitamin A. In: Biogenesis of Natural Compounds. 2nd. Edn., pp. 641–678. Ed. by P. BERNFELD. Oxford: Pergamon Press 1967.
CHRISTOMANOS, A.: Purple pigment and protein in the threads of the sea anemone *Adamsia rondeleti*. Nature (London) 171, 886–887 (1953).
CHU, E. J.: A simple qualitative test to distinguish between protoporphyrin IX or its esters and porphyrins containing no vinyl group. J. Biol. Chem. 166, 463–464 (1946).
CHURCH, A. H.: Researches on turacin, an animal pigment containing copper. Phil. Trans. Roy. Soc. London 159, 627–636 (1870).
CLARKE, B.: Divergent effects of natural selection on two closely related polymorphic snails. Heredity 14, 423–443 (1960).
CLARK, R. B.: The eyes and photonegative behaviour of *Nephthys* (Annelida, Polychaeta). J. Exp. Biol. 33, 461–477 (1956).
CLEFFMAN, G.: Agouti pigment cells *in situ* and *in vitro*. Ann. N. Y. Acad. Sci. 100, 749–761 (1963).
CLOUD, P. E.: Color-patterns in terebratuloids. Amer. J. Sci. 239, 905–907 (1941).
COCKAYNE, E. A.: The distribution of fluorescent pigments in the Lepidoptera. Trans. Entomol. Soc. London, (1924), 1–19.
COHEN, J.: Feathers and Patterns. Advan. Morphog. 5, 1–38 (1966).
COLLINS, R. E.: Transport of gases through haemoglobin solutions. Science, 133, 1593–1594 (1961).
COLLINSON, E., DAINTON, F. S., HOLMES, B.: Inactivation of ribonuclease in dilute aqueous solution: Inactivation by OH-radicals. Nature (London) 165, 267–269 (1950).
COMFORT, A.: The pigmentation of Molluscan shells. Biol. Rev. Cambridge Phil. Soc. 26, 285–301 (1951).
COMMONER, B., TOWNSEND, J., PAKE, J. E.: Free radicals in biological material. Nature (London) 174, 689–691 (1954).
CONANT, J. B., CHOW, B. F., SCHOENBACH, E. B.: The oxidation of haemocyanin. J. Biol. Chem. 101, 463–473 (1933).
CONN, H. J.: Biological Stains. 6th. Edn. Geneva and New York: Biotech. Publications 1953.
COOHILL, T. P., BARTELL, C. K., FINGERMAN, M.: Relative effectiveness of u.v. and visible light in

eliciting pigment dispersion directly in melanophores of the fiddler crab *Uca pugilator*. Physiol. Zool **43**, 232–239 (1970).

Cook, S. F.: The action of potassuum cyanide and potassium ferricyanide on certain respiratory pigments. J. Gen. Physiol. **11**, 339–348 (1928).

Cope, F. W., Sever, R. J., Polis, B. D.: Reversible free radical generation in the melanin granules of the eye by visible light. Arch. Biochem. Biophys. **100**, 171–177 (1963).

Cormier, M. J.: Luminescence in Annelids. In: Chemical Zoology, Vol. IV, pp. 467–479. Ed. by M. Florkin and BT. Scheer. New York – London: Academic Press. 1969.

Cornman, I.: Sperm activation by *Arbacia* egg extracts with special reference to echinochrome. Biol. Bull. **80**, 202–207 (1941).

Costello, D. P.: Experimental studies of germinal localisation in *Nereis* I: The development of isolated blastomeres. J. Exp. Zool. **100**, 19–66 (1945).

Costello, L. C., Smith, W., Fredricks, W.: The comparative biochemistry of developing *Ascaris* eggs VI: Respiration and terminal oxidation during embryonation. Comp. Biochem. Physiol. **18**, 217–224 (1966).

Cott, H. B.: Adaptive Coloration in Animals. London: Methuen 1940.

Crane, F. L.: Distribution of ubiquinones. In: Biochemistry of Quinones, pp. 183–206. Ed. by R. A. Morton. London – New York: Academic Press 1965.

Crescitelli, F.: A note on the absorption spectra of the blood of *Eudistylia gigantea* and of the pigment of the red corpuscles of *Cucumaria miniata* and *Molpadia intermedia*. Biol Bull. **88**, 30–36 (1945).

Crocker, A. C.: Pigmentation in the lipidoses. In: Pigments in Pathology, pp. 287–306. Ed. by M. Wolman. New York: Academic Press 1969.

Gromartie, R. I. T., Harley-Mason, J.: Melanin and its precursors X: The oxidation of methylated 5,6-dihydroxyindoles. Biochem. J. **66**, 713–720 (1957).

Crozier, G. F.: Tissue carotenoids in prespawning and spawning sockeye salmon (*Onchorhynchus nerka*). J. Fish. Res. Bd. Canada, **27**, 973–975 (1970).

Crozier, W. J.: The orientation of a holothurian by light. Amer. J. Physiol. **36**, 8–20 (1914).

Crozier, W. J.: Correspondence of skin-pigments in related species of Nudibranchs. J. Gen. Physiol. **4**, 303–304 (1922).

Cunningham, J. T., MacMunn, C. A.: On the coloration of the skins of fishes, especially of Pleuronectidae. Phil. Trans. Roy. Soc. London B, **184**, 765–812 (1893).

Czeczuga, B.: Comparative studies on the occurrence of carotenoids in the host-parasite system with reference to *Argulus foliaceus* and its host *Gasterosteus foliaceus*. Acta Physiol. Pol. **19**, 185–194 (1971).

Dales, R. P.: Feeding mechanisms and structure of the gut in *Owenia fusiformis* Della Chiaje (= *Ammochares*). J. Mar. Biol. Ass. U. K. **36**, 81–89 (1957).

Dales, R. P.: The nature of the pigments in the crowns of sabellid and serpulid Polychaetes. J. Mar. Biol. Ass. U. K. **42**, 259–274 (1962a).

Dales, R. P.: The coelomic and peritoneal cell system of some sabellid Polychaetes. Quart. J. Micr. Sci. **102**, 327–346 (1962b).

Dales, R. P.: Pigments in the skin of the Polychaetes *Arenicola*, *Abarenicola*, *Dodecacaria* and *Halla*. Comp. Biochem. Physiol. **8**, 99–108 (1963).

Dales, R. P.: The coelomocytes of the terebellid polychaete, *Amphitrite johnstoni*. Quart. J. Micr. Sci. **105**, 263–279 (1964).

Dales, R. P.: Iron compounds in the heart-body of the terebellid polychaete, *Neoamphitrite figulus*. J. Mar. Biol. Ass. U. K. **45**, 345–351 (1965)

Dales, R. P., Kennedy, G. Y.: On the diverse colours of *Nereis diversicolor*. J. Mar. Biol. Ass. U. K. **33**, 699–708 (1954).

Dalgleish, C. E.: Isolation and examination of urinary metabolites containing an aromatic system. J. Clin. Pathol. **8**, 73–78 (1955).

Dartnall, H. J. A.: Visual pigments of the Coelacanth. Nature (London). **239**, 341–342 (1972).

Dartnall, H. J. A., Tansley, K.: Physiology of vision: Retinal structure and visual pigments. Ann. Rev. Physiol. **25**, 433–458 (1963).

Dausend, K.: Über die Atmung der Tubificiden. Z. Vergl. Physiol. **14**, 557–608 (1931).

Davenport, H. E.: The haemoglobins of *Ascaris lumbricoides*. Proc. Roy. Soc. London B, **136**, 255–270 (1949).

Davis, N.S., Silverman, G.J., Masurovsky, E.B.: Radiation-resistant pigmented coccus from isolated haddock tissue. J. Bacteriol. **86**, 294–298 (1963).
Davis, R.B., Schwartz, S.: Effect of 'haematoporphyrin' on the adhesiveness and aggregation of rabbit platelets. Nature (London) **214**, 186–187 (1967).
Davson, H.: Textbook of General Physiology. 4th. Edn. London: Churchill 1970.
Davson, H., Danielli, J.F.: Permeability of Natural Menbranes. Cambridge: University Press 1943.
Dawes, B.: The Trematoda. Cambridge: University Press 1946.
Dawson, A.B.: The coloured corpuscles of the blood of the purple sea-spider, *Anoplodactylus*. Biol. Bull. **66**, 62–68 (1934).
Dawson, C.R., Tarpley, W.B.: On the pathway of the catechol-tyrosinase reaction. Ann. N.Y. Acad. Sci. **100**, 937–950 (1963).
Dawson, J.W., Gray, H.B., Hoenig, H.E., Rossman, C.R., Schredder, J.M., Wang, R.-H.: A magnetic susceptibility study of haemerythrin using an ultrasensitive magnetometer. Biochemistry **11**, 461–465 (1972).
Deanin, G.G., Steggerda, F.R.: Use of the spectrophotometer for measuring melanin dispersion in the frog. Proc. Soc. Exp. Biol. Med. **67**, 101–104 (1948).
Deisseroth, A., Dounce, A.L.: Catalase. Physiol. Rev. **50**, 319–375 (1970).
De Jong, H.G.B.: La Coacervation, Les Coacervates et leur Importance en Biologie. Paris: Hermann et Cie 1936.
De Kok, A., Veeger, C., Hemmerich, P.: The effect of light on flavins and flavoproteins in the presence of α-keto acids. In: Plavins and Flavoproteins, pp. 63–81. Ed. by H. Kamin. Baltimore: University Park Press and London: Butterworth 1971.
Delkeskamp, E.: Über den Porphyrinstoffwechsel bei *Lumbricus terrestris* L. Z. Vergl. Physiol. **48**, 332–340 (1964).
De Moss, R.D., Happel, M.E.: Biosynthesis of violacein. Bacteriol. Proc. (**1955**), 138–139 (1955).
Dennell, R.: The hardening of insect cuticles. Biol. Rev. Cambridge Phil. Soc. **33**, 178–196 (1958).
Denton, E.J.: The buoyancy of fish and cephalopods. Prog. Biophys. **11**, 177–234 (1961).
Denton, E.J.: Reflections in fishes. Sci. Amer. **224** (1), 65–72 (1971).
Deol, M.S.: The relationship between abnormalities of pigmentation and of the inner ear. Proc. Roy. Soc. London B **175**, 201–217 (1970).
De Phillips, H.A., Nickerson, K.W., Johnson, M., Van Holde, K.E.: Physical studies of haemocyanins IV: Oxygen-linked dissociation of *Loligo pealei* haemocyanin. Biochemistry **8**, 3665–3672 (1969).
De Phillips, H.A., Nickerson, K.W., Van Holde, K.E.: Oxygen binding and sub-unit equilibria of *Busycon* haemocyanin. J. Mol. Biol. **50**, 471–479 (1970).
Derrien, E., Turchini, J.: Nouvelles observations de fluorescence rouge chez les animaux. C.R. Soc. Biol. **92**, 1030–1031 (1925).
Desselberger, H.: Über das Lipochrom der Vogelfeder. J. Ornithol. **78**, 328–376 (1930).
Deuel, H.J.: The Lipids. New York: Interscience 1957.
Dewey, V.C., Kidder, G.W.: 2,4-diamino-5,6,7,8-tetrahydroquinazoline, a new pteridine antagonist. Biochem. Biophys. Res. Comm. **34**, 495–502 (1969).
Dhéré, C., Baumeler, C.: Sur la rufine, pigment tégumentaire de l'*Arion rufus*. C.R. Soc. Biol. **99**, 492–496 (1928).
Dhéré, C., Baumeler, C.: Recherches sur la rufescine, pigment de la coquille de l'*Haliotis rufescens*. Arch. Int. Physiol. **32**, 55–79 (1930).
Dhéré, C., Baumeler, C., Schneider, A.: Spectrochimie de la rufine et ses dérivés. C.R. Soc. Biol **99**, 722–725 (1928).
Dickens, F., Neil, E.: Oxygen in the Animal Organism. (Int. Union Biochem. Symp. No. 31). Oxford: Pergamon Press 1964.
Dickerson, R.E.: The structure and history of an ancient protein. Sci. Amer. **226** (4), 58–72 (1972).
Dimelow, E.J.: Pigments present in the arms and pinnules of the Crinoid *Antedon bifida* (Pennant). Nature (London) **182**, 812 (1958).
Dimroth, O., Kämmerer, H.: Carminic acid. Ber. Deut. Chem. Ges. **53**B, 471–480 (1920)
Dingle, J.T., Lucy, J.A.: Vitamin A, carotenoids and cell function. Biol. Rev. Cambridge Phil. Soc. **40**, 422–461 (1965).
Diwany, H.F.: Étude histologique de l'embryotrophe et du tube digestif de quelques invertébrés. Thèsis Fac. Sci. Paris: Le François 1919.

Dixon, G.Y., Dixon, A.F.: Report on the invertebrate fauna near Dublin. Proc. Roy. Irish Acad. Ser. 3, II, pp. 19–33 (1891).
Dixon, K.C.: Hydrophobic lipids in injured cells. Histochem. J. 2, 151–187 (1970).
Dixon, M., Webb, E.C.: Enzymes. 2nd. Edn. London: Longmans 1964.
Djangmah, J.S.: The effects of feeding and starvation on copper in the blood and hepatopancreas and on blood-proteins of *Crangon*. Comp. Biochem. Physiol. 32, 709–31 (1970).
Doan, C.A., Wilson, H.E., Wright, C.S.: Folic Acid (*L. casei* factor) essential pan-haematopoietic factor: Experimental and clinical studies. Ohio State Med. J. 42, 139–144 (1946).
Domagk, G.: Grundlagen und Probleme einer Chemotherapie des Krebses. Krebsarzt. 12, 1–4 (1957).
Donnasson, J., Faure-Fremiet, E.: Sur le pigment de *Fabrea salina*. C.R. Soc. Biol. 71, 515–517 (1911).
Donnellon, J.A.: An experimental study of the clot formation in the perivisceral fluid of *Arbacia*. Physiol. Zool. 11, 389–397 (1938).
Dorner, M., Reich, E.: Oxidative phosphorylation and some related phenomena in pigment granules of mouse melanomas. Biochim. Biophys. Acta 48, 534–546 (1961).
Douzou, P.: Reversible molecular reactions and biochemical mechanisms. In: Electronic Aspects of Biochemistry, pp. 347–364. Ed. by B. Pullman. New York: Academic Press 1964.
Drabkin, D.L.: Metabolism of the hemin chromoproteins. Physiol. Rev. 31, 345–431 (1951).
Drabkin, D.L.: The forgotten cell. In: Haems and Haemoproteins, pp. 599–604. Ed. by B. Chance, R.W. Estabrook, and T. Yonetani. New York-London: Academic Press 1966.
Drach, P.: Mue et cycle d'intermue chez les Crustacés Decapodes. Ann. Inst. Oceanog. 19, 103–391 (1939).
Dresel, E.I.B.: Passage of haemoglobin from blood into eggs of *Daphnia*. Nature (London) 162, 736–737 (1948).
Driesch, H.: Studien über das Regulationsvermögen bei Organismen 2: Quantitative Regulationen bei der Reparation der *Tubularia*. Wilhelm Roux' Arch. Entwicklungsmech. Organismen 9, 103–136 (1900).
Driesch, H.: Studien über das Regulationsvermögen bei Organismen 7: Zwei neue Regulationen bei *Tubularia*. Wilhelm Roux' Arch. Entwicklungsmech. Organismen 14, 532–538 (1902).
Drochmans, P.: The fine structure of melanin granules (the early, mature and compound forms). In: Structure and Control of the Melanocyte, pp. 90–95. Ed. by G. Della Porta and O. Mühlbock. Berlin – Heidelberg – New York: Springer 1966.
Drummond, J.C., Gilding, H.P., Macwalter, R.J.: The fate of carotene injected into the circulation. J. Physiol. (London) 82, 75–78 (1934).
Dubois, R.: Sur le vénin de la glande à pourpre des *Murex*. C.R. Soc. Biol. 55, 81. (1903).
Dubois, R.: Nouvelles recherches sur la pourpre du *Murex brandaris*. Action de lumières colorées, teinture purpuro-photographies. C.R. Soc. Biol. 62, 718–720 (1907).
Du Boy, H.G., Showacre, J.L., Messelbach, M.L.: Enzymic and other similarities between melanosome granules and mitochondria. Ann. N.Y. Acad. Sci. 100, 569–583 (1963).
Duchateau, G., Florkin, M., Jeuniaux, C.: Composante aminoacide des tissues chez les Crustacés. I: Composante aminoacide des muscles de *Carcinus maenas* L. lors du passage de l'eau de mer à l'eau saumatre et au cours de la mue. Arch. Int. Physiol. Biochem. 67, 489–500 (1959).
Duewell, H., Human, J.P.E., Johnson, A.W., Macdonald, S.F., Todd, A.R.: Colouring matters of the Aphididae. Nature (London) 162, 759–761 (1948).
Durivault, A.: Sur la nature des pigments de l'*Alcyonaria palmatum* Pallas. C.R. Soc. Biol. 126, 787–789 (1937).
Dürken, B.: Über die Wirkung verschiedenfarbiger Umgebung auf die Variation von Schmetterlingspuppen. Z. Wiss. Zool. 116, 587–626 (1916).
Dürken, B.: Über die Wirkung farbigen Lichtes auf die Puppen des Kohlweisslings (*Pieris brassicae*) und das Verhalten der Nachkommen. Arch. Mikrosk. Anat. Entwicklungsmech. 99, 222–389 (1923).
Dürken, B.: Experimental Analysis of Development. Transl. by H.G. and A.M. Newth. London: Allen and Unwin 1932.
Dustmann, J.H.: Die redox pigment von *Carausius morosus*. Z. Vergl. Physiol. 49, 28–57 (1964).
Dustmann, J.H.: Pigment studies on several eye-colour mutants of the honey bee, *Apis mellifera*. Nature (London) 219, 950–952 (1968).
Ebrey, T.G., Cone, R.A.: Melanin a possible pigment for photostable electrical responses of the eye. Nature (London) 213, 360–362 (1967).

EDGREN, R. A.: Factors controlling color change in the tree frog. *Hyla versicolor* Wied. Proc. Soc. Exp. Biol. Med. **87**, 20–23 (1954).
EDSER, E.: Light for Students. London: Macmillan 1902.
EDWARDS, N. A., HASSALL, K. A.: Cellular Biochemistry and Physiology. London: McGraw-Hill 1971.
EHRENSVARD, G., GATENBECK, S.: Die metabolische Herkunft polyzyklischer Chinone. In: Symposia on Biochemistry B. 1: Natural pigments and their biosynthesis (17th. Int. Congr. Pure Appl. Chem. Munich, 1959: Main Lectures, Vol. II, pp. 99–111. 1960.
EKMAN, B.: Oxydation zyklischer Verbindungen durch Vitamin C. Acta Physiol. Scand. **8**, Suppl. 22, 196 pp. 1944.
ELEY, D. D.,: The kinetics of haemoglobin reactions. Trans. Faraday Soc. **39**, 172–181 (1943).
ELEY, D. D., WILLIS, M. R.: Photoconduction and the degradation of organic molecules. In: Pigments, pp. 66–85. Ed. by D. PATTERSON. Amsterdam-London-New York: Elsevier Publ. Co. 1967.
ELIAS, H.: Chromatophores as evidence of phylogenetic evolution. Amer. Nat. **76**, 405–414 (1942).
ELLENBY, C.: Haemoglobin in the 'chromotrope' of an insect parasitic Nematode. Nature (London) **202**, 615–616 (1964).
ELLERTON, H. D., CARPENTIER, D. E., VAN HOLDE, K. E.: Physical studies of haemocyanins V: Characterisation and subunit structure of the haemocyanin of *Cancer magister*. Biochemistry **9**, 2225–2232 (1970).
ELLIOT, W. B.: Discussion. In: Haems and Haemoproteins, p. 189. Ed. by B. CHANCE, R. W. ESTABROOK, and T. YONETANI. New York-London: Academic Press 1966.
ELVIDGE, J. A., LEVER, A. B. P.: Manganese phthalocyanine as an oxygen carrier. Proc. Chem. Soc. London **1959**, 195.
EMERSON, R.: On the behaviour of nickel carbonate in relation to photosynthesis. J. Gen. Physiol. **13** 159–168 (1929).
ENAMI, M.: Melanophore responses in an Isopod Crustacean, *Ligia exotica*. Jap. J. Zool. **9**, 497–515 (1941).
ENAMI, M.: Chromatophore activator in the central nervous organs of *Uca dubia*. Proc. Imp. Acad. Japan **19**, 693–697 (1943 a).
ENAMI, M.: Interspecificity of the pigmentary hormones as tested upon *Uca dubia*. Proc. Imp. Acad. Japan **19**, 698–702 (1943 b).
ENAMI, M.: Studies on the controlling mechanism of black chromatophores in the young of the fresh water crab *Sesarmia haematocheir* II: Hepatopancreas principle antagonising the chromatophorotropic principle from the ganglionic tissue. Jap. J. Ecol. **4**, 1 (1950).
ENAMI, M.: The sources and activities of two chromatophorotropic hormones in crabs of the genus *Sesarmia* I: Experimental analyses; II: Histology of incretory elements. Biol. Bull. **100**, 28–43; **101**, 241–258 (1951).
ENDEAN, R.: Studies on the blood and tests of some Australian ascidians I: The blood of *Pyura stolonifera* (Heller). Anst. J. Mar. Freshwater Res. **6**, 35–59 (1955).
EPHRUSSI, B., HEROULD, J. L.: Studies on the eye-pigments of *Drosophila* II: Effects of temperature on the red and brown pigments in the mutant 'blood' (w^{bl}). Genetics **30**, 62–70 (1945).
EPSTEIN, E., NABORS, M. W., STOWE, B. B.: Origin of indigo of woad. Nature (London) **216**, 547–549 (1967).
EPSTEIN, S. S., SMALL, M., FALK, H. L., MANTEL, N.: Association between photodynamic and carcinogenic actions in polycyclic compounds. Cancer Res. **24**, 855–862 (1964).
ERGENE, S.: Untersuchungen über Farbanpassung und Farbwechsel bei *Acrida turrita*. Z. Vergl. Physiol. **32**, 530–551 (1950).
ERGENE, S.: Farbanpassung entsprechend der jeweiligen Substratfärbung bei *Acrida turrita*. Z. Vergl. Physiol. **34**, 69–74 (1952).
ERGENE, S.: Homochrome Farbanpassungen bei *Mantis religiosa*. Z. Vergl. Physiol. **35**, 36–41 (1953).
ERGENE, S.: Über die Faktoren, die Grünfärbung bei *Acrida* bedingen. Z. Vergl. Physiol. **37**, 221–229 (1955).
ERGENE, S.: Homochromere Farbwechsel bei geblindeten *Oedaleus decorus*. Z. Vergl. Physiol. **38**, 311–316 (1956).
ERHARDT, F., OSTROY, S. E. and ABRAHAMSON, E. W.: Protein configuration changes in the photolysis of rhodopsin I: The thermal decay of cattle lumirhodopsin *in vitro*. Biochim. Biophys. Acta **112**, 256–264 (1966).
'ESPINASSE, P. G.: The developmental anatomy of the brown leghorn breast feather and its reactions to oestrone. Proc. Zool. Soc. London A **109**, 247–288 (1939).

Ewer, R. F., Fox, H. M.: On the function of chlorocruorin. Proc. Roy. Soc. London B **129**, 137–153 (1940).
Ewer, R. F., Fox, H M.: The function of chlorocruorin. Pubbl. Sta. Zool. Napoli **24**, 197–200 (1953).
Falk, J. E.: Porphyrins and Metalloporphyrins (BBA Library No. 2). Amsterdam – London – New York: Elsevier Publ. Co. 1964.
Falk, S., Rhodin, J.: Mechanism of pigment migration within teleost melanophores. In: Electron Microscopy, pp. 213–215. Ed. by F. S. Sjöstrand and J. Rhodin. Stockholm: Almqvist and Wiksell 1957.
Fan, C. C., York, J. L.: Implications of histidine at the active site of haemerythrin. Biochem. Biophys. Res. Comm. **36**, 365–372 (1969).
Fan, C. C., York, J. L.: Role of tyrosine in the haemerythrin active site. Biochem. Biophys. Res. Comm. **47**, 472–476 (1972).
Fänge, R.: Gastrotricha, Kinorhyncha, Rotatoria, Kamptozoa, Nematomorpha, Nemertina and Priapula. In: Chemical Zoology, Vol. III, pp. 593–609. Ed. by M. Florkin and B. T. Scheer. New York – London: Academic Press 1969.
Fatt, I., La Force, R. C.: Theory of oxygen transport through haemoglobin solutions. Science **133**, 1919–1921 (1961).
Fauvel, P.: Classe des Annélides Polychètes. In: Traité de Zoologie, Tome V. 1. pp. 13–196. Ed. by P.-P. Grasse. Paris: Masson et Cie 1959.
Fawcett, D. W.: Differences in physiological activity in brown and white fat as revealed by histochemical reactions. Science **105**, 123 (1947).
Feigl, F.: Spot-tests in Organic Analysis 7th. Edn. Amsterdam – London – New York: Elsevier Publ. Co. 1966.
Feitknecht, W.: The theory of colour in inorganic substances. In: Pigments, pp. 1–25. Ed. by D. Patterson. Amsterdam – London – New York: Elsevier Publ. Co. 1967.
Fell, H. B., Mellanby, E.: Metaplasia produced in cultures of chick ectoderm by high vitamin A. J. Physiol. London **119**, 470–488 (1953).
Felsenfield, G., Printz, M. P.: Specific reactions of hydrogen peroxide with the active site of haemocyanin: The formation of methaemocyanin. J. Amer. Chem. Soc. **81**, 6259–6264 (1959).
Feltwell, J., Valadon, L. R. G.: Plant pigments identified in the Common Blue butterfly. Nature (London) **225**, 969. (1970).
Fernlund, P., Josefsson, L.: Crustacean color-change hormones: Amino acid sequence and chemical synthesis. Science **177**, 173–175 (1968).
Ferrell, R. E., Kitto, G. B.: Properties of *Dendrostomum pyroides* haemerythrin. Biochemistry **9**, 3053–3062 (1970).
Fieser, L. F., Fieser, M.: Reduction potential of various naphthoquinones. J. Amer. Chem. Soc. **57**, 491–494 (1935).
Figge, F. H. J.: Indophenol depigmentation of normal and hypophysectomised Amphibia. J. Exp. Zool. **78**, 471–481 (1938).
Finar, I. L.: Organic Chemistry. Vol. I. 3rd. Edn. London: Longmans 1959.
Finar, I. L.: Organic Chemistry. Vol. II. 3rd. Edn. London: Longmans 1964.
Fingerman, M.: Physiology of the black and red chromatophores of *Callinectes sapidus*. J. Exp. Zool. **133**, 87–106 (1956a).
Fingerman, M.: The physiology of the melanophores of the Isopod *Ligia exotica*. Tulane Stud. Zool. **3**, 139–148 (1956b).
Fingerman, M.: Black pigment concentrating factor in the fiddler crab. Science **123**, 585–586 (1956c).
Fingerman, M.: Endocrine control of the red and white chromatophores of the dwarf crawfish, *Cambarellus shufeldti*. Tulane Stud. Zool. **5**, 137–150 (1957a).
Fingerman, M.: Physiology of the red and white chromatophores of the dwarf crayfish, *Cambarellus shufeldti*. Physiol. Zool. **30**, 142–154 (1957b).
Fingerman, M.: The Control of Chromatophores. Oxford: Pergamon Press 1963.
Fingerman, M: Neurosecretory control of pigmentary effectors in Crustaceans. Amer. Zool. **6**, 169–179 (1966).
Fingerman, M.: Comparative physiology: Chromatophores. Ann. Rev. Physiol. **32**, 345–372 (1970).
Fingerman, M., Aoto, T.: Chromatophorotropins in the crayfish, *Orconectes clypeatus* and their relationship to long-term background adaptation. Physiol. Zool. **31**, 193–208 (1958).
Fingerman, M., Fitzpatrick, C.: An endocrine basis for the seasonal difference in melanin dispersion in *Uca pugilator*. Biol. Bull. **110**, 138–143 (1956).

Fingerman, M., Mobberley, W. C.: Investigation of the hormones controlling the distal retinal pigment of the prawn *Palaemonetes*. Biol. Bull. 118, 393–406 (1960).
Fingerman, M., Lowe, M. E., Sundarara, B. I.: Dark-adapting and light-adapting hormones controlling the distal retinal pigment of the prawn, *Palaemonetes vulgaris*. Biol. Bull. 116, 30–36 (1959).
Fischer, E.: Sur les fonctions de l'hepatopancréas des Crustacés: Les pigments d'excretion. C. R. Soc. Biol. 96, 850–852 (1927).
Fischer, P.-H.: Sur le role de la glande purpurigène des *Murex* et des Pourpres. C. R. H. Acad. Sci. 180, 1369–1371 (1925).
Fisher, E. R.: Pigmentation of the intestinal tract. In: Pigments in Pathology, pp. 489–506. Ed. by M. Wolman. New York – London: Academic Press 1969.
Fisher, J. W.: The structure and physiology of erythropoietin. In: The Biological Basis of Medicine, Vol. III, pp. 41–79. Ed. by E. E. Bittar and N. Bittar. New York: Academic Press 1969.
Fisher, L. R.: Vitamins. In: Physiology of Crustacea, Vol. I, pp. 259–289. Ed. by T. H. Waterman. New York – London: Academic Press 1960.
Fisher, L. R., Kon, S. K.: Vitamin A in Invertebrates. Biol. Rev. Cambridge Phil. Soc. 34, 1–36 (1959).
Fisher, R. A.: The Genetical Theory of Natural Selection. Oxford: Clarendon Press 1930. 2nd. Edn. New York: Dover Publ. Inc. 1958.
Fitzpatrick, T. B.: Discussion. In: Structure and Control of the Melanocyte, p. 216. Ed. by G. Della Porta and O. Mühlbock. Berlin-Heidelberg-New York: Springer 1966.
Fitzpatrick, T. B., Brunet, P. C. J., Kukita, A.: The nature of hair pigment. In: The Biology of Hair Growth pp. 255–303. Ed. by W. Montagna. New York: Academic Press 1958.
Fitzpatrick, T. B., Quevedo, W. C., Levine, A. L., McGovern, V. J., Mishima, Y., Oettle, A. G.: Terminology of Vertebrate melanin-containing cells. Science 152, 88–89 (1965).
Flatmark, T.: Multiple molecular forms (isocytochromes) of beef heart cytochrome c. In: Haems and Haemoproteins, pp. 411–414. Ed. by B. Chance, R. W. Estabrook, and T. Yonetani. New York – London: Academic Press 1966.
Flesch, P.: Epidermal iron-pigments in red species. Nature (London) 217, 1056–1057 (1968).
Flesch, P.: The red pigmentary system and its relation to black melanin genesis. J. Soc. Cosmet. Chem. 21, 77–83 (1970).
Flesch, P., Rothman, S.: Role of sulphydryl compounds in pigmentation. Science 108, 505–506 (1948).
Flickinger, R. A.: Study of the metabolism of Amphibian neural crest cells during their migration and pigmentation in vitro. J. Exp. Zool. 112, 465–484 (1949).
Florey, E.: Ultrastructure and function of Cephalopod chromatophores. Amer. Zool. 9, 429–442 (1969).
Florkin, M.: La fonction respiratoire du milieu interieur dans la série animale. Ann. Physiol. et Physicochim. Biol. 10, 599–684 (1934).
Florkin, M.: Biochemical Evolution. Trans. S. Morgulis. New York: Academic Press 1949.
Florkin, M.: Unity and Diversity in Living Organisms. Trans. T. Wood. Oxford: Pergamon Press 1960.
Florkin, M.: Biochemical evolution. I. C. S. U. Rev. World Sci. 5, 202–209 (1963).
Florkin, M.: Molecular Approach to Phylogeny. Amsterdam – London – New York: Elsevier Publ. Co. 1966.
Florkin, M.: Evolution as a biochemical process. In: Biochemical Evolution and The Origin of Life, pp. 366–380. Ed. by E. Schoffeniels. Amsterdam: North Holland Publ. Co. 1971.
Flux, J. E. C.: Colour change of mountain hare (*Lepus timidus scoticus*) in north-east Scotland. J. Zool. 162, 345–358 (1970).
Folkers, K.: Unidentified vitamins and growth-factors. In: The Biochemistry and Physiology of Growth, pp. 72–90. Ed. by A. K. Parpart. Princeton: University Press. 1949.
Fontaine, M.: Spectrographie de l'absorption et de fluorescence de la fabréine. C. R. H. Acad. Sci. 198, 1077–1079 (1934).
Fontaine, M.: Sur la teneur en flavine de divers organes le l'*Anguilla* C. R. H. Acad. Sci. 204, 1367–1368 (1937).
Fontaine, M.: Propriétiés physiques et chimiques – repartition role physiologique. Les flavines. 16th Int. Congr. Physiol. Zurich, 35 pp. (Quoted from D. L. Fox, 1953).
Fontaine, M., Gourevitch, A.: Sur les variations de la teneur en flavine pendant le développement des oeufs et embryons des Sélaciens. C. R. Soc. Biol. 123, 443–444 (1936).

FONTAINE, M., RAFFY, A., COLLANGE, S.: Teneur en vitamine B_2 (riboflavine) de quelques Crustacés (Cirripedes et Isopodes) Relation avec la biologie de l'espèce considerée. C.R. Soc. Biol. **137**, 27–29 (1943).

FORD, E.B.: The inheritance of dwarfing in *Gammarus chevreuxi*. J. Genet. **20**, 93–102 (1929).

FORD, E.B.: Studies on the chemistry of pigments in the Lepidoptera with special reference to their bearing on systematics 3: The red pigments of the Papilionidae. Proc. Roy. Entomol. Soc. London A **19**, 92–106 (1944).

FORD, E.B.: A murexide test for the recognition of pterins in intact insects. Proc. Roy. Entomol. Soc. London A. **22**, 72–76 (1947a).

FORD, E.B.: Studies on the chemistry of pigments in the Lepidoptera with special reference to their bearing on systematics 5: *Pseudopontia paradoxa* Felder. Proc. Roy. Entomol. Soc. London A, **22**, 77–78 (1947b).

FORD, E.B.: The genetics of polymorphism in the Lepidoptera. Advan. Genet. **5**, 43–87 (1953).

FORD, E.B.: Ecological Genetics. 3rd. Edn. London: Methuen 1971.

FORD, E.B., HUXLEY, J.S.: Mendelian genes and rates of development in *Gammarus chevreuxi*. J. Exp. Biol. **5**, 112–134 (1927).

FORREST, H.S.: Aspects of biosynthesis of pteridines. In: Symposia on Biochemistry, B. 1: Natural Pigments and their Biosynthesis. (17th. Int. Congr. Pure Appl. Chem. Munich, 1959) Main Lectures, Vol II, pp. 40–51. 1960.

FORREST, H.S., MITCHELL, H.K.: Pteridines in Drosophila I: Isolation of a yellow pigment; II Structure of the yellow pigment. J. Amer. Chem. Soc. **76**, 5656–5658; 5658–5662 (1954).

FORREST, H.S., MITCHELL, H.K.: Pteridines from *Drosophila* III: Isolation and identification of three more pteridines. J. Amer. Chem. Soc. **77**, 4865–4869 (1955).

FOSTER, F.W.: Colour change in *Fundulus* with special reference to the colour changes of the iridosomes. Proc. Nat. Acad. Sci. U. S. **19**, 535–540 (1933).

FOSTER, F.W.: Colour change in *Fundulus* with special reference to the colour changes of the iridosomes. Proc. Nat. Acad. Sci. U. S. **19**, 535–540 (1933).

FOSTER, M.: Manometric and histochemical demonstration of tyrosinase in foetal guinea pig skin. Proc. Soc. Exp. Biol. Med. **79**, 713–715 (1952).

FOX, D.L.: Biochromes. Science **100**, 470–471 (1944).

FOX, D.L.: Animal Biochromes and Structural Colours. Cambridge: University Press 1953.

FOX, D.L.: Pigments of plant origin in animal phyla. In: Comparative Biochemistry of Photoreactive Systems, pp. 11–31. Ed. by M.B. ALLEN. New York – London: Academic Press 1960.

FOX, D.L., CRANE, S.C.: The pigments of the Spotted Octopus and the Opalescent Squid. Biol. Bull. **82**, 284–291 (1942).

FOX, D.L., HAXO, P.T.: Pigments and algal commensalism in the blue oceanic siphonophore, *Velella lata*. Proc. 15th. Int. Congr. Zool. London, pp. 280–282 (1959).

FOX, D.L., MILLOTT, N.: A biliverdin-like pigment in the skull and vertebrae of the ocean skipjack, *Katsuwonus pelamis* (L.). Experientia **10**, 185. (1954a).

FOX, D.L., MILLOTT, N.: The pigmentation of the jelly-fish, *Pelagia noctiluca* (Forskal) var. *panopyra* Péron and Lesueur. Proc. Roy. Soc. London B **142**, 392–408 (1954b).

FOX, D.L., PANTIN, C.F.A.: The colours of the plumose anemone, *Metridium senile* (L.). Phil. Trans. Roy. Soc. London B **230**, 415–450 (1941).

FOX, D.L., PANTIN, C.F.A.: Pigments in the Coelenterata. Biol. Rev. Cambridge Phil. Soc. **19**, 121–134 (1944).

FOX, D.L., SCHEER, B.T.: Comparative studies of the pigments of some Pacific Coast echinoderms. Biol. Bull. **80**, 441–455 (1941).

FOX, D.L., UPDEGRAFF, D.M.: Adenochrome, a glandular pigment from the branchial hearts of the octopus. Arch. Biochem. **1**, 339–356 (1944).

FOX, D.L., CRANE, S.C., MCCONNAUGHEY, B.H.: A biochemical study of the marine annelid worm *Thoracophelia mucronata*; its food biochromes and carotenoid metabolism. J. Mar. Res. **7**, 567–585 (1948).

FOX, H.M.: Chlorocruorin: a pigment allied to haemoglobin. Proc. Roy. Soc. London B **99**, 199–220 (1926).

FOX, H.M.: The oxygen-affinity of chlorocruorin. Proc. Roy. Soc. London B **111**, 356–363 (1932).

FOX, H.M.: Chlorocruorin and haemoglobin. Nature (London) **160**, 825 (1947a).

FOX, H.M.: *Daphnia* haemoglobin. Nature (London) **160**, 431–432 (1947b).

FOX, H.M.: On chlorocruorin and haemoglobin. Proc. Roy. Soc. London B **136**, 378–388 (1949).

Fox, H. M.: Oxygen affinities of respiratory blood pigments in *Serpula*. Nature (London) **168**, 112. (1951).

Fox, H. M.: Haemoglobin and biliverdin in parasitic Cirripede Crustacea. Nature (London) **171**, 162–163 (1953).

Fox, H. M.: The effect of oxygen on the concentration of haem in invertebrates. Proc. Roy. Soc. London B **143**, 203–214 (1955).

Fox, H. M.: Respiratory pigments. In: Colour and Life, pp. 55–59. Ed. by W. B. Broughton. London: Institute of Biology 1964.

Fox, H. M., Phear, E. A.: Factors influencing haemoglobin synthesis by *Daphnia*. Proc. Roy. Soc. London B **141**, 179–189 (1953).

Fox, H. M., Vevers, H. G.: The Nature of Animal Colours. London: Sidgwick and Jackson. 1960.

Fox, S. W.: Production of spherules from synthetic proteinoid and hot water. Science, **129**, 1221–1223 (1959).

Fox, S. W.: How did Life begin? Science **132**, 200–208 (1960).

Fox, S. W. (Ed.): The Origins of Prebiological Systems. New York – London: Academic Press 1965a.

Fox, S. W.: Discussion. In: The Origin of Prebiological Systems, p. 455 ff. Ed. by S. W. Fox. New York – London: Academic Press 1965b.

Frank, W., Fetzer, U.: Anhäufung von Carotenoiden in *Filaria* aus einem Chamäleon. Z. Protistenkunde **30**, 199–206 (1968).

Franz, V.: Zur Anatomie, Histologie und funktionellen Gestaltung des Selachierauges. Jena Z. Naturwiss. **40**, 697–840 (1905).

Freeman, A. R., Connell, P. M. and Fingerman, M.: An electrophoretic study of the red chromatophore of the prawn *Palaemonetes:* Observations on the action of red pigment concentrating hormone. Comp. Biochem. Physiol. **26**, 1015–1029 (1968).

Freeman, H. C.: Crystal structure studies of cupric-peptide complexes. In: The Biochemistry of Copper, pp. 77–113. Ed. by J. Peisach, P. Aisen, and W. E. Blumberg. New York: Academic Press 1966.

Freeman, H. C.: Discussion. In: The Biochemistry of Copper, p. 581. Ed. by J. Peisach, P. Aisen, and W. E. Blumberg. New York: Academic Press 1966.

Friede, R. L.: The relation of the formation of lipofuscin to the distribution of oxidative enzymes in the human brain. Acta Neuropathol. **2**, 113–125 (1962).

Frieden, E., Eaton, D. N., Tripp, M. J., Eaton, J. E.: Functional relations of iron and copper metalloproteins in development. Fed. Proc. Fed. Amer. Soc. Exp. Biol. **31**, 487 (Abst. 1539: Biochem.) (1972).

Friedheim, E. A. H.: Sur deux ferments respiratoires accessoires d'origine animale. C.R. Soc. Biol. **111**, 505–507 (1932).

Friedheim, E. A. H.: Sur la fonction respiratoire du pigment rouge de *Penicillium phoeniceum*. C. R. Soc. Biol. **112**, 1030–1032 (1933).

Fries, E. F. B.: Some neurohumoral evidence for double innervation of xanthophores in killifish (*Fundulus*). Biol. Bull. **82**, 261–272 (1942).

Fries, E. F. B.: Iridescent white reflecting chromatophores (antaugophores, iridoleucophores) in certain teleost fishes, particularly in *Bathygobius*. J. Morphol. **103**, 203–253 (1958).

Fujii, R.: Chromatophores and pigments. In: Fish Physiology, Vol. III, pp. 307–353. Ed. by W. S. Hoar and D. J. Randall. New York – London: Academic Press 1969.

Fujii, R., Novales, R. R.: Cellular aspects of the control of physiological colour change in fishes. Amer. Zool. **9**, 453–464 (1969).

Fukushi, Y., Seki, T.: Differences in amino acid compositions of pupal sheaths between wild and black pupa strains in some species of insects. Jap. J. Genet. **40**, 203–208 (1965).

Fürth, O. von, Schneider, H.: Über tierischen Tyrosinasen und ihre Beziehungen zur Pigmentbildung. Beitr. Chem. Physiol. Pathol. **1**, 229–242 (1902).

Fuseau-Braesch, S.: Etude biologique et biochimique de la pigmentation d'un insecte, *Gryllus bimaculatus* de Geer (Gryllide, Orthoptère) Bull. Biol. France Belg. **94**, 526–627 (1960).

Fuseau-Braesch, S.: Contribution à l'étude de l'homochromie chez *Locusta migratoria*. In: Colloq. Int. C. N. R. S.: Sur l'Effet de Groupe Chez les Animaux, No. 173 1967.

Gaffron, H.: Some effects of derivatives of vitamin K on the metabolism of unicellular Algae. J. Gen. Physiol. **28**, 259–268 (1945).

Gaffron, H.: The origin of Life. In: Evolution After Darwin, Vol. I, pp. 39–84. Ed. by S. Tax. Chicago: University Press 1960.

GAFFRON, H.: On dating stages in Evolution. Comp. Biochem. Physiol. 4, 205–216 (1962).
GAFFRON, H.: The role of light in Evolution: Transition from a one-quantum to a two-quantum mechanism. In: The Origins of Prebiological Systems, pp. 437–460. Ed. by S. W. Fox. New York – London: Academic Press 1965.
GALSTON, A. W., BAKER, R. S.: Inactivation of enzymes by visible light in the presence of riboflavin. Science 109, 485–486 (1949).
GAMBLE, F. W.: Platyhelminthes and Mesozoa. In: Cambridge Natural History, Vol. II, pp. 3–98. Ed. by S. F. HARMER and A. E. SHIPLEY. London: Macmillan 1896.
GAMBLE, F. W., KEEBLE, F. W.: *Hippolyte varians:* a study in colour change. Quart. J. Microsc. Sci. 43, 589–698 (1900).
GARBETT, K., DARNALL, D. W., KLOTZ, I. M., WILLIAMS, R. J. P.: Spectroscopy and structure of haemerythrin. Arch. Biochem. Biophys. 135, 429–434 (1969).
GATES, M.: Chemistry of the pteridines. Chem. Rev. 41, 63–96 (1947).
GEDIGH, P.: Pigmentation caused by inorganic materials. In: Pigments in Pathology, pp. 1–32. Ed. by M. WOLMAN. New York: Academic Press 1969.
GEISSMAN, T.: The biosynthesis of phenolic plant products. In: Biogenesis of Natural Compounds. 2nd. Edn., pp. 743–799. Ed. by P. BERNFELD. Oxford: Pergamon Press 1967.
GEORGE, P., HANANIA, G. I. H., EATON, W. A.: Effect of electrostatic environment on redox potentials. In: Haems and Haemoproteins, pp. 267–271. Ed. by B. HANCE, R. W. ESTABROOK and T. YONETANI. New York – London: Academic Press 1966.
GEROULD, J. H.: Blue-green caterpillars: the origin and ecology of a mutation in haemolymph colour in *Colias (Eurymus) philodice* J. Exp. Zool. 34, 385–415 (1921).
GESCHWINDT, I. I.: Chemistry of the melanocyte-stimulating hormone. In: Structure and Control of the Melanocyte, pp. 28–44. Ed. by G. Della PORTA and O. MÜHLBOCK. Berlin – Heidelberg – New York: Springer 1966.
GEYER, K.: Untersuchungen über die chemische Zusammensetzung der Insektenhämolymphe und ihre Bedeutung für die geschlechtliche Differenzierung. Z. Wiss. Zool. 105, 349–499 (1913).
GHIRETTI, F.: Molluscan haemocyanins. In: Physiology of the Mollusca, Vol. II, pp. 233–248. Ed. by K. M. WILBUR and C. M. YONGE. New York – London: Academic Press 1966.
GHIRETTI, F. (Ed.): Physiology and Biochemistry of Haemocyanins. London – New York: Academic Press 1968.
GIERSBERG, H.: Über den morphologischen Farbwechsel der Stabheuschrecke *Dixippus (Carausius) morosus.* Z. Vergl. Physiol. 7, 657–695 (1928).
GIERSBERG, H.: Die Färbung der Schmetterlinge I. Z. Vergl. Physiol. 9, 523–552 (1929).
GIESE, A. C.: An intracellular photodynamic sensitiser in *Blepharisma.* J. Cell. Comp. Physiol. 28, 119–127 (1946).
GIESE, A. C.: Some properties of a photodynamic pigment from *Blepharisma.* J. Gen. Physiol. 37, 259–269 (1953).
GIESE, A. C.: Cell Physiology. 3rd. Edn. Philadelphia: W. B. Saunders. 1968.
GIESE, A. C., GRAINGER, R. M.: Studies on the red and blue forms of the pigment of *Blepharisma.* Photochem. Photobiol. 12, 489–503 (1970).
GILBERT, J. S., THOMPSON, G. A.: Alpha tocopherol-control of sexuality and polymorphism in the rotifer *Asplanchna.* Science, 159, 734–736 (1968).
GILCHRIST, B. M.: Haemoglobin of *Artemia.* Proc. Roy. Soc. London B, 143, 136–146 (1954).
GILCHRIST, B. M., GREEN, J.: The pigments of *Artemia.* Proc. Roy. Soc. London B 152, 118–136 (1960).
GILCHRIST, B. M., GREEN, J.: Bile pigment in *Chirocephalus diaphanus* Prevost (Crustacea: Anostraca). Comp. Biochem. Physiol. 7, 117–125 (1962).
GILCHRIST, B. M., LEE, W. L.: Carotenoid pigments and their possible role in reproduction in the sand crab, *Emerita analoga* (Stimpson, 1857). Comp. Biochem. Physiol. 42, 263–294 (1972).
GILI, D., KILPONEN, R. G., RIMAI, L.: Resonance Raman scattering of laser radiation by vibrational modes of carotenoid pigment molecules in intact plant tissues. Nature (London) 227, 743–744 (1970).
GILMOUR, D.: The Metabolism of Insects. Edinburgh: Oliver and Boyd 1965.
GIRARDIE, A.: Fonctions de la pars intercerebralis chez *Locusta migratoria* L. C.R.H. Acad. Sci. 254, 2669–2670 (1962).
GLASER, R., LEDERER, E.: Echinochrome et spinochrome; dérivés méthylés; distribution; pigments associés. C.R.H. Acad. Sci. 208, 1939–1942 (1939).

Glass, B.: In pursuit of a gene. Science 126, 683–689 (1957).
Glass, B.: Summary. In: Light and Life, pp. 817–911. Ed. by W. E. McElroy and B. Glass. Baltimore: The Johns Hopkins Press 1961.
Glassman, E.: Tyrosinase produced quinones and the disappearance of kynurenine in larval extracts of *Drosophila melanogaster*. Arch. Biochem. Biophys. 67, 74–89 (1957).
Glossman, H., Horst, J., Plagens, U., Braunitzer, G.: Isoelektrisches Fokussieren von Hämoglobin der Insekten *Chironomus thumni*. Hoppe Seyler's Z. Physiol. Chem. 351, 342–348 (1970).
Glover, J.: Biosynthesis of biologically active quinones and related compounds. In: Biochemistry of Quinones, pp. 207–260. Ed. by R. A. Morton. London – New York: Academic Press 1965.
Goldberg, E. D., McBlair, W., Taylor, K. M.: The uptake of vanadium by Tunicates. Biol. Bull. 101, 84–94 (1951).
Goldman, A. S.: Synthesis of pigment during the reconstitution of *Tubularia*, Biol. Bull. 105, 450–465 (1953).
Goldschmidt, R.: Physiologische Theorie der Vererbung. Berlin: Springer 1927.
Goldschmidt, R.: Physiological Genetics. New York: McGraw-Hill 1938.
Goldsmith, T. H.: The nature of the retinal action potential and the spectral sensitivities of u.v. and green receptor systems of the compound eye of the worker honeybee. J. Gen. Physiol. 43, 775–799 (1960).
Gomperts, B. D.: Biochemistry of red blood corpuscles in health and disease. In: Biological Basis of Medicine. Vol. 3, pp. 81–127. Ed. by E. E. Bittar and N. Bittar. New York: Academic Press 1969.
Goodrich, E. S.: The study of nephridia and genital ducts since 1895. Quart. J. Microsc. Sci. 86, 115–392 (1945).
Goodrich, H. B., Green, J. M.: An experimental analysis of the development of a colour pattern in the fish *Brachydanis albolineatus*. Blyth. J. Exp. Zool. 141, 15–45 (1959).
Goodrich, H. B., Hill, G. A., Arrick, M. S.: The chemical identification of gene-controlled pigments in *Platypoecilus* and *Xiphophorus* and comparison with other tropical fish. Genetics 26, 573–586 (1941).
Goodwin, T. W.: Carotenoids and Reproduction. Biol. Rev. Cambridge Phil. Soc. 25, 391–413 (1950a).
Goodwin, T. W.: Biochemistry of locusts 4: Insectorubin metabolism in the desert locust (*Schistocerca gregaria* Forsk.) and the African migratory locust (*Locusta migratoria migratoria* K and F). Biochem J. 47, 554–562 (1950b).
Goodwin, T. W.: Carotenoid metabolism during development of lobster eggs. Nature (London) 167, 559. (1951).
Goodwin, T. W.: The biochemistry of locust pigmentation. Biol. Rev. Cambridge Phil. Soc. 27, 439–460 (1952a).
Goodwin, T. W.: Comparative Biochemistry of Carotenoids. London: Chapman and Hall 1952b.
Goodwin, T. W.: The biosynthesis and function of the carotenoid pigments. Advan. Enzymol. 21, 295–368 (1959).
Goodwin, T. W.: Carotenoids: Structure, distribution and function. In: Comparative Biochemistry, Vol. IV, pp. 643–675. Ed. by M. Florkin and H. S. Mason. New York: Academic Press 1962.
Goodwin, T. W.: Biosynthesis of Vitamins and Related Compounds. New York: Academic Press 1963.
Goodwin, T. W.: The chemistry of animal colours. In: Colour and Life, pp. 25–40. Ed. by W. B. Broughton. London: Institute of Biology 1964.
Goodwin, T. W., Srisukh, S.: The carotenoids of the locust integument. Nature (London) 161, 525–526 (1948).
Goodwin, T. W., Srisukh, S.: The carotenoids of locusts. 1st. Int. Congr. Biochem. Abst. No. 402, p. 65 (1949).
Goodwin, T. W., Srisukh, S.: Biochemistry of locusts 3: The redox pigment present in the integument and eyes of the desert locust (*Schistocerca gregaria* Forsk.), the African migratory locust (*Locusta migratoria migratorioides* R. and F.) and other insects. Biochem. J. 47, 549–562 (1950).
Goodwin, T. W., Srisukh, S.: Biochemistry of locusts 6: The occurrence of flavin in the eggs and of a pterin in the eyes of the African migratory locust (*Locusta migratoria migratorioides* Forsk). Biochem. J. 49, 84–87 (1951).
Goodwin, T. W., Taha, M. M.: The carotenoids of the gonads of the limpets *Patella vulgata* and *Patella depressa*. Biochem. J. 47, 244–248 (1950)
Gordon, A. S.: Haemopoietine. Physiol. Rev. 39, 1–40 (1959).

Gordon, C. and Sang, J.H.: The gene 'antennaless' (Drosophila). Proc. Roy. Soc. London B 130, 151–184 (1941).
Gordon, M.: Physiological genetics of fishes. In: Physiology of Fishes, Vol. II, pp. 431–501. Ed. by M.E. Brown. New York: Academic Press 1957.
Goto, T., Hama, T.: Über die fluoreszierenden Stoffe aus der Haut eines Frosches Rana nigromaculata I: Isolierung und Eigenschaften des 'Ranachrom'. Proc. Japan. Acad. 34, 724. (1958).
Gould, D.C., Ehrenberg, A.: Electron spin resonance in copper-protein studies. In: Physiology and Biochemistry of Haemocyanins, pp. 95–112. Ed. by F. Ghiretti. London – New York: Academic Press 1968.
Granick, S.: Ferritin I: Physical and chemical properties of horse spleen ferritin. J. Biol. Chem. 146, 451–461 (1942).
Granick, S.: Evolution of heme and chlorophyll. In: Evolving Genes and Proteins, pp. 67–88. Ed. by V. Bryson and H.J. Vogel. New York – London: Academic Press 1965.
Granick, S. Sassa, S: Delta-Ala synthetase and the control of heam and chlorophyll synthesis. In: Metabolic Regulation, pp. 77–141. Ed. by H.J. Vogel. New York – London: Academic Press 1971.
Granit, R.: The photopic spectrum of the pigeon. Acta Physiol. Scand. 5, 118–124 (1942).
Granit, R.: 'Red' and 'green' receptors in the retina of Tropidonotus. Acta Physiol. Scand. 5, 108–113 (1943).
Granit, R.: Receptors and Sensory Perception. New Haven Connecticut: Yale University Press 1955.
Grant, P.: Informational molecules and embryonic development. In: The Biochemistry of Animal Development, Vol. I, pp. 483–593. Ed. by R. Weber. New York – London: Academic Press 1965.
Grant, W.C.: Special aspects of the metamorphic process: Second metamorphosis. Amer. Zool. 1, 163–171 (1961).
Gray, C.H.: Bile Pigments in Health and Disease. Springfield, Illinois: C.C. Thomas 1961.
Gray, H.: Textbook of Anatomy. 28th. Edn. Ed. by T.B. Johnston and J. Whillis. London – New York – Toronto: Longmans, Green 1944.
Green, D.E., Brierly, G.P.: The role of coenzyme Q in electron transfer. In: Biochemistry of Quinones, pp. 405–431. Ed. by R.A. Morton. London – New York: Academic Press 1965.
Green, J.: Carotenoids in Daphnia. Proc. Roy. Soc. London, B, 147, 392–401 (1957).
Green, J.: Carotenoid pigment in Spirobacillus cienkowskii Metchnikoff, a pathogen of Cladocera. Nature (London) 183, 56–57 (1959).
Green, J.: Biliverdin in the eyes of Polyphemus pediculus (L.) (Crustacea, Cladocera). Nature (London) 189, 227–228 (1961).
Green, J.: Pigments of the hydracharine, Eylais extendus. Comp. Biochem. Physiol. 13, 469–472 (1964).
Green, J.: Chemical embryology of the Crustacea. Biol. Rev. Cambridge Phil. Soc. 40, 580–600 (1965).
Green, J., Dales, R.P.: Biliverdin in the eggs of Nereis fucata. Nature (London) 181, 1412–1413 (1958).
Green, J., McHale, D.: Quinones related to vitamin E. In: Biochemistry of Quinones, pp. 261–285. Ed. by R.A. Morton. London – New York: Academic Press 1965.
Green, J.P., Neff, M.R.: A survey of the fine structure of the integument of the fiddler crab. Tissue and Cell 4, 137–171 (1972).
Green, M.M.: Homologous eye-colour mutants in the honey bee and Drosophila. Evolution 9, 215–216 (1955).
Greenstein, J.P.: Biochemistry of Cancer. 2nd. Edn. New York: Academic Press 1954.
Griffith, L.A.: Detection and identification of the polyphenoloxidase substrate of the banana. Nature (London) 184, 58–59 (1959).
Griffiths, A.B.: Sur la composition de la pinnaglobuline: Une nouvelle globuline. C.R.H. Acad. Sci. 114, 840–842 (1892).
Griffiths, M., Sistrom, W.R., Cohen-Bazire, G., Stanier, R.Y.: Function of carotenoids in photosynthesis. Nature (London) 176, 1211–1214 (1955).
Grob, E.C.: Die Biogenese der Carotenoide bei pflanzlichen Organismen. In: Symposia on Biochemistry B. 1: Natural Pigments and Their Biogenesis (17th. Int. Congr. Pure Appl. Chem. Munich, 1959) Main Lectures, Vol. II, pp. 52–72 (1960).
Groudinsky, O., Jacq, C., Labeyrie, F., Lederer, F., Monteilhet, C., Minio, M., Pajot, P.,

Risler, J.: New developments concerning the quaternary structure of yeast L-lactate-cytochrome c oxidoreductase (cytochrome b_2). In: Flavins and Flavoproteins. pp. 581–598. Ed. by H. Kamin. Baltimore: University Park Press and London: Butterworth 1971.

Gruber, M.: Structure and function of *Helix pomatia* haemocyanin. In: Physiology and Biochemistry of Haemocyanin, pp. 49–59. Ed. by F. Ghiretti. London – New York: Academic Press 1968.

Guillory, R. J., Racker, E.: Oxidative phosphorylation in brown adipose tissue mitochondria. Biochim. Biophys. Acta 153, 490–493 (1968).

Gurd, F. N. R., Banaszak, L. J., Veros, A. J., Clark, J. F.: Chemical modification of sperm whale myoglobin in the crystalline state. In: Hemes and Hemoproteins, pp. 221–240. Ed. by B. Chance, R. W. Estabrook, and T. Yonetani. New York – London: Academic Press 1966.

Gurr, E.: A fresh look at dyes. Chem. in Britain, 3, 301–4 (1967).

Gurvich, A. G.: Die Mitogenetische Strahlung. Jena: G. Fischer 1959.

Gutensohn, W.: Chemische und biochemische Untersuchungen zum Abbau von Tetrahydroneopterin. Thesis: Ludwig-Max Universität, München 1968.

Hack, M. H., Helmy, F. M.: Comparative chemical studies of the liver of various vertebrates. Acta Histochem. 19, 316–328 (1964a).

Hack, M. H., Helmy, F. M.: An analysis of melanoprotein from *Amphiuma* liver and from a human liver melanoma. Proc. Soc. Exp. Biol. Med. 116, 348–350 (1964b).

Hackman, R. H.: The chemistry of insect cuticle. In: Physiology of Insects. Vol. III, pp. 471–506. Ed. by M. Rockstein. New York: Academic Press 1964.

Haddow, A., Alson, L. A., Roe, E. M. F., Rudall, K. M., Timmis, G. M.: Artificial production of coat colour in the albino rat. Nature (London) 155, 379–381 (1945).

Haddow, A., Ross, W. C. J., Timmis, G. M.: The chain of unexpected discovery. Perspect. Biol. Med. 15, 177–217 (1972).

Hadley, C. E.: Color changes in two Cuban lizards. Bull. Mus. Comp. Zool. Harvard 69, 107–113 (1929).

Hadley, M. E., Goldman, J. M.: Physiological colour changes in reptiles. Amer. Zool. 9, 489–504 (1969).

Hadorn, E.: Developmental Genetics and Lethal Factors. Trans. U. Mittwoch. London: Methuen 1961.

Haggag, G. M., El-Duweini, A. K.: Main nitrogenous constituents of the excreta and tissues of earthworms. Proc. Egypt. Acad. Sci. 13, 1–5 (1959).

Haldane, J. B. S.: The Causes of Evolution. London: Longmans Green 1932.

Hale, F.: Pigs born without eyeballs. J. Hered. 24, 105–106 (1933).

Hall, D. O., Cammick, R., Rao, K. K.: Role for ferredoxins in the origin of Life and biological evolution. Nature (London) 233, 136–138 (1971).

Hama, T.: The relation between the chromatophores and pterin compounds. Ann. N. Y. Acad. Sci. 100, 977–984 (1963).

Hamilton, W. J., Heppner, F.: Radiant solar energy and function of black homoiotherm pigmentation: An hypothesis. Science 155, 196–197 (1967).

Harant, M., Grassé, P-P.: Classe des Annelides achètes ou Hirudinées ou Sangsues. In: Traité de Zoologie. Tome V. 1. pp. 471–593. Ed. by P.-P. Grassé. Paris: Masson et Cie. 1959.

Harborne, J. B.: Chemical colour in plants. In: Colour and Life. Ed. by W. B. Broughton. London: Institute of Biology 1964.

Harborne, J. B.: Flavonoid pigments. In: Plant Biochemistry, pp. 618–640. Ed. by J. Bonner and J. E. Varner. New York – London: Academic Press 1965.

Harbury, H. A., Myer, Y. P., Murphy, A. J., Vinogradov, S. N.: Optical rotatory dispersion studies of cytochrome *c*. In: Hemes and Hemoproteins, pp. 415–426. Ed. by B. Chance, R. W. Estabrook and T. Yonetani. New York – London: Academic Press 1966.

Hardy, Sir A. C.: The Open Sea: World of Plankton. London: Collins 1956.

Hare, T. J.: Chemoreception. In: Fish Physiology, Vol. V, pp. 19–120. Ed. by W. S. Hoar and P. J. Randall. New York: Academic Press 1971.

Harmsen, R.: Effect of atmospheric O_2 pressure on the biosynthesis of simple pterins in pierid butterflies. J. Insect Physiol. 15, 2239–2244 (1969).

Harris, J. E., Mason, P.: Vertical migration in eyeless *Daphnia*. Proc. Roy. Soc. London B 145, 280–290 (1956).

Hartenstein, R.: Nitrogen metabolism in non-insect Arthropods. In: Comparative Biochemistry

of Nitrogen Metabolism, Vol. I, pp. 299–385. Ed. by J.W. CAMPBELL. London – New York: Academic Press 1970.
HARTMANN, M.: Sex problems in algae, fungi and Protozoa. Amer. Zool. **89**, 321–346 (1955).
HARTMANN, M., SCHARTAU, O., KUHN, R., WALLENFELS, K.: Über die Sexualstoffe der Seeigel. Naturwissenschaften **27**, 433. (1939).
HARTMANN, M., MEDEM, F. Graf, KUHN, R., BILLIG, H. J.: Untersuchungen über die Befruchtungstoffe der Regenbogenforelle. Z. Naturforsch. 2B, 330–349 (1947).
HARRISON, K.: A theory of oxidative phosphorylation. Nature (London) **181**, 1131 (1958).
HARVEY, E. N.: The mechanism of membrane formation and other early changes in developing sea urchins' eggs as bearing on the problem of artificial parthenogenesis. J. Exp. Zool. **8**, 355–376 (1910).
HARVEY, E. N.: Living Light. Princeton: University Press 1940.
HARVEY, E. N.: Bioluminescence. New York: Academic Press 1952.
HARVEY, E. N.: Bioluminescence. In: Comparative Biochemistry, Vol. II, pp. 545–591. Ed. by M. FLORKIN and H.S. MASON. New York: Academic Press 1960.
HAUGHTON, T. M., KERKUT, G. A., MUNDAY, K. A.: The oxygen dissociation and alkaline denaturation of haemoglobin from two species of earthworm J. Exp. Biol. **35**, 360–368 (1958).
HAUROWITZ, F.: Die Gleichgewichte zwischen Hämoglobin und Sauerstoff. Hoppe Seyler's Z. Physiol. Chem. **254**, 266–274 (1938).
HAUROWITZ, F., WAELSCH, H.: Über die chemische Zusammensetzung der Quelle, *Velella spirans*. Hoppe Seyler's Z. Physiol. Chem. **161**, 300–317 (1926).
HAWK, P. B., OSER, B. L., SUMMERSON, W. H.: Practical Physiological Chemistry 11th. Edn. London: J. and A. Churchill 1947.
HECHT, S.: Sensory equilibrium and dark adaptation in *Mya arenaria*. J. Gen. Physiol. **1**, 545–558; The nature of the latent period in the photic response of *Mya arenaria*. ib. 657–666; The effect of temperature on the latent period in the photic response of *Mya arenaria*. ib. 667–685 (1919).
HECHT, S.: The photochemical nature of the photosensory process. J. Gen. Physiol. **2**, 229–246; Intensity and the process of photoreception ib. 337–347; The dark adaptation of the human eye. ib. 499–517 (1920).
HECHT, S.: Time and intensity in photosensory stimulation. J. Gen. Physiol. **3**, 367–373; The relation between wavelength of light and its effect on the photosensory process. ib. 375–390 (1921).
HECHT, S.: Intensity discrimination and the stationary state. J. Gen. Physiol. **6**, 355–373 (1924).
HECHT, S.: The effect of exposure period and temperature on the photosensory process in *Ciona*. J. Gen. Physiol. **8**, 291–301 (1926).
HECHT, S.: The kinetics of dark adaptation. J. Gen. Physiol. **10**, 781–809 (1927).
HECHT, S.: The relation of time, intensity and wavelength in the photosensory system of *Pholas*. J. Gen. Physiol. **11**, 657–672 (1928).
HEDGES, E. S.: Liesegang's Rings and Other Periodic Structures. London: Chapman and Hall 1932.
HEILBRUNN, L. V.: The effect of anaesthetics on the surface precipitation reaction. Biol. Bull. **66**, 264–275 (1934).
HEILBRUNN, L. V.: An Outline of General Physiology. 3rd. Edn. Philadelphia – London: W. B. Saunders 1952.
HEIMAN, V., WILHELM, L. A.: The transmission of xanthophylls in feeds to the yolk of the egg(hen). Poultry Sci. **16**, 400–403 (1937).
HELMY, M. H., HACK, F. H.: Comparative histochemistry of liver (Amphibia and rat) and the melanin pigment cell system of Amphibian liver. Acta Histochim. **18**, 194–212 (1964).
HELMY, M. H., HACK, F. H.: Melanin pigment cell system of ovary of frog and toad and of Bidder's organ of the toad. Acta Histochim. **22**, 324–332 (1965).
HEMMERICH, P., SPENCER, J. Interaction of flavin with molybdenum V- and VI- and iron II- and III- redox systems. In: Flavins and Flavoproteins, pp. 82–98. Ed. by E.C. SLATER. Amsterdam-London-New York: Elsevier Publ. Co. 1966.
HEMMERICH, P., GHISLE, S., HARTMANN, U., MÜLLER, R.: Chemistry and molecular biology of flavin in the "fully reduced state". In: Flavins and Flavoproteins, pp. 83–105. Ed. by H. KAMIN. Baltimore: University Park Press and London: Butterworth 1971.
HEMMING, F. W., PENNOCK, J. F.: Vitamins and ubiquinone status in animals. In: Biochemistry of Quinones, pp. 287–315. Ed. by R.A. MORTON. London – New York: Academic Press 1965.
HEMMINGSEN, E., SCHOLANDER, P. F.: Specific transport of oxygen through haemoglobin solutions. Science, **132**, 1379–1381 (1960).

HEMPEL, K.: Investigation on the structure of melanin in malignant melanoma with ^{3}H and ^{14}C-DOPA labelled at different positions. In: Structure and Control of the Melanocyte, pp. 162–175. Ed. by G. Della PORTA and O. MÜHLBOCK. Berlin-Heidelberg – New York: Springer 1966.

HEMPEL, K., DIENNEL, M. 1962. Quoted from ROBSON and SWAN, 1966.

HEMPELMANN, F.: Chromatophoren bei *Nereis.* Z. Wiss. Zool. **152**, 353–383 (1939).

HENDRICKS, S.B.: How light interacts with living matter. Sci. Amer. **219** (3), 175–186 (1968).

HENKE, K.: Einfache Grundvorgänge in der tierischen Entwicklung I: Über Zellteilung, Wachstum und Formbildungen der Organeentwicklung der Insekten. Naturwissenschaften **34**, 149–157, 180–186 (1947).

HENKE, K.: Einfache Grundvorgänge in der tierischen Entwicklung II. Über die Entstehung von Differenzierungsmustern. Naturwissenschaften **35**, 176–181, 203–211, 239–250 (1948).

HENZE, M.: Untersuchungen über das Blut der Ascidien I: Die Vanadiumverbindung der Blutkörperchen. Hoppe Seyler's Z. Physiol. Chem. **72**, 494–501 (1911).

HENZE, M.: Über das sogenannte "Pinnaglobuline". Hoppe Seyler's Z. Physiol. Chem. **162**, 136–138 (1927).

HEPPLESTON, A.G.: Pigmentation and disorders of the lung. In: Pigments in Pathology, pp. 33–73. Ed. by M. WOLMAN. New York – London: Academic Press 1969.

HESLOP, R.B., ROBINSON, P.L.: Inorganic Chemistry. Amsterdam – London – New York: Elsevier Publ. Co. 1960.

HILL, R.: The chemical nature of haemochromogen and its carbon monoxide compound. Proc. Roy. Soc. London B **100**, 419–430 (1926).

HILLMAN, P., HOCHSTEIN, S., MINKE, B.: A visual pigment with two physiologically active stable states. Science, **175**, 1486–1488 (1972).

HINDE, R., SMITH, D.C.: Persistence of functional chloroplasts in *Elysia viridis* (Opisthobranchia, Saccoglossa). Nature, New Biol. (London) **239**, 30–31 (1972).

HIROMI, K., STURTEVANT, J.M.: Cytochrome b$_2$. In: Flavins and Flavoproteins, pp. 283–305. Ed. by E.C. SLATER. Amsterdam – London – New York: Elsevier Publ. Co. 1966.

HIROMI, K., STURTEVANT, J.M.: Free energy levels of heme and flavin of cytochrome b$_2$. In: Flavins and Flavoproteins. pp. 59–70. Ed. by K. YAGI. Tokyo: University Press and Baltimore: University Park Press 1969.

HITTELMAN, K.J., LINDBERG, O.: Fatty acid uncoupling in brown fat mitochondria. In: Brown Adipose Tissue, pp. 245–262. Ed. by O. LINDBERG. New York: American Elsevier Publ. Co. Inc. 1970.

HOAR, W.S.: Phototactic and pigmentary responses of sockeye salmon smolts following injury to the pineal organ. J. Fish. Bd. Canada **92**, 178–185 (1955).

HOAR, W.S.: General and Comparative Physiology. Englewood Cliffs, N.J.: Prentice Hall Inc. 1966.

HOCHSTEIN, P., COHEN, G.: The cytotoxicity of melanin precursors. Ann. N.Y. Acad. Sci. **100**, 876–886 (1963).

HODGES, R.B.: Vitamins: Fat-soluble. In: Encyclopaedia of the Biological Sciences, 2nd. Edn. pp. 984–986. Ed. by P. GRAY. New York: Reinhold 1970.

HODGSON, G.W., PETERSON, E., KVENVOLDEN, K.A., BUNNENBERG, E., HALPERN, B., PONNAMPERUMA, C.: Search for porphyrins in lunar dust. Science, **167**, 763–765 (1970).

HOGBEN, L.T.: The pigmentary effector system VII: The chromatic function in elasmobranch fishes. Proc. Roy. London B **120**, 142–158 (1936).

HOGBEN, L.T., MIRVICH, L.: The pigmentary effector system V: The nervous control of excitement pallor in reptiles. J. Exp. Biol. **5**, 295–308 (1938).

HOHORST, H.J., STRATMANN, D.: Oxidative phosphorylation by mitochondria of brown adipose tissue of the guinea pig. Abstracts 4th. Meeting Fed. Eur. Biochem. Socs Oslo, p. 109 (No. 435) 1967.

HOLLANDE, A.C.: Coloration vitale du corps adipeux d'un insecte avec la nourriture. Arch. Zool. Exp. Gen. **51**, Notes et Rev. 53–58 (1913).

HOLLANDE, A.C.: Les cérodécytes ou oenocytes des insectes considérés au point de vue biochimique. Arch. Anat. Microsc. **16**, 1–66 (1914).

HOLLANDE, A.C.: La cellule pericardiale des insectes. Arch. Anat. Microsc. **18**, 85–307 (1923).

HOLMES, W.: The colour changes of Cephalopods. Endeavour **14**, 78–82 (1955).

HONIG, B., KARPLUS, M.: Implications of torsional potential of retinal isomers for visual excitation. Nature (London) **229**, 558–560 (1971).

HOPKINS, F.G.: The pigments of the Pieridae: A contribution to the study of excretory substances which function in ornament. Phil. Trans. Roy. Soc, London B **186**, 661–682 (1895).

Hopkins, F. G.: A contribution to the chemistry of pterins. Proc. Roy. Soc. London B **130**, 359–379 (1942).
Horowitz, N. H.: The partition of nitrogen in the developing eggs of *Urechis caupo*. J. Cell. Comp. Physiol. **14**, 189–195 (1939).
Horowitz, N. H.: A respiratory pigment from the eggs of a marine worm. Proc. Nat. Acad. Sci. U.S. **26**, 161–163 (1940).
Horowitz, N. H., Baumberger, J. P.: Studies on the respiratory pigment of *Urechis* eggs. J. Biol. Chem. **141**, 407–415 (1941).
Horstadius, S.: The Neural Crest. London: Oxford University Press 1949.
Hoshi, T.: Studies on the physiology and ecology of plankton XII: Changes in oxygen consumption of the daphnia, *Simocephalus vetulus* with a decrease of oxygen concentration. Sci. Rep. Tohoku Imp. Univ. Ser. IV, **23**, 27–34; XIII: haemoglobin and its role in the respiration of the daphnia, *Simocephalus vetulus*. ib. 35–58 (1957).
Houchin, O. B.: Vitamin E and muscle degeneration in the hamster. Fed. Proc. Fed. Amer. Socs. Exp. Biol. **1**, 117–118 (1942).
Howard, H. W.: The genetics of *Armadillidium vulgare* Latr. III: Dominant and recessive genes for red body colour. J. Genet. **51**, 259–269 (1952–3).
Hsiao, S. C. T.: Melanosis in the common cod, *Gadus callanas* L., associated with trematode infection. Biol. Bull. **80**, 37–44 (1941).
Hsu, W.-J., Chichester, C. O., Davies, B. H.: Metabolism of β-carotene and other carotenoids in the brine shrimp *Artemia*. Comp. Biochem. Physiol. **32**, 69–79 (1970).
Hubbard, R.: The thermal stability of rhodopsin and opsin. J. Gen. Physiol. **42**, 259–280 (1958).
Hubbard, R., St. George, R. C. C.: The rhodopsin system of the squid. J. Gen. Physiol. **41**, 501–528 (1958).
Hubbard, R., Brown, P. K., Kropf, A.: Action of light on visual pigments: Vertebrate lumi- and meta- rhodopsins. Nature (London) **183**, 442–446 (1959).
Hüber, R., Epp, O., Steigemann, W., Formanek, H.: The atomic structure of erythrocruorin in the light of the chemical sequence and its comparison with myoglobin. Europ. J. Biochem. **19**, 42–50 (1971).
Huettner, A. F.: Comparative Embryology of Vertebrates. 2nd. Edn. New York: Macmillan 1949.
Hull, D., Hardman, M. J.: Brown adipose tissue in new-born mammals. In: Brown Adipose Tissue, pp. 97–115. Ed. by O. Lindberg. New York: Elsevier Publ. Co. 1970.
Human. P. J. E., Johnson, A. W., MacDonald, S. F., Todd, A. R.: Colouring matters of the Aphididae. II. Matters from *A. Fabae*. J. Chem. Soc. Lond. (**1950**), 477–485 (1950).
Hundley, J. M.: Achromotrichia due to copper deficiency. Proc. Soc. Exp. Biol. Med. **74**, 531–536 (1950).
Hüttel, R.: Die chemische Untersuchung des Schreckstoffes aus Elritzenhaut. Naturwissenschaften **29**, 333–334 (1941).
Huxley, J. S.: Threat and warning coloration in birds, with a general discussion of the biological functions of colour. Proc. 8th. Int. Congr. Ornithol. Oxford, pp. 430–455. Ed. by F. C. R. Jourdain. Oxford: University Press 1934.
Huxley, J. S., de Beer, G. R.: Elements of Experimental Embryology. Cambridge: University Press 1934.
Huxley, J. S., Wolsky, A.: Structure of normal and mutant eyes in *Gammarus chevreuxi*. Nature (London) **129**, 242–243 (1932).
Hydén, H., Lindström, B.: Microspectrographic studies on the yellow pigment in nerve cells. Discuss. Faraday. Soc. **9**, 436–441 (1950).
Hyman, L. H.: The Invertebrates: Protozoa through Ctenophora. New York – London: McGraw-Hill Book Co. Inc. 1940
Hyman, L. H.: The Invertebrates, Vol. II: Platyhelminthes and Rhynchocoela. New York-Toronto-London: McGraw-Hill Book Co. Inc. 1951.
Hyman, L. H.: The Invertebrates, Vol. IV: Echinodermata. New York – Toronto – London: McGraw-Hill Book Co. Inc. 1955.
Hyman, L. H.: The Invertebrates, Vol. V: Smaller Coelomate Groups. New York – Toronto – London: McGraw-Hill Book Co. Inc. 1959.
Hynes, H. B. N., Nicholas, W. L.: The resistance of *Gammarus* spp. to infection by *Polymorphus minutus* (Goeze, 1782) (Acanthocephala). Ann. Trop. Med. Parasitol. **52**, 376–383 (1958).

IGA, T.: Studies on melanomas. Bull. Shimane Univ. Nat. Sci. **14**, 85–91 (1965).
IHNAT, M., BERSOHN, R.: ^{1}H-nuclear magnetic resonance study of the Cu(II)-carnosine complex in aqueous solution. Biochemistry **9**, 4555–4566 (1970).
INAGAMI, K.: Mechanism of formation of red melanin in the silkworm. Nature (London) **174**, 1105 (1954).
INGRAM, V.M.: The evolution of a protein. Fed. Proc. Fed. Amer. Socs. Exp. Biol. **21**, 1053–1057 (1962).
INTERNATIONAL CRITICAL TABLES of numerical data: Physics, Chemistry and Technology (National Research Council of the U.S.A.). E.P. CARR and M.L. SHERRILL: Bibliography of absorption spectra of solutions: Vol. **5**, 326–358; Organic solutions, ib. 331–358 (1929) W.C. HOLMES: The absorption spectra of dyes: Vol. **7**, 173–211 (1930)
ISAKA, S.: Inhibitory effect of xanthopterin upon the formation of melanin *in vitro*. Nature (London) **169**, 74–75 (1952).
ISSIDORIDES, M.R.: Neuronal vascular relationships in the zona compacta of normal and Parkinsonian substantia nigra. Brain Res. **25**, 289–299 (1971).
ITO, Y., NOZAWA, Y., KAWAI, K.: Effect of the naphthoquinone xanthomegnin from *Microsporidium cookei* on the respiration of rat liver mitochondria. Experientia **26**, 826–827 (1970).
JAKOBI, H. 1957. Quoted from GILBERT and THOMPSON, 1968.
JACOBSON, W.: The role of *Leuconostoc citrovitrorum* factor (LCF) in cell division and the mode of action of folic acid antagonists on normal and leucaemic cells. J. Pathol. Bacteriol. **64**, 245–247 (1952).
JACOBSON, W., SIMPSON, D.M.: The fluorescence spectra of pterins and their possible use in the elucidation of the anti-pernicious anaemia factor. Biochem. J. **40**, 3–14 (1946).
JAFFE, H.H., ORCHIN, M.: Theory and Application of U.V. Spectroscopy. New York: Wiley 1962.
JANSEN, R.: Der Farbwechsel von *Piscicola geometra* L. I: Beschreibung des Farbwechsels und seiner Elemente. Z. Morphol. Ökol. Tiere **24**, 327–341 (1932).
JENKINS, F.A., WHITE, H.E.: Fundamentals of Optics. New York: McGraw-Hill Book Co. Inc. 1957.
JENSEN, R.A.: Metabolic interlock: Regulatory interactions existing between biochemical pathways. J. Biol. Chem. **244**, 2816–2823 (1969).
JIRSA, M.: The bile pigments. In: Pigments in Pathology, pp. 151–190. Ed. by M. WOLMAN. New York – London: Academic Press 1969.
JOHANSEN, K., MARTIN, A.W.: Circulation in a giant earthworm, *Glossoscolex giganteus* II: Respiratory properties of the blood and some patterns of gas exchange. J. Exp. Biol. **45**, 165–172 (1966)
JOHNSON, A., JONES, O.T.G.: Enzymic formation of haems and other metalloporphyrins. Biochim. Biophys. Acta, **93**, 171–173 (1964).
JOHNSON, F. 1970. Quoted from C. Sagan: Life beyond the Earth. In: Exobiology, pp. 465–477. Ed. by C. PONNAMPERUMA. Amsterdam – London: North Holland Publ. Co. 1972.
JOHNSON, F.H.: Introduction. In: Bioluminescence in Progress, pp. 3–21. Ed. by F.H. JOHNSON and Y. HANEDA. Princeton, N.J.: University Press 1966.
JOHNSON, L.P.: A study of *Euglena rubra* Hardy 1911. Trans. Amer. Microsc. Soc. Menasha, **58**, 42–48 (1939).
JOHNSON, L.P., JAHN, T.: Causes of green-red color change in *Euglena rubra*. Physiol. Zool. **15**, 89–94 (1942).
JOHNSON, M.B.: Formation of an ommatin pigment from tyrosine by u.v. irradiation. Nature (London) **190**, 924–925 (1961).
JOLY, P.: Determinisme endocrine de la pigmentation chez *Locusta migratoria* L. C.R. Soc. Biol. **145**, 1362–1364 (1951).
JONES, F.: The colour and constitution of organic molecules. In: Pigments pp. 26–49. Ed. by D. PATTERSON. Amsterdam – London – New York: Elsevier Publ. Co. 1967.
JONES, J.D.: Observations on the respiratory physiology and on the haemoglobin of the polychaete genus *Nephthys (N. hombergi)*. J. Exp. Biol. **32**, 110–125 (1955).
JONES, J.D.: Functions of respiratory pigments in invertebrates. In: Problems in Biology, pp. 9–90. Ed. by G.A. KERKUT. Oxford: Pergamon Press 1963.
JONES, O.T.G.: Biosynthesis of chlorophylls. In: Porphyrins and Related Compounds, pp. 131–145. Ed. by T.W. GOODWIN. London – New York: Academic Press 1968.
JØRGENSEN, C.K.: Symmetry and chemical bonding in copper-containing chromophores. In: The

Biochemistry of Copper, pp. 1–14. Ed. by J. Peisach, P. Aisen, and W.E. Blumberg. New York – London: Academic Press 1966.

Jori, G., Galiazzo, G., Scoffoni, E.: Photodynamic action of porphyrins on amino acids and proteins I: Selective photooxidation of methionine in aqueous solutions. Biochemistry 8, 2868–2875 (1969).

Joshi, S., Brown, R.R.: Pigment formation from hydroxykynurenine by a rat liver system. Fed. Proc. Fed. Amer. Socs Exp. Biol. 18, 355. (Biochem. Abst. No. 1008) (1959)

Jucci, C.: Ulteriori ricerche sul pigmento dei bozzolo di bachi da seta di razza verde giapponese. Boll. Soc. Ital. Biol. Sper. Napoli, 7, 163–165; Nuovo ricerche sui pigmenti dei bozzoli e della uova in varie razza di bachi da seta I: Flavoni. ib. 573–575 (1932).

Judd, D.B.: Basic correlates of the visual stimulus. In: Handbook of Experimental Psychology, pp. 811–867. Ed. by S.S. Stevens. New York: Wiley 1951.

Juhn, M., Fraps, L.M.: Developmental analysis in plumage I: The individual feather methods. Physiol. Zool. 9, 293–318; II: Plumage configuration and the mechanism of feather development. ib. 319–377 (1936).

Junge, H.: Über grüne Insekten-Farbstoffe. Hoppe Seyler's Z. Physiol. Chem. 268, 179–186 (1941).

Just, E.E.: Basic Methods for Experiments on Eggs of Marine Animals. Philadelphia: Blakiston 1939.

Kahn, J.S., Chang, I.C.: A soluble protein-chlorophyll complex from spinach chloroplasts III: Determination of molecular weight and comparison of complex isolated from different sources. Photochem. Photobiol. 4, 733–738 (1965).

Kalmus, H.: Physiology and ecology of cuticle colour in insects. Nature (London) 148, 428–429 (1941a).

Kalmus, H.: The resistance to desiccation of *Drosophila* mutants affecting body-colour. Proc. Roy. Soc. London B, 130, 185–201 (1941b).

Kalmus, H., Satchell, J.E., Bowen, J.C.: On the colour forms of *Allolobophora chlorotica* Sav. Ann. Mag. Nat. Hist. 12th Series 8, 795–800 (1955).

Kamen, M.D., Dus, K.M., Flatmark, T., de Klerk, H.: Cytochrome c. In: Electron and Coupled Energy Transfer in Biological Systems, Vol. IA, pp. 243–324. Ed. by T.E. King and M. Klingenberg. New York: Dekker 1971.

Kamin, H. (Ed.): Flavins and Flavoproteins. Baltimore: University Park Press and London: Butterworth 1971.

Kampa, E.M.P. Euphausiopsin, a new photosensitive pigment from the eyes of Euphausid Crustaceans. Nature (London) 175, 996–998 (1955).

Karkhanis, Y.D., Cormier, M.J.: Isolation and properties of *Renilla reniformis* luciferase, a low molecular weight energy-conversion enzyme. Biochemistry 10, 317–326 (1971).

Karlson, P.: Introduction to Modern Biochemistry. Trans. C.H. Doering. 3rd. Edn. New York: Academic Press 1968.

Karlson, P., Bückmann, D.: Experimentelle Auslösung der Umfärbung bei *Cerura*-Raupen durch Prothorakdrüsen Hormone. Naturwissenschaften 43, 44–45 (1956).

Karrer, P., Jucker, E.: Carotenoids. Trans. E. Braude. Amsterdam – London – New York: Elsevier Publ. Co. 1950.

Kasha, M.: Relation between exciton bands and conduction bands in molecular lamellate systems. In: Biophysical Science, pp. 162–169. Ed. by J.L. Oncley. New York: John Wiley and Sons Inc. 1959.

Kasha, M.: The nature and significance of $n \to \pi^*$ transitions. In: Light and Life, pp. 31–68. Ed. by W.E. McElroy and B. Glass. Baltimore: Johns Hopkins Press 1961.

Katagiri, M., Takemori, S.: Reaction mechanism of flavin-containing oxygenases. In: Flavins and Flavoproteins, pp. 447–461. Ed. by H. Kamin. Baltimore: University Park Press and London: Butterworth 1971.

Kaufman, S.: Nature of the primary oxidation product formed from tetrahydropteridines during phenylalanine hydroxylation. J. Biol. Chem. 236, 804–810 (1961).

Kaufman, S.: Pteridine cofactors. Ann. Rev. Biochem. 36, 171–184 (1967).

Kautsky, H., Müller, G.O.: Luminescenzumwandlung durch Sauerstoffnachweis geringster Sauerstoffmengen. Z. Naturforsch. 2A, 167–172 (1947).

Kawaguti, S.: Haemerythrin found in the blood of *Lingula*. Mem. Fac. Sci. Agric. Taihohu, 23, 95–98 (1941).

Keilin, D.: On cytochrome, a respiratory pigment common to animals, yeast and higher plants. Proc. Roy. Soc. London, B 98, 312–339 (1925).

KEILIN, D.: Comparative study of turacin and haematin. Proc. Roy. Soc. London, B, 100, 129–151 (1926).
KEILIN, D., RYLEY, J.F.: Haemoglobin in Protozoa. Nature 172, 451 (1953).
KEILIN, D., SMITH, E. L: Direct perception of pigment in nerve tissue of the human retina. Nature (London) 143, 333. (1939).
KEILIN, D., WANG, Y.L.: Haemoglobin of *Gastrophilus* larvae: Purification and properties. Biochem. J. 40, 855–866 (1946).
KEILIN, J.: Helicorubin and cytochrome h. Biochem. J. 64, 663–676 (1956).
KEILIN, J.: Properties of helicorubin and cytochrome h. Nature (London) 180, 427–429 (1957).
KEILIN, J.: Binding of nitrogenous bases to myoglobin. In: Hemes and Hemoproteins, pp. 173–188. Ed. by B. CHANCE, R.W. ESTABROOK, and T. YONETANI. New York – London: Academic Press 1966.
KEILIN, J.: Helicorubin and cytochrome h – spectroscopic properties, amino acid composition and reactivity with oxidases. In: Structure and Functions of Cytochromes, pp. 691–700. Ed. by K. OKUNUKI, M.D. KAMEN, and I. SEKUZU. Tokyo: University Press and Baltimore-Manchester: University Park Press 1968.
KEILIN, J., MCCOSKER, P.J.: Reactions between uroporphyrin and copper and their biological significance. Biochim. Biophys. Acta, 52, 424–435 (1961).
KELLER, R., WUTHRICH, R., DEBRUNNER, P.G.: Proton magnetic resonance reveals high spin Fe II in ferrocytochrome P_{450cam} from *Pseudomonas putida*. Proc. Nat. Acad. Sci. U. S. 69, 273–275 (1972).
KENNEDY, D.: Neural photoreception in a Lamellibranch Mollusc. J. Gen. Physiol. 44, 277–299 (1960).
KENNEDY, G.Y.: A porphyrin pigment in the integument of *Arion ater* (L.). J. Mar. Biol. Ass. U. K. 38, 27–32 (1959).
KENNEDY, G.Y.: Pigments of Annelida, Echiuroidea, Sipunculoidea, Priapuloidea and Phoronidea. In: Chemical Zoology, Vol. IV, pp. 311–376. Ed. by M. FLORKIN, and B.T. SCHEER. New York – London: Academic Press 1969.
KENNEDY, G.Y.: Harderoporphyrin: a new porphyrin from the Harderian gland of the rat. Comp. Biochem. Physiol. 36, 21–36 (1970).
KENNEDY, G.Y., Dales, R.P.: Functions of the heart body in Polychaetes. J. Mar. Biol. Ass. U. K. 37, 15–31 (1958).
KENNEDY, G.Y., NICOL. J.A.C.: Pigments of *Chaetopterus variopedatus* (Polychaeta). Proc. Roy. Soc. London B 150, 509–538 (1959).
KENNEDY, G.Y., VEVERS, H.G.: The biology of *Asterias rubens* V: A porphyrin pigment in the integument. J. Mar. Biol. Ass. U. K. 32, 235–247 (1953).
KENNEDY, G.Y., VEVERS, H.G.: The occurrence of porphyrins in certain marine invertebrates. J. Mar. Biol. Ass. U. K. 33, 663–676 (1954).
KENNEDY, G.Y., VEVERS, H.G.: Porphyrin pigments in the Tectibranch *Akera bullata* O.F. Müller. J. Mar. Biol. Ass. U. K. 35, 35–39 (1956).
KENNEDY, J.S.: Phase transformation in locust biology. Biol. Rev. Cambridge Phil. Soc. 31, 349–370 (1956).
KERR, W.E., LAIDLAW, H.H.: General genetics of the bee. Advan. Genet. 8, 109–153 (1956).
KERTESZ, D.: The copper of polyphenol oxidase. In: The Biochemistry of Copper, pp. 359–369. Ed. by J. PEISACH, P. AISEN, and W.E. BLUMBERG. New York – London: Academic Press 1966.
KETTLEWELL, H.B.D.: Further selection experiments on induction of melanin in the Lepidoptera. Heredity 10, 287–301 (1956).
KETTLEWELL, H.B.D.: The phenomenon of industrial melanism in Lepidoptera. Ann. Rev. Entomol. 6, 245–262 (1961).
KETTLEWELL, H.B.D.: A twelve year survey of the frequencies of *Biston betularia* (L.) and its melanic forms in Great Britain. Entomol. Rec. 77, 195–218 (1965).
KEY, K.H.L., DAY, M.F.: A temperature controlled physiological colour change in the grasshopper *Koscuiscola tristis* Sjost (Orthoptera, Acrididae). Aust. J. Zool. 2, 309–363 (1954).
KIDDER, G.W.: Nitrogen distribution, nutrition and metabolism. In: Chemical Zoology, Vol. I, pp. 93–159. Ed. by M. FLORKIN and B.T. SCHEER. New York – London: Academic Press 1967.
KIERKEGAARD, P.C., NORRESTAM, R., WERNER, P., CSOREGH, I., GLEHN, M., KARLSSON, R., LEIJONMARCK, M., LONNQUIST, O., STENSLAND, B., TILLBERG, O., TORBJORNSSON, L.: X-ray structure investigations of flavin derivatives. In: Flavins and Flavoproteins, pp. 1–22. Ed. by H. KAMIN. Baltimore: University Park Press and London: Butterworth 1971.

Kikkawa, H.: Biochemical genetics of *Bombyx mori.* Advan. Genet. **5**, 107–140 (1953).
Kikuyama, S., Miura, K., Yasumasu, I.: Role of pituitary gland in melanin synthesis in the tadpole skin. Endocrinol. Jap. **16**, 275–278 (1969)
Kim, K., Tchen, T.T., Hu, F.: Studies on ACTH-induced melanocyte formation in cultures of goldfish caudal fin. Ann. N.Y. Acad. Sci. **100**, 708–718 (1963).
Kimura, E., Hosoya, Y.: Further studies on cone substances. Jap. J. Physiol. **6**, 1–11 (1956).
King, M.E.: Regulation of cytochrome biosynthesis in some eukaryotes. In: Metabolic regulation, pp. 55–76. Ed. by H.J. Vogel. New York – London: Academic Press 1971.
King, R.L., Taylor, A.B.: *Malpighamoeba locustae* n. sp., Protozoan parasite of malpighian tubules of grasshoppers. Trans. Amer. Microsc. Soc. **55**, 6–10 (1936).
King, T.E., Klingenberg, M.: Electron and coupled Energy Transfer in Biological Systems. Vol. I/A. New York: M. Dekker Inc. 1971
Kinosita, H.: Electrophoretic theory of pigment migration within the fish melanophore. Ann. N.Y. Acad. Sci. **100**, 992–1004 (1963).
Kirby-Smith, J.S., Blum, H.S., Grady, H.G.: Penetration of u.v. radiation into skin as a factor in carcinogenesis. J. Nat. Cancer Inst. **2**, 403–412 (1942).
Kit, S.: Intermediary metabolism of building blocks involved in growth. In: Fundamental Aspects of Normal and Malignant Growth, pp. 1–136. Ed. by W. W. Nowinsky. Amsterdam – London – New York: Elsevier Publ. Co. 1960.
Klapper, M.H., Klotz, I.M.: Effect of bound ligands on the quaternary structure of haemerythrin. In Physiology and Biochemistry of Haemocyanin. Ed. by F. Ghiretti. London – New York: Academic Press 1968.
Kleinholz, L.H.: Crustacean eyestalk hormone and retinal pigment migration. Biol. Bull. **70**, 159–184 (1936).
Kleinholz, L.H.: Studies in reptilian colour change. J. Exp. Biol. **15**, 474–499 (1938).
Klotz, I.M., Klotz, T.A.: Oxygen-carrying proteins: A comparison of the oxygenation reactions in haemocyanin and haemerythrin with that in haemoglobin. Science **141**, 477–480 (1955).
Klüver, H.: On naturally occurring porphyrins in the central nervous system. Science **99**, 482–484 (1944a).
Klüver, H.: Porphyrins in the nervous system and behaviour. J. Psychol. **17**, 209–227 (1944b).
Klüver, H.: Functional differences between the occipital and temporal lobes with special reference to the interrelations of behavior and extracerebral mechanisms. In: Cerebral Mechanisms in Behavior, pp. 147–199. Ed. by L.A. Jeffress. New York: John Wiley and Sons Inc. London: Chapman and Hall Ltd. 1951.
Knowles, Sir F.G.W.: Pigment movement after sinus gland removal in *Leander aspersus.* Physiol. Comp. Oecol. **2**, 289–296 (1952).
Knowles, Sir F.G.W.: Endocrine activity in the Crustacean nervous system. Proc. Roy. Soc. London B **141**, 248–267 (1953).
Knowles, Sir F.G.W.: Neurosecretion in the tritocerebral complex of Crustaceans. Pubbl. Sta. Zool. Napoli **24**, Suppl. 74–78 (1954).
Knowles, Sir F.G.W.: Some problems in the study of colour change in Crustaceans. Ann. Sci. Nat. Zool. 11e Série **18**, 315–324 (1956).
Kobayashi, K., Forrest, H.S.: Isolation and identification of a new pteridine, neopterinyl 3'-β-D-glucuronic acid from *Bacillus subtilis.* Comp. Biochem. Physiol. **33**, 201–207 (1970).
Kokolis, N., Ziegler, I.: Wiedererscheinen von Tetrahydrobiopterin in der Regenerationknospe von *Triturus*-arten. Z. Naturforsch. **23 B**, 850–855 (1968).
Koller, G.: Farbwechsel bei *Crangon vulgaris.* Verh. Deut. Zool. Ges. **30**, 128–132 (1925).
Koller, G.: Über Chromatophorensysteme, Farbensinn und Farbwechsel bei *Crangon vulgaris.* Z. Vergl. Physiol. **5**, 191–246 (1927).
Koller, G.: Versuche über die inkretorischen Vorgänge beim Garneelenfarbwechsel. Z. Vergl. Physiol. **8**, 601–612 (1928).
Kon, S.K.: Vitamins A and B_{12} and some comment on refection. Proc. Roy. Soc. London B **156**, 351–365 (1962).
Konings, W.N., Siezen, R.J., Gruber, M.: Association-dissociation behaviour of *Helix pomatia* haemocyanin. Biochim. Biophys. Acta, **194**, 276–285 (1969).
Koschara, W., Haug, H.: Über die physiologische Bedeutung des Uropterins. Hoppe Seyler's Z. Physiol. Chem. **259**, 97–112 (1939).
Kosower, E.M.: The role of charge-transfer complexes in flavin chemistry and biochemistry. In:

Flavins and Flavoproteins, pp. 1–14. Ed. by E.C. SLATER. Amsterdam – London – New York: Elsevier Publ. Co. 1966.

KRAMER, S.: Color changes correlated with parental behaviour in cichlid fishes. Anat. Rec. **138**, 362–363 (1960).

KRASNOVSKY, A.A.: The models of the evolution of photochemical electron transfer. In: Chemical Evolution and the Origin of Life, pp. 279–287. Ed. by C. PONNAMPERUMA. Amsterdam: North Holland Publ. Co. 1972.

KRIEBEL, M.E., FLOREY, E.: Nervous control of chromatophores in Cephalopods. J. Invest. Dermatol. **54**, 90 (1970).

KRISHNAN, G.: Histochemical studies on the nature and formation of the egg capsules of the shark *Chiloscyllium griseum*. Biol. Bull. **117**, 298–307 (1959).

KRITSKY, M.S.: Participation of flavins in photobiological processes in contemporary organisms. In: Chemical Evolution and the Origin of Life, pp. 313–315. Ed. by C. PONNAMPERUMA. Amsterdam: North Holland Publ. Co. 1972.

KROPF, A., BROWN, P.K., HUBBARD, R.: Lumi- and meta- rhosopsins of squid and octopus. Nature (London) **183**, 446–448 (1959).

KROPP, B.: The pigment of *Velella spirans* and *Fiona marina*. Biol. Bull. **60**, 120–123 (1931).

KRUGER, P., KERN, H.: Die physikalische und physiologische Bedeutung des Pigments bei Amphibien und Reptilien. Pfluger's Arch. Gesamte Physiol. Menschen Tiere **202**, 119–138 (1924).

KRUKENBERG, C.F.W.: Vergleichend-Physiologische Studien an den Küsten der Adria; Experimentelle Untersuchungen, Erste Reihe. Heidelberg: Carl Winter's Universitäts Buchbehandlung 1880.

KRUKENBERG, C.F.W.: Vergleichend- Physiologische Studien, Zweite Reihe: Abt. 3: Die Pigmente, ihre Eigenschaften, ihre Genese und ihre Metamorphosen bei den wirbellosen Tieren. Heidelberg: Carl Winter 1882.

KUBIŠTA, V.: Flavones in *Helix pomatia*. Experientia **6**, 100 (1950).

KUBO, M.: The oxygen equilibrium of haemerythrin and the Bohr effect. Bull. Chem. Soc. Japan **26**, 189–192 (1953).

KÜHN, A.: Zur Genetik und Entwicklungsphysiologie des Zeichnungsmusters der Schmetterlinge. Nach. Ges. Wiss. Göttingen **6**, 312–335 (1932).

KÜHN, A.: Entwicklungsphysiologie. Berlin: Springer 1955.

KÜHN, R., WAGNER-JAUREGG, T.: Über die aus Eiklar und Milch isolierten Flavine. Ber. Deut. Chem. Ges. **66**, 1577–1582 (1930).

KÜHN, R., WALLENFELS, K.: Echinochromes as prosthetic groups of high molecular symplexes in the eggs of *Arbacia pustulosa*. Ber. Deut. Chem. Ges. **73 B**, 458–464 (1940).

KÜHN, R., LEDERER, E., DEUTSCH, A.: Astacin aus den Eiern der Seespinne. Hoppe Seyler's Z. Physiol. Chem. **220**, 229–235 (1933).

KÜHN, R., WALLENFELS, K., WEYGAND, F., MOLL, T., HEPDING, L.: Zur Specifität des Vitamins K. Naturwissenschaften **27**, 518–519 (1939).

KUNTZ, A.: The histological basis of adaptive shades and colours in the flounder, *Paralichthys albiguttatus*. Washington Bull. Bur. Fish. **35**, 1–29 (1916).

LABEYRIE, F.: Discussion. In: Flavins and Flavoproteins, p. 22. Ed. by M. KAMIN. Baltimore: University Park Press and London: Butterworth 1971.

LA DU, B.N., ZANNONI, V.G.: Ochronosis. In: Pigments in Pathology, pp. 465–488. Ed. by M. WOLMAN. New York – London: Academic Press 1969.

LAMBOOY, J.P.: Riboflavin and related flavins. In: Comprehensive Biochemistry, **11**, pp. 23–35. Ed. by M. FLORKIN and E.H. STOTZ. Amsterdam – London – New York: Elsevier Publ. Co. 1963.

LAMOLA, A.A.: Electronic energy transfer in solution: Theory and applications. In: Energy Transfer and Organic Photochemistry, pp. 17–132. Ed. by A.A. LAMOLA and J.J. TURRO. New York: Wiley (Interscience Publishers) 1969.

LANKESTER, E.R.: Blue stentorin, – the colouring matter of *Stentor coeruleus*. Quart. J. Microsc. Sci. **13**, 139–142 (1873).

LAND, E.H.: Colour vision and the natural image. Proc. Nat. Acad. Sci. U.S. **45**, 115–129; 636–644 (1959) see also Sci. Amer. **200** (5), 84–99 (1959).

LASCELLES, J.: Tetrypyrrole Biosynthesis and its Regulation. New York: Benjamin 1964.

LASCELLES, J.: Regulation of haem and chlorophyll biosynthesis. In: Porphyrins and related Compounds, pp. 49–59. Ed. by T.W. GOODWIN New York – London: Academic Press 1968.

Lawrence, W.J.C., Price, J.R.: The genetics and chemistry of flower colour variation. Biol. Rev. Cambridge Phil. Soc. 15, 35–58 (1940).
Lea, A.J.: A neutral solvent for melanin. Nature (London) 156, 478. (1945).
Leclerq, J.: Mis en évidence de la nature pterinique des pigments jaunes des Hyménoptères adultes. Bull. Ann. Soc. Entomol. Belg. 87, 64–74 (1951).
Lederer, E.: Les Carotenoides des Animaux. Paris: Hermann et Cie 1935.
Lederer, E.: Recherches sur les caroténoides des Invertébrés. Bull. Soc. Chim. Biol. Paris 20, 567–610 (1938).
Lederer, E.: L'isolement et la constitution chimique de la bonelline, pigment verte de *Bonellia viridis*. C.R.H. Acad. Sci. 209, 528–530 (1939).
Lederer, E.: Les pigments des Invertébrés. Biol. Rev. Cambridge Phil. Soc. 15, 273–306 (1940).
Lederer, E., Glaser, R.: Sur l'échinochrome et le spinochrome. C.R.H. Acad. Sci. 207, 454–456 (1938).
Lederer, E., Huttrer, C.: Pigments from the secretion of *Aplysia* (Sea Hare or Sea Slug). Trans. Mem. Soc. Chim. Biol. (1942), 1055–1061.
Lederer, E., Tixier, R.: Sur les porphyrins de l'ambergris. C.R.H. Acad. Sci. 225, 531–532 (1947).
Lederer, E., Teissier, G., Huttrer, C: Sur l'isolement et la constitution chimique de la calliactine, pigment de l'anémone de mer, *Sagartia parasitica* (= *Calliactis effoeta*). Bull. Soc. Chim. France 7, 608–615 (1940).
Lee, D.L., Smith, M.H.: Haemoglobins of parasitic animals. Exp. Parasitol 16, 392–424 (1965).
Leermakers, P.A.: Fundamental concepts. In: Energy Transfer and Organic Photochemistry, pp. 1–16. Ed. by A.A. Lamola and N.J. Turro. New York: Wiley (Interscience) 1969.
Lehmann, F.E.: Einführung in die Physiologische Embryologie. Basel: Birkhauser 1945.
Lehmann, F.E.: Plasmatische Eiorganisation und Entwicklungsleistung beim Keim von *Tubifex* (Spiralia). Naturwissenschaften 43, 289–296 (1956).
Lehmann, F.E.: Action of morphostatic substances and the role of proteases in regenerating tissues and in tumour cells. Advan. Morphog. 1, 153–187 (1961).
Lehmann, F.E.: Die Physiologie der Mitose. Ergeb. Biol. 27, 116–161 (1964).
Lemberg, R.: The disintegration of haemoglobin in the animal body. In: Perspectives in Biochemistry, pp. 137–149. Ed. by J. Needham and D.E. Green. Cambridge: University Press 1937.
Lemberg, R.: Animal pigments. Ann. Rev. Biochem. 7, 421–448 (1938).
Lemberg, R., Legge, J.W.: Haematin Compounds and bile Pigments. New York and London: Interscience Publ. 1949.
Lemberg, R., Gilmour, M.V., Cutler, M.E.: Recent studies on cytochrome oxidase. In: Structure and Function of Cytochromes, pp. 54–65. Ed. by K. Okunuki, M.D. Kamen and I. Sekuzu. Tokyo: University Press and Baltimore – Manchester: University Park Press 1968.
Lenel, R.: Sur l'absorption des pigments caroténoides du crabe *Carcinus maenas* Pennant par son parasite *Sacculina carcini* Thompson C.R.H. Acad. Sci. 238, 948–949 (1954).
Lepkovsky, S., Wang, W., Koike, T., Dimick, M.K.: The oxygen uptake and phosphorus uptake by particulate suspensions (mitochondrial) from liver, brown and white fatty tissues. Fed. Proc. Fed. Amer. Socs Exp. Biol. 18, 272 (1959).
Lerner, A.B.: Metabolism of phenylalanine and tyrosine. Advan. Enzymol. 14, 73–128 (1953).
Lerner, A.B.: Vitiligo. J. Invest. Dermatol. 32, 285–310 (1959).
Lerner, A.B., Case, J.D.: Pigment cell regulatory factors. J. Invest. Dermatol. 32, 211–221 (1959).
Lester, R.L., Crane, F.L.: The natural occurrence of coenzyme Q and related compounds. J. Biol. Chem. 234, 2169–2175 (1959).
Letellier, A.: Recherches sur la pourpre produite par le *Purpura lapillus*. C.R.H. Acad. Sci. 109, 82–84 (1889).
Lettvin, J.Y., Platt, J.R., Wald, G., Brown, K.T.: General Discussion: Early receptor potential. Cold Springs. Harbor Symp. Quant. Biol. 30, 501–504 (1965).
Lewin, R.: Vitamin E: The rise of a new theory. New Sci. 50, 165–167 (1971).
Lewis, J.S., Prinn, R.G.: Chemistry and photochemistry of the atmosphere of Jupiter. In: Exobiology, pp. 123–142. Ed. by C. Ponnamperuma. Amsterdam: North Holland Publ. Co. 1972.
Lewis, M.R.: The disappearance of the pigmentation of the eye and the skin of tadpoles (*Rana sylvatica*) that develop in solutions of indophenol dyes. J. Exp. Zool. 64, 37–71 (1932).
Lewisohn, R., Laszlo, D., Leuchtenberger, R., Leuchtenberger, C.: The action of xanthopterin on tumor growth. Proc. Soc. Exp. Biol. Med. 56, 144–145 (1944).

L'Helias, C.: Le rôle des ptérines intermédiares photosensibles et thermosensibles dans la genèse des hormones du complex endocrinien de l'insecte. Ann. Biol. 3e Série 37, 367–392 (1961).

L'Helias, C.: Corrélations entre les pterines et la photopériodisme dans la regulation du cycle sexuel chez les pucerons. Bull. Biol. France Belg. 96, 187–198 (1962).

L'Helias, C.: Etude du facteur inducteur de tumeurs (ADN) induit par le déséquilibre du rapport ptérines/hormones de croissance. Bull. Biol. France Belg. 98, 511–542 (1964).

L'Helias, C.: Growth and development. In: Chemical Zoology V, A, pp. 343–400. Ed. by M. Florkin and B.T. Scheer. New York – London: Academic Press 1970.

Liaaen-Jensen, S.: Developments in the carotenoid field. Experientia 26, 697–710 (1970).

Lillie, F.R., Juhn, M.: The physiology of development of feathers I: Growth rate and pattern in the individual feather. Physiol. Zool. 5, 124–184 (1932).

Lillie, R.D.: Histochemistry of melanins. In: Pigments in Pathology, pp. 327–351. Ed. by M. Wolman. New York – London: Academic Press 1969.

Lindberg, O. (Ed.): Brown Adipose Tissue. New York: American Elsevier Publ. Co. Inc. 1970.

Linschitz, H., Abrahamson, E.W.: Kinetics of porphyrin-catalysed chemiluminescent decomposition of peroxides and the mechanism of photosensitised oxidation. Nature (London) 172, 909–910 (1953).

Linzen, B.: Über Ommochrome XV: Über die Identifizierung des 'Urechochroms' als Xanthommatin. Hoppe Seyler's Z. Physiol. Chem. 314, 12–14 (1959).

Linzen, B.: Ommidine, ein neuer Typus von Ommochromen aus Orthopteren. Z. Naturforsch. 21 B, 1038–1047 (1966).

Linzen, B.: Zur Biochemie der Ommochrome. Naturwissenschaften 54, 259–267 (1967).

Linzen, B., Bückmann, D.: Biochemische und histologische Untersuchungen zur Umfärbung der Raupe von *Cerula vinula* L. Z. Naturforsch. 16 B, 6–18 (1961).

Linzen, B., Hendricks-Hertal, U.: Kynurenin-3-Hydroxylase in der Metamorphose von *Bombyx mori rb:* Organ- und stadien-spezifischer Wechsel der Enzymaktivität. Wilhelm Roux' Arch. Entwicklungsmech. Organismen 165, 26–34 (1970).

Lipmann, F.: Metabolic generation and utilisation of phosphate bond energy. Advan. Enzymol. 1, 99–162 (1941).

Little, C.C.: The Inheritance of Coat Colour in Dogs. Ithaca, N.Y.: Comstock Publ. Assoc. (Cornell U.P.), and London: Constable and Co. Ltd. 1957.

Little, J.E., Sproston, T.J., Foote, M.W.: Synthesis and antifungal action of 2-methyl mercapto-1,4-napthoquinone. J. Amer. Chem. Soc. 71, 1124–1125 (1949).

Llewellyn, J.: Observations on the food and the gut pigment of the polyopisthocotylea (Trematoda, Monogenea). Parasitology 44, 428–437 (1954).

Lockwood, A.P.M.: Aspects of the Physiology of Crustacea. Edinburgh – London: Oliver and Boyd 1968.

Loconti, J.D., Roth, L.M.: Composition of the odorous secretion of *Tribolium castaneum*. Ann. Entomol. Soc. Amer. 46, 281–289 (1953).

Loewe, R., Linzen, B., von Stachelberg, W.: Die gelösten Stoffe in der Hämolymphe einer Spinne. Z. Vergl. Physiol. 66, 27–34 (1970).

Lönnberg, E., Hellström, H.: Zur Kenntnis der Carotenoide bei marinen Evertebraten. Ark. Zool. 23 A, (15) 1–74 (1931).

Lontie, R., Witters, R.: *Helix pomatia* haemocyanins. In: Biochemistry of Copper, pp. 455–463. Ed. by J. Peisach, P. Aisen, and W.E. Blumberg. New York – London: Academic Press 1966.

Loomis, W.F.: Skin-pigment regulation of vitamin D biosynthesis in Man. Science 157, 501–506 (1967).

Ludwig, M.L., Anderson, R., Apgar, P., Le Quesne, M.: A preliminary crystallographic study of Clostridial flavodoxin. In: Flavins and Flavoproteins, pp. 171–184. Ed. by M. Kamin. Baltimore: University Park Press and London: Butterworth 1971.

Lundstrom, H.M., Bard, P.: Hypophyseal control of cutaneous pigmentation in an elasmobranch fish. Biol. Bull. 62, 1–9 (1932).

Lumry, R.: Fundamental problems in the physical chemistry of proteins. In: Electron and Coupled Energy Transfer in Biological Systems, pp. 1–116. Ed. by T.E. King and M. Klingenberg. New York: Dekker 1971.

Lwoff, A.: Le cycle du pigment carotenoide chez *Idya furcata* (Bahd) (Copépode Harpacticide): Nature, origine, evolution du pigment et des reserves ovulaires au cours de la segmentation. Bull. Biol. France Belg. 61, 193–240 (1927).

Lwoff, A.: L'Evolution Physiologique. Paris: Hermann et Cie 1944.

Lyman, C. P.: Control of coat colour in the varying hare. Bull. Mus. Comp. Zool. Harvard 93, 393–461 (1943).

Maccoll, A.: Colour and constitution. Quart. Rev. Chem. Soc. London 1, 16–58 (1947).

Macdonald, R. A.: Human and experimental haemochromatosis and haemosiderosis. In: Pigments in Pathology, pp. 115–149. Ed. by M. Wolman. New York – London: Academic Press 1969.

Macdonald, S. F.: Colouring matter of the Aphididae Part XI: Pigments from *Hamamelistes* species. J. Chem. Soc. (1954), 2378–2381 (1954).

Macfarlane, R. G.: Introduction. In: The Haemostatic Mechanism in Man and Other Animals, pp. 1–6. Ed. by R. G. Macfarlane. London: Academic Press, for the Zoological Society of London 1970.

Maclean, N., Jurd, R. D.: The control of haemoglobin synthesis. Biol. Rev. Cambridge Phil. Soc. 47, 393–437 (1972).

Mac Munn, C. A.: Studies on Animal Chromatology. Proc. Birmingham Nat. Hist. Soc. 3, 351–407 (1883).

Mac Munn, C. A. Observations on the chromatology of *Actinia*. Phil. Trans. Roy. Soc. London B 176, 641–643 (1885).

Mac Munn, C. A.: On the presence of haematoporphyrin in the integument of certain invertebrates. J. Physiol. London 7, 240–252 (1886).

Mac Munn, C. A.: Contributions to animal chromatology. Quart. J. Microsc. Sci. 30, 51–96 (1890).

Mac Rae, E. K.: The occurrence of porphyrin in the planarian. Biol. Bull. 110, 69–76 (1956).

Mac Rae, E. K.: Localisation of porphyrin fluorescence in planarians. Science 134, 331–332 (1961).

MacRae, E. K.: Chromatographic identification of porphyrins in several species of Turbellaria. Experientia 19, 77–78 (1963).

Mager, A.: Über die Zusammensetzung des Eridictyolglykosides. Hoppe Seyler's Z. Physiol. Chem. 274, 109–115 (1942).

Mahdihassan, S.: Two symbionts of *Psylla mali*. Nature (London) 159, 749 (1947).

Mahler, H. R., Cordes, E. H.: Biological Chemistry. London – New York: Harper and Row 1966.

Malkin, R., Rabinowitz, J. C.: Non-heme iron electron-transfer proteins. Ann. Rev. Biochem. 36, 113–148 (1967).

Mangum, C. P., Dales, R. P.: Products of haem-synthesis in Polychaetes. Comp. Biochem. Physiol. 15, 237–252 (1965).

Manunta, C.: La determinazione colorimetrica del continuto in pigmenti carotenoide e flavone della uova di varie razze ed incroce di bacchi da seta. Boll. Soc. Ital. Biol. Sper. 8, 1278–1292 (1933).

Manunta, C.: Permeabilita intestinale e permeabilita ghiandolare per le xantofilla e per le carotine in varie razza di bacchi da seta. Atti Soc. Nat. Mat. Modena 66, 104–113 (1935).

Manunta, C.: I flavoni nel vitello e nei gusci delle uova di bachi a pelle gialla. Boll. Soc. Ital. Biol. Sper. 11, 50–51 (1936).

Manunta, C.: Sui pigmenti della uova e della pelle dei camaleonte. Boll. Soc. Ital. Biol. Sper. 12, 33–34 (1937a).

Manunta, C.: Sulla differente natura dei pigmente che colorando la pelle e le uova di due varieta di *Carassius* I, II: Boll. Soc. Ital. Biol. Sper. 12, 628–630 (1937b).

Manunta, C.: I carotenoidi del bozzolo nella F_2 dell'incrocio fra razze 'limone' e 'rosa' die bacchi da seta. Boll. Soc. Ital. Biol. Sper. 12, 31–32 (1937c).

Manunta, C.: L'analisi cromatografica dei carotenoidi della foglia di gelso e del sangue, della ghiandole e della uova die varie razze di bacchi da seta. Boll. Soc. Ital. Biol. Sper. 12, 626–628 (1937d).

Manunta, C.: Astaxanthin in insects and other terrestrial Arthropods. Nature (London) 162, 298. (1948).

Manwell, C.: Oxygen equilibrium of *Phascolosoma agassizii* hemerythrin. Science 127, 592–593 (1958a).

Manwell, C.: The oxygen-respiratory pigment equilibrium of the haemocyanin and haemoglobin of the Amphineuran Mollusc, *Cryptochiton stelleri*. J. Cell. Comp. Physiol. 52, 341–352 (1958b).

Manwell, C.: Comparative physiology: Blood pigments. Ann. Rev. Physiol. 22, 191–244 (1960a).

Manwell, C.: Heme-heme interactions in the oxygen equilibrium of some Invertebrate myoglobins. Arch. Biochem. Biophys. 89, 194–201 (1960b).

Manwell, C.: Respiratory functions of body fluids. In: Comparative Animal Physiology, pp. 198–237. By C. L. Prosser and F. A. Brown. Philadelphia – London: W. B. Saunders Co. 1961.

MANWELL, C.: Chemistry, genetics and function of Invertebrate respiratory pigments: Configurational changes and allosteric effects. In: Oxygen in the Animal Organism, pp. 49–116. Ed. by F. DICKENS and E. NEIL. Oxford: Pergamon Press 1964.
MARAK, G. E., GALLIK, G. J., CORNESBY, R. A.: Light sensitive pigments of insect heads. Opthalmol. Res. 1, 65–71 (1970).
MARGOLIN, P.: Regulation of tryptophan synthesis. In: Metabolic Regulation, pp. 389–446. Ed. by H. J. VOGEL. New York – London: Academic Press 1971.
MARGULIS, L.: Early cellular evolution. In: Exobiology, pp. 342–368. Ed. by C. PONNAMPERUMA, Amsterdam: North Holland Publ. Co. 1972.
MARICIC, S. L.: Discussion. In: Hemes and Hemoproteins, pp. 282–284. Ed. by B. CHANCE, R. W. ESTABROOK, and T. YONETANI. New York – London: Academic Press 1966.
MARKERT, C. L., SILVERS, W. K.: The effects of genotype and cell environment on melanoblast differentiation in the house mouse. Genetics 41, 429–450 (1956).
MARRIAN, G. F.: A note on haemerythrin. J. Exp. Biol. 4, 357–364 (1926).
MARSDEN, C. D.: Brain melanin. In: Pigments in Pathology, pp. 395–420. Ed. by M. WOLMAN. New York – London: Academic Press 1969.
MARSHALL, S. M., NICHOLLS, A. G., ORR, A. P.: On the biology of *Calanus finmarchicus* V: Seasonal distribution, size, weight and chemical composition in Loch Striven in 1933 and their relation to the phytoplankton. J. Mar. Biol. Ass. U.K. 19, 793–819 (1934).
MARSHALL, S. M., ORR, A. P.: The Biology of a Marine Copepod, *Calanus finmarchicus*. Edinburgh – London: Oliver and Boyd 1935.
MARTELL, A. E., CALVIN, M.: Chemistry of Metal Chelate Compounds. Englewood Cliffs: Prentice Hall 1952.
MARXER, A.: Über ein carcinostatisch wirksames Hydrochinonderivat. Experientia 11, 184–186 (1955).
MASON, H. S.: The chemistry of melanins III: Mechanism of oxidation of dihydroxyphenylalanine by tyrosinase. J. Biol. Chem. 172, 83–97 (1948).
MASON, H. S.: Discussion. In: Oxygen in The Animal Organism, pp. 117–119. Ed. by F. DICKENS and E. NEIL. Oxford: Pergamon Press 1964.
MASON, H. S., PETERSON, E. W.: Melanoproteins I: Reactions between enzyme-generated quinones and amino acids. Biochim. Biophys. Acta 111, 134–146 (1965).
MASON, S. F.: Molecular electronic absorption spectra. Rev. Chem. Soc. London 15, 287–371 (1961).
MASSEY, V., PALMER, G., BALLOU, D.: On the reaction of reduced flavins and flavoproteins with molecular oxygen. In: Flavins and Flavoproteins, pp. 349–361. Ed. by H. KAMIN. Baltimore: University Park Press and London: Butterworth 1971.
MAST, S. O.: Changes in shade, color and pattern in fishes and their bearing on the problem of adaptation and behaviour with special reference to the flounders *Paralichthys* and *Ancylopsetta*. Bull. U.S. Bur. Fish. 34, 173–238 (1916).
MATSUMOTO, J.: Studies on fine structure and cytochemical properties of erythrophores in swordtail *Xiphophorus helleri* with special reference to their pigment granules (pterinosomes). J. Cell. Biol. 27, 493–504 (1965).
MATSUMOTO, J.: The pterinosome: A subcellular unit in the brightly coloured pigmentation of the lower Vertebrates. J. Invest. Dermatol. 54, 92. (1970).
MATSUMOTO, J., BAGNARA, J. T., TAYLOR, J. D.: Isolation and separation of pteridines from animals. In: Experiments in Physiology and Biochemistry, Vol. 4, pp. 289–364. Ed. by G. A. KERKUT. London – New York: Academic Press 1971.
MATTHEW, R., MASSEY, V.: Free and complexed forms of old yellow enzyme. In: Flavins and Flavoproteins, pp. 329–348. Ed. by H. KAMIN. Baltimore: University Park Press and London: Butterworth 1971.
MATTHEWS, S. A.: Observations on pigment migration within the fish melanophore. J. Exp. Zool. 58, 471–486 (1931).
MATTY, A. J., SHELTWAY, M. J.: The relation of thyroxine to skin purines in *Salmo irideus*. Gen Comp. Endocrinol. 9, 473. (1967).
MAXIMOW, A. A., BLOOM, W.: A Textbook of Histology. 4th. Edn. Philadelphia – London: W. B. Saunders Co. 1942.
MAYER, F., COOK, A. H.: The Chemistry of Natural Colouring Matters. New York: Reinhold Publ. Co. 1943.
MAYER, H. D., MELNICK, J. L.: Studies on the acridine orange staining of the purified RNA viruses. Virology 14, 74–82 (1961).

Mazza, F. P., Stolfi, G.: Ricerche sul pigmento di *Halla parthenopaea* Costa. Arch. Sci. Biol. Napoli **16**, 183–197 (1931).
Mc Elroy, W. D., Green, A.: The enzymic basis of light emission. In: Enzymes: Units of Biological Structure and Function, pp. 369–380. Ed. by O. H. Gaebler. New York: Academic Press 1956.
Mc Elroy, W. D., Hastings, J. W.: Biochemistry of firefly luminescence. In: Luminescence of Biological Systems, pp. 161–198. Ed. by F. H. Johnson. Washington, D. C.: Amer. Ass. Advan. Sci. 1955
Mc Ginnety, J. A., Payne, N. C., Ibers, J. A.: The role of the metal atom in the reversible uptake of molecular oxygen. J. Amer. Chem. Soc. **91**, 6301–6310 (1969).
Mc Kusick, V. A.: Human Genetics. 2nd. Edn. Englewood Cliffs: Prentice Hall 1969.
Mc Whinnie, M. A., Sweeney, T. M.: The demonstration of two chromatophorotropically active substances in the land Isopod *Trachelipus rathkei*. Biol. Bull. **108**, 160–174 (1955).
Medawar, P. B.: Old age and natural death. Modern Quarterly **2**, 30–49 (1946).
Meikle, J. E. S., Mc Farlane, J. E.: The role of lipid in the nutrition of the house cricket *Acheta domestica* L. (Orthoptera: Gryllidae). Can. J. Zool. **43**, 87–98 (1965).
Meinschein, W. G., Barghoorn, E. S., Schopf, J. W.: Biological remnants in a Precambrian sediment. Science **145**, 262–263 (1964).
Mell, R.: Die Trockenzeitform als Hemmungserscheinung (*Diagora nigrivena* (Leech) als Trockenzeitform von *Hestina assimilis* L.). Biol. Zentralbl. **51**, 187–194 (1931).
Menaker, M.: Non-visual light reception. Sci. Amer. **226** (3), 22–29 (1972).
Menke, H.: Periodische Bewegungen und ihre Zusammenhänge mit Licht und Stoffwechsel. Pflüger's Arch. Gesamte Physiol. Menschen Tiere **140**, 37–91 (1911).
Merker, E.: Die Empfindlichkeit feuchthautiger Tiere im Licht II: Warum kommen Regenwürmer in Wasserlachen um und warum verlassen sie bei Regen ihre Wohnrohre? Zool. Jahrb. Abt. Allg. Zool. **42**, 487–555 (1926).
Merker, E., Gilbert, H.: Die Widerstandsfähigkeit von Süßwasserplanarien in ultraviolettreichem Licht. Zool. Jahrb. Abt. Allg. Zool. Physiol. Tiere **50**, 441–556 (1932).
Mervyn, L.: The metabolism of the cobalamins. In: Comprehensive Biochemistry, Vol. 21, pp. 153–177. Ed. by M. Florkin and E. H. Stotz. Amsterdam – London – New York: Elsevier Publ. Co. 1971.
Metcalf, R. L.: Isolation of a red-fluorescent pigment, lampyrine, from the Lampyridae. Ann. Entomol. Soc. Amer. **36**, 37–40 (1943).
Metcalf, R. L.: A study of the metabolism of chlorophyll in the Squash Bug, *Anasa tristis* De Geer. Ann. Entomol. Soc. Amer. **38**, 397–402 (1945).
Metcalf, R. L., Patton, R. L.: A study of riboflavin metabolism in the American roach by fluorescence microscopy. J. Cell. Comp. Physiol. **19**, 373–376 (1942).
Metzler, D. E.: Discussion. In: Flavins and Flavoproteins, p. 320. Ed. by E. C. Slater. Amsterdam – London – New York: Elsevier Publ. Co. 1966.
Michaelis, L., Coryell, C. D., Granick, S.: Ferritin III: The magnetic properties of ferritin and some other colloidal ferric compounds. J. Biol. Chem. **148**, 463–480 (1943).
Miles, N. A., Postman, W. M., Heggie, R.: Evidence for the presence of vitamin A and carotenoids in the olfactory area of the steer. J. Amer. Chem. Soc. **61**, 1929–1930 (1939).
Millar, I. T., Springall, H. D.: A Shorter Sidgwick's Organic Chemistry of Nitrogen. Oxford: Clarendon Press 1969.
Miller, P. L.: Possible function of haemoglobin in *Anisops*. Nature (London) **201**, 1052. (1964).
Millot, J.: Le pigment purique chez les Vertébrés inferieurs. Bull. Biol. France Belg. **57**, 261–363 (1923).
Millott, N.: Naphthoquinone pigment in the tropical sea urchin, *Diadema antillarum* Philippi. Proc. Zool. Soc. London **129**, 263–272 (1957).
Millott, N., Jacobson, F. W.: Phenolases in the echinoid, *Diadema antillarum* philippi. Nature (London) **168**, 878 (1951).
Millott, N., Jacobson, F. W.: The occurrence of melanin in the sea urchin, *Diadema antillarum* Philippi. J. Invest. Dermatol. **18**, 91–95 (1952).
Millott, N., Yoshida, M.: Reactions to shading in the sea urchin *Psammechinus miliaris* (Gmelin). Nature (London) **178**, 1300 (1956).
Millott, N., Yoshida, M.: The spectral sensitivity of the echinoid, *Diadema antillarum* Philippi. J. Exp. Biol. **34**, 394–401 (1957).
Mitchell, J. S., Marrian, D. H.: Radiosensitisation of cells by a derivative of 2-methyl-1,4-naphtho-

quinone. In: Biochemistry of Quinones, pp. 503–541. Ed. by R. A. MORTON, London – New York: Academic Press 1965.

MITCHELL, P.: Chemiosmotic coupling in oxidative and photosynthetic phosphorylation. Biol. Rev. Cambridge Phil. Soc. 41, 445–502 (1966).

MOEWIS, F.: Untersuchungen über die Sexualität und Entwicklung von Chlorophyceen. Arch. Protistenk. 80, 469–526 (1923).

MOEWIS, F.: Zur Physiologie und Biochemie der Selbststerilität bei *Forsythia*. Biol. Zentralbl. 69, 181–197 (1950).

MØLLER, K.M.: On the nature of stentorin, with an appendix on the photodynamic action of the pigment by K.M. MØLLER and A.H. WHITELEY. C.R. Trav. Lab. Carlsberg 22, 471–498 (1960).

MONAGHAN, B., SCHMITT, O.: The absorption spectrum of medullated and of non-medullated nerve. Proc. Soc. Exp. Biol. Med. 28, 705–708 (1931).

MOORE, H.B.: A comparison of the biology of *Echinus esculentus* in different habitats. J. Mar. Biol. Ass. U.K. 21, 711–719 (1937).

MOORE, R.E., SINGH, H., SCHEUER, P.J.: Isolation of eleven new spinochromes from Echinoids of the genus *Echinothrix*. J. Org. Chem. 31, 3645–3650 (1966).

MOORE, T.: The interrelations of vitamins. Vitam. Horm. 3, 1–21 (1945).

MORERE, J.-L.: La carotene: substance indispensable pour la nutrition de *Plodia interpunctella*. C.R.H. Acad. Sci. 272D, 2229–2231 (1971).

MORGAN, L.R., SINGH, R., FISETTE, R.J.: Relationship of oxygen consumption and cytochrome oxidase-succinic dehydrogenase activities in the *Amphiuma means*. Comp. Biochem. Physiol. 20, 343–349 (1967).

MORGAN, T.H.: Regeneration in *Tubularia*. Wilhelm Roux' Arch. Entwicklungsmech. Organismen, 11, 346–381 (1901a).

MORGAN, T.H.: Regeneration. New York: The Macmillan Co.; London: Macmillan and Co. 1901b.

MORI, Y., MATSUMOTO, J., HAMA, T.: On the properties of cyprinopourpre A_2, a pterin isolated from the skin of Cyprinidae and its relation to ichthyopterin or 7-hydroxypterin. Z. Vergl. Physiol 43, 531–543 (1960).

MORPURGO, G., WILLIAMS, R.J.P.: The state of copper in biological systems. In: Physiology and Biochemistry of Haemocyanin, pp. 115–130. Ed. by F. GHIRETTI. London – New York: Academic Press 1968.

MORRIS, S.J., THOMSON, R.H.: The flavonoid pigments of the Marbled White butterfly (*Melanargia galathea* Saltz). J. Ins. Physiol. 9, 391–399 (1963); The flavonoid pigments of the Small Heath butterfly, *Coenonympha pamphilus* L. ib. 10, 377–383 (1964).

MORRISON, M., ROMBAUTS, W.A., SCHROEDER, W.A.: Lactoperoxidase. In: Hemes and Hemoproteins, pp. 345–348. Ed. by B. CHANCE, R.W. ESTABROOK, and T. YONETANI. New York – London: Academic Press 1966.

MORTENSON, L.E., VALLENTINE, R.C., CARNAHAN, J.E.: An electron-transport factor from *Clostridium pasteurianum*. Biochem. Biophys. Res. Comm. 7, 448–460 (1962).

MORTON, R.A.: Spectroscopy of quinones and related substances. In: Biochemistry of Quinones, pp. 23–65. Ed. by R.A. MORTON. London – New York: Academic Press 1965.

MORTON, R.K., ARMSTRONG, J.McD.: Flavin-heme interaction in crystalline cytochrome $b_2L(+)$-lactate dehydrogenase of yeast. In: Intracellular Respiration (Proc. 5th. Int. Cong. Biochem. Moscow 1961), pp. 213–224. Ed. E.C. SLATER. Oxford: Pergamon Press 1963.

MOSELEY, H.N.: On the colouring matters of various animals and especially of deep sea forms dredged by H.M.S. CHALLENGER. Quart. J. Microsc. Sci. 17, 1–23 (1877).

MOSS, T.H., MOLESKI, C., YORK, J.L.: The magnetic susceptibility evidence for a binuclear iron complex in hemerythrin. Biochemistry 10, 840–842 (1971).

MOTE, M.I., GOLDSMITH, T.H.: Compound eyes: Localisation of two colour receptors in the same ommatidium. Science 171, 1254–1255 (1971).

MOYER, F.H.: Genetic effects on melanosome fine structure and ontogeny in normal and malignant cells. Ann. N.Y. Acad. Sci. 100, 584–606 (1963).

MOYER, F.H.: Genetic variations in the function, structure and ontogenesis of mouse melanin granules. Amer. Zool. 6, 43–66 (1966).

MÜLLER, F., HEMMERICH, P., EHRENBERG, A.: On the molecular and submolecular structure of flavin free radicals and their properties In: Flavins and Flavoproteins, pp. 107–122. Ed. by H. KAMIN. Baltimore: University Park Press and London: Butterworth 1971.

MULLINS, D. E., CAHRAN, D. G.: Nitrogen excretion in cockroaches: uric acid not a major product. Science 177, 699–701 (1972).

MUNTZ, W. R. A.: Vision in frogs. Sci. Amer. 210 (3), 111–119 (1964).

MUNZ, F. W.: Vision: Visual pigments. In: Fish Physiology, Vol. V, pp. 1–32. Ed by W. S. HOAR and D. J. RANDALL. New York – London: Academic Press 1971.

MURRAY, J. (Ed.): Narrative of the Cruise of H. M. S. Challenger, With a General Account of the Scientific Results of the Expedition, London: H. M. Stationery Office 1885.

MURRAY, J. D.: A simple method for obtaining approximate solutions for a class of diffusion kinetics enzyme problems: I: General class and illustrative examples. Math. Biosci. 2, 379–411; II: Further examples and non-symmetric problems. ib. 3, 115–133 (1968).

NAKAMURA, T., MASON, H. S.: An electron spin resonance study of copper valence in oxyhaemocyanin. Biochem. Biophys. Res. Comm. 3, 297–299 (1960).

NAPOLITANO, L., MC NARY, J. E., KLOEP, L. P.: The release of free fatty acids from brown and white adipose tissues after incubation with ACTH or epinephrine. Metab. Clin. Exp. 14, 1076–1083 (1965).

NARA, S., YASUNOBU, K. T.: Some recent advances in the field of amine oxidases. In: The Biochemistry of Copper, pp. 423–441. Ed. by J. PEISACH, P. AISEN, and W. E. BLUMBERG. New York – London: Academic Press 1966.

NARDI, G., ZITO, R.: Spectrophotometric titration of haemocyanin from *Octopus vulgaris*. Pubbl. Sta. Zool. Napoli 36, 507–509 (1968).

NEEDHAM, A. E.: Microanatomical studies on *Asellus aquaticus*. Quart. J. Microsc. Sci. 84, 49–72 (1942).

NEEDHAM, A. E.: Formation of melanin in regenerating limbs of a Crustacean. Nature (London) 164, 717–718 (1949).

NEEDHAM, A. E.: Regeneration and Wound Healing. London: Methuen 1952.

NEEDHAM, A. E.: Origination of Life. Quart. Rev. Biol. 34, 189–209 (1959).

NEEDHAM, A. E.: Properties of the connective tissue pigment of *Lithobius forficatus* (L.). Comp. Biochem. Physiol. 1, 72–100 (1960).

NEEDHAM, A. E.: The Growth Process in Animals. London: Pitman and Sons 1964.

NEEDHAM, A. E.: Body pigment of *Polycelis*. Nature (London) 206, 209–210 (1965a).

NEEDHAM, A. E.: The Uniqueness of Biological Materials. Oxford: Pergamon Press 1965b.

NEEDHAM, A. E.: Regeneration in the Arthropods and its endocrine control. In: Regeneration in Animals and Related Problems, pp. 283–323 Ed. by V. KIORTSIS and H. A. L. TRAMPUSCH. Amsterdam: North Holland Publ. Co. 1965c.

NEEDHAM, A. E.: The chlorogogen pigment of earthworms. Life Sci. 5, 33–39 (1966a).

NEEDHAM, A. E.: The tissue pigments of some fresh water leeches. Comp. Biochem. Physiol. 18, 427–461 (1966b).

NEEDHAM, A. E.: Integumental pigments of the millipede, *Polydesmus angustus* (Latzel). Nature (London) 217, 975–977 (1968a).

NEEDHAM, A. E.: Distribution of protoporphyrin, ferrihaem and indoles in the body-wall of *Lumbricus terrestris* L. Comp. Biochem. Physiol. 26, 429–442 (1968b).

NEEDHAM, A. E.: Uniqueness of biological materials. J. Brit. Interplanet. Soc. 21, 26–37 (1968c).

NEEDHAM, A. E.: The integumental pigments of some Isopod Crustacea. Comp. Biochem. Physiol. 35, 509–534 (1970a).

NEEDHAM, A. E.: Nitrogen metabolism in Annelida. In: Comparative Biochemistry of Nitrogen Metabolism, Vol. I, pp. 207–297. Ed. by J. W. CAMPBELL. London – New York: Academic Press 1970b.

NEEDHAM, A. E.: Haemostatic mechanisms in the Invertebrata. In: The Haemostatic Mechanism in Man and Other Animals, pp. 19–44. Ed. by R. G. MACFARLANE. London: Academic Press for the Zoological Society of London 1970c.

NEEDHAM, A. E.: Integumental pigments of the Amphipod *Jassa*. Nature (London) 228, 1336–1337 (1970d).

NEEDHAM, A. E., BRUNET, P. C. J.: The integumental pigment of *Asellus*. Experientia 13, 207–209 (1957).

NEEDHAM, J.: Chemical Embryology. Cambridge: University Press 1931.

NEEDHAM, J.: Biochemistry and Morphogenesis. Cambridge: University Press 1942.

NEEDHAM, J., NEEDHAM, D. M.: On phosphorus metabolism in embryonic life. I: Invertebrate eggs. J. Exp. Biol. 7, 317–348 (1930).

NEUBERT, G.: 1956. Quoted from BUTENANDT et al., 1960a.
NEUMANN, D.: Morphologische und experimentelle Untersuchungen über die Variabilität der Farbmuster auf der Schale von *Theodoxus fluviatilis* L. Z. Morphol. Ököl. Tiere **48**, 349–411 (1959).
NEUWIRT, J., PONKA, J.P., BOROVA, J.: Role of haem in regulation of γ-ALA and haem synthesis in rabbit reticulocytes. Eur. J. Biochem. **9**, 36–41 (1969).
NEVILLE, A.C.: Daily growth layers in animals and plants. Biol. Rev. Cambridge Phil. Soc. **42**, 421–441 (1967a).
NEVILLE, A.C.: A dermal light sense influencing skeletal structure in locusts. J. Ins. Physiol. **13**, 933–939 (1967b).
NICHOLAS, D.J.D.: Role of metals in enzymes with special reference to flavoproteins. Nature (London) **179**, 800–804 (1957).
NICHOLAS, G., FUSEAU-BRAESCH, S.: Etude de quelques facteurs contrôlant l'homochromie chez *Locusta migratoria migratorioides* (Orthoptère). C.R. Soc. Biol. **162**, 1091–1094 (1968).
NICOLAUS, R.A. 1962. Quoted from BLOIS, 1965.
NICOLAUS, R.A.: Melanins. Paris: Hermann et Cie 1968.
NICHOLLS, E.M.: DOPA and the red, brown and black pigments of hair and feathers. J. Invest. Dermatol. **53**, 302–309 (1969).
NICHOLLS, P.: Cytochrome oxidase, an allosteric enzyme. In: Structure and Function of Cytochromes, pp. 76–86. Ed. by K. OKUNUKI, M.D. KAMEN, and I. SEKUZU. Tokyo: University Press; Baltimore-Manchester: University Park Press 1968.
NICKERSON, B.: Pigmentation of the hoppers of the desert locust in relation to phase coloration. Anti-locust Bull. No. 24. 37 pp. London: Anti-locust Res. Centre 1956.
NICKERSON, M.: Relation between black and red melanin pigments in feathers. Physiol. Zool. **19**, 66–77)1941).
NICOL, J.A.C.: Some aspects of photoreception and vision in fishes. Advan. Mar. Biol. **1**, 171–208. (1963).
NICOL, J.A.C.: Special effectors: Luminous organs, chromatophores, pigments and poison glands. In: Physiology of Mollusca, Vol. I, pp. 353–381. Ed. by K.M. WILBUR and C.M. YONGE. New York: Academic Press 1964.
NICOL, J.A.C.: Biology of Marine Animals. 2nd. Edn. London: Pitman and Sons 1967.
NIXON, M.: Feeding Mechanisms and Growth in *Octopus vulgaris*. Thesis, PhD. Degree: University of London 1968.
NIYAGI, K., HARTMAN, G.R., RUNGE, W., WATSON, C.J.: Studies on the affinity of nuclear material for uroporphyrin. Biochem. Med. **4**, 391–402 (1970).
NORMAN, C., GOLDBERG, E.: Effect of light on motility, life span and respiration of bovine spermatozoa. Science **130**, 624–625 (1959).
NORRIS, D.M.: Quinol stimulation and quinone deterrence of gustation by *Scolytus multistriatus*. Ann. Entomol. Soc. Amer. **63**, 476–478 (1970).
NORRIS, D.M., FERKOVICH, S., BAKER, J., ROZENTHAL, J., BERG, T.: Energy transduction in quinone inhibition of insect feeding. J. Ins. Physiol. **17**, 85–97 (1971).
NORTH, W.J.: Sensitivity to light of the sea-anemone, *Metridium senile* (L). In: Comparative Biochemistry of Photoreactive Systems, pp. 295–301. Ed. by M.B. ALLEN. New York: Academic Press 1960.
NORTH, W.J., PANTIN, C.F.A.: Sensitivity to light of the sea-anemone, *Metridium senile* (L) Adaptation and action spectra. Proc. Roy. Soc. London B **148**, 385–396 (1958).
NOVAK, V.J.A.: Morphogenetic analysis of the effects of juvenile hormone analogues and other morphogenetically active substances on embryos of *Schistocerca*. J. Embryol. Exp. Morphol. **21**, 1–21 (1969).
NOVALES, R.R.(Ed.): Cellular aspects of the control of colour changes. Amer. Zool. **9**, 427–540 (1969).
OBIKA, M., MATSUMOTO, J., TAYLOR, J.D.: Some aspects of the control of melanisation in the goldfish. J. Invest. Dermatol. **54**, 94. (1970).
OHUYE, T.: A note on the formed elements in the coelomic fluid of a Brachiopod, *Terebratalia corcanica*. Sci. Rep. Tohoku University IV, **11**, 231–238 (1936).
OKADA, Y.K.: Etudes sur la régénération chez les coelentérés. Arch. Zool. Exp. Gen. **66**, 497–551 (1927).
OKAMURA, M.Y., KLOTZ, I.M., JOHNSON, C.E., WINTER, M.R.C., WILLIAMS, R.J.P.: The states of iron in haemerythrin: A Mössbauer study. Biochemistry **8**, 1951–1958 (1969).

Okay, S.: 1946. Quoted from Prosser and Brown, 1961.
Okunuki, K., Kamen, M.D., Sekuzu, I. (Eds.): Structure and Function of Cytochromes. Tokyo: University Press; Baltimore – Manchester: University Park Press 1968.
Onslow, H.: A contribution to our knowledge of the chemistry of coat colour in animals and of dominant and recessive whiteness. Proc. Roy. Soc. London B **89**, 36–58 (1917).
Oparin, A.I.: The Origin of Life on Earth. Trans. Ann Synge. Edinburgh: Oliver and Boyd 1957.
Oparin, A.I. Life, its Nature, Origin and Development. Edinburgh: Oliver and Boyd 1961.
Oro, J.: Stages and mechanisms of prebiological organic synthesis. In: The Origins of Prebiological Systems, pp. 137–171. Ed. by S.W. Fox. New York – London: Academic Press 1965.
Ortiz, E., Williams-Ashman, H.G.: Identification of skin pteridines in the pasture lizard, *Anolis pulchellus*. Comp. Biochem. Physiol. **10**, 181–190 (1963).
Osaki, S., Mc Dermott, J.A., Johnson, D.A., Frieden, E.: The inhibition of the ascorbate oxidising activity of serum ceruloplasmin by citrate and apotransferrin. In: The Biochemistry of Copper, pp. 559–569. Ed. by J. Peisach, P. Aisen, and W.E. Blumberg. New York – London: Academic Press 1966.
Osaki, S., Mc Dermott, J.A., Frieden, E.: Re-examination of the oxidising activity and the physiological role of ceruloplasmin (ferroxidase). In: Physiology and Biochemistry of Haemocyanin pp. 25–36. Ed. by F. Ghiretti. London – New York: Academic Press 1968.
Osawa, H., Sato, M., Natori, S., Ogawa, H.: Occurrence of rhodoquinone-9 in the muscle of *Ascaris*. Experientia **25**, 484–485 (1969).
Osborne, C.M.: The experimental production of melanin pigment on the lower surface of summer flounders (*Paralichthys dentatus*. Proc. Nat. Acad. Sci. U.S. **26**, 155–161 (1940).
Oshima, C., Seki, T., Ishizaki, H.: Studies on the mechanism of pattern formation in the elytra of lady beetles. Genetics **41**, 4–20 (1956).
Oshima, K., Johnson, C.L., Gorbman, A.: Relations between prolonged hypothyroidism and electrophysiological events in the trout *Salmo gairdnerii:* Effects of replacement dosages of thyroxine Gen. Comp. Endocrinol. Suppl. **3**, 529–541 (1972).
Oster, G.: The chemical effects of light. Sci. Amer. **219** (3), 158–170 (1968).
Otsu, T.: Component amino acids of chromatophore – concentrating hormones from Decapod Crustacea. Naturwissenschaften **52**, 187–188 (1965).
Ottolenghi-Nightingale, E.: Induction of melanin synthesis in albino mouse skin by DNA from pigmented mice. Proc. Nat. Acad. Sci. U.S. **64**, 184–189 (1969).
Padlan, E.A., Love, W.E.: Structure of the haemoglobin of the marine Annelid worm, *Glycera dibranchiata*, at 5.5 Å resolution. Nature (London) **220**, 376–378 (1968).
Palmer, G., Muller, F., Massey, V.: Electron paramagnetic resonance studies on flavoprotein radicals. In: Flavins and Flavoproteins, pp. 127–150. Ed. by H. Kamin, Baltimore: University Park Press 1971.
Palmer, L.S.: The physiological relation of plant carotenoids to the carotenoids of the cow, horse, sheep, goat, pig and hen. J. Biol. Chem. **27**, 27–32 (1916).
Palmer, L.S.: Carotenoids and Related Pigments: The Chromolipids, New York: Chemical Catalog Co. 1922.
Palmer, L.S., Eccles, C.H.: The fate of carotin and xanthophylls during digestion. J. Biol. Chem. **17**, 237–243 (1914).
Palmer, L.S., Kempster, H.L.: Relation of plant carotenoids to growth, fecundity and reproduction in fowls. J. Biol. Chem. **39**, 299–312 (1919a).
Palmer, L.S., Kempster, H.l.: Yellow pigmentation and fecundity: The physiological relation between fecundity and the natural yellow pigmentation of various breeds of fowl. J. Biol. Chem. **39**, 313–329 (1919b).
Palmer, L.S., Kempster, H.L.: Effects of pigments on egg yolks: The influence of specific feeds and certain pigments on the colour of the egg yolk and body fat of fowls. J. Biol. Chem. **39**, 331–337 (1919c).
Palmer, M.F., Chapman, G.: The state of oxidation of haemoglobin in the blood of living *Tubifex*. J. Zool. London **167**, 203–209 (1970).
Pant, K.P., Mukerjee, D.P.: The melanising activity of the semen of buffalo bulls and its relation to the properties of live spermatozoa. J. Reprod. Fertil. **25**, 103–105 (1971).
Pantin, C.F.A.: Physiological adaptation. J. Linn. Soc. Zool. London **37**, 705–711 (1932).

Parker, G.H.: The influence of light and heat on the movements of the melanophore pigment in lizards. J. Exp. Zool. 3, 401–414 (1906).
Parker, G.H.: Color changes due to erythrophores in the squirrel fish *Holocentrus*. Proc. Nat. Acad. Sci. U.S. 23, 206–211 (1937).
Parker, G.H.: The colour changes in lizards, particularly in *Phrynosoma*, J. Exp. Biol. 15, 48–73 (1938).
Parker, G.H.: Animal color changes and their neurohumors. Quart. Rev. Biol. 18, 205–227 (1943).
Parker, G.H.: Animal Color Changes and Their Neurohumours. Cambridge: University Press 1948.
Parker, G.H., Rosenblueth, A.: The electric stimulation of the concentrating (adrenergic) and the dispersing (cholinergic) nerve fibres of the melanophores in the catfish. Proc. Nat. Acad. Sci. U.S. 27, 198–204 (1941).
Passama-Vuillaume, M.: Sur la pigmentation verte de *Mantis religiosa* (L.) C.R.H. Acad. Sci. 258, 6549–6552 (1964).
Passama-Vuillaume, M.: Etude du pigment vert chez *Locusta migratoria* (L.) normal et albinos. Bull. Soc. Zool. 90, 485–501 (1965).
Passama-Vuillaume, M.: Sur la pigmentation brune de trois Orthoptères: *Mantis religiosa* (L.), *Sphodromantis viridis* (F.) et *Locusta migratoria* (L.). C.R.H. Acad. Sci. 262D, 1597–1599 (1966).
Passama-Vuillaume, M., Barbier, M. Sur la biosynthèse de la biliverdin IXα par la mante *Mantis religiosa* et le criquet *Locusta migratoria*. C.R.H. Acad. Sci. 263, 924–925 (1966).
Pasteels, J.J.: Recherches sur la morphogenèse et le determinisme des segmentations inégales chez les Spiralia I: *Aplysia*, II: *Myzostoma*, III: *Chaetopterus*. Arch. Anat. Microsc. Morphol. Exp. 30, 161–197 (1934).
Patterson, D., Pilling, B.: Photoconductivity measurements on a zwitterionic dye. Nature (London) 201, 294–295 (1964).
Pautsch, F.: The colour change of the zoea of the shrimp, *Crangon crangon* L. Experientia 9, 274–276 (1953).
Payne, N.M.: Hydroid pigments I: General discussion and pigments of the Sertulariidae. J. Mar. Biol. Ass. U.K. 17, 739–749 (1931).
Pearson, P.B., Burgin, C.J.: The pantothenic acid content of 'royal jelly'. Proc. Soc. Exp. Biol. Med. 48, 415–417 (1941).
Pease, M.S.: Yellow fat in rabbits, a linked character? Z. Indukt. Abstam. Vererbungsl. Suppl. 2, 1153–1156 (1928).
Peisach, J., Aisen, P., Blumbach, W.E.: The Biochemistry of Copper. New York – London: Academic Press 1966.
Penrose, L.S.: Outline of Human Genetics. London: Heinemann 1959.
Penzer, G.R., Radda, G.K.: The Chemistry and biological functions of isoalloxazines (flavins). Quart. Rev. Chem. Soc. London 21, 43–65 (1967).
Peredelsky, A.A., Blacher, L.J.: 1929. Quoted from Fox and Vevers, 1960.
Perkins, E.B., Snook, T.: The movement of pigment within the chromatophores of *Palaemonetes*. J. Exp. Zool. 61, 115–128 (1932).
Pershin, J.C.: Photolabile pigments in Invertebrates. Science 114, 120–121 (1951).
Perutz, M.F.: Stereochemistry of cooperative effects in haemoglobin. Nature (London) 228, 726–739 (1970).
Perutz, M.F.: Haemoglobin: Genetic abnormalities. New Sci. 50, 762–765 (1971).
Perutz, M.F.: Nature of haem-haem interactions. Nature (London) 237, 495–499 (1972).
Perutz, M.F., Rossman, M.G., Cullis, A.F., Muirhead, H., Will, G., North, A.C.T.: Structure of haemoglobin: A three dimensional Fourrier synthesis at 5.5Å resolution, obtained by X-ray analysis. Nature (London) 185, 416–422 (1960).
Petersen, S., Gauss, W., Urbschat, E.: Synthese einfacher Chinon-Derivate mit fungiziden, bacteriostatischen oder cytostatischen Eigenschaften. Angew. Chem. 67, 217–240 (1955).
Phear, E.A.: Gut haems in the Invertebrates. Proc. Zool. Soc. London 125, 383–406 (1955).
Phillips, D.C.: The three dimensional study of an enzyme molecule. Sci. Amer. 215 (5), 78–90 (1966).
Phillips, J.P., Forrest, H.S.: Terminal synthesis of xanthommatin in *Drosophila melanogaster* II: Enzymic formation of the phenoxazone nucleus. Biochem. Genet. 4, 489–498 (1970).
Phillips, U., Haldane, J.B.S.: Relative sexuality in unicellular Algae. Nature 143, 334. (1939).
Phillips, W.D., Knight, E., Blomstrom, D.C.: Fe^{37} Mössbauer spectroscopy and some biological applications. In: Non-Heme Iron Proteins: Role in Energy Conversion, pp. 69–85. Ed. by A. San Pietro. Yellow Springs, Ohio: Antioch Press 1965.

PICKFORD, G.E.: Histological and histohemical observations upon an aberrant Annelid, *Poeobius meseres* Heath. J. Morphol. 80, 287–319 (1947).

PIRENNE, M.H., MARRIOT, F.H.C.: Light sensitivity of the aquatic flatworm, *Dendrocoelum lacteum*. Nature (London) 175, 642. (1955).

PIRIE, A.: Crystals of riboflavin making up the tapetum lucidum in the eye of the lemur. Nature (London) 183, 985–986 (1958).

PIRIE, A., VAN HEYNINGEN, R.: Biochemistry of The Eye. Oxford: Blackwell Scientific Publications 1956.

PITOTTI, M.: La distribuzione delle ossidasi e perossidasi nelle uova di *Myzostoma, Beroe e Nereis*. Pubbl. Sta. Zool. Napoli 21, 93–100 (1947).

PITTENDRIGH, C.S.: Adaptation, natural selection and behaviour. In: Behaviour and Evolution, pp. 390–416. Ed. by A. ROE and G.G. SIMPSON. New Haven, Connecticut: Yale University Press 1958.

PLAUT, G.W.E.: Biosynthesis of riboflavin. In: Comprehensive Biochemistry, Vol. 21, pp. 11–45. Ed. by M. FLORKIN and E.H. STOTZ. Amsterdam – London – New York: Elsevier Publ. Co. 1971.

POLONOVSKI, M., BUSNEL, R.-G., PESSON, M.: Propriétiés biochimiques des ptérines. Helv. Chim. Acta 29, 1328–1333 (1946).

PORRA, R.J., ROSS, B.D.: Haem synthase and cobalt porphyrin synthase in various microorganisms. Biochem. J. 94, 557–562 (1965).

PORTA, E.A., HARTROFT, W.S.: Lipid pigment in relation to ageing and dietary factors. In: Pigments in Pathology, pp. 191–235. Ed. by M. WOLMAN. New York – London: Academic Press 1969.

PORTER, G.: The triplet state in fluid solution. In: Light and Life; pp. 69–77. Ed. by W.D. MCELROY and B. GLASS. Baltimore: Johns Hopkins Press 1961.

PORTER, H.: The tissue copper proteins, cerebrocuprein, erythrocuprein, hepatocuprein and neonatal hepatic mitochondrocuprein. In: The Biochemistry of Copper, pp. 159–174. Ed. By J. PEISACH, P. AISEN, and W.E. BLUMBERG. New York – London: Academic Press 1966.

PORTER, J.W., LINCOLN, R.E.: Lycopersicon selections containing a high percentage content of carotenes and colourless polyenes II: The mechanism of carotene biosynthesis. Arch. Biochem. 27, 390–403 (1950).

POTTS, W.T.W.: Excretion in Molluscs. Biol. Rev. Cambridge Phil. Soc. 42, 1–41 (1967).

POULTON, E.B.: An inquiry into the cause and extent of a special colour relation between certain exposed lepidopterous pupae and the surfaces which immediately surround them. Proc. Roy. Soc. London B 42, 94–108 (1887).

POULTON, E.B.: The Colours of Animals. London: Paul Trench, Trubner and Co. Ltd. 1890.

PREISLER, P.W.: Oxidation-reduction potentials and the possible respiratory significance of the pigment of the Nudibranch, *Chromodoris zebra*. J. Gen. Physiol. 13, 349–359 (1930).

PRENANT, A.: Sur l'origine mitochondriale des grains de pigment. C. R. Soc. Biol. 74, 926–929 (1913).

PRESSMAN, B.C., LARDY, H.A.: Effect of surface active agents on the latent ATPase of mitochondria. Biochim Biophys. Acta 21, 458–466 (1956).

PROSSER, C.L. BROWN, F.A. Jr.: Comparative Animal Physiology. Philadelphia – London: W.B. Saunders Co. 1961.

PROTA, G.: Structure and biogenesis of phaeomelanin. J. Invest. Dermatol. 54, 95. (1970).

PROTA, G. SCHERILLO, G., NICOLAUS, R.A.: On the structure of trichosiderin. Rend. Accad. Sci. Fis. Nat. (Soc. Naz. Sci. Lett. Arti Napoli) 35, 1–4 (1968).

PROTA, G., D'AGOSTINO, M., MISURACA, G.: Isolation and characterisation of Hallachrome, a red pigment from the sea worm, *Halla parthenopeia*. Experientia 27, 15–16 (1971).

PRUSINER, S., CANNON, B., LINDBERG, O.: Mechanisms controlling oxidative metabolism in brown adipose tissue. In: Brown Adipose Tissue pp. 283–318. Ed. by O. LINDBERG. New York: American Elsevier Publ. Co. Inc. 1970.

PRYOR, M.G.M.: On the hardening of the ootheca of *Blatta orientalis* Proc. Roy. Soc. London B, 128, 378–393; On the hardening of the cuticle of insects. ib. 393–407 (1940).

PRYOR, M.G.M.: Tanning of blowfly puparia. Nature (London), 175, 600. (1955)

PRYOR, M.G.M.: Sclerotisation. In: Comparative Biochemistry, Vol. IV, pp. 371–396. Ed. by M. FLORKIN and H.S. MASON. New York: Academic Press 1962.

PULLMAN, B.: Electronic factors in Biochemical Evolution. In: Exobiology, pp. 136–169. Ed. by C. PONNAMPERUMA. Amsterdam: North Holland Publ. Co. 1972.

Pullman, A., Pullman, B.: The band structure of melanins. Biochim. Biophys. Acta **54**, 384–385 (1961).
Pullman, B., Pullman, A.: Electronic levels and biochemical evolution. Nature (London) **196**, 1137–1142 (1962).
Pullman, B., Pullman, A. (Eds.): Quantum Biochemistry. New York: Interscience 1963.
Pullman, M.E., Schatz, G.: Mitochondrial oxidation and energy coupling Ann. Rev. Biochem. **36**, 539–610 (1967).
Purrmann, R.: Xanthin als Pigmentbestandteil der Flügel von Pieriden. Hoppe Seyler's Z. Physiol. Chem. **260**, 105–107 (1939).
Quevedo, W.C., Youle, M.C., Rovee, D.T., Bienicki, T.C.: The developmental fate of melanocytes in murine skin. In: Structure and Control of the Melanocyte, pp. 228–241. Ed. by G. Della Porta and O. Mühlbock. Berlin – Heidelberg – New York: Springer 1966.
Raff, R.A., Mahler, H.R.: The non-symbiotic origin of mitochondria. Science, **177**, 577–582 (1972).
Ramsey, W.N.M.: Plasma bilirubin in the horse. Biochem. J. **39**, 32–33 (1945).
Rangier, M.: L'acide urique urinaire et l'urochrome. Bull. Soc. Chim. Biol. **17**, 502–518 (1935).
Raper, H.S.: The aerobic oxidases. Biol. Rev. Cambridge Phil. Soc. **8**, 245–282 (1928).
Raper, J.R.: Hormones and sexuality in lower plants. Symp. Soc. Exp. Biol. **11**, 143–165 (1957).
Rapoport, E.H.: Gloger's Rule and pigmentation of Collembola. Evolution, **23**, 622–626 (1969).
Rappaport, H., Nakai, T., Shubik, P., Swift, H.: Carcinogen-induced melanotic tumors in the Syrian hamster. Ann. N.Y. Acad. Sci. **100**, 279–296 (1963).
Rasool, S.I.: Planetary atmospheres. In: Exobiology, pp. 369–399. Ed. by C. Ponnamperuma. Amsterdam: North Holland Publ. Co. 1972.
Rebell, G., Lamb, J.H., Mahvi, A., Lee, H.R.: The identification of L-kynurenine as the cause of fluorescence of the hair of the laboratory rat. J. Invest. Dermatol. **29**, 471–477 (1957).
Redfield, A.C.: The haemocyanins. Biol. Rev. Cambridge Phil. Soc. **9**, 175–212 (1934).
Redfield, A.C., Goodkind, R.: The significance of the Bohr Effect in the respiration and asphyxiation of the squid, *Loligo pealii*. J. Exp. Biol. **6**, 340–349 (1929).
Redfield, A.C., Ingalls, E.N.: The effect of salts and H-ion concentration upon the oxygen dissociation constant of the haemocyanin of *Busycon canaliculatum*. J. Cell Comp. Physiol. **1**, 253–275 (1932).
Redfield, A.C., Ingalls, E.N.: The oxygen dissociation curves of some bloods containing haemocyanin. J. Cell. Comp. Physiol. **3**, 169–202 (1933).
Redmond, J.R.: The respiratory function of haemocyanin in Crustacea. J. Cell. Comp. Physiol. **46**, 209–247 (1955).
Redmond, J.R.: The respiratory function of haemocyanin. In: Physiology and Biochemistry of Haemocyanin, pp. 5–23. Ed. by F. Ghiretti. London – New York: Academic Press 1968.
Reed, N., Fain, J.N: Hormonal regulation of the metabolism of free brown fat cells. In: Brown Adipose Tissue, pp. 207–224. Ed. by O. Lindberg. New York: American Elsevier Publ. Co. Inc. 1970.
Rembold, H., Hanser, G.: Über den Weiselzellenfuttersaft der Honigbiene V: Untersuchungen über die Bildung des Futtersaftes der Ammenbiene. Hoppe Seyler's Z. Physiol. Chem. **319**, 206–212; VI: Der Stoffwechsel des Biopterins in der Honigbiene. ib. 213–219 (1960).
Rho, J.H., Bauman, A.J., Yen, T.F., Bonner, J.: Fluorimetric examination of a lunar sample. Science, **167**, 754–755 (1970).
Rich, A.R.: The formation of bile pigment from haemoglobin in tissue cultures. Johns Hopkins Hosp. Bull. **35**, 415–416 (1964).
Richards, A.G.: The Integument of Arthropods. Minneapolis: University of Minnesota Press. 1951.
Richards, O.W., Davies, R.G.: A General Textbook of Entomology (Imms), 9th. Edn. London: Methuen 1957.
Riggs, A., Herner, A.E.: The hybridisation of donkey and mouse haemoglobins. Proc. Nat. Acad. Sci. U.S. **48**, 1664–1670 (1962).
Riley, V., Pack, G.T.: Enzymic, metabolic, electron microscopic and clinical characteristics of a human malignant blue naevus. In: Structure and Control of the Melanocyte, pp. 184–199. Ed. by G. Della Porta and O. Mühlbock. Berlin – Heidelberg – New York: Springer 1966.
Riley, V., Hobby, G., Burke, D.: Oxidising enzymes of mouse melanomas, their inhibition, enhan-

cement and chromatographic separation. In: Pigment Cell Growth, pp. 231–266. Ed. by M. Gordon. New York: Academic Press 1953.
Rill, R.L., Klotz, I.M.: Tyrosine ligands to iron in haemerythrin. Arch. Biochem. Biophys. 136, 507–514 (1970).
Rimington, C.: Porphyrins. Endeavour 14, 126–135 (1955).
Rimington, C., Kennedy, G.Y.: Porphyrin structure, distribution and metabolism. In: Comparative Biochemistry, Vol. IV, pp. 557–614. Ed. by M. Florkin and H.S. Mason. New York: Academic Press 1962.
Rimington, C., Holliday, E.R., Jope, E.M.: Indigoid pigments derived from pathological urine. Biochem. J. 40, 669–674 (1946).
Risinger, G.E., Durst, H. Du P., Hsieh, H.H.: Reductive conversion of thiochrome to thiamine. Nature (London) 210, 94–95 (1966).
Robbins, S.L.: Pathology. Philadelphia: W.B. Saunders 1967.
Robertson, J.D.: Ionic regulation in the crab, *Carcinus maenas* (L.) in relation to the moulting cycle. Comp. Biochem. Physiol. 1, 183–212 (1960).
Robson, N.C., Swan, G.A.: Studies on the structure of some synthetic melanins. In: Structure and Control of The Melanocyte, pp. 155–162. Ed. by G. Della Porta and O. Mühlbock. Berlin-Heidelberg – New York: Springer 1966.
Roche, J.: Electron microscope studies on high molecular weight erythrocruorin (Invertebrate haemoglobin) and chlorocruorin of Annelids. In: Studies in Comparative Biochemistry, pp. 62–80. Ed. by K.A. Munday. Oxford: Pergamon Press 1965.
Roels, O.A.: Present knowledge of vitamin E. Nutr. Rev. 25, 33–37 (1967).
Rook, A., Wilkinson, D.S., Ebling, F.J.G.: Textbook of Dermatology. Oxford: Blackwell 1968.
Root, R.W.: The respiratory functions of the blood of marine fishes Biol. Bull. 61, 426–456 (1931).
Roots, B.I.: Some observations on the chlorogogenous tissue of earthworms. Comp. Biochem. Physiol. 1, 218–226 (1960).
Rosenberg, J.C., Assimacopoulos, C., Rosenberg, S.A.: The malignant melanoma of hamsters III: Effects of sex and castration on the growth of the transplanted tumor. Ann. N.Y. Acad. Sci. 100, 297–304 (1963).
Rosenberg, J.L.: Photochemistry and luminescence. In: Physical Techniques in Biological Research, Vol. I, pp. 1–49. Ed. by G. Oster and A.W. Pollister. New York: Academic Press 1955.
Rosenheim, O., Webster, F.A.: The stomach oil of the fulmar petrel (*Fulmarus glacialis*). Biochem. J. 21, 111–118 (1927).
Rostand, J., Tetry, A.: An Atlas of Human Genetics. Trans. K. Mc. Whirter. London: Hutchinson 1965.
Roth, L.M., Eisner, T.: Chemical defences in the Arthropoda. Ann. Rev. Entomol. Palo Alto 7, 107–136 (1962).
Roth, L.M., Stay, B.: The occurrence of p-quinones in some Arthropods with emphasis on the quinone secreting tracheal glands of *Diploptera punctata* (Blattaria). J. Ins. Physiol. 1, 305–318 (1958).
Rothemund, P.: The chemiluminescence of the chlorophylls and of some other porphyrin metal complex salts. J. Amer. Chem. Soc. 60, 2005. (1938).
Rothschild, Lord: Pertilisation. London: Methuen 1956.
Rothschild, M.: Change of pelage in the stoat, *Mustela erminea* L. Nature (London) 149, 78 (1942); 154, 180–181 (1944).
Rowlands, A.: The influence of water and light upon the pigmentary system of the common frog, *Rana temporaria*. J. Exp. Biol. 27, 446–460 (1950).
Rowlands, A.: The influence of water and light upon the colour change of sightless frogs (*R. temporaria*). J. Exp. Biol. 29, 127–136 (1952).
Rowley, R.R., Williams, J.S.: Unique coloration of two Mississippian Brachiopods. J. Washington Acad. Sci. 23, 46–58 (1933).
Rubey, W.W.: Development of the hydrosphere and atmosphere with special reference to the probable composition of the early atmosphere. In: The Crust of The Earth: Geol. Soc. Amer. Special Papers 62, 631–650 (1955).
Rüdiger, W.: Gallenfarbstoffe bei wirbellosen Tieren. Naturwissenschaften 57, 331–337 (1970).
Rushton, W.A.H.: Kinetics of cone pigments measured objectively on the living human fovea. Ann. N.Y. Acad. Sci. 74, 291–304 (1958).

Russell, E.S.: A quantitative histological study of the pigment found in the coat colour mutants of the house mouse II: Estimates of the total volume of pigment. Genetics 33, 228–236 (1948).

Ryland, J.S.: Embryo colour as a diagnostic character in Polyzoa. Ann. Mag. Nat. Hist. Series 13, 1, 552–556 (1958).

Salaque, A., Barbier, M., Lederer, E.: Sur la biosynthèse de l'echinochrome A par l'oursin *Arbacia pustulosa*. Bull. Soc. Chim. Biol. 49, 841–848 (1967).

Salt, G.: Experimental studies in insect parasitism VIII: Host reactions following artificial parasitisation. Proc. Roy. Soc. London B 144, 380–398 (1955).

Salt, G.: Experimental studies in insect parasitism X: The reactions of some endopterygote insects to an alien parasite. Proc. Roy. Soc. London B 147, 167–184 (1957).

Sandeen, M.I.: Chromatophorotropins in the central nervous system of *Uca pugilator* with special refernece to their origins and actions. Physiol. Zool. 23, 337–352 (1950).

Sang, J.A.: The biochemical basis of melanotic tumour expression in *Drosophila*. Heredity 24, 515. (1969).

Sannasi, A.: Quinone tanning in Reptilia and Aves. Experientia 25, 968–969 (1969).

San Pietro, A. (Ed.): Non-Heme Iron Proteins: Role in Energy Conversion. Yellow Springs Ohio: Antioch Press 1965.

Sarkar, B., Kruck, T.P.A.: Copper amino acid complexes in human serum. In: The Biochemistry of Copper, pp. 183–196. Ed. by J. Peisach, P. Aisen, and W.E. Blumberg. New York – London: Academic Press 1966.

Sato, M., Ozawa, H.: Occurrence of ubiquinone and rhodoquinone in the parasitic nematodes, *Metastrongylus elongatus* and *Ascaris lumbricoides* var. *suis*. J. Biochem. (Japan) 65, 861–867 (1969).

Saxen, L.: Tetrycycline pigments. In: Pigments in Pathology, pp. 75–91. Ed. by M. Wolman. New York – London: Academic Press 1969.

Scharrer, E.: Die Lichtempfindlichkeit blinder Elritzen. Z. Vergl. Physiol. 7, 1–38 (1928).

Scheer, B.T.: Some features of the metabolism of the carotenoid pigments of the California sea mussel, *Mytilus californicus*. J. Biol. Chem. 136, 275–299 (1940).

Scheinberg, I.H.: Ceruloplasmin: A review. In: The Biochemistry of Copper, pp. 513–524. Ed. by J. Peisach, P. Aisen, and W.E. Blumberg. New York – London: Academic Press 1966.

Schejter, A.: Discussion. In: Hemes and Hemoproteins, pp. 391–394. Ed. by B. Chance, R.W. Estabrook, and T. Yonetani. New York – London: Academic Press 1966.

Schildknecht, H.: Evolutionary peaks in the defensive chemistry of insects. Endeavour 20, 136–141 (1971).

Schlieper, C.: Der Farbwechsel von *Hyperia galba* (Zugleich der Nachweis des Einflusses taktiler Reize auf den Farbwechsel eines Krebses). Z. Vergl. Physiol. 3, 547–557 (1926)

Schmidt, F.: Beobachtung bei Aufzucht von Hermelinen, *Mustela erminea aestiva* Kerr 1792. Säugetierk. Mitt. 2, 166–174 (1954).

Scholander, P.F.: Oxygen transport through haemoglobin solutions. Science 131, 585–590 (1960).

Schreiber, G.: Studi sul pigmento cromolipoide l'apparato fenestrato e la respirazione di supplemento del sistema nervoso. Pubbl. Sta. Zool. Napoli 10, 151–195 (1930).

Schubert, J., Westfall, W.M.: Effects of ionising radiation and peroxides on haemocyanin. Nature (London) 195, 1096–1097 (1962).

Schulman, M.P., Wald, G.: The valence of copper in haemocyanins. Biol. Bull. 101, 239–240 (1951).

Schulze, P.: Neue Beiträge zu einer Monographie der Gattung *Hydra*. Arch. Biontol. Berlin 4, 39–119 (1917).

Schwartz, V.: Zur Phänogenese der Flügelzeichnung von *Plodia interpunctella*. Z. Indukt. Abstam. Vererbungsl. 85, 51–96 (1953).

Schwink, I.: Über den Nachweis eines Redox-Pigments (Ommochrome) in der Haut von *Sepia officinalis*. Naturwissenschaften 40, 365. (1953).

Schwink, I.: Vergleich des Redox-Pigments aus Chromatophoren und Retina von *Sepia officinalis* mit Insektenpigmenten der Ommochromgruppe. Verh. Deut. Zool. Ges. (Zool. Anz. 19, Suppl. 71–75) (1956).

Searle, A.G.: Comparative Genetics of Coat Colour in Mammals. London: Logos Press 1968.

Seyffert, R., Grassl, M., Lynen, F.: The sidechain of cytohaemin. In: Hemes and Hemoproteins, pp. 45–52. Ed. by B. Chance, R.W., Estabrook, and T. Yonetani. New York – London: Academic Press 1966.

Seiji, M., Shimao, K., Birkbeck, M.S.C. Fitzpatrick, T.B.: Subcellular localisation of melanin biosynthesis. Ann. N.Y. Acad. Sci. **100**, 497–533 (1963).
Sereni, E.: Sulla innervazione dei Cefalopoda. Boll. Soc. Ital. Biol. Sper. Napoli **3**, 707–711 (1928).
Sexton, W.A.: Chemical Constitution and Biological Activity. 2nd. Edn. London: Spon. 1953. 3rd. Edn. 1963.
Shaklai, N., Daniel, E.: Fluorescence properties of haemocyanin from *Levantina hierosolima*. Biochemistry **9**, 564–568 (1970).
Shemin, D.: The succinate-glycine cycle: The role of δ-ALA in porphyrin synthesis. In: Porphyrin Biosynthesis and Metabolism, pp. 4–42. Ed. by G.E.W. Wolstenholme and E.C.P. Millar. London: J. and A. Churchill 1955.
Sheppard, P.M.: Natural Selection and Heredity. London: Hutchinson 1958.
Shimomura, O., Johnson, F.H.: Structure and light emitting moiety of aequorin. Biochemistry **11**, 1602–1608 (1972).
Shiota, T.: The biosynthesis of folic acid and 6-substituted pteridine derivatives. In: Comprehensive Biochemistry, **21**, 111–152. Ed. by M. Florkin and E.H. Stotz. Amsterdam – London – New York: Elsevier Publ. Co. 1971.
Shugar, H.J., Rossman, G.R., Thibeault, J., Gray, H.B.: Simultaneous pair electronic excitations in a binuclear iron (III) complex. Chem. Phys. Lett. **6**, 26–28 (1970).
Shugar, H.J. et al., 1972. In Press.
Sick, K.: Haemoglobin-polymorphism in fishes. Nature (London) **192**, 894–896 (1961).
Siegelman, W.H., Chapman, D.J., Cole, W.J.: The bile pigments of plants. In: Porphyrins and Related Compounds, pp. 107–120. Ed. by T.W. Goodwin. London – New York: Academic Press 1968.
Sieker, L.C., Adman, E., Jensen, L.H.: Structure of the Fe-S complex in bacterial ferredoxin. Nature (London) **235**, 40–42 (1972).
Singh, H., Seshadi, T.R., Subramanian, G.B.V.: A note on the colouring matter of the lac larvae. Tetrahedron Letters **10**, 1101–1108 (1966).
Singh, H., Moore, R.E., Scheuer, P.J.: The distribution of quinone pigments in Echinoderms. Experientia **23**, 624–626 (1967).
Slater, E.C. (Ed.): Flavins and Flavoproteins. Amsterdam – London – New York: Elsevier Publ. Co. 1966.
Smith, C.L.: Reproduction in female Amphibia. Mem. Soc. Endocrinol. G.B. **4**, 39–56 (1955).
Smith, D.C.: The effects of temperature changes upon the chromatophores of Crustaceans. Biol. Bull. **58**, 193–202 (1930).
Smith, H.G.: The receptive mechanism of the background response in chromatic behaviour of Crustacea. Proc. Roy. Soc. London B **125**, 250–263 (1938).
Smith, M.H.: The pigments of Nematoda and Acanthocephala. In: Chemical Zoology, Vol. III, pp. 501–520. Ed. by M. Florkin and B.T. Scheer. New York – London: Academic Press 1969.
Smith, M.H., Lee, D.L.: Metabolism of haemoglobin and haematin compounds in *Ascaris lumbricoides*. Proc. Roy. Soc. London B **157**, 234–257 (1963).
Smith, M.H., George, P., Preer, J.R.: Preliminary observations on isolated *Paramecium* haemoglobin. Arch. Biochem. Biophys. **99**, 313–318 (1962).
Smith, R.E.: Thermoregulation by brown adipose tissue in cold. Fed. Proc. Fed. Amer. Socs. Exp. Biol. **21**, 221 (1962).
Smith, R.E., Roberts, J.C., Hittelman, K.J.: Non-phosphorylating respiration of mitochondria from brown adipose tissue of rats. Science **154**, 653–654 (1966).
Snell, P.S., Bischitz, P.G.: The melanocytes and melanin in human abdominal wall skin: a survey made at different ages in both sexes and during pregnancy. J. Anat. London **97**, 361–376 (1963).
Snyder, F., Cornatzer, W.E.: Vanadium inhibition of phospholipid synthesis and sulphydryl activity in rat liver. Nature (London) **182**, 462. (1958).
Song, P.-S.S.: Chemistry of flavins in their excited states. In: Flavins and Flavoproteins, pp. 37–61. Ed. by H. Kamin. Baltimore: University Park Press and London: Butterworth 1971.
Sorby, H.C.: 1878. Quoted from Fox and Vevers, 1960, p. 33.
Spek, J.: Zustandsänderungen der Plasmakolloide bei Befruchtung und Entwicklung des *Nereis* Eies. Protoplasma **9**, 370–427 (1930).
Stadtman, T.C.: Vitamin B_{12}. Science **171**, 859–867 (1971).
Stanier, R.: Carotenoid pigments: Problems of synthesis and function. Harvey Lect. **54**, 219–255 (1960).

STEDMAN, E., STEDMAN, E.: Haemocyanin, Part I: The dissociation curves of the oxyhaemocyanin in the blood of some Decapod Crustacea. Biochem. J. **19**, 544–551 (1925).
STEDMAN, E., STEDMAN, E.: Haemocyanin, Part II: The influence of hydrogen ion concentration on the dissociation curve of the oxyhaemocyanin from the blood of the common lobster, *Homarus vulgaris*. Biochem. J. **20**, 938–948 (1926).
STEFANIS, C. N., ISSIDORIDES, M. R.: Aberrant *m*-RNA in Parkinsonism: Support for hypothesis from studies with chloramphenicol. Nature (London) **225**, 962–963 (1970).
STEINACH, E.: Studien über die Hautfärbung und über den Farbwechsel der Cephalopoden. Pfluger's Arch. Gesamte Physiol. Menschen Tiere **87**, 1–37 (1901).
STEINBERG, M. S.: Studies on the mechanism of physiological dominance in *Tubularia*. J. Exp. Zool. **127**, 1–26 (1954).
STEINBEEG, M. S.: Cell movements, rate of regeneration and the axial gradient in *Tubularia*. Biol. Bull. **108**, 219–234 (1955).
STEPHENS, G. C., FRIEDL, F., GUTTMAN, B., Electrophoretic separation of chromatophorotropic principles of the fiddler crab *Uca*. Biol. Bull. **111**, 312–313 (1956).
STEPHENSON, E. M., STEWART, C.: Animal Camouflage. Harmondsworth: Penguin Books 1946.
STERN, K. G.: The relationship between prosthetic group and protein carrier in certain enzymes and biological pigments. Cold Springs Harbor Symp. Quant. Biol. **6**, 286–300 (1938).
STERN, R. G., SALOMON, K.: Ovoverdin, a pigment chemically related to visual purple. Science **86**, 310–311 (1937).
STERN, R. G., SALOMON, K.: On ovoverdin, the carotenoid-protein pigment of the egg in the lobster. J. Biol. Chem. **122**, 461–475 (1938).
STEVEN, D. M.: The dermal light sense. Biol. Rev. Cambridge Phil. Soc. **38**, 205–240 (1963).
STOFFOLANO, J. G.: Darkening of the anal organ of the larva of *Musca autumnalis* and *Orthellia caesarion*. J. Invert. Pathol. **17**, 3–8. (1971).
STOVES, J. L.: The structural significance of iron and sulphur in natural melanins. Research **6**, 29S–30S (1953).
STRAIN, H. H., MANNING, W. M., HARDIN, G.: Xanthophylls and carotenes of Diatoms, Brown Algae, Dinoflagellates and Sea Anemones. Biol. Bull. **86**, 169–196 (1944).
STREHLER, B. L., SHOUP, C. S.: The chemiluminescence of riboflavin. Arch. Biochem. Biophys. **47**, 8–15 (1953).
STREHLER, B. L., ARNOLD, W.: Light production by green plants. J. Gen. Physiol. **34**, 809–820 (1951).
STREHLER, B. L., SHOUP, C. S.: The chemiluminescence of riboflavin. Arch. Biochem. Biophys. **47**, 8–15 (1953).
STURGEON, P., SHODEN, A.: Haemosiderin and ferritin. In: Pigments in Pathology, pp. 93–114. Ed. by M. WOLMAN. New York – London: Academic Press 1969.
SUMNER, F. B.: Evidence for the protective value of changeable coloration in fishes. Amer. Nat. **69**, 245–266 (1935).
SUMNER, F. B.: Quantitative changes in pigmentation resulting from visual stimuli in fishes and Amphibia. Biol. Rev. Cambridge Phil. Soc. **15**, 351–375 (1940).
SUMNER, F. B: Vision and guanine production in fishes. Proc. Nat. Acad. Sci. U.S. **30**, 285–294 (1944).
SUTHERLAND, M. D., WELLS, J. W.: Pigments of marine animals IV: The anthraquinoid pigments of the crinoids *Ptilometra australis* Wilton and *Tropiometra afra* Hartlaub. Aust. J. Chem. **20**, 515–553 (1967).
SUZUKI, K., KIMURA, T.: An iron protein as a component of steroid 11β-hydroxylase complex. Biochem. Biophys. Res. Comm. **19**, 340–345 (1965).
SUZUKI, S. K.: Egg pigments of the Polychaetes, *Lumbriconereis japonica* and *Neanthes diversicolor*. Sci. Rep. Tôhoku Imp. Univ. IV, **33**, 175–178 (1968).
SWAN, G. A.: Chemical structure of melanins. Ann. N.Y. Acad. Sci. **100**, 1005–1019 (1963).
SWAN, G. A.: Further studies on the structure of DOPA melanin. J. Invest. Dermatol. **54**. 98. (1970).
SWAN, G. A., WRIGHT, D.: A study of melanin formation by use of 2-(3,4-dihydroxy[3-^{14}C] phenyl)-, 2-(3,4-dihydroxy[4-^{14}C]-phenyl)- and 2-(3,4-dihydroxy[5-^{14}C] phenyl) -ethylamines. J. Chem. Soc. (1956), 1549–1557 (1956).
SZABO, G.: Tyrosinase in the epidermal melanocytes of white human skin. Amer. Med. Ass. Arch. Derm. Syphilis **76**, 324–329 (1957).
SZABO, G., GERALD, A. B., PETHACK, M. A., FITZPATRICK, T. B.: Ultrastructure of retinal colour-differences in Man. J. Invest. Dermatol. **54**, 98. (1970).
SZENT-GYORGYI, A.: Introduction to a Submolecular Biology. New York: Academic Press 1960.

Szodoray, L., Nagy-Vezekenyi, C.: Histochemical investigations of melanotic tumors. In: Structure and Control of the Melanocyte, pp. 205–210. Ed. by G. Della Porta and O. Mühlbock. Berlin-Heidelberg – New York: Springer 1966.

Szutka, A.: Probable synthesis of porphin-like substances during chemical evolution. In: Origins of Prebiological Systems. pp. 243–254. Ed. by S.W. Fox. New York – London: Academic Press 1965a.

Szutka, A.: Discussion. In: Origins of Prebiological Systems, p. 457. Ed. by S.W. Fox. New York-London: Academic Press 1965b.

Takasada, H., Hamaguchi, K.: Circular dichroism of the haemocyanin of the octopus. J. Biochem. Tokyo 63, 725–729 (1968).

Tanaka, M., Benson, A.M., Mower, H.F., Yasunobu, K.T.: A proposed structure of *Clostridium pasteurianum* ferridoxin, In: Non-Haem Iron Proteins: Role in Energy Conversion, pp. 221–224. Ed. by A. San Pietro. Yellow Springs Ohio: Antioch Press 1965.

Tardent, P.: Regeneration in the Hydrozoa. Biol. Rev. Cambridge Phil. Soc. 38, 293–333 (1963).

Tartar, V.: The Biology of *Stentor*. Oxford: Pergamon Press 1961.

Taylor, J.D.: Some aspects of iridophore reflecting platelets. J. Invest. Dermatol. 54, 98 (1970).

Teissier, G.: Changements de coloration des embryons de *Clava squamata* au cours de l'ontogénèse: Interpretations chimique et physiologique. Trav. Sta. Zool. Wimereux 9, 233–238 (1925).

Teissier, G.: Origine et nature du pigment carotenoide des oeufs de *Daphnia pulex* de Geer. C.R. Soc. Biol. 109, 813–815 (1932).

Thathachari, Y.T.: X-ray diffraction studies on melanins. J. Invest. Dermatol. 54, 99 (1970).

Thayer, G.H.: Concealing Coloration in the Animal Kingdom. New York: The Macmillan Co. 1918.

Thomas, G.J.: Infra red and Raman spectroscopy. In: Physical Techniques in Biological Research, 2nd. Edn., Vol. I A, pp. 277–346. Ed. by G. Oster. New York – London: Academic Press 1971.

Thomas, M.: Melanins. In: Modern Methods of Plant Analysis, pp. 661–675. Ed. by K. Paech and M.V. Tracey. Berlin – Göttingen – Heidelberg: Springer 1955.

Thompson, D'A.W.: On Growth and Form. 2nd Edn. Cambridge: University Press 1942.

Thompson, S.W.: Lipogenic pigments related to treatment with exogenous lipid. In: Pigments in Pathology, pp. 237–286. Ed. by M. Wolman. New York – London: Academic Press 1969.

Thompson, W.R., Meinwald, J., Aneshansley, D., Eisner, T.: Flavonols: Pigments responsible for ultraviolet absorption in nectar guide of flower. Science 177, 528–530 (1972)

Thomson, R.H.: Naturally Occurring Quinones. London: Butterworth 1957.

Thomson, R.H.: Quinones: Structure and distribution. In: Comparative Biochemistry, Vol. III, pp. 631–725. Ed. by M. Florkin and H.S. Mason. New York – London: Academic Press 1962.

Thomson, R.H.: Discussion. In: Colour and Life, p. 40. Ed. by W.B. Broughton. London: Institute of Biology 1964.

Threlfall, G.: Cell proliferation in the rat kidney induced by folic acid. Cell and Tissue Kinetics 1, 385–392 (1968).

Tietz, A., Lindberg, M., Kennedy, E.P.: A new pteridine-requiring enzyme system for the oxidation of glyceryl ethers. J. Biol. Chem. 239, 4081–4090 (1964).

Timon-David, J.: Pigments des insectes. Anneé Biol. Paris 51, 237–271 (1947).

Tixier, R.: Contribution à l'étude de quelques pigments pyrroliques naturels des coquilles de Mollusques, de l'oeuf d'emu et du squelette du corail bleu (*Heliopora coerulea*). Ann. Inst. Océanog. 22, 343–397 (1945).

Tixier, R., Lederer, E.: Sur l'haliotivioline, pigment principal des coquilles de' *Haliotis cracherodii*. C.R.H. Acad. Sci. 228, 1669–1671 (1949).

Tollin, G.: Discussion. In: Flavins and Flavoproteins, p. 36. Ed. by M. Kamin. Baltimore: University Park Press and London: Butterworth 1971.

Torri, S.: Identity of *Ascaris* blue with an hydrolysis product of folic acid. Igahu Seibutsugahu 38, 113–115 (Chem. Abst. 52, 6434c) (1956).

Trager, W., Subbarow, Y.: The chemical nature of growth factors required by mosquito larvae. Biol. Bull. 75, 75–84 (1938).

Tregear, R.T.: Physical Functions of The Skin. London: Academic Press 1966.

Triebs, A.: Chlorophyll-and Hämin-Derivate in bituminösen Gesteinen, Erdölen, Kohlen, Phosphoriten. Liebig's Ann. 517, 172–196 (1935).

TURING, A.M.: The chemical basis of morphogenesis. Phil. Trans. Roy. Soc. London B **237**, 37–72 (1952).
TURNER, W.J.: Studies on porphyria I. Observations on the fox-squirrel, *Sciurus niger.* J. Biol. Chem. **118**, 519–530 (1937).
TWEEDELL, K.S.: Cytological studies during germinal vesicle breakdown of *Pectinaria gouldii* with vital dyes, centrifugation and fluorescence microscopy. Biol. Bull. **123**, 424–449 (1962).
TYLER, A.: Crystalline echinochrome and spinochrome: their failure to stimulate the respiration of eggs and sperm of *Stronglocentrotus.* Proc. Nat. Acad. Sci. U.S. **25**, 523–528 (1939).
UDA, M.: On the relations between blood-colour and cocoon-colour in silkworms, with special reference to Mendel's law of heredity. Genetics **4**, 395–416 (1919).
UMEBACHI, Y.: 1962. Quoted from BRUNET, 1965.
UMEBACHI, Y., TAKAHASHI, H.: Kynurenine in the wings of the papilionid butterflies. J. Biochem. Tokyo, **43**, 73–81 (1956).
UNDERWOOD, H., MENAKER, M.: Extraretinal light perception: Entrainment of the biological clock controlling lizard locomotor activity. Science **170**, 190–193 (1970).
UNGER, H.: Neurohormone bei Seesternen *Marthasterias glacialis.* Symp. Biol. Hungary **1**, 203– (1960).
URBAN, P.F., KLINGENBERG. M.: On the redox potential of ubiquinone and cytochrome *b* in the respiratory chain. Eur. J. Biochem. **9**, 519–525 (1969).
VANDEL, A.: Sur l'existence de charactères sexuels secondaires non encore signalés chez les espèces du genre *Spiloniscus* (Isopodes terrestres: Trichoniscidae) et sur une nouvelle espèce de ce genre: *Spiloniscus darwini* n. sp. Arch. Zool. Exp. Gen **79**, Notes Rev. pp. 89–94 (1938).
VAN DER LEK, B.: Photosensitive melanophores: some aspects of the light-induced pigment migrations in the tail fin melanophores of the larvae of the clawed toad, *Xenopus laevis* (Daud.). Rotterdam: Bronder Offset 1967.
VAN DER LEUN, J.C.: Delayed pigmentation and u.v. erythrema. In: Recent Progress in Photobiology, pp. 387–388. Ed. by E.J. BOWEN. Oxford: Blackwell.
VAN DE VEERDONK, F.C.G., KONIJN, T.M.: The role of adenosine-3',5'-cyclic monophosphate and catecholamines in the pigment migration process in *Xenopus laevis.* Acta Endocrinol. **64**, 364–376 (1970).
VAN DUIJN, P.: Arenicochrome, a new pigment from *Arenicola marina* L. Rec. Trav. Chim. Pays. Bas. **71**, 585–594; Arenicochrome: degradation to arenicochromine, a sulphur-free pigment. ib. 595–600. (1952).
VAN GANSEN, P.S.: Les cellules chlorogogènes des Lombriciens. Bull. Biol. France Belg. **90**, 335–356 (1956).
VAN VEEN, A.G., MARTENS, W.K.: Die Giftstoffe der sogenannten Bongkrekvergiftungen auf Java. Rec. Trav. Chim. Pays Bas, **53**, 257–266; Das Toxoflavin, der gelbe Giftstoff der Bongkrek. ib. 398–404 (1934).
VERNBERG, F.J.: Comparative physiology: latitudinal effects on physiological properties of animal populations. Ann. Rev. Physiol. **24**, 517–546 (1962).
VERNE, J.: Les Pigments dans L'Organisme Animale. Paris: Gaston Doin et Cie 1926.
VERNET, G.: Les pigments de *Lineus ruber* (O.F. MÜLLER) (Heteronemertes, Lineidae). C.R.H. Acad. Sci. **263**, 191–194 (1966).
VEVERS, H.G.: The biology of *Asterias rubens* L.: Growth and reproduction. J. Mar. Biol. Ass. U.K. **28**, 165–187 (1949).
VEVERS, H.G.: Pigmentation of the Echinoderms. Proc. 16th. Int. Congr. Zool. **3**, 120–122 (1963).
VEVERS, H.G., MILLOTT, N.: Carotenoid pigments of the integument of the starfish *Marthasterias glacialis* L. Proc. Zool. Soc. London, **129**, 75–80 (1957).
VIEHOEVER, A., COHEN, I.: The responses of *Daphnia magna* to vitamin E. Amer. J. Pharmacol. **110**, 297–315 (1938).
VILLELA, G.G.: Adenochrome-like pigment of the Polyzoan, *Bugula neritina* (L.). Proc. Soc. Exp. Biol. Med. **68**, 531–533 (1948a).
VILLELA, G.G.: Biocromos (pigmentos) de Invertebrades. I: Briozoarios. Mem. Inst. Osw. Cruz **46**, 459–471 (1948b).
VILLELA, G.G., PRADO, J.L.: Riboflavin in the blood-plasma of some Brazilian snakes. J. Biol. Chem. **157**, 693–697 (1945).

VISCONTINI, M., HADORN, E., KARRER, P.: Fluoreszierende Stoffe aus *Drosophila melanogaster,* die roten Augenfarbstoffe. Helv. Chim. Acta 40, 579–585 (1957).
VISCONTINI, M., MATTERN, G.: Hydroxylierung der Tryptophans mittels Tetrahydropterin unter physiologischen Bedingungen. Helv. Chim. Acta 50, 1664–1665 (1967).
VISCONTINI, M., STERLIN, M.: Isolierung, Struktur und Synthese von Pterinen aus *Ephestia kuhniella* Zeller. Hel. Chim. Acta 44, 1783–1785 (1961).
VISCONTINI, M., STERLIN, M.: Fluoreszierende Stoffe aus *Ephestia kuhniella* Zeller: Synthese von Erythropterin, Ekapterin und Lepidopterin. Helv. Chim. Acta 46, 51–56 (1963).
VLES, F., VELLINGER, E.: Recherches sur le pigment de l'oeuf d'*Arbacia* envisagé comme indicateur de pH intracellulaire. Bull. Inst. Oceanog. No. 513, pp. 1–32 (1928).
VOET, D., RICH, A.: An H-bonded complex between riboflavin and an adenosine derivative, and its possible relation to FAD function. In: Flavins and Flavoproteins, pp.24–35. Ed. by H. KAMIN. Baltimore: University Park Press and London: Butterworth 1971.
VOGEL, A.I.: Textbook of Practical Organic Chemistry. London: Longmans. 1956.
VOINOV, V.: La pigmentogénèse chez *Glossiphonia paludosa* (Carena) C.R. Soc. Biol. 99, 1081–1083 (1928).
VÖLKER, O.: Über fluoreszierende, gelbe Federpigmente bei Papageien, eine neue Klasse von Federfarbstoffen. J. Ornith. Leipzig 85, 136–146 (1937).
VÖLKER, O.: Zur Kenntnis des Porphyrins in Vogelfedern. Hoppe Seyler's Z. Physiol. Chem. 258, 1–5 (1939).
VOLKER, O.: Über das Vorkommen von Protoporphyrin in den Eischalen der Vögel. Hoppe Seyler's Z. Physiol. Chem. 273, 277–282 (1942).
VON GELDERN, C.E.: Color changes and structure of the skin of *Anolis carolinensis.* Proc. Calif. Acad. Sci. 10, 77–117 (1921).
VORONTZOVA, M.A.: Morphogenetische Analyse der Färbung bei weißen axolotl. Wilhelm Roux' Arch. Entwicklungsmech. Organismen 115, 93–109 (1929).
VUILLAUME, M.: Pigmentations et variations pigmentaires de trois insectes: *Mantis religiosa* (L.), *Sphodromantis viridis* (F.) et *Locusta migratoria* (L.). Bull. Biol. France Belg. 102, 147–232 (1968).
VUILLAUME, M.: Les Pigments des Invertébrés: Biochimie et Biologie des Colorations. Paris: Masson 1969.
WADDINGTON, C.H.: An Introduction to Modern Genetics. London: Allen and Unwin 1939.
WADDINGTON, C.H.: The biological foundations of measurements of growth and form. Proc. Roy. Soc. London B 137, 509–515 (1950).
WADDINGTON, C.H.: Principles of Embryology. London: Allen and Unwin 1956.
WADDINGTON, C.H.: New Patterns in Genetics and Development. New York: Columbia Univ. Press 1962.
WAGNER-JAUREGG, T.: Riboflavin: Chemistry. In: The Vitamins, Vol. III, pp. 301–332. Ed. by W.H. SEBRELL and R.S. HARRIS. New York: Academic Press Inc. 1954.
WAKIL, S.J.: Enzymic synthesis of fatty acids. Comp. Biochem. Physiol. 4, 123–158 (1962).
WALD, G.: Pigments of the retina II: sea robin, sea bass and scup. J. Gen. Physiol. 20, 45–56 (1936).
WALD, G.: Vitamins A in Invertebrate eyes. Amer. J. Physiol. 133, 479–480 (1941).
WALD, G.: The chemical evolution of vision. Harvey Lectures 41, 117–160 (1945).
WALD, G.: The significance of Vertebrate metamorphosis. Science 128, 1481–1490 (1958).
WALD, G.: The photoreceptor process in vision. In: Handbook of Physiology, Section I, Volume I, pp. 671–692. Ed., by J. FIELD. Washington, D.C.: American Physiological Society 1959.
WALD, G.: The distribution and evolution of visual systems, In: Comparative Biochemistry Vol. I, pp. 311–345. Ed. by M. FLORKIN and H.S. MASON 1960.
WALD, G.: Life in the second and third periods, or why phosphorus and sulphur for high energy bonds? In: Horizons in Biochemistry, pp. 127–142. Ed. by M. KASHA and B. PULLMAN. New York: Academic Press 1962.
WALD, G. and BURG, S.P.: The vitamin A of the lobster. J. Gen. Physiol. 40, 609–626 (1957).
WALD, G., BROWN, P.K., BROWN, P.S.: Visual pigments and depth of habitat of marine fishes. Nature (London) 180, 969–971 (1957).
WALKER, A.G., RADDA, G.K.: Photoreactions of retinol and derivatives sensitised by flavins. Nature (London) 215, 1483. (1967).
WALLS, G.L.: The Vertebrate Eye and its Adaptive Radiation. (Cranbrook Inst. Sci. Bull. No. 19). Michigan: Bloomfield Hills 1942. New York: Hafner Publ. Co. 1963.

WALSHE, B.M.: On the function of the haemoglobin in *Chironomus* after oxygen lack. J. Exp. Biol. **24**, 329–342 (1947).
WALSHE-MAETZ, B.M.: Le métabolisme de *Chironomus plumosus* dans des conditions naturelles. Physiol. Comp. et Oecol. **3**, 135–154 (1953).
WANG, J.H.: Transport of oxygen through haemoglobin solutions. Science **133**, 1770–1771 (1961).
WANG, J.H.: Discussion. In: Hemes and Hemoproteins, p. 287. Ed. by B. CHANCE, R.W. ESTABROOK, and T. YONETANI. New York – London: Academic Press 1966.
WARING, H.: The coordination of Vertebrate melanophore responses. Biol. Rev. Cambridge Phil. Soc. **17**, 120–150 (1942).
WARING, H.: Colour Change Mechanisms of Cold-Blooded Vertebrates. New York – London: Academic Press 1963.
WATERMAN, T.H.: Light sensitivity and vision. In: Comparative Physiology of Crustacea, Vol. II, pp. 1–64. Ed. by T.H. WATERMAN. New York – London: Academic Press 1960.
WATSON, H.C.: Discussion. In: Hemes and Hemoproteins, p. 275. Ed. by B. CHANCE, R.W. ESTABROOK, and T. YONETANI. New York – London: Academic Press 1966.
WATT, W.B.: Adaptive significance of pigment polymorphism in *Colias* butterflies II: Thermoregulation and photoperiodically controlled melanin variation in *C. eurytheme*. Proc. Nat. Acad. Sci. U.S. **63**, 767–774 (1969).
WEBB, D.A.: Observations on the blood of certain ascidians with special reference to the biochemistry of vanadium. J. Exp. Biol. **16**, 499–523 (1939).
WEBER, G.: Fluoressence parameters and photosynthesis. In: Comparative Biochemistry of Photoreactive Systems, pp. 395–411. Ed. by M.B. ALLEN. New York – London: Academic Press 1960.
WEBER, G.: Intramolecular complexes of flavins. In: Flavins and Flavoproteins, pp. 15–21. Ed. by E.C. SLATER. Amsterdam – London – New York: Elsevier Publ. Co. 1966
WEBER, R.E. Relations between function and molecular properties of Annelid haemoglobins: Interaction between haems in the haemoglobin of *Arenicola marina* L. Comp. Biochem. Physiol. **35**, 179–189 (1970).
WEBER, W.: Mehrfachinnervation von Tintenfisch-chromatophoren. Naturwissenschaften **55**, 348–349 (1968).
WEBER, W.: Zur Ultra Structur der Chromatophorenmuskelzellen von *Loligo*. Z. Zellforsch. Mikrosc. Anat. **108**, 446–456 (1970).
WEGLICKI, W.B., REICHEL, W., NAIR, P.P.: Accumulation of lipofuscin like pigment in rat adrenal as a function of vitamin E deficiency. J. Gerontol. **23**, 469–475 (1968).
WEINSTEIN, P.P.: The *in vitro* cultivation of helminths with reference to morphogenesis. In: Biology of Parasites, pp. 143–154. Ed. by E.J.L. SOULSBY. New York – London: Academic Press 1966.
WEISS, U., ALTLAND, H.W.: Red pigments of *Dactylonotus rudbeckiae* and *D. ambrosiae* (Homoptera, Aphididae). Nature (London) **207**, 1295–1297 (1965).
WEISSKOPF, V.F.: How light interacts with matter. Sci. Amer. **219** (3) 60–71 (1968).
WEISZ, P.B.: On the mitochondrial nature of the pigmented granules in *Stentor* and *Blepharisma*. J. Morphol. **86**, 177–184 (1950).
WELLNER, D.: Flavoproteins. Ann. Rev. Biochem. **36**, 669–690 (1967).
WELLNER, D., MEISTER, A.: New symbols for the common amino acid residues of peptides and proteins. Science **151**, 77–78 (1966).
WELLS, J.W.: Supposed colour markings in Ordovician Trilobites from Ohio. Amer. J. Sci. New Haven **240**, 710–713 (1942).
WENIG, K.: The chemical and physical basis of the luminescence of the earthworm *Eisenia submontana* (In Czech with English summary). Vest. Ceskosl. Zool. Spolec. Praha **10**, 293–355 (1946).
WEST, E.S. and TODD, W.R.: Textbook of Biochemistry. 2nd. Edn. New York: Macmillan 1957.
WETTERBERG, L., GELLER, E., YUWILER, A.: Harderian gland: An extraretinal photoreceptor influencing the pineal gland in neonatal rats? Science **167**, 884–885 (1970).
WHELAND, G.W.: Resonance in Organic Chemistry. New York: Wiley 1955.
WHYTE, L.P.: Melanin a naturally occurring cation-exchange material. Nature (London) **182**, 1427–1428 (1958).
WIELAND, H., LIEBIG, R.: Ergänzende Beiträge zur Kenntnis der Pteridine: Über die Pigmente der Schmetterlingsflügel XIV. Liebig's Ann. Chem. **555**, 146–156 (1944).
WIGGLESWORTH, V.B.: The fate of haemoglobin in *Rhodnius prolixus* (Hemiptera) and other blood-sucking Arthropods. Proc. Roy. Soc. London B **131**, 313–339 (1943).
WIGGLESWORTH, V.B.: Control of Growth and Form. New York: Cornell Univ. Press 1959.

WILHELM, R.: Nervös beeinflußte Oszillationen der Melanophoren bei Fischen. Verh. Deut. Zool. Ges. (Zool. Anz. 33, Suppl.) 273–278. (1969).
WILLIAMS, B.G., NAYLOR, E.: Synchronisation of the locomotor tidal rhythm of *Carcinus*. J. Exp. Biol. 51, 715–725 (1969).
WILLIAMS, R.H.: Textbook of Endocrinology. 4th. Edn. Philadelphia: W.B. Saunders 1968.
WILLIAMS, R.J.P.: Some theoretical problems concerning iron porphyrins. In: Hemes and Hemoproteins, pp. 557–576. Ed. by B. CHANCE, R.W. ESTABROOK, and T. YONETANI. New York – London: Academic Press 1966 a.
WILLIAMS, R.J.P.: Reactivity at the iron-haem centre. ib. pp. 585–588. 1966 b.
WILLIAMS, R.J.P.: The haemoprotein P-450 reactions. In: Structure and Function of Cytochromes, pp. 645–648. Ed. by K. OKUNUKI, M.D. KAMEN, and T. SEKUZU. Tokyo: Univ. Press and Baltimore – Manchester: Univ. Park Press 1968.
WILLIAMS, R.T.: *In vitro* studies on the environmental biology of *Goniodes colchici* (Mallophaga). III: Effects of temperature and humidity on the uptake of water vapour. J. Exp. Biol. 55, 553–568 (1971).
WILSON, C.W.: A study of the boric acid colour reaction of flavone derivatives. J. Amer. Chem. Soc. 61, 2303–2306 (1939).
WILT, F.H.: The differentiation of visual pigments in metamorphosing larva of *Rana catesbiana*. Develop. Biol. 1, 199–233 (1959).
WITH, T.K.: Uroporphyrin from turacin: a simplified method. Nature (London) 179, 824. (1957).
WITH, T.K.: Bile Pigments. Trans. J.P. Kennedy. New York: Academic Press 1968.
WITTENBERG, J.B.: Oxygen transport: a new function proposed for myoglobin. Biol. Bull. 117, 402–403 (1959).
WITTENBERG, J.B.: Facilitated diffusion of oxygen through haemerythrin solutions. Nature (London) 199, 816–817 (1963).
WITTENBERG, J.B.: Myoglobin-facilitated oxygen diffusion: Role of myoglobin in O_2 entry into muscles. Physiol. Rev. 50, 559–636 (1970).
WITTERS, R., LONTIE, R.: Stability regions and amino acid composition of Gastropod haemocyanin. In: Physiology and Biochemistry of Haemocyanin, pp. 61–73. Ed. by F. GHIRETTI. London – New York: Academic Press 1968.
WOLKEN, J.J.: Photoreceptors: Comparative studies. In: Comparative Biochemistry of Photoreactive Systems, pp. 145–167. Ed. by M.B. ALLEN. New York: Academic·Press 1960.
WOLKEN, J.J., SCHEER, I.J.: The eye-pigment of the cockroach. Exp. Eye Res. 2, 182–188 (1963).
WOLVEKAMP, H.P., KRUYT, W.: Experiments on the carbon-dioxide transport by the blood of the edible snail (*H. pomatia* L.), the common crab (*Cancer pagurus* L.) and the common lobster (*Homarus vulgaris* M.E.). Arch. Neerl. Physiol. 28, 620–629 (1947).
WOLVEKAMP, H.P., VREEDE, M.C.: On the gas-binding properties of the blood of the lugworm (*Arenicola marina* L.). Arch. Neerl. Physiol. 25, 265–276 (1941).
WOLVEKAMP, H.P., BAERENDS, G.P., KOK, B., MOMMAERTS, W.F.H.M.: Oxygen and carbon-dioxide binding properties of the blood of the cuttle fish (*Sepia officinalis* L.) and the common squid (*Loligo vulgaris* Lam.). Arch. Neerl. Physiol. 26, 203–211 (1942).
WOOD, E.J.: The copper proteins. Chest Piece (J. Med. Stud. Ass. Univ. Malta), March 1968, pp. 5–13 (1968).
WOOD, E.J., BANNISTER, W.H.: Photooxidation of haemocyanin in the presence of methylene blue. Biochem. J. 104, 42–43 P (1967).
WOOD, E.J., BANNISTER, W.H.: The effect of photooxidation and histidine reagents on *Murex truncatulus* haemocyanin. Biochim. Biophys. Acta 154, 10–16 (1968).
WOOD, E.J., SALISBURY, C.M., BANNISTER, W.H.: The binding of copper in haemocyanin. Biochem. J. 108, 26–27 P (1968).
WOODS, M., BUCK, D., HUNTER, J.: The ontogenetic status of melanin granules. Ann. N.Y. Acad. Sci. 100, 534–539 (1963).
WRIGHT, R.H.: The Science of Smell. London: Allen and Unwin 1964.
WRIGHT, S.: The physiology of the gene. Physiol. Rev. 21, 487–527 (1941).
WYMAN, J., ALLEN, D.W.: The problem of heme interactions in haemoglobin and the basis of the Bohr effect. J. Polymer Sci. 7, 499–518 (1959).
YAGI, K. (Ed.).: Flavins and Flavoproteins. Tokyo: Univ. Press and Baltimore: Univ. Park Press 1969.

Yamamoto, T.-O.: Photochemical phenomena in the egg of a polychaete worm, *Ceratocephale osawai*. J. Fac. Sci. Univ. Tokyo (Zool). **4**, 99–110 (1935).

Yamamoto, T.-O.: 1938. Quoted from J. Needham, 1942.

Yamashita, O., Hasegawa, K.: The response of the pupal ovaries of the silkworm to the diapause hormone, with special reference to their physiological age. J. Ins. Physiol. **12**, 325–330; Further studies on the mode of action of the diapause hormone on the silkworm, *Bombyx mori* L. ib. 957–962 (1966).

Yamazaki, I., Yokota, K., Tamura, M.: Horse-radish peroxidase compound III. In: Hemes and Hemoproteins, pp. 319–326. Ed. by B. Chance, R. W. Estabrook, and T. Yonetani. New York – London: Academic Press 1966.

Yinon, U.: Adaptation in the compound eye and interaction with screening pigments. J. Exp. Biol. **56**, 119–128 (1972).

Yonetani, T., Schleyer, H., Chance, B., Ehrenberg, A.: The chemical nature of complex ES of cytochrome c peroxidase. In: Hemes and Hemoproteins, pp. 293–305. Ed. by B. Chance, R. W., Estabrook, and T. Yonetani. New York – London: Academic Press 1966.

York, J. L., Bearden, A. J.: Active site of hemerythrin: Iron electronic states and the binding of oxygen. Biochemistry **9**, 4549–4554 (1970).

Yoshida, M.: On the light response of the chromatophore of the sea urchin, *Diadema setosum* (Leske). J. Exp. Biol. **33**, 119–123 (1956).

Yoshida, M.: Ocellar pigment of the anthomedusan, *Spirocodon saltatrix*. Bull. Mar Sta. Asamushi Tohoku Univ. **13**, 215–219 (1968).

Yoshida, M., Millott, N.: Light sensitive nerve in an Echinoid. Experientia **15**, 13–14 (1959).

Young, J. Z.: The photoreceptos of lampreys II: The functions of the pineal complex. J. Exp. Biol. **12**, 254–270 (1935).

Young, J. Z.: The Life of Vertebrates. Oxford: Clarendon Press 1950. 2nd. Edn. 1962, 1969.

Zagalsky, P. F., Cheesman, D. F., Ceccaldi, H. J.: Studies on carotenoid containing lipoproteins isolated from the eggs and ovaries of certain marine invertebrates. Comp. Biochem. Physiol. **22**, 851–871 (1967).

Zetterberg, G.: Mutagenic effects of u. v. and visible light. In: Photophysiology, Vol. 2, pp. 245–281. Ed. by A.C. Giese. New York – London: Academic Press 1964.

Ziegler, I.: Genetic aspects of ommochrome and pterin pigments. Advan. Genet. **10**, 349–403 (1961).

Ziegler, I., Jaenicki, L.: Zur Wirkungsweise des white-allels bei *Drosophila*. Z. Indukt. Abstam. Vererb. **90**, 53–61 (1959).

Ziegler-Gunder, I.: Pterine: Pigmente und Wirkstoffe im Tierreich. Biol. Rev. Cambridge Phil. Soc. **31**, 314–348 (1956).

Zimmerman, W. F., Goldsmith, T. H.: Photosensitivity of the circadian rhythm and of visual receptors in carotenoid-depleted *Drosophila melanogaster*. Science **171**, 1167–1169 (1971).

Zoond, A. and Eyre, J.: Studies on reptilian colour-response I: The bionomics and physiology of the pigmentary activity of the chamaeleon. Phil. Trans. Roy. Soc. London B **223**, 27–55 (1934).

Zuckerkandl, E.: La variation au cours du cycle d'intermue des fractions proteiques de l'hemolymph de *Maia squinado* separés par electrophorese. C.R. Soc. Biol. **150**, 39–41 (1956).

Zuckerkandl, E.: Le teneur en cuivre de l'hépatopancréas de *Maia squinado* aux divers stades d'intermue. C.R. Soc. Biol. **151**, 676–679; L'economie du cuivre chez *Maia squinado* au cours du cycle d'intermue. ib. 850–853 (1957).

Zuckerkandl, E., Pauling, L.: Molecular disease, evolution and genic heterogeneity. In: Horizons in Biochemistry, pp. 189–225. Ed. by M. Kasha and B. Pullman. New York: Academic Press 1962.

Subject and Systematic Index

Absorption bands 6, 16, 19, 20, 191, 194, 253
– coefficient 19
– difference method 168
– of heat 160, 348–349
– spectrum 14, 31, 35, 39, 44, 45, 49, 50, 51, 56–58, 60, 61, 63, 65, 67, 69, 70, 72, 74–77, 163, 166–169, 172, 174, 175, 183, 184, 187, 190, 192, 207, 213, 235, 338
Acanthaster naphthoquinones 42
Acanthocephala 99, 256, 257, 259
Acarina 91, 108, 114, 349
Acetate units in biosynthesis 282–284, 286, 296
N-acetyl dopamine 41
N-acetyl dopamine-3,4-quinone 285
N-acetyl-5-methoxytryptamine 143
Acetyl choline and cholinergic mechanisms 143, 148, 150, 153
Acetylenic bonds 13, 33, 34
Achaeta 268
Acrida 133, 134
Acridine (phenazine) 66, 67, 200, 249
Acridioxanthin 64
Acriflavin 208, 209
Actinia 32, 103, 114, 302
Actiniaria 303
Actiniochrome 74
Actinioerythrin 31, 34
Actiniohaematin 103
Actinomycetes 187
Actinomycin 64, 160
Action at a distance 62
– potential 152, 168, 176
– spectrum 142, 168, 172, 174, 175, 206
Activating agents 9, 47, 179
Activation energy 8, 9, 12, 125, 179, 228
Adamsia 74, 79
Adaptation to background 131 ff., 202
Adaptive features 114, 133, 136, 138, 140, 155, 176, 215, 216, 220, 228, 229, 270, 293–296, 345 ff.
– radiation 197, 294, 295, 320, 339
Addison's disease 144, 254
Adenine 55, 72, 73, 190
Adenochrome 75, 101, 106, 114, 184
Adenosine monophosphate 209
– triphosphate (ATP) 181, 184–186, 232
Adrectal (hypobranchial, purple) gland 59, 115, 159, 249
Adrenal gland 37, 49, 51, 101, 111, 197, 251, 254, 263

Adrenalin (epinephrine) 99, 101, 143, 148, 153, 155, 198, 254, 289
Adrenocorticotropic hormone (ACTH) 144, 254, 310
Adrenodoxin 51, 195
Adsorption of biochromes 80, 103, 112, 118, 125
Advantages of variety in colour pattern 347, 348
Adventitious chromes 102, 103
Aequidans 264
Aequorea 207
Aetioporphyrins 336
Age changes in zoochromes 198, 202, 217, 243, 244, 255, 272, 275, 276, 305
Agouti 244, 317, 326, 328, 330, 343
Agriolimax 255
Akera 88
β-alanine 328
Alarm substance 177
Albedo 136, 137, 154, 348
Albino 253, 313, 314, 317, 319, 325, 326, 328, 349
Albino locus 325, 330
Aleyonaria 74, 95
Aldehyde oxidase 51
Alimentary system 51, 54, 73, 75, 79, 87, 88, 107, 110, 276, 306, 312, 353
Alizarin 102
Alkaloids 28
Alkaptonuria 89, 112, 249, 250, 255
Alleles at specific loci 326, 327
Allenic compounds 33, 34
Alligator 216
Allolobophora 93, 103, 114, 217, 330
Allomelanin 60
Alloxan 67
Alloxanthine 34
Alloxazine 71
Alma 221, 351
Alopecia 251
Amine oxidase 49, 50, 197
D-Amino acid oxidase 71
Aminohydroxypteridine (AHP) 66, 68 ff., 291, 292
Aminohydroxypteridine-6-COOH 292
δ-aminolaevulinic acid (δ-ALA) 53, 286, 308
δ-aminolaevulinic acid synthase 307, 308
Aminopyrazine ring 207
Ammonium sulphide 338
Amoebae 123

Amoeboid chromatophores 138, 148
Ampharetidae 87, 211
Amphibia 82, 122, 138, 142, 143, 165, 173, 212, 216, 265, 266, 275, 303, 335, 353
Amphineura 87
Amphitrite 105, 272
Amphoteric biochromes 77
Anabiosis 219
Anaemia 251–253, 257, 310, 326
Anaerobic respiration 187, 202, 205, 206, 301
Analogue 223, 236, 238, 249, 273, 276
Anasa 97, 107, 108, 115, 306
Anatomical relationships of zoochromes 335
Androgen 268, 310
Anemonia 17, 19, 74
Angular annelation 200
Anionic dyes 80
Anisops 87, 99, 221
Annelation 14, 20, 44, 45, 200, 201
Annelida 74, 86–88, 92, 138, 208, 243, 295, 303, 309
Anodonta 99
Anolis 112, 113, 141, 264
Anomalous properties of oxygen-transport zoochromes 228 ff., 235, 238
Anomura 92, 145, 146
Anoplodactylus 75
Anoxic conditions 218, 219, 221, 227
Antagonistic mechanisms 143, 144, 153, 155
Antedon 45, 47, 78, 82, 105, 118, 174
Antedon naphthoquinones 45, 47, 78
Antennaless mutant 251
Anthelminth chromes 159
Anthocyanins 14, 32, 38, 77, 78, 175, 304, 323, 329, 338
Anthomedusae 165, 172
Anthoxanthin 38
Anthracene 14
Anthranilic acid 283
Anthraquinone 8, 40 ff., 63, 67, 108, 114, 115, 183, 200, 245, 284, 297, 336
– dimers 285
Antibiotics 112, 159, 160, 200, 248
Anti-buffer action 220
Anticoagulant action 248, 253
Antiferromagnetic coupling 234, 236
Antimitotic agent 272
Antioxidant chromes 35, 37, 156, 160, 197, 206, 251, 270, 349
Antiradiation chromes 156 ff., 273, 341, 348, 349, 353
Apatesis (deceitful coloration) 131
Aphidae 268
Aphins 17, 29, 33, 183, 200, 248, 285, 352
Aphrodite 87, 96, 99, 108
Aplysia (sea hare) 78, 81, 87, 88, 97, 99, 156, 159, 174, 175, 306
Aplysina 74
Aplysioazurin 78
Aplysiorhodin 78, 88
Aplysioverdin 53
Aplysioviolin 78, 247
Apogenin 329
Aposematic (warning) colour schemes 131, 135
Apposition and superposition modes of perception 170
Aquatic animals 135, 140, 141, 160, 173
Aquofuscins 35, 37
Araneida 91
Arbacia 77, 78, 200, 271, 272, 274, 284
Arca 87
Arenicochrome 19, 43, 47, 77, 78, 118, 183, 201, 218
Arenicochromine 43
Arenicola 43, 51, 87, 106, 116, 159, 216, 219, 221, 231
Argulus 256
Argynnis 264
Armadillidium 324
Arion 73, 75, 79, 87, 157, 159
Aromatic structures 14, 21, 40, 44, 60, 62, 108, 127, 177, 241, 284, 296, 337
Artemia 87, 265, 266, 271, 301, 303, 309
Arthropoda 41, 49, 63, 74, 75, 85, 88, 89, 90, 142, 154, 156, 159, 170, 173, 245, 256, 296, 303, 330
Ascaris 40, 86, 87, 96, 107, 187, 217, 221, 223, 351
Ascaris blue 89, 91
Ascidians 73, 75, 90, 103, 108, 115, 118, 173, 266
Ascorbic acid 205
– oxidase 49, 195, 201
Asellus 99, 111, 257, 259, 317
Asplanchna 268
Astacene 19
Astacus 32
Astaxanthin 25, 29, 31, 32, 103, 108, 176, 183, 264, 265, 271
Asterias 31, 93, 244, 302, 306
Asterina 32
Asteroid chromatocytes 122, 148, 149, 153
Asteroidea 82, 88, 167, 175, 303
Astropecten 92
Attacus 305
Audouinia 104
Aurellia 95
Aurone 39, 284
Austerity (biochemical) 25, 28, 53, 80, 295, 339
Autonomic nervous system 143, 452
Auto-catalysis 290, 298
Autoregulation 182
Auxillary genes 328
Auxochrome groups 11, 14, 28, 32, 47, 56, 80
Axial organ 82
Azurin 49

Bábak's law   140
Bacillus subtilis   68
Background response   136, 154
Bacteria   49, 156, 189, 190, 200, 202, 208, 220, 339
Bacteriostatic chromes   159, 249, 272
Bacterium pyocyaneum   67
Bacterium violaceum   60
Balanced polymorphism   327
Balanus   169, 175
Banana   61
Barometer crab   156
Bathochromic shift and colour   11–16, 20, 21, 31, 39, 40, 44, 46, 47, 56, 61, 63, 65, 67–69, 163, 177, 182, 184, 207, 288, 296, 316
Beaver   86
Bee *(Apis)*   169, 172, 268, 273, 328, 330
Beeswax   97
Benzene   14
Benzimidazole   55, 66
Benzochromone   38
Benzoporphyrins   338
Benzoquinone   37, 40 ff., 81, 85, 158, 199, 206, 282, 284, 297
o-Benzoquinone   40, 42, 44, 85, 127, 159
p-Benzoquinone   40, 42, 44, 45, 127, 159
Benzoyl alcohol-3, 4-quinone   285
Berberine   67
Beroe   274
Beryx   90, 172
Betacyanin   59, 60, 81
Biladiene   32, 55, 56
Bilamonene   56, 58, 88
Bilane   56, 58
Bilatriene   15, 56, 102, 194, 272
Bile fluid and chromes   88, 123, 241, 242
Bilifuscin   88
Bilin   18, 29, 32, 52 ff., 54, 57, 58, 76, 81, 88, 103, 108–110, 159, 173, 206, 253, 308, 352
Biliprotein   30
Bilirubin   29, 52, 56, 58, 88, 103, 111, 112, 201, 241
Biliverdin IXα   52, 57–8, 93, 103, 112, 115, 202, 247, 253, 264, 265, 306
Biliverdin IXγ   57, 88, 92, 93
Binding of chrome to matrix   328, 329
Binucleate complex   236
Biochemical evolution   329, 354
– lesion   251, 254, 255, 323 ff., 328
– thermogenesis   198
Biochromatology VII,   3, 4, 37
Biochromes and fertilisation   271, 272
– as oxidation products   37, 242
Biological rationale   139, 140, 147, 160, 167, 214, 309, 313, 350
– significance   200, 203, 249
Bioluminescence   90, 125, 131, 156, 206–209, 345

Biophysical properties VII,   129, 179
Biopterin   68, 69, 181, 204, 268, 290, 292
Biosynthesis of zoochromes   93, 156, 273, 275
– carotenoids   205, 270, 281, 282, 301
– chromones   282–284
– flavonoids   82, 282–284
– haemoglobin   196, 252, 270, 309
– indigotins   59
– indolic melanins   52, 61, 62, 122, 137, 144, 195, 204, 254, 288, 289, 311, 313
Biosynthesis of isoalloxazines   249, 292, 293
– of ommochromes   63, 65, 89, 144, 199, 204, 245, 276, 289, 290, 350
– of porphyrins and derivatives   87, 88, 106, 115, 248, 286–288, 295, 306–308, 309–311
– of pterins   68, 81, 144, 204, 246, 249, 290–292
– purines   144, 273, 290
– Quinones   41, 284, 285
Biosynthetic pathways   63, 106, 199, 205, 281 ff., 323, 347
– pathways that continue as physiological pathways   163, 289, 291
Bipalium   96
Bipyrryl compounds   56
Birds   75, 86, 88, 89, 97, 101, 114, 131, 137, 138, 155, 157, 165, 174, 198, 212, 219, 232, 243, 249, 252, 265, 268, 271, 303, 304, 324, 325, 330
Birefringence   115, 118
Bisglycinato-copper hydrate   48
Biston   133, 134, 324
Biuret compounds   48, 49
Bivalvia   82, 86, 106, 173, 303, 305
Bixanthone, bixanthylene   8
Black and white colour schemes   132, 136, 139, 154
Black sea bream   140, 264
Blattella   302
Bleaching of chromes   35, 36, 39, 62, 126, 165, 182, 247
Blepharisma   43, 86, 201, 248
Blood cell chromes   75, 103, 211 ff.
– chromes   43, 49, 50, 73, 75, 79, 86, 92, 103, 109, 110, 201, 211 ff., 251, 305, 323, 351, 352
– clotting   103, 105, 199, 251
Blue mutant of *Colias*   323
Blushing (flushing)   141, 264
Body fluid   61, 111, 211 ff.
– wall   75, 87, 90, 103, 221
Bohr effect   212, 214 ff., 224, 225 ff., 230, 237, 238
Bombardier beetle *(Brachynus)*   159, 347
Bombyx   78, 242, 265, 267, 270, 298, 323
Bond energy   8
– strength   30, 229, 235
Bonding/antibonding orbital   10, 11, 202

407

Bone 87, 102, 113, 251, 352
– marrow 257
Bonellin 91, 247
Bothrops 305
Botryoidal tissue 36, 106, 115
Brachiopoda 86, 91, 105, 114, 212, 224, 336
Brachydanis 275, 336
Brachyura 92, 139, 146, 147, 264
Brain 204, 268, 276
Branchial heart appendage 75, 101, 106
Branchiomma 219
Branchiostoma (Amphioxus) 174, 175
Bomine derivative of polycyclic aromatic compounds 59, 350
N-Bromosuccinimide 281
Brown atrophy of heart 244
– bodies of Ectoprocta 246
– fat 36, 102, 197, 245, 252
Brown locus 326, 328, 330
Bruise 111
Buffering action of haemoglobin 215, 216
Bugula 75, 91, 105, 118, 184, 271
Bull 249, 263, 267
– frog 212
Bunsen-Roscoe reciprocity relation 173, 176
Bursicon 144
Busycon 225, 237
Butein 329

Caffeine 81
Calanus 256
Calcium 15, 39, 47, 62, 102, 173, 218
Calliactis 74, 77, 79, 184
Callinectes 140
Calliphora 169
Calorescence 160
Cambarellus 145, 146
Camouflage VII, 3, 20, 125, 131, 154, 155, 273, 342, 343
Cancer 31, 237
Canthaxanthin 32, 265, 304
Carausius 122, 139, 140, 144, 147, 148
Carbamino compound 215
Carbohydrate as conjugant 29ff., 194
Carbon dioxide/oxygen interactions 214, 225, 226, 235, 248
– dioxide transport 214, 238
– monoxide affinity 219, 351
– monoxide binding 195, 218, 223, 272
Carbonic anhydratase 215
6-Carboxy-isoxanthopterin 68
6-Carboxy-pterin (AHP-6-COOH) 68, 70
Carcinogenesis 248, 249, 253
Carcinostatic agents 272, 273
Carcinus 31, 32, 108, 114, 132, 133, 256, 265, 266, 302–304, 324
Carmine 42, 80, 125, 126
Carminic acid 19, 40, 43, 45

Carnivore/herbivore contrasts 82, 202, 302, 303
Carotenaemia 251
Carotenase 302
Carotene 18, 33 ff., 206, 303, 337
α-Carotene 35
β-Carotene 31, 32, 33, 82, 163, 256, 264, 265, 312
γ-Carotene 82
Carotene/Oxycarotenoid contrasts 302, 303, 349
Carotenoic acid 18, 32, 33 ff.
Carotenoid 17, 19, 20, 28, 29, 33 ff., 36, 81, 82, 109, 110, 114, 156, 163 ff., 169, 174, 176, 183, 203, 245, 247, 250, 251, 256, 263, 264, 266, 268, 270, 301–304, 312, 323, 339, 342, 349
Carotenol, carotenone 33
Carotenoprotein 29–32, 34, 82, 126, 197, 282, 324
Carrier rhythm 140
Cartilage 112, 250, 254
Caste determination 273
Catabolism of biochromes 54, 55, 71, 202, 302,
Catalase 50, 54, 86, 159, 187, 192–195, 205, 238, 249, 309
Cataract (eye) 251
Catecholamine (see adrenalin)
Catechol 41, 61
Catfish 143
Cation exchange 62, 80
Caudina 87
Cavernicolous animals 302, 348
Central nervous system 142 ff., 145–147, 174
Centropristis 90
Cepaea 324
Cephalopoda 61, 82, 86, 88, 89, 92, 106, 108, 111, 138, 140, 142, 151–153, 154–156, 163, 165, 167, 211, 212, 216, 225 ff., 302, 330
Ceratocephale 265, 270
Ceratotrichia 102
Cercopithecus 134
Cerebrocuprein 49, 100, 196
Cerianthus 74
Ceroid 18, 111, 124
Ceruloplasmin 49, 50, 103, 195, 196, 213, 235
Cerumen chrome 18, 37, 75, 160
Cerura 276, 350
Cestoda 256
Chaetodipteus 154
Chaetognatha 86, 91
Chaetopterus 54, 74, 89, 103, 108
Chain of mediation in chromatic response 141 ff., 154, 313, 314
– link in molecule 39, 60, 67, 207, 284
– shortening 281, 287, 291
Chalcone 38, 284
Chamaeleon 132, 141, 256, 265, 304,
Chaoborus 140

Charge-complementarity 191
Charge number 228, 239
– transfer 13, 15, 21, 48, 62, 72, 173, 181, 191, 192, 249
Chelicerata 86, 88, 91, 159, 227
Chelonia 165, 212
Chemical classes of biochrome 3, 4, 25 ff.
– defence 40, 60, 91, 158 ff., 193, 199, 201, 352
– excitation 80, 258
– reactivity 15, 44, 58, 62, 125, 126
– tests for biochromes 35, 36, 39, 46, 56, 57, 58, 62, 71, 72, 76
Chemi-electric transduction 163, 176
Chemiluminescence 9, 208, 353
Chemi-osmotic coupling 79, 181
Chemoautotrophe 340
Chemo-photic transduction 206
Chemosensory function 267, 351
Chilopoda 92
Chinchilla allele 326
Chirocephalus 32, 267
Chironomus 87, 107, 212, 218, 219, 309
Chloasma 254
Chloragogen 124, 243
Chloragosome 124
Chloride shift 215
Chlorin 55
Chlorochromin 266
Chloroferrihaem 106
Chloroflavin 72
Chlorohaem 54, 87, 286
Chlorohaemoglobin 19, 88, 93, 115, 211, 212, 217–219, 223, 224, 233, 272, 286, 351
Chlorophyll 9, 19, 29, 37, 52, 53, 55, 57, 81, 132, 156, 181, 197, 206, 284, 287, 307 ff., 339, 340
Chloroplast 49, 123, 157, 197, 199, 342, 345
Chloroporphyrin 88, 92, 286
Choanichthyes 266
Choleglobin 54
Choletelin 88
Chondrostei 266
Chondrostoma 330
Choroid (eye) 62, 90, 101, 170, 173, 203
Chromaffin properties 80, 101, 248
Chromagogen tissue 73, 74, 76, 82, 90, 106, 116, 305
Chroman 37, 38
Chromasome 43, 64, 118 ff., 122, 126, 138 ff., 148 ff., 151, 158, 170, 171, 345
Chromatic aberration 170
– properties 37, 44, 80, 158, 228, 345, 347
Chromatium 195
Chromatocyte 97, 100, 101, 111, 113, 136 ff., 170, 174, 329
Chromatographical methods 35, 58, 77
Chromatophore 136 ff., 147 ff., 151, 175, 336

Chromatosome 118
Chromatropism 133
Chromes and first organismus 338 ff.
– as by-products 242, 299, 338
– as end-products of biosynthetic paths 242
Chromium 238, 239
Chromobacterium 67
Chromocyte 95, 96, 98, 102
Chromodoris 75, 79, 184, 197
Chromogenes and taxonomy 329, 330
Chromogenesis VII, 136 ff., 144, 218, 223, 274, 276, 279 ff., 307 ff., 313, 315 ff., 323 ff.
Chromogenic colour change 136 ff., 141 ff., 144, 154, 156, 202, 264, 276, 315, 342
– pathways as redox pathways 281, 337, 339
– mutants 325, 326
Chromomotor colour change 114, 136, 138 ff., 141 ff., 145, 146, 264, 315, 342, 346
Chromone 38, 40, 183, 199, 297, 304
Chromopathies 247 ff.
Chromophore 5, 11, 13, 25–28, 29, 44, 47, 48
Chromoprotein 28, 29 ff., 47, 74, 80, 211 ff., 329
– enzyme 34, 49 ff., 182 ff., 192 ff., 210, 238
Chromorhizon 150
Chromosome 118
Chrysoaphin 18, 44, 45, 47, 78
Chrysopsin 167
Chrysopterin 18, 68, 70, 292
Cichlid fishes 140
Cilia and biochromes 270, 273, 274
Ciliata 82, 86, 87, 123, 223
Cinnabar 113
Cinnamic acid 283
Ciona 173, 175, 266, 274
Cionus 39, 78
Circadian cycle 139, 173–175
Circular dichroism (CD) 20, 234, 237
Circulatory system 105
Circumoesophageal commissure 145, 146
Circumvention mechanisms 312, 313
Cirripedia 205, 217, 305
Cis/trans configuration 16, 33, 60, 163
Cladocera 220, 309
Classification of animals 83–85
– of colour schemes 131
Clathrina 96, 173
Clava 32, 265, 271
Clostridium 51, 195, 287
Cloud of fire 209
Cnidaria 74, 82, 95, 342, 347
Coal 336, 339
Cobalt 53, 55, 173, 201, 223, 228, 238, 239, 287, 310
– complex 223, 228
Cobaltodihistidine 223
Cobamide 53, 55, 209, 247, 251, 252, 287, 298, 304, 305, 307 ff.

409

Cobrinic acid   53, 55
Coccidae   29, 80, 86, 108, 244, 245, 265
Cochineal   29, 80
Coelentera   82, 86, 95
Coelomocyte   75, 82, 105, 110, 114, 118, 184, 200, 211
Coelomic fluid   105, 110, 211, 212, 220, 221, 224
Coenzyme biochromes   29, 40, 43, 50, 51, 71, 89, 123, 182 ff., 209
Cofactor   30, 71, 72
Coiled coil structure   120
Coleoptera   82
Collagen   112, 113
Colias   160, 323
Colour   4, 5 ff.
– blindness   6, 169, 326
– change   101, 111, 129, 136 ff., 154 ff., 320
– circle   5, 6, 168
– morph   323
– pattern   132, 155, 315 ff.
– pattern determined by muscles   316
– vision   6, 137, 168 ff., 342
Coloured complex   188, 194
Colourless biological materials   7, 8, 9
Compensatory properties   139, 281
Competition between functions   156, 263
– gene loci   329
– pathways   93, 298, 301
– tissues   245, 263, 267, 304
Competitive inhibition   200, 223
Complementary colour   6, 151
Compound eye   171
Comproportionation   188, 191
Conalbumen   51
Condensed ring systems   21, 28, 44, 46, 67, 296
Conjugants of biochromes 9, 16, 17, 20, 21, 29 ff., 36, 43, 45, 47 ff., 53 ff., 58, 72, 80, 113, 126, 159, 163, 165, 241
Conjugated double bond system   12, 14–16, 20, 21, 28, 33, 37, 46, 47, 52, 55, 59, 61, 65, 207, 281, 296
– pterin   71, 89
Conjugation step   287, 289
Connections between chromogenic pathways   297
Connective tissue biochromes   91, 101, 102, 110, 114, 184, 250, 254
Control mechanism   244, 245, 253, 268, 310
– for colour change   140 ff., 147, 243, 244, 347
– of body temperature by biochromes   135, 197, 198, 349, 350, 353
– of chromogenesis   276, 296, 298, 307 ff., 337, 347
– of supply of biochromes   301 ff.
Controlled instability   236

Convergence between chromogenic pathways   297
Convergent evolution   238, 295, 329, 345, 351
Co-ordination complex   228 ff., 238
– number   229, 235, 237, 239
Co-oxidation reaction   204
Copepoda   157, 207
Copper   15, 17, 30, 31, 47, 62, 196, 227, 235, 245, 289, 347
– complex   111
Copper-phaeophorbide   88
Copper porphyran   49, 88, 201
– protein   29, 40 ff., 80, 81, 86, 186, 193, 195, 235
– uroporphyrin-III   88
Coprohaem   87, 88, 91, 243, 286
Copronigrin   88
Coproporphyrin   53, 57, 78, 87, 88, 91, 243, 286
– I   53, 243, 286
– III   53, 286
Corethra   138, 144
Corpus allatum   144, 311
– cardiacum   144, 311
– luteum   36, 101, 301
Corrin   30, 55, 66, 201, 294, 308
Corrosive biochrome   46, 159
Corymorpha   271
Cosmic rays   8, 158
Cotton effect   20
Cottus   102
Coupling of phosphorylation with oxidation   188, 197, 198
Countershading   131, 137, 139, 141, 154, 243, 316
Courtship colour pattern   131, 138, 140
Covalent bonding   8, 15, 79
p-Coumaric acid   38, 204, 282, 283
Cow   156, 187, 263, 301, 351
Crago (Crangon)   139, 141, 145–147, 155, 156, 226, 351
Cramasie   80, 348
Crinoidea   42, 43, 86, 336
Crocetin   34
Crocodile   165, 212, 216
Crossaster   31
Cross-over point   228
Crustacea   82, 86, 87, 89, 90, 102, 108, 137, 138, 142, 144–147, 163, 165, 167, 175, 212, 216, 224, 226, 227, 243, 244, 351
Crustacocyanin   32, 115
Crypsis (camouflage)   131, 138, 156, 343
Crystalline form   115, 116, 124, 351
– state   114–116, 151, 172, 223, 230, 234, 253, 276, 306, 327, 350
Cryptochiton   212, 225
Cryptoxanthine   265
Cuckoo wrasse   140, 264

Cucumaria 87, 212
Cuff-link dimer 59, 67, 68
Cuprous compounds 236
Curie scalar/vector problem 186
Cuvierian organs 73, 75, 77
Cyanein 31
Cyanidin 329
Cyanophyceae 347
Cyanopsin 32, 167
Cyclic AMP 150, 151, 198, 313, 314
– compound 14 ff., 28 ff., 207
– process 163–165, 194, 196, 232, 242, 263, 293, 312, 317, 337, 340
Cyclops 32, 268
Cyclostoma 138, 143, 165, 173, 212
Cymatogaster 303
Cypridina 125, 207
Cypris 74, 184
Cysticerus 256
Cysteine 51, 52, 158, 233, 234, 236, 255
Cytochrome 53, 54, 99, 102, 125, 191, 192, 197, 198, 203, 242, 274, 309, 339, 351
Cytochrome a 37, 53, 185, 186, 284
– a_2 187
– a_3 185, 186
– b 54, 185, 186, 191, 193
– b_1 187, 309
– b_2 71, 187
– c 54, 77, 183, 186, 187, 191, 192, 193, 205, 221, 329
– c_1 185, 186
– c peroxidase 125
– f 181
– h 201
– oxidase 50, 118, 185, 192, 219, 223, 228, 235, 274, 340, 347
– P_{450} 192
Cytosol-dissolved biochromes 114

D*actylopius* 114
Daphnia 32, 87, 93, 96, 173, 175, 212, 218, 243, 256, 265, 266, 268, 269, 270, 272, 303, 309, 312
Daphnis 74, 79
Dark adapted eye 171
Dazzle pattern 132
Deep sea animals 209, 348
– tissues of body 4, 73, 82, 96, 200, 248, 323, 343, 349, 352
Defence functions 156 ff., 245
Deficiency of biochromes 111, 197, 199, 204, 248, 250 ff., 253, 258, 273, 302, 326, 346
– mutant 323, 325, 329
Degeneration of spinal cord 252
Dehydrogenation reaction 184, 189, 190, 281, 287, 296
Delayed light 206

Delocalisation of electrons 12, 13, 15, 16, 21, 48, 62, 65, 228
Delphinidin 329
Demyelination 251
Denaturation 29, 58, 64, 235, 236, 306
Dendrostomum 224
Deoxynucleotides as conjugants 51, 71
Depigmentation process 102, 112
Derepression 314
Dermal light sense 142, 173 ff., 351
– melanocyte 120
Desaturation 202, 281, 296, 347
Destruction of biochromes 63, 101, 108, 276, 287, 306, 313, 314, 323
Detoxication 241, 245
Deuteranopic colour blindness 169
Deuteroporphyrin 53, 54
Diabetes and carotenaemia 251
Di- and tri-methyl p-benzoquinones 159
Diadema 89, 96, 105, 139, 141, 174
Diamagnetism 190, 229, 234, 235, 236, 240
Diamine oxidase 49
Dianthrone 43
Diaphorase 189, 193
Diaptomus 32
Diazine compounds 67
Dibenzpyrazine 67
Dibromoindigotin 12, 59, 67, 97, 115, 159, 350
Dibiuretocuprate anion 48
Dichromatism 17, 19, 207
Dicysteinyl DOPA 59, 327
6,7-diethyllumazine 204
Dietary biochromes 42, 80, 250, 252
Dielectric properties 21
Differential synthesis of biochromes 308, 325
Diffusing carrier mechanism for oxygen 221–223
Diffusion of biochromes 102, 109
Digestive fluid 54, 77, 88, 241, 242, 346
7,8-di-H-neopterin 291, 292
7,8-di-H-pterin 68, 292
Di-H-xanthommatin 64, 203
7,8-di-H-xanthopterin 292
3,4-dihydroxybenzyl alcohol 283, 285
5,6-dihydroxyindole 63, 288
5,6-dihydroxyindolyl-indol-5,6-quinone 62, 288
5,6-dihydroxyindole-5,6-quinone-2-COOH 288
Diketocarotenoid 32
Dilute locus 326, 329, 330
Dilution locus 325
Dimer 8, 41, 43, 59, 62, 284, 285
Dimethylbenzanthracene 254
1,1-dimethylindigotin 59
6,7-dimethyllumazine 68, 291, 292
Dinoflagellata 123, 256

2,3diphosphoglyceric acid 232
Diplopoda 92
Diploptera 159
Dipole induced-dipole interaction 62
— moment 14, 15, 21, 33, 44
— moment of transition 16
Dipnoi 165, 166, 212
Diptera 144, 172, 328
Dipyrimidine biochrome of *Sulphomonas* 66
Dipyrryl compound 88
Direct action on biochromes 141, 312, 313, 315
Display of chromasomes 122, 136, 138, 141, 145–147, 148, 151, 160
Disruptive colour pattern 131
Distribution of zoochromes 23 ff., 81 ff., 95 ff.
Diurnal cycle 139, 154, 170
DNA 71, 160, 249
Dominance and chromogenes 325, 329
— and oxidised biochromes 327
Dominant mutants 324, 325, 345
Dominant spotting mutant 326
Dominant white genotype 244, 245, 253, 325
Dominator component in retina 168
DOPA (Dihydroxyphenylalanine) 61, 204, 254, 285, 288, 289, 298
Dopachrome 288, 289
Dopamine 41, 99, 285
Dopamine β-hydroxylase 49, 50
Dopaquinone 183, 199, 288, 290
d-orbital 47, 50
Doris 75, 79
Drach stages of moult cycle 226
Dragonfly 103
Dreissenia 99
Drosophila 107, 118, 122, 158, 172, 175, 251, 255, 257, 289, 319, 326, 328, 330
Drosopterin 68, 69, 122, 292
Dubin-Johnson syndrome 111, 123
Duck 173, 212
Dust cell 99, 100
Dustbin treatment 269, 270
Dye 8, 14, 59, 63, 67, 80, 103, 248, 269
Dynamic stability 127
Dystrophy of tissue 199, 251
Dytiscus 39

Earthworm 29, 175, 205, 217, 220, 243, 249, 306
Ecdysone 144, 311
Echinarachnius 78
Echinenone 82, 304
Echinochrome 29, 40, 42, 45, 77, 78, 159, 174, 183, 199, 213, 271, 272, 335, 339, 351
Echinochromoprotein 29, 30
Echinoderma 42, 75, 82, 86, 88, 89, 102, 105, 138, 173, 245, 284, 342
Echinoidea 82, 86, 157, 265, 266, 303
Echinus 108, 159, 200

Echiura 65, 82, 87, 91, 303
Eclipse plumage 264
Ecological aspects of visual effect functions 131 ff., 140, 157, 165–167
Ecological factors and chromatology 187, 205, 238, 257, 258, 269, 305, 347
Economy in biosynthesis 284, 293, 295
Ectoprocta 75, 86, 91, 105, 266, 274, 346
Edentata 330
Eel 212, 275
Effects of biochromes on embryogenesis 270–274
Efficient biological mechanisms 220–223, 232, 235, 242, 243, 351
Egg 31, 32, 74, 75, 77, 79, 87, 89, 107, 114, 123, 156, 199, 203, 252, 263, 264, 265, 266, 269, 302, 306, 312, 346
Egg-case of dogfish 86
Eisenia 66, 90, 159, 208, 353
Ekapterin 68, 69, 70, 292
Elasmobranchi 138, 142, 143, 165, 170, 212, 265, 273
Elastic fibre 102, 112, 253
Electric vector 16
Electrical asymmetry of molecule 44
— conductivity 350
— properties of biochromes 126, 350
Electrochemical gradient 185
Electromgnetic radiation 9, 10, 127, 156, 206
Electron acceptor 33, 62, 71, 72, 188, 197, 202, 240
— donor 33, 47, 48, 71, 72, 188, 197, 204, 239, 240
— microscopy 111, 113 ff.
— paramagnetic resonance (EPR) 12, 195, 235, 236
— poor region of molecule 202
— spin resonance (ESR) 12, 173, 203
— transfer 126, 165, 175, 181, 182 ff., 186, 203
— transfer system 182 ff., 298, 347
— transition 8, 9, 10–13, 15, 47, 48, 190, 228, 239
Electronic basis of colour 9 ff.
— oscillations 9
Electrophoretic potential 148, 150, 151
— theory of chromasome movement 150
Electrophotic transduction 209
Electrotonic coupling 151
Embelin 159
Emerita 267, 271
Emerson's systems I and II 181
Emotional colour change 141, 264
Endocrine organ 37, 101, 109, 110, 252
Endogenous modification of dietary chromes 302–304, 305, 307
Endogenously synthesised zoochromes 4, 103, 204, 251, 252–255, 270, 271, 276, 285, 290, 295, 301, 306–314, 323, 337

Endoplasmic-Golgi cisternal system 101, 120, 123
Endothelial tissues 103–105
Energy increment of electrons 9, 12, 15, 21, 62, 190
– level overlap 228
Ensis 87
Enthalpy change between oxidised and reduced states 191, 228
Entomostracan crustacea 87, 92
Entrainment 139, 173
Entropy considerations 17, 230
Ephestia 270, 318, 330
Epidermal melanocyte 101
– secretion 76, 82, 110
Epidermis 73, 74, 97–99, 109, 110, 114, 122, 147, 148, 245
Epigamic mechanism 131, 134
Episematic (attractive) scheme 131
Episomatic scheme 131, 134, 256
Ergosterol 157
Ericymba 133
Eriocheir 146, 265
Eriphia 31
Erpobdella 90
Erythraemia 248, 313, 350
Erythroaphin 18, 40, 44, 45, 78
Erythroblastosis foetalis 253
Erythrocuprein 49, 103, 196
Erythrocyte of blood 19, 114, 215
– of integument 122
Erythromelanin 18, 60 ff., 78, 88, 255, 327
Erythrophore 98, 122, 143, 145, 146, 147, 151, 154
Erythropoiesis 310
Erythropoietin 310–1
Erythropterin 18, 68, 70, 292
Erythrosome 143, 150
Escherichia coli 187
Essential oil 28
Ethological aspects of visual effect functions 131 ff.
2-Ethyl-1, 4-benzoquinone 159
Ethylenic compound 12, 16, 33, 44
Eudryas 305
Euglena 123, 157, 171, 349
Eulalia 74, 79
Eumelanin 18, 61, 325, 326, 327, 328
Eupagurus 31, 256
Eupolymnia 221
Evolution of biochromes 82, 93, 132, 142, 167, 195, 221, 232, 244, 269, 275, 277, 315, 323, 329, 330, 333, 335 ff., 341–343, 345, 353
Evolutionary sequence 281, 282, 284, 327
– sophistry 191
Excess mutant 325
Excess of biochrome due to parasite 255–258

– of dietary chrome 251
– of endogenous chrome 252–255
Exchange diffusion 223
Excitation of electrons 31, 33, 56
Excited states of electrons 9, 16, 21, 177
Exciton 13
Excretion 4, 91, 111, 123, 241 ff., 268, 276, 277, 290, 296, 302, 305, 306, 346, 350, 352
Excretory organs and tissues 105, 106, 243, 302
Exogenous biochrome 4, 90, 102, 103, 280, 301–307, 312, 323, 343, 349
Exoskeleton 30–32, 95, 97, 109, 110, 114, 158, 173, 175, 201, 226, 245, 302
Extension locus 325, 326, 330
Extension of pigment locus 325
Extracellular biochrome 73, 111 ff.
– biochrome in lipid droplets 112
Extraction technique 26, 27, 28, 30, 58, 64, 76, 77
Extraterrestrial chrome 338
Extravasal tissue 106
Exuvia 245
Eye 89–91, 93, 100, 101, 123, 142, 145, 146, 151, 154, 157, 204, 245, 274, 323, 328, 330
Eylais 31

Fabrein 18, 45
Facilitated diffusion 222
Faecal chrome 59, 88, 241 ff.
Farnesyl compound 53, 54
Fasting/feeding differences in biochromes 241, 242, 244, 245, 302, 304, 306, 351
Fat body 43, 102, 105, 106, 276, 312
– tissue 92, 102, 109, 110, 197 ff., 303, 323
Feather 19, 60, 75, 88, 97, 99, 114, 120, 122, 158, 198, 244, 267, 317, 324
Feedback control 246, 252, 254, 295, 307, 308, 347
Ferredoxin 48, 49, 50, 51, 76, 181, 184, 194, 195, 212, 339
Ferric hydroxide 50, 113
Ferrihaem 33, 106, 257, 272
Ferritin 50, 51, 106, 111
Ferroxidase 196
Ferroverdin 73
Ferryl ion 228
Fertility 252, 267, 268, 270, 302
Ficulina 32
Filograna 271
Filtering chrome 137, 172
Fiona 31
Firefly 207 ff.
First excited state 10, 11, 77
Fishes 89, 115, 142, 173, 215, 243, 255, 303, 353
Five fold symmetry 237
Flabelligeridae 87, 211, 266
Flamingo 302

413

Flavanone 38
Flavilium 38, 39
Flavin-adenine dinucleotide (FAD) 71, 89, 181, 185, 187, 190, 274, 347
Flavin mononucleotide (FMN) 71, 72, 89, 181, 187, 190, 274
Flavodoxin 51
Flavonoid 18, 29, 36, 37, 38 ff., 60, 76, 78, 81, 82, 159, 183, 198, 199, 265, 268, 269, 297, 303, 323, 329, 339
Flavoprotein (FP) 20, 32, 125, 126, 185, 186, 188–191, 193–195, 198, 347
Flight muscle 103, 203
Flip-flop switch, intramolecular 231
Flower 14, 38, 39, 325, 329, 348
Fluorescence 9, 11, 17 ff., 29, 32, 35, 36, 39, 44, 56, 57, 61, 65, 69, 72, 74, 75, 76, 77, 190, 207, 237, 338, 340
Fluorescence and chemiluminescence 270 ff.
Folic pterin 68, 292
Formylkynurenine 64, 289
Formyl transfer 204, 298
Fortuitity in biological acquisition 109, 160, 172, 174, 226, 240, 242, 244, 270, 293, 294, 316, 345, 349
Fossil biochrome 53, 91, 125, 336
Franck-Condon principle 17
Free radical 12, 13, 18, 20, 21, 46, 61, 62, 72, 173, 188, 191, 193–5, 203, 204, 206, 347, 350
Freshwater Teleostei 165
Fringelite 41, 43, 284, 336
Frog 99, 114, 138, 155, 160, 169, 216, 353
Fucoxanthin 34, 35
Fulmar petrel 108
Functional hypertrophy and atrophy 137, 140
– significance of biochromes vii, 60, 62, 72, 95, 109, 187, 237, 238, 242, 335, 345 ff.
Fundulus 139, 141, 143, 264, 302
Fungi 49, 81, 187
Fungicides and fungistats 159, 272
Fuscin 18, 25, 33, 35 ff., 76, 81, 82, 99, 101, 107, 108, 124, 197, 198, 242, 244, 245, 251, 252, 354
Fusitriton 212

Galactose oxidase 49
Galago 90
Gallstone 102
Gallus (fowl) 267, 273, 317
Gambusia 141
Gamma (γ) rays 8
Gammarus 32, 92, 256, 330
Gastrophilus 87, 99, 114, 216, 217, 219
Gastropoda 54, 76, 82, 86–88, 92, 106, 198, 212, 216, 225, 241, 303, 319, 350
Gastrotricha 91
Geeldikkop 248
Generator potential 163, 168

Genes controlling chromoproteins 327, 328
Genetic (innate) agencies 249, 252, 253
– determination of colour scheme 136, 155, 313, 315, 316, 319, 320, 323 ff.
– homology 329
Genistein 268
Geographical variant 324
Geometrinae 74, 79
Geoteuthis 336
Gephyrea 302
Geranylgeraniol 281
Gill 89, 99, 259
Girella 136, 302
Glaucoma 123
Glial tissue 100, 110
Globins 211 ff., 327
Glomerid millipede 91
Glossiphonia 118
Glossoscolex 217, 224
Glucocorticoid 310
Gluconic acid as conjugant 29, 68, 202, 205, 253
Glycera 87
Glycine for biosynthesis 286, 290
Glycolipoprotein 31, 32
Glycolytic respiration 200, 255
Glycoprotein 31, 34
Glycose derivatives as conjugants 28, 29, 33, 39, 43
Glycosidation 29, 33, 39, 43, 44, 63
Gmelin test 56, 58, 202
Gnathostome haemoglobin 216, 217, 230
Goat 302, 303
Goldfish 137, 169, 273, 309
Golfingia 224, 233
Gonad 31, 158, 251, 263
Gonadal hormone 263, 268
Gradients, ontogenetic 318, 324
Gram-negative and positive bacteria 81, 187
Granular biochromes 113 ff., 118
Graphite 61
Green gland 243
Gregaria phase (migratoria) 135, 156, 311
Grey and white horse 255
Greying of hair (achromotrichia) 251, 289, 350
Ground state of electron 9–11, 77, 177, 228
Growth effect 251, 252, 273, 317
Gryllus 311, 312, 351
Guanine 67, 91, 170, 249, 290
Guanophore 122
Guanosine diphosphate 287
Guanosine-7-triphosphate 291, 292
Guanosine-9-triphosphate 291, 292
Guanosome 113
Guinea pig 102, 108, 302
Gut (see alimentary system)

Haem   17, 18, 30, 33, 36, 53 ff., 86, 87, 109, 110, 156, 210, 250, 257, 264, 306–308, 339
– oxygenase   287
Haematein   39
Haematin   39, 241   (see also protoferrihaem)
Haematochrome   123
Haematoporphyrin   30, 53, 57, 268
Haematoxylin   39, 80, 126
Haemerythrin   18, 50, 78, 86, 105, 108, 114, 212 ff., 223, 224, 233 ff., 351
Haemochromatosis   252
Haemochrome   58, 103, 108, 242, 243, 272, 306
Haemocuprein   196
Haemocyanin   17, 20, 30, 49, 50, 86, 114, 190, 196, 211 ff., 216, 218, 224–227, 235–237, 247, 350
Haemocyte   115, 118
Haemoferrin   50
Haemoglobin   9, 19, 29, 81, 86, 87, 96, 114, 116, 117, 125, 126, 132, 191, 192, 211 ff., 229 ff., 233, 234, 243, 252, 253, 268, 272, 323, 327, 329, 330, 347, 351
– variants   327
Haemophilus   269
Haemopoiesis   196, 311
Haemoprotein   29, 54, 86, 87, 103, 123, 174, 188, 193, 223, 308
Haemorrhage   310
Haemosiderin   50, 76, 111, 196, 252
Hair   30, 51, 60, 62, 65, 89, 90, 99 126, 158, 244, 276, 317, 326, 328–330
Halichondria   74, 79
Haliotis   76, 78, 79, 88
Halla   42, 86
Hallachrome   42, 47, 78, 183, 200
Hallorange   42
Halochromy   65
Halogen   12, 239, 350
Halogenated amino acid   95
Halogenation reaction   194
Hamster   254, 268
Hardening agent   158, 198, 348
Harderian gland   54, 174
Harmathoe   108
Harmonia   298
Hawkmoth   276
Heart   90, 103, 187, 191, 194, 252
– body (Polychaeta)   104–106
Heat-radiation   160, 353
Heat of oxygenation   228, 238
Heavy metal   14, 46, 62, 69, 113, 123, 289
Helicorubin   77, 87, 183, 201, 241, 242
Helix   18, 29, 39, 76, 78, 82, 97, 108, 159, 183, 198, 212, 225, 237
Helmichthydes   330
Hemichorda   91, 105, 106
Hemiptera   42, 82
Hen   92, 216, 266, 302

Henricia   31
Hepatocuprein   49, 108, 196
Hepatopancreas (see liver)
Hermione   108
Herring   107
Hestina   137, 313, 324
Heterocope   32
Heterocyclic ring   14, 47 ff., 66 ff., 127, 207
Heterocypris   306
Heterometrus   227
Heterotrophe's biochrome   217
Heterotylenchus   256
Hibernation gland   197
High altitude   135, 136, 160, 348, 349
– latitude   135, 136, 160, 348, 349
– oxidation state of biochromes   337
High-spin (spin-free) state   77, 79, 192, 228, 234
Higher Metazoa   211, 328
Himalayan rabbit   138, 313, 316
Hippolyte   32, 140, 157
Hirudinea (leeches)   36, 61, 62, 63, 74, 79, 102, 105, 106, 139, 142, 154, 157, 159, 243, 272, 342
Hirudo   244, 306
Histidine   29, 30, 192, 229, 233, 236, 238, 248
Holocentrus   143
Holothuria   19, 37, 73, 75, 89, 90, 96
Holothuroidea   31, 92, 100, 175, 265, 303
Homogentisic acid   250
Homoiothermic vertebrates   100, 135, 160, 216, 350
Homology   330, 336, 339, 345
Homophany   135
Hoplias   141
Hormonal control of biochromes   141 ff., 268, 310, 311, 318, 324
Horse   103, 216, 219, 241, 254, 325
Hot spring   349
Hue   6, 16, 21, 168
Huffner unit   211, 216, 224
Humidity and integumental colour   135, 137, 140, 142, 155, 157, 313
Hyas   134
Hybrid prosthetic group   186, 187, 191, 347
Hydra   97, 271, 302
Hydracharina   349
Hydractinia   271
Hydrangea   14, 39
Hydrocarbon   10, 11, 336, 337
Hydrogen peroxide   159, 187, 193, 194, 199, 205, 208, 218, 235, 247, 249
Hydrophilic properties   45, 57, 60, 72, 186, 191
Hydrophobic properties   45, 57, 60, 72, 186, 191
3-Hydroxyanthranilic acid   64, 65, 290
7-Hydroxybiopterin   68, 69
3-Hydroxykynurenine   63, 64, 204, 289, 298
Hydroxyquinone   91, 200, 272
Hydrozoa   265, 271, 301

Hygroperceptor 313
Hyla 140
Hynobius blue 68, 69
Hyperconjugation 15, 207
Hyperia 140
Hypericin 43, 156, 248
Hyperoxia 310
Hypophysis 37, 174
Hypothalamus 142, 275
Hypoxia 310
Hypsochromic shift and colour 13, 19, 56, 70, 163, 209, 288, 316

Ichthyopterin 18, 68, 69, 183, 204, 292
Idotea 32, 139, 140
Idya 31, 256, 301
Imidazole structure 49, 50, 66, 67
Improbable structure 164, 192, 291
– step 289, 294
Inclusion, cellular 113, 120, 124
Independence between chromogenic genes 325
  components of development 319
  control mechanisms of types of chromatophores 147
Independent sensori-motor systems 141, 142
India ink 113
Indican 59
Indigofera 59
Indigo green 59
Indigotins 28, 59 ff., 88, 106, 202, 249, 266
Indirect autocatalysis 290
– response 247, 312, 313
Indirubin (indigo red) 59, 60, 106
Indium 239
Indole-5, 6-quinone 61, 288
Indoleacetic acid 249
Indolic biochromes 59 ff.
– melanins 60 ff.
– monomers 60, 61
Indolylpyrrylmethine 60
Indophenol 273
Industrial melanism in moths 133, 324
Infra red radiation 7, 50, 160, 349
Ink of molluscs 61, 62, 88, 97, 156, 159, 209, 245, 313, 336
Inositol hexaphosphate 232
Insecta 68, 82, 86, 89, 131, 137, 138, 142, 158, 163, 165, 167, 169, 175, 204, 245, 251, 255, 266, 274–276, 303, 323, 324, 326, 330, 348, 350
Integumental zoochrome VII, 31, 32, 37, 64, 74–76, 79, 82, 88, 89, 92, 93, 95, 97 ff., 109, 110, 131 ff., 151, 173, 249, 263, 269, 275, 295, 303, 306, 315 ff., 323 ff., 341 ff., 346, 349
Intensity of light 16, 46, 56, 151
Intensive/extensive properties 220
Interaction between chromogenic genes 329
– members of different chromogenic pathways 297, 298

– ontogentetic patterns 318
– sub-units of molecule 216, 222, 230, 234, 237
– whole molecules 231
Intercellular transfer of zoochromes 111, 122
Interconversions between biochromes 297
Intermediary compounds in pathways 20, 53, 62, 72, 163, 187, 201, 281, 282, 285, 287, 289, 298
– forms in a single reaction 41, 190, 193–195, 199
– states of a molecule 231 ff.
Intermeshing of chromogenic pathways 255, 289–290
Intersystem crossing 11, 12
Internal clock 139
– quenching of fluorescence 61, 65
Inter-relationships between chromogenic pathways 296–299
Intracellular biochrome 73, 74, 75, 95 ff., 106, 111 ff.
Intramolecular interactions 214, 230
Invertebrates 61, 82, 85, 88, 89, 99, 101, 102, 105, 113, 165, 167, 168, 173, 204, 216, 218, 221, 241, 250, 265, 270, 275, 303
Iodinin 66, 67, 200
Iodopsin 31, 167
Ion-exchange properties 80
Ionic properties 8, 79, 239 (see also pH)
Ionising radiation 30, 156, 193, 350
Ionone ring 33, 165, 281
Iridescence 151
Iridophore 122, 144, 330
Iridosome 115
Iris of eye 92, 100, 328
Iron 15, 17, 47, 50 ff., 53, 56, 62, 73, 76, 113, 173, 188, 205, 251, 252, 257, 287, 289, 306, 308, 347
– flavoprotein 51
Iron-protein 29, 49, 50 ff., 80, 81, 193, 228
Isatis 59
Ischaemia 310
Isoalloxazine 28, 63, 67, 71 ff., 190, 245, 317, 339
Isodrosopterin 69, 292
Isoelectric point 28, 56, 57, 62, 221
Isoeleutherin 41
Isoflavone 39, 268
Isoguanine 67
Isolated leg: colour change 139
Isologous compounds 93, 336
Isopentenyl unit 86, 281, 282, 284
Isopoda 145, 146, 205, 305, 324
Isorhodoptilometrine 43
Isosepiapterin 68, 69, 292
Isoxanthopterin 68, 70, 292, 298
Iulid millipedes 91, 159
Ixodes 114

Jablonski diagram   10, 11
Janickina (Paramoeba)   258
Janthina   19, 75, 79
Janus heads   271
Jassid Amphipoda   92
Jaundice (Icterus)   96, 102, 112, 245, 251, 253
Jupiter   338

Keber's organ   106
Keratin   62, 157, 158
Keratinocyte   111, 114
Kermesic acid   8, 43, 45
Kidney   37, 49, 90, 106, 189, 197, 305, 311
Kidney of accumulation   243
Kinetic instability   126
– stability   164, 191
Kingfisher   135
Koscuiscola   140
Kupffer cell   103, 111, 124
Kynurenine   18, 63–65, 204, 289
– hydroxylase   330
Kynurenic acid   64, 89

Lability of biochrome   79, 163, 228 ff., 236, 315
Laboratory-synthesised chrome   77, 242, 295, 338
Laccaic acid   43, 47, 78
Laccase   49
Lacciferinic acid   64
Lacertilia   212, 353
Lactam-lactim forms of purine   66
Lactic dehydrogenase   189, 202
Lactocuprein   49
Lactoperoxidase   54
Ladybird beetle   303
Lagopus   138
Lambert and Beer's relation   176
Lampetra   139, 165, 175
Lamprey   216, 230, 237
Lampyrid beetle   74, 90
Lampyrine   19, 74, 79, 107, 125
Lanaurin   88
Lanigerin   18
Latia   207, 208
Latitude and colour schemes   135, 136
Latreutes   157
Leander   145, 146, 157
Lebistes   136, 141
Leghaemoglobin   219, 223
Lemon locus of *Drosophila*   326
Lepadogaster   89
Lepas   32
Lepidoptera   32, 37, 68, 74, 79, 82, 88, 89, 92, 133 ff., 137, 160, 276, 312, 318, 328
Lepidopterin   68, 70, 292
Leptocoris   97
Lernaeocera   217
Lethal mutant   324

Leucochloridium   257, 346
Leucoflavin   71, 72
Leucophore   98, 145, 146, 148, 149, 151
Leucopterin   18, 68, 70, 292
Leucosome   122
Liesegang's rings   316
Ligand   15, 16, 29, 30, 47, 49, 50, 55
– field   21, 47, 48, 186, 228 ff., 234, 235, 351
Light-adapted eye   171
Light as triggering agent   136, 137, 141, 154, 156
– perception   VIII, 8, 100, 142, 163 ff., 351
Light-sensitivity   46, 126, 164, 167
Ligia   145
Lignin   61
Limulus   86, 91, 99, 145, 165, 167, 168, 170, 212, 227
Lineus   89, 159
Lingula   212, 224, 234
Lipid phase   25, 26, 28, 30, 80, 192
Lipidosis   252
Lipophore   122
Lipochrome   33
Lipofuscin   36, 105, 113, 114, 250
Lipomelanin   111, 123
Lipoprotein   29, 31, 32, 34, 163, 186, 264
Liquid crystalline state   125
Lithobius   19, 47, 78, 79, 97, 99, 183, 201
Litmus   39
Littoral fauna   139, 140
Littorina   97
Liver   37, 49, 50, 51, 103, 108, 110, 111, 123, 124, 189, 190, 196, 202, 205, 220, 227, 241, 248, 251, 252, 257, 271, 301, 302, 305, 324, 349
Lizard   114, 141, 157, 165, 268, 353
Loading pressure of oxygen (P_{50})   212, 214 ff., 238
Lobster (*Homarus*)   32, 99, 226, 265
Local control   155
Locus coeruleus   99
Locust   31, 32, 135, 156, 173, 265, 267, 273, 275
Locusta   257, 312
Loligo   92, 225, 237
Lone pair electrons   11, 12
Low-spin (spin-paired) state   77, 79, 191, 228, 240
Lower taxa: chromatic contrasts   73, 92, 93, 335
– vertebrates   122, 243, 296, 306
Loxorhynchus   226
Luciferase   208 ff.
Luciferescein   208
Luciferin   207 ff.
Luidia   92
Lumbriconereis   74, 266
Lumbricus   73, 74, 93, 217
Lumichrome   18, 66, 71, 206, 247, 249
Lumiflavin   18, 71, 206, 247
Lumirhodopsin   165

Luminous bacteria 208
Lunar periodicity 139, 140
Lung 99, 100, 111
Lutein 31, 32, 33, 265
Luteolin 39
Lycopene 303
Lymnaea (Limnaea) 87, 89, 100, 221
Lymph organ 103, 113
Lyochrome 71
Lysosome 123

Macrobrachium 140
Macrocorixa 107
Macrocytic (hyperchromic, pernicious) anaemia 108, 251, 252
Macropipus (Portunus) 139, 265
Macropodus 141
Macrura 92, 139, 146, 147
Maesoquinone 159
Magelona 86, 212
Magnesium 15, 53, 248, 287, 306, 338
Maia 226, 352
Main chromogenic pathways 296
Malacostracan crustacea 92
Malaria parasite 257
Malonate 285, 286
Malpighamoeba 256
Malpighian tubule 106, 118, 243, 276, 305
Mammal 36, 37, 49–51, 59, 60, 75, 82, 86–88, 99, 101, 119–122, 137, 138, 155, 157, 165, 170, 174, 198, 204, 212, 216, 217, 219, 241, 249, 252, 268, 274, 303, 309, 316, 325–327, 329, 330, 339, 351
Man 90, 101, 108, 122, 135, 136, 141, 144, 157, 168, 169, 198, 202, 215, 216, 219, 225, 241, 244, 249–251, 255, 258, 290, 313, 323, 326–328, 342, 349, 350
Mandrill 141
Manganese 53, 201, 213, 238, 239, 289
Mantid 133
Marbled White butterfly 29
Marginalin 38, 39, 284
Marine Elasmobranchi 165
– Teleostei 165
Marphysa 108
Mars 338
Marthasterias 93, 144, 163
Mastigoproctus 86
Matrix material 64, 80, 118–120, 122, 126, 328, 329
Mechanical properties of biochromes 30, 62, 65, 158
Mechanisms of colour-change 141 ff.
Meconium 245, 276, 350
Megalomma 106
Meganyctiphanes 207
Melanargia 304
Melanin 7, 12, 13, 17, 18, 25, 28–30, 36, 46, 51, 60 ff., 80, 88, 109–111, 120, 157, 158, 160, 170, 172, 174, 177, 202, 203, 245, 247, 249, 250, 253, 256, 265, 273, 275, 276, 297, 313, 326, 328, 330, 341, 349, 350
Melanising activity 107, 109
Melanochrome intermediaries 288
Melanocyte 98, 101, 111, 119–121, 254, 276
Melanogenesis 136, 144, 276
Melanoma 144, 202, 253–255, 256, 268
Melanophore 143, 145, 146
Melanophore-stimulating hormone (MSH) 143, 144, 148, 150, 155, 254, 313
Melanoplus 273
Melanoprotein 52, 62, 202
Melanosis coli 124
Melanosome 62, 101, 114, 118, 120, 144, 173, 202, 203
Melatonin 143, 144
Mellitic acid 47
Membrane phenomena involving biochromes 30, 111, 176, 177, 199, 241, 246, 302
Mermis 268
Menadione (vitamin K_3) 181
Mesobilirhodin 56
Mesobilirubin 56, 58
Mesobiliverdin 32
Mesobiliviolin 56
Mesohaem IX 193
Mesoporphyrin 53, 55
Mesopyrrochlorin 53
Metabolic chromoprotein 125, 182 ff.
– cycle 337, 341
– depression 251, 310
– functions of biochromes VII, 3, 12, 17, 20, 21, 28, 50, 90, 93, 182 ff., 241 ff., 264
– interlock 296
– load 93
– product 103, 113, 132, 242–246
– rate affecting biochromes 203, 312
Metals 16, 29, 39, 57, 73, 127
Metal co-factors 30, 72, 188, 191
– salt of biochrome 14, 39, 46, 57, 71
Metallobilin 54
Metalloporphyrin 47
Metalloprotein 25, 28, 29, 47 ff., 73, 86, 238, 265, 339, 341
Metamorphosis 144, 245, 273, 275, 295, 313, 336, 350
Metarhodopsin 77, 78, 165, 167
Metazoa 95, 174
Methaemerythrin 224, 234
Methaemocyanin 200
Methaemoglobin 108, 193, 223
Methine bridge 54–56, 58, 60, 202, 287
Methoxy group as substituent 42
2-Methyl-1,4-benzoquinone (*p*-toluiquinone) 159
Methyl transfer 201, 204

Methylene blue 67, 111, 200, 205, 249
– bridge 287
Metridene 31
Metridium 88, 114, 118, 174, 175, 324, 348
Micro- and macro- pattern 132, 315
Micrococcus 51
Microcytic (hypochromic) anaemia 251, 252
Microhyla 306
Micromethod 141, 143, 147 ff., 168, 169
Micro-organism 28, 50, 55, 59, 64, 81, 187, 189, 190, 284
Microsome 190, 252, 308
Microsporidium 258
Microtubule theory of chromasome movement 150, 151
Migratoria phase of locust 135, 144, 160, 311
Mimicry 131, 324, 348
Miniature potential 152
Minnow 151
Mitochondria 49, 99, 103, 113, 118, 120, 122, 123, 125, 186, 188, 194, 196, 198, 199, 203, 252, 274, 345
Mitogenetic radiation 353
MK9 187
Model of biological molecule 223, 228, 238
Modulation 140, 307, 310
– of genetic mechanism by physiological response 315, 320, 324
Modulator component of retinal response 168
Molecular activation 7, 9 ff., 181
– analogue 223, 229, 233
– asymmetry 12, 14, 16, 207, 294
– basis of colour 4, 5 ff.
– basis of oxygen-transport 228 ff.
– conformation changes 125, 192, 230 ff., 232 ff., 234, 237
– dissociation 237
– engineering 290
– orbital approach 9 ff., 62, 71, 203
– polymorphism 225
– rotation 7, 9, 16, 126, 222
– refraction 20
– size 10, 16, 21, 45, 61, 63
– strain 192, 194, 229, 233, 234
– sub-unit 231
– symmetry 15, 20, 52, 56, 286
– vibration 7, 9–11, 16, 177
– weight 50, 51, 61, 63, 73, 146, 163, 224, 231, 237
Mollusca 49, 75, 87–89, 97, 106, 224, 225, 243, 244
Molpadia 75, 184, 211
Molybdenum 188, 238
Monochromatic cell 122
Monomer 62, 231, 234, 289
– interaction(n) 212, 214, 230, 231, 237
6-Monomethylrhodocomatuline 43
Monoplacophora 106

Morphogenetic abnormalities 251, 252, 271, 275
– action of biochromes 270 ff., 351
– colour-change 264, 311, 315
– tide 319
Mosquito 274
Motile chromasomes 122, 123, 126
Moult cycle 226, 227, 352
Moulting (ecdysis) and colour-change 137, 158, 244, 264
Mountain hare 138
Mouse 268, 313, 325, 326, 328
Murex 59, 67, 159, 266, 350
Murexide test 67, 71
Muscle 37, 87, 103, 110, 111, 174, 197–199, 222, 251, 274, 316, 317, 353
Muscular effector organ 151, 248
Mustela 138
Mustelus 141, 142
Mutagenesis 248
Mutant 107, 255, 323, 326
Mutation 8, 93, 351, 352
Mutual extinction of chromogenic waves 319
– stimulation of biosynthesis 308
Mya 175, 176
Myoglobin 87, 103, 125, 217, 219, 221, 222, 230, 341
Myohaematin 91
Myoid of rod and cone 170, 171
Myoneme 148
Myriapoda 91
Mysis 149
Mystacoceti 303
Mytiloxanthin 78, 265
Mytilus 82, 265, 267, 304
Myxicola 106
Myxine 175
Myxothexa 123
Myzostoma 266, 274

n → π^* transition 11, 72, 190
NADH dehydrogenase 51
Naevi 254
Namakochrome 42
Naphthalene 14
Naphthaquinone 40 ff., 81, 102, 157, 159, 176, 183, 187, 245
Natural selection VII, 8, 57, 71, 93, 127, 132 ff., 168, 192, 203, 206, 217, 243–245, 250, 281, 283, 286, 287, 293, 315, 317, 319, 325, 336, 337, 342, 345, 346
Neanthes 14, 266
Negative feedback 308
Nematoda 87, 89, 91, 111, 158, 211, 212, 256, 265
Nemertina 87, 91, 92, 99, 103, 111
Neodrosopterin 68, 69, 292
Neoplasia 249, 254
Neopterin 68, 290, 292
Neotenin 144, 204, 275, 311

Nephridial gland   106
Nephridium   101, 243, 274
Nephrops   31
Nephthys   105, 174, 218, 220, 221
Nereis   96, 105, 107, 108, 139, 165, 167, 264, 266, 270, 274
Nerve cell   18, 37, 49, 82, 87, 91, 99, 100, 110, 174, 198, 245
– plexus   142, 174
Neural control mechanism   141, 151–153, 154–156, 311
– crest   101, 109, 110, 120–122, 202, 254, 316
– tissue   99–100, 145–147, 197, 251, 252, 352
Neuro-hormonal control mechanism   142, 307, 314
Neurosensory agent   144, 145–147
Neutral density   221
– mutations   223
N-heterocyclic ring   14, 80, 127
NH_2-group   46, 47, 61, 71, 127, 159, 289
Nickel   238, 336
Nicotinamide   206, 209
Nicotinamide-adenine dinucleotide (NAD)   18, 181, 184, 185, 188, 191, 204
– dinucleotide phosphate (NADP)   287
Nidamental gland   107
Night blindness   251
Niobium   73, 205
Nipple   37, 264
Nitrate reduction   188, 189, 193, 194
Nitrogen   7, 11, 13, 14, 21, 28, 47ff., 50, 53, 60, 65, 229
– economy by plants   28, 33, 61, 81, 341
Nitrogen-fixing bacteria   223
Nocardia   55
Non-bonding(n) orbital   10, 15
Non-radiative transition   11
Non-stoichiometric complex   50, 51, 61
Nor-carotenoid   34
Nucella   59, 263, 267
Nucleic acid   7, 8, 17, 29, 243, 248, 253, 267, 273, 276, 296, 306, 307, 338, 350
Nucleotide   55, 71
Number of chromogenic loci   328–329
Nuptual colour-change   137, 138, 140, 264
Nymphaline butterflies   109

Ochronosis   112, 250
Octahedral ligand field   50, 229, 233
Octopus   75, 92, 153, 225, 236, 241, 242, 302, 303
Odontosyllis   207
Oedaleus   132
Oedipoda   31, 32, 133
Offensin   158
OH-group   14, 15, 29, 39, 40, 43, 47, 71, 127, 185
Olfactory sense and biochromes   176, 177

Oligochaeta   54, 65, 212
Oligomer   61–63, 89, 211
Ommatidia   63, 101, 149, 154, 170, 171
Ommatin   63, 172
Ommatin D   63, 203, 264
Ommatophore   152, 153
Ommidin   64
Ommin   63, 64
Ommin A   63–65
Ommochromasome   122, 147, 151, 152
Ommochrome (Phenozazone)   29, 61, 63 ff., 68, 69, 110, 118, 137, 138, 157, 159, 172, 203 244, 245, 255, 257, 266, 273, 275, 276, 294, 297, 324, 326, 328, 330, 335, 341, 350
Onchorhynchus   267
Ontogenesis and biochromes   269–277, 315 ff., 323 ff.,
Onychophora   86
Oochrome   265, 266, 270, 271
Operator gene   314
Ophiocomina naphthaquinones   42
Ophiuroidea   303
Opsin   163
Optical asymmetry   20
– exaltation   20
– properties   5, 17 ff., 31
– rotary dispersion (ORD)   20, 234
Orchestia   176
Orconectes   145
Organ of Bellonci   146
Orsellinic acid   41, 284
Orthellia   256
Orthopterid insects   88, 157, 202, 306, 312
Ortho-substituted aromatic compounds   40 ff., 67
Oryzias   148, 256
Outstanding adaptations   346, 347
Ovary   31, 32, 37, 107, 110, 263 ff.
Ovorubin   30, 114, 265
Ovoverdin   29, 30, 32, 265, 270
Owenia   108
Oxazine   63, 65, 289
Oxidase   189, 190
Oxidation of biochromes   36, 39, 43, 44, 47, 56, 58, 62, 65, 70, 72, 73, 302
– state of metal   49–51, 186, 188, 191, 193–197, 201, 205, 228 ff., 234, 235, 238, 239
Oxidative pathways of chromogenesis   296, 299, 303, 304, 327
– phosphorylation   188, 202
– sulphuration   203
Oxidising action of biochromes   46, 159, 172
Oxonium cation   14, 38, 39, 46, 65
Oxycarotenoid   33, 82, 265
Oxygen as controlling agent   308, 309 ff.
– in molecule of biochrome   7, 11, 13, 14, 17, 21, 28, 30, 60, 65, 229
Oxygen-affinity   217 ff., 223, 234, 235, 257

Oxygen-binding   186, 192, 195, 200, 228, 238, 351
Oxygen-bridge   234, 236
Oxygen-capacity   212, 214 ff., 351
Oxygen-consumption   197, 199, 202–204, 218, 226, 298
Oxygen-containing side-chain   14, 80
Oxygen-equilibrium curve   214 ff., 230
Oxygen-gradient   220, 221
Oxygen-pocket   192, 230, 232
Oxygen-pressure   211, 214, 310, 350
Oxygen-saturation   214, 218
Oxygen-store   219–221
Oxygen-transporting zoochromes   VII, 29, 32, 50, 54, 77, 93, 105, 111, 184, 192, 201, 205, 211 ff., 295, 341, 347, 351
Oxygenation cycle   217, 228 ff., 236, 237, 351
Oxysome   125, 186, 187
Ozone   9

Palaemon   132
Palaemonetes   139, 140, 145, 146, 151
Palinurus   31
Palomino horse   325
Pancreas   252
Pandalus   146
Pantothenic acid   209, 251, 252, 268, 299
Panulirus   226, 302
Papilio   324
Papiliochrome   18, 64, 65, 89, 92, 245, 294
Paracentrotus   78, 272
Para-disubstituted aromatic ring   40 ff.
Paradoxical effects   32, 160, 190, 191, 205, 218,
– properties   229, 235, 237, 251, 267, 304, 309
Paralichthys   139
Parallel evolution   221, 236, 295, 329, 335, 341
Parallels between molecular and anatomical evolution   282
– biosynthetic pathways   283, 284, 287
Paramagnetic state   17, 21, 46, 50, 62, 190, 196, 228, 229, 234, 235,
Paramecium   123
Parascaris   217
Parasites and biochromes   82, 99, 123, 140, 217, 250, 252, 255–258, 269
Parathyroid   101
Parenchymal chromes   95
Parietal eye   173
Parkinson's paralysis agitans   252, 254
Partially characterised zoochromes   75, 76, 90, 97
Patchy distribution of biochromes   92, 155
Patchy whiteness   254
Patella   75, 266
Pathological conditions   102, 111–113, 123, 245, 247 ff., 325
Pattern-discrimination   341, 342
Pauli principle   12

Peacock   135
Pearl   97
Pecten   31, 265
Pecten of eye   86
Pectenoxanthin   265
Pectinaria   266
Pediculus   107, 265
Pelagia   74, 111
Pelargonidin   329
Pelobates   302
Peltogaster   217
Penicillium   284
Penis   264
Pentacarboxylic porphyrin   54
Pentacene   14
Peobius   87, 211
Perca   141
Perceptor chrome   163 ff., 169, 172, 175, 176
Perceptor organ   163 ff., 311
Perceptor potential   169, 176
Perhydroxyl ion   228, 236
Pericardial tissue   105, 106, 202, 306
Perinereis   77, 79, 266, 274
Periodicity in patterns   317 ff.
Periplaneta   176, 305
Peritoneum   102, 105, 106, 110, 353
Perivisceral melanin   352, 353
Pernicious anaemia   108, 251, 252
Perosis   252
Peroxidase   54, 86, 193, 194, 201, 238, 309
Peroxisome   193
Peroxo ion   228, 236
Perylene   285
Petromyzon   165
pH   14, 25, 26, 28, 31, 39, 46, 50, 56, 62, 63, 65, 70, 72, 73, 74–80, 167, 183, 184, 201, 205, 212, 214, 216, 231, 235, 236, 238, 319
pH colour change   46, 47, 70 ff.
pH indicator   77 ff., 274
Phaeodium   123
Phaeomelanin   18, 60 ff., 88, 325–328
Phaeophorbide   53, 55, 57, 88, 97, 103, 108, 115, 306
Phaeoporphyran   55, 248
Phaeoporphyrin   53, 55
Phagocytosis of chromes   103, 105, 111, 113, 257
Phagosome   123, 124
Phasmid   144
Pheasant   97, 324
Phenanthraquinone   46, 200, 272
Phenocopy   313, 316
Phenol oxidase   274
Phenolic melanin   41, 60, 81, 250, 285
Phenoxazine   63, 64, 67, 200
Phenthiazine   200
Phenylalanine   250, 254, 283, 284, 296
Phenylalanine hydroxylase   204

421

Phenylketonuria 254
Phenylpropane 61
Pheromone 177
Phobocampe 276
Phoronida 87, 105, 106, 274
Phosphatidyl ethanolamine 165
Phosphorescence 9, 11, 17 ff., 208
Photinus 208
Photoactivation 8, 9, 157, 163, 172, 173, 176, 182, 201, 202, 204–206, 208, 270, 337–339, 340, 346
Photochemoelectrical transduction 163 ff., 177
Photochromatism 20
Photoconduction 13
Photodecomposition 30, 46, 247, 248
Photodynamic properties of biochromes 46, 53, 57, 72, 156, 157, 197, 201, 247–249, 298, 313, 343
Photoelectric effect 72, 163, 167, 168, 209
Photoexcitation reaction 13, 29, 52, 62, 80, 126, 201, 208
Photogenic organ 209
Photokinesis 175
Photolysis 71, 72, 206, 247
Photomechanical transduction 142
Photon 7, 9
Photooxidation 183, 204, 208, 247, 270
Photoperception 34, 137, 163 ff., 172 ff., 268, 270, 315, 341, 342, 346
Photoperiod control of reproduction 138, 268
Photopsin 163 ff.
Photoreduction 197, 206
Photosensitisation by chromes 8, 156, 175, 176, 206, 247 ff., 268, 273
Photosensitised neurological response 248, 249
Photosensitivity of biochromes 62, 70, 72, 172, 175
Photosynthesis 3, 8, 11, 123, 157, 172, 181, 182, 186, 194, 339, 340, 345, 347
Phoxinus 139
Phrynosoma (horned 'toad') 139, 140, 141, 160
Phthalocyanin 59, 126, 201
Phycocyanin 32, 81, 88, 103, 347
Phycoerythrin 32, 81, 88, 103, 347
Phyllodoce 74, 79
Phylloerythrin 53, 55, 108, 248
Physiological functions of zoochromes 129 ff., 213 ff., 354
Physiologically active stable state 167, 172
Physiology, Biophysics and Biochemistry 129, 179
Phytochrome 81, 173
Phytoene 281
Phytoflagellata 123, 342
π-orbital 10, 15, 21, 62, 71
$\pi \to \pi^*$ transition 11, 72, 188, 189
Pied piper bicoloration 324
Pieridae 68, 88, 92, 115, 246, 264, 292, 313

Piezoelectric flow of electrons 192
Pig 301, 302, 349,
Pigeon 92, 172
Pigment VII, 4, 28, 138, 144
– epithelium of eye 100, 120, 171
Pike 112, 253
Pila 225, 270
Pilocarpine 153
Pineal organ 101, 142, 143, 173, 174
Pink-eyed locus 325, 326, 330
Pinnipedia 212
Pisaster 31
Pituitary gland 101, 143, 144
Petyriasis versicolor 258
Placenta 107, 115, 241
Planarity of molecule 16, 20, 21, 30, 33, 49–50, 59–61, 73, 190, 191, 281
Planktonic organisms 275, 343, 349
Planorbis 87, 212, 218, 219, 221, 309
Plants 28–9, 32–34, 38, 43, 49, 50, 59, 60, 66, 68, 77, 81, 89, 187, 189, 190, 284, 339, 352
Plasma chromes 31, 32, 75, 251
Plastocyanin 49, 181, 195
Plastoquinone 181, 188, 199
Platyfish 330
Platyhelmia 87, 88, 95, 111, 159, 173
Pleiotropic gene 326
– pathway 284
Pleurobranchus 90
Pleuroploea 225
Plistophora 257
Plodia 271
Plusia 32
Plymouth Rock fowl 317
Pogonophora 87
Poikilothermic vertebrates 89, 90, 100, 138, 139
Poikilotherms: colour schemes 135, 160, 313
Poised molecular state 228, 240
Polar bear 108, 132, 251
– groups of molecule 28, 30, 39, 57, 61
– solvent 35, 39, 57, 58, 65, 72, 80
Polarity of molecule 14, 15, 21, 44, 58, 60, 65, 67
Polycelis 74, 90
Polychaeta 43, 54, 87, 88, 102, 154, 212, 265, 303, 351
Polycheria 42
Polychromatism 19
Polychrome cell 122, 123
Polydesmus 87, 91
Polycyclic compound 14
– quinone 8, 19, 40 ff., 248, 284, 335, 336
Polymer 211, 228
Polymerisation 46, 61–63, 101, 216, 229, 281, 284, 285, 288, 296, 338
Polymorphism 133, 160, 323–325, 347, 348
Polymorphus 256

Polyphemus 93, 173
Polyphenol oxidase 41
Polyphyletic origin 342
Polyplacophora 86
Pomacea 30, 31, 114, 265
Pomatoceros 74
Population-density and integumental coloration 156, 311, 312
Porania 31
p-orbital 47
Porifera 74, 82, 86, 88, 89, 95, 111, 303, 342
Porphobilinogen unit 53, 286
Porphyran 49–51, 52 ff., 58, 339
– analogue 223
Porphyranoprotein 12, 52 ff., 54, 235
Porphyria 245, 248, 352
– cutanea tarda 248
Porphyrin 8, 15–19, 28, 29, 52 ff., 77, 81, 87, 109, 110, 157, 174, 175, 201, 202, 206, 245, 247, 250, 263, 265, 272, 273, 336–338, 352
Porphyrinogen 287
Porphyropsin 31, 163 ff., 175
Porpita 32
Positive feedback 308
Post-commissural organ 145, 146
Potassium salt 29, 148, 150, 199, 215
Preadaptation 217, 221
Prebiological evolution of organic chromes 337, 338
– synthesis of chromes 338
Prebiological excess of chromes 337–339
Precambrian biochrome 336
Predator-prey relation 131, 133 ff.
Prelumirhodopsin 165
Premelanosome 110, 120, 121, 328
Prenyl side-chain 40, 55, 282, 284, 296
Presetting of chemical system 163, 177
Priapula 82, 86, 108, 212, 224, 303
Priapulus 108, 115, 224, 242
Primality of chromogenic pathways 293, 294
Primary colour-change response 141, 147, 151, 163, 174, 342
– structure of protein 29, 230, 233
Primates 303
Prionotus 90
Problems concerning integumental colour-pattern 135, 154 ff., 157, 158, 159, 161, 347 ff.
– concerning other biochromes 218, 221, 228, 234, 235, 245, 255, 264, 267, 275, 286, 304, 307, 347 ff., 351 ff.
– of chromogenesis 286, 287
Proctoeces 219
Prodigal use of exogenously acquired biochromes 251, 295, 302, 306, 307
Prodigiosin 52, 57, 183, 201, 287
Prolactin 144, 275
Proportionate atrophy 244, 302

Prosthetic group 31, 32, 47, 50, 71, 182, 238, 239
Protective functions 131 ff., 156 ff., 218–220, 247, 270, 273, 341
Protein 7, 9, 17, 21, 29 ff., 32, 43, 46, 47, 248, 350
– as conjugant 29 ff., 43, 44, 49 ff., 61, 62, 68, 70, 72, 73, 163, 213, 230, 327, 328
Proteinoid 338
Prothoracic gland 144, 276
Protoaphin 40, 44, 45, 47, 78, 103
Protocatechuic acid 41, 283, 285
Protochorda 86
Protoferrihaem 30, 36, 97, 102, 241, 286, 306
Protoferrihaemochrome 107
Protoferrohaem 52, 106, 211
Protoferrohaemoglobin 211
Protoferrohaemochrome 108
Protohaem IX, 54, 71, 187, 193, 286
Protohaemoglobin 86, 87, 92, 272, 286
Protoporphyrin IX 53, 57, 58, 78, 87, 92, 93, 248, 268, 286
Protopterus 165, 212
Prototropic (intramolecular) reaction 58, 202
Protozoa 82, 88, 89, 123, 174, 175, 204, 252, 273, 290, 335, 343
Pseudaposematic scheme 131
Pseudemys 165
Pseudepisomatic scheme 131
Pseudocoele 111, 211
Pseudoporphyran 48, 59, 126, 201, 233, 236
Pseudosematic scheme 131
Psittaci 19, 75
Psylla 74
Pteridine 68, 127
Pterin 17, 18, 28, 29, 35, 36, 39, 67, 68 ff., 77, 89, 109, 110, 122, 137, 169, 170, 172, 177, 183, 203, 204, 245, 247, 265, 268, 273, 275, 276, 326, 328, 330, 335, 338, 339, 350
Pterin substituents at positions 6 and 7 291
Pterinophore 122, 330
Pteropods 138
Pterorhodin 66, 68, 292
Pteroylglutamic acid 66, 68 ff., 81, 89, 204, 209, 251, 273, 290, 292, 298, 305
Ptilometric acid 43
Puberty and zoochromes 263, 264
Pugmoth caterpillar 82, 303, 304
Pulmonata 76, 82
Purine 4, 28, 66, 67 ff., 72, 81, 107, 122, 170, 243, 247, 267, 273, 276, 306, 307, 326, 330, 350, 353
Purinotely 243, 306
Purple bacteria 340
Purpura 59
Purpuric acid 67
Pycnogonida 75
Pyocyanin 67, 200

423

Pyrazine 67
Pyridazine 67
Pyridine compounds 60, 65, 66 ff., 72, 338
Pyridinoprotein system 125, 184, 185, 193
Pyridoxal 197, 205, 249
Pyrilium 38, 39
Pyrimidine 67, 273
Pyrrolic compounds 52 ff., 183
Pyura 205

Qualitative chromopathies 249, 250
Quanta of energy 7–11, 21, 165, 181, 342, 350
Quantitative differences between alleles 326
Quaternary structure of protein 230, 308
Quenching of fluorescence 9, 12, 17, 21, 29, 32, 56, 61, 70, 72, 247, 343, 352
Quinhydrone 41, 46, 61, 62, 72, 188, 199, 203, 206, 285
Quinol 41, 46, 159, 176, 206
Quinoline 67
Quinone 14, 17, 30, 40 ff., 60, 77–79, 101, 109, 110, 159, 176, 183, 187, 199–201, 203, 206, 265, 271, 272, 293, 335, 339, 352
m-Quinone 44, 67, 72
o-Quinone 40 ff., 285
p-Quinone 40 ff., 159
Quinonoid properties 38, 44, 60, 65, 71, 77, 79, 80, 158, 208

Rabbit 107, 219, 302, 323
Radiationless transfer of energy 9
Radiative transfer of energy 9
Radiolaria 123
Radula 87, 97, 212
Raman scattering 20
Rana 93, 100, 140, 144, 149, 212, 303, 313, 330, 352
Ranachrome 3, 68, 69, 292
Range of colours in each taxon 82 ff.
– of colours in each tissue 95 ff., 109
– of colours in each class of biochrome 31, 32, 35, 36, 37, 44, 64, 93
Raper pathway of melanogenesis 288
Rat 65, 174, 188, 197, 204, 241, 317, 351
rb mutant of *Bombyx* 298
Reaction of chromes with c-H_2SO_4 35, 36, 46, 65, 71, 75, 76, 77
Recapitulation 123, 167, 272, 275, 335, 343
Recessive allele 323, 327, 328, 330
Reciprocity of responses of dark and light chromes 137
Recruitment 152
Red nucleus 100
Redox agent 33, 71, 72, 181, 197 ff., 199, 172, 337, 339, 345, 351
– colour change 46, 183, 184
– criteria 183, 184
– cycle 165, 194, 197, 200, 204, 205

– enzyme 37, 50, 54, 181 ff.
– indicator 182
– potential 32, 46, 51, 71, 72, 181–184, 186, 187, 195, 199, 201, 203, 210, 229, 239, 255, 347
– properties of biochromes vii, 37, 39, 46, 50, 65, 71, 72, 74, 75, 77, 179, 181 ff., 197 ff., 238 ff., 242, 273, 351
– reaction 80, 163, 179, 197
Reduction of biochrome 36, 39, 46, 47, 50, 56, 58, 62, 65, 70, 72, 73, 336
Redundancy 237
Reflected light 136, 154
Reflecting biochromes 89, 91, 92, 99, 101, 115, 151, 170, 243, 349
Regeneration 143, 271, 275, 335
– (molecular) 163, 164, 165, 167, 168
Reindeer 132
Release-recapture technique 134
Renal erythropoietic factor 310, 311
Reniera 34, 82
Renilla 207
Repression of biosynthesis 308
Reproduction and biochromes 175, 262–269, 346
Reproduction and biochromes 175, 262–269,
Reptantia 145
Reptilia 89, 90, 138, 142, 143, 160, 232, 243
Reseda 39
Resonance (molecular) 8, 9, 15, 16, 21, 33, 39, 41, 44, 55, 61, 71, 72, 177, 188, 290
Resonance forms of indolic melanin 59
– *o*- and *p*- benzoquinones 41
Respiration 3, 20, 54, 71, 72, 181 ff., 197, 200, 203, 205, 267, 340, 345
Respiratory enzyme system 125, 185, 252
– metabolism 202, 251, 252, 340
– organ 74, 99, 110
Response to incident light 136, 154, 155
– to reflected light 136, 154, 155
Retention purinotely 306
Reticulo-endothelial system 50, 111
Retina 119, 120, 203, 253
– biochromes 90, 92, 119, 120, 163 ff., 170 ff., 275, 302
Retinal (retinene) 3, 282, 345, 346
– I 31, 163 ff., 175, 197, 275
– II 31, 32, 163 ff., 275
– field 154, 155
– isomerase 165
– melanosome 120
Retinol (vitamin A) 34, 163, 165, 251
Retinomotor movements 171
Retinopsin 34, 100, 163 ff., 342, 349
Retinulae 169
Retraction of zoochrome 140, 146, 147
Retroperitoneal chrome 105, 157, 353
Re-use of material 163, 164, 241, 242
Reversal of countershading 137, 316

Reversed (negative) Bohr effect 214 ff., 220, 225, 227, 230
Reversibility of redox reaction 183, 184, 197, 199–201, 205, 231 ff., 240, 291
Rhabdite 97, 118, 159
Rhizocephala 241
Rhizostoma 31
Rhodium 239
Rhodnius 51, 96, 103, 107, 108, 202, 242, 266, 306
Rhodocomatuline 43
Rhodoflavin 72
Rhodommatin 63
Rhodopseudomonas 307, 308
Rhodopsin 29–31, 33, 163 ff., 175
Rhodoptilometrine 43
Rhodoquinone–9 40, 45, 91
Rhynchodemus 74, 79
Rhythmic and cyclic colour-change 138, 139, 143, 151, 152
Ribitylflavin(RF) 7, 12, 17, 18, 20, 29, 32, 37, 66, 67, 71 ff., 77, 89, 90, 109, 110, 118, 159, 169, 170, 172, 175, 183, 187, 188–191, 200, 201, 205, 206, 208, 209, 243–245, 247, 249, 251, 266, 269, 273, 289, 291, 298, 305, 339
m-RNA 313
Rodentia 65, 89, 90, 102, 158, 244, 317
Rods and cones 13, 163, 165, 171, 251
Root effect 214, 215, 221, 231
– nodule 219, 223
Rotation of leucosome 151
Rotifera 86, 91
Royal jelly 268
Rubredoxin 51, 195
Rufescin 76, 79
Rufin 73, 76, 79, 159

Sabella 223, 266
Sabellaria 108
Sabellidae 87, 89, 97, 105, 211
Sable 132
Sacculina 256, 303
Sacculus of chromatophore 151
Sagitta 258
Salamandrid urodele 92, 118, 313
Salinity and haemoglobin production 309
– pigment pattern 319
Salivary gland 108, 242, 306
Salmon 103, 141, 170, 171, 197, 267, 268, 275, 330
Samia 65
Sargasso shrimp 157
Saturation/desaturation of bonds 56, 58, 67
Scalar process 175, 186, 328
Scale insect 64
Scavenging function 100, 105, 123, 124, 196
Sceleporus 173
Schistocerca 275, 303, 311

Scintillation 151
Sclerotic of eye 112
Sclerotin 61
Scolopus 87
Scolytus 176
Scotopsin 163 ff., 167. 175
Screening (masking) chrome 64, 100, 142, 145, 146, 156 ff., 170 ff., 341, 352,
Scyliorhinus 141
Sea anemone 131, 348
– otter 102, 304
Seasonal colour-change 137, 138, 263, 264, 315, 324, 338
Sebastodes 214
Secondary chromopathy 247, 251–253
Secondary colour-change response 141, 142 ff., 147, 154
Secreted chrome 59, 97, 108, 158–160, 198
Segregation of chromes 277
Selection for minimal production of chromes 243, 293, 336
– simplicity 317
– of chromes by tissue 109
Selective/indiscriminate treatment of cartenoids 303, 304
Selenium 124, 199
Self-regulating mechanism 220, 222, 281
Self-perpetuation 338
Sematic (advertising) scheme 4, 131, 151
Semen 107, 249, 267
Semiconduction 13, 62, 125
Semiquinone radical (see Quinhydrone)
Sense organ 100, 101, 110, 163 ff.
Sensory perception 163 ff.
Separation method 35, 39, 57, 58
Sepia 152, 153, 225, 264, 313
Sepia locus 330
Sepiapterin 68, 69, 122, 204, 292
Septosaccus 266
Sequestration of chromes 111, 114, 211, 337
Serpulidae 211, 265
Serpulimorph 87, 88, 93, 106, 108, 211, 224, 272
Serratia marcescens 52
Sertularella 36, 82
Sertularia 78, 271
Sertulariid 29, 82, 92, 304
Sesarmia 146
Sewage microorganism 55
Sex-determination 268
Sex differences in chromes and metabolism 203, 244, 245, 263 ff.
Shade 6, 21
Shell (egg) 29, 87, 102, 249, 263, 266, 319
– (molluscan) 29, 30, 75, 76, 78, 87, 92, 97, 249
Shikimic acid 38, 282–284, 296
Shikimic-prephenic pathways 283
Short-lived animal 244–246

425

Siamese cat 138, 316
Sickle cell anaemia 253, 327
Side chain 21, 28–30, 33, 37, 39, 40, 42–44, 46, 53, 54, 57, 58, 60, 61, 63, 65, 68, 284, 285, 347
Siderophilin 51
σ -orbital 10, 15
Sigmoid curve 214, 217
Silk 78, 97, 350
Silkworm 103, 108, 265
Silver 113
Similarities between chromogenic pathways 296
Simocephalus 32, 218, 264, 272, 309
Simulidae 96
Singlet excited state 11, 12, 21, 72, 188
Sinus gland 145, 146, 147
Siphonosoma 224
Siphons 173
Sipuncula 82, 86, 101, 108, 118, 198, 212, 233, 234, 274, 303
Sipunculus 198, 224
Six-fold symmetry 237
Skate 212, 216
Skeletal tissue 74, 87, 95, 102, 109, 110, 113, 158
Skin 112, 151, 153, 252, 267, 328, 349
Slowly healing lesion 159, 248, 249
Small Tortoiseshell butterfly 156
Smoke screen 159
Snake 92, 165, 170
Sol-gel change 148
Solar radiation 6, 7, 21
Solaster 31
Solid state 4, 13, 16, 18, 67, 72, 111, 114, 115, 125, 126, 201, 222, 223
Solitaria/gregaria colour changes 135, 144, 155, 156, 311
Solubility properties of biochromes 4, 20, 26, 27, 28, 33, 35–37, 44, 45, 57, 61, 62, 65, 67, 69–71, 72, 74, 75, 80, 113, 114, 230, 275, 351
Somatic mutation 225, 325
Sophisticated mechanism 152, 161, 231 ff., 245, 257, 293, 294, 319, 329, 345–347
Sophisticated structure 245, 339, 345
Soret absorption peak 56, 213
Spadella 258
Sparrow 264
Spatiotemporal wave 319
Spectral display 6
Spermatogenesis 267
Spherulitic structure 118
Sphinx 32
Spinal cord 99, 174, 176, 249
Spinochrome 42, 45, 183
Spiralian taxonomic group 87, 88, 89, 90, 245, 335, 350
Spirobacterium 256
Spirocodon 89, 172

Spirographis 212, 223, 224
Spleen 50, 113, 252, 257
Splitting of orbital energy levels 15
Spontaneity of biological processes 13, 29, 30, 58, 65, 164, 182, 187, 200, 202, 204, 205, 230, 231, 237, 281, 283, 290, 293, 316, 330, 337, 338
Sporadic demand for chromes 270
Sprue 249, 251
Squalus 142, 330
Square conformation 236
Squid 77, 156, 209
Squilla 145, 146
Stabilisation 30, 33, 48, 58, 67, 77, 125, 158, 173, 188, 191, 202, 203, 205, 229, 230, 232, 234, 236
Stability 44, 47, 52–54, 60, 62, 64, 71, 72, 75, 126, 127, 163, 169, 190, 196, 229, 239, 240, 290, 336–338
Standard respiratory pathways 184 ff., 206
Starling 264
Static (genetic) colour scheme 136, 141, 148, 151, 315, 320
Steady state 220, 337
– state auto/heterotrophe system 340
Stellacyanin 49, 50
Stenotomus 90
Stentor 86, 118, 248
Stentorin 18, 19, 45, 47, 118
Stentorol 18, 45
Stercobilin 56
Steric molecular aspects 45, 51, 61, 62, 72, 73, 125, 163, 190–192, 201, 230 ff.
Steric hindrance 73, 164
Steroid 190, 204, 205, 255
Stickleback 140, 151, 256, 264
Stigma 172
Stoichiometric conjugation 29, 34, 51, 62, 64, 72
– reaction 205, 224, 225, 228, 234–236, 238
Stokes law 17, 20
Storage excretion 243
Storage function for zoochromes 50, 51, 90, 93, 102, 105, 106, 115, 123, 124, 125, 196, 204, 227, 243 ff., 276, 295, 305, 306, 312, 343, 346, 349, 352
Stratum corneum 349, 350
Strobinin 18
Strongylus 217, 221, 351
Structural colour 4, 16, 67, 69, 99, 136, 328
Styela 265, 266, 274
Styelopsis 79
Strongylus 217, 221, 351
Subchromatic molecule 207, 208
Sub-particles of electron-transfer system 185
Substantia ferruginea 99
– nigra 99, 252, 254
Succinate 185, 286
Succinea 257

Succinic dehydrogenase 51
Sudan III incorporation into egg 269
Sulphomonas 67
Sulphonium cation 65
Sulphur in biochromes 51, 63
Sulphuric ester 63, 203
Sulphydryl group 46, 47, 49, 61, 158, 159, 165, 176, 185, 196, 199, 201, 205, 236, 238, 289, 327
Sumner's relation 136, 176
Suntan 157, 313
Super-ring of porphyrin 52, 53, 57, 58
Surface effect of chromes 199
Susceptibility to disease, and deficiency of chromes 151, 325
Sweat 36, 37, 67, 97, 112, 160, 245, 251
Swim-bladder of fishes 215, 221
Swordtail 330
Symbiont 285, 292, 305
Sympetrum 203, 267
Synaptic transmitter 150, 152, 153
Synergism 187, 312, 329
Synkavit 188
Systemic control mechanism 198, 308, 312
– response 174

Tactile stimuli 142
Talitrus 176
Tanning process 29, 30, 41, 43, 46, 65, 158
Tanymastix 31, 32
Tanytarsus 218
Tapetum 90, 170
Tardigrada 86
Taste sense and chromes 176
Tattooing 113
Tautomerism 67, 69
Taxonomic distribution VII, 81 ff., 122, 147, 173, 271, 303
– variations 187, 213, 230
Taxonomic distribution VII, 81 ff., 122, 147,
Tealia 74
Teeth 87, 102, 113, 158
Teleonomy vii, 135, 243, 281, 287, 290, 293
Teleostei 82, 90, 102, 103, 107, 122, 131, 138, 142, 143, 147, 151, 152, 165, 166, 212, 216, 265, 330, 352
Temperature control of integumental colour pattern 135, 137, 139, 140, 142, 155, 157, 313, 316, 319, 330, 349
Terebella 87, 105
Terebellidae 108
Terminal respiratory enzyme 185, 187, 202, 204, 238
Terestrial poikilotherm 216, 349
Ternary quinone 40 ff., 81, 85, 86
Tertiary response 154
– structure of proteins 29, 230, 308
Testies 37, 107, 110, 353
Tetracene 14

Tetracycline 113
Tetra-H-biopterin 68, 69, 204, 249, 268, 292, 298, 336
Tetra-H-pteridine 71
Tetralin hydroperoxide 208
Tetrameres 221
Tetrapyrroles 52 ff.
Tettigonia 32
Thalassema 87, 96, 99
Theodoxus 319, 320
Thermal activation 8, 228
– energy 7, 163, 176, 228, 349
– regulation 136, 160
Thermochromatic response 8, 20
Thermodynamic considerations 9, 164, 177, 230
– improbability 164, 177
– stability 15, 21, 53, 126
Thermogenic function 198, 216
Thermosensitisation 273
Theromyzon 275
Thiamine 209
Thiazine 63, 65, 67
Thiazole 67
Thiochrome 67, 209
Thomisia 132
Thoracophelia 82, 223, 227, 303
Threshold intensity for photostimulation 165, 167, 168, 176
Thyone 87
Thyroid gland 101, 198, 275
Thyroxin 12, 144, 310
Tide cycle 139, 140
Tiger 316
Tineola 302, 305
Tint 7, 21
Tissue-distribution of chromes 43, 95 ff.
Tissue spaces 111, 112
Titanium 69, 238
α-Tocopherol 37, 38, 111, 124, 197, 199, 209, 251, 268, 282
Tocopheroquinone 37, 183, 199
Tocopherylphosphate 199
Tortoise 167
Tortoiseshell cat 329, 341
Touraco 3, 57, 88
Toxicity of chromes 108, 159, 199, 251
Toxiflavin 66, 67
Tracheal end cell 99, 114, 201, 221
Transduction of energy 163, 165, 342
Transferrin 51, 111, 113, 196
Transition-dipole of molecule 16
Transition state 209
Transitional metal 10, 15, 17, 21, 47, 71, 79, 228 ff., 238–240
Translation of m-RNA 308
Transliteration of DNA 160, 308
Transmitted light 5, 6, 16, 19, 70, 151, 160, 166, 268

Transparent, colourless organisms   343
Transport of biochromes in body   50, 103, 111, 263, 264, 267, 269–274, 276, 304, 305, 306, 312, 314, 323
– by chromes   126
Transverse motif in colour patterns   155, 315 ff.
Travisia   105
Trematoda   88, 107, 131, 256, 265
Tribolium   159, 302
Tricarboxylic porphyrin   54
Trichosiderin   51, 60
Trilobita   336
Triops   74, 79. 87, 212, 265
4-Triphosphoribosamine-2, 5-diamino-6-hydroxypyrimidine   292
Triple bond   13, 33
Triplet excited state   11, 12, 17, 21, 172, 188
Tritanopic colour blindness   169
Tritonia   87
Triturus   336
Trout   270
Tryptophan   17, 49, 59, 60, 63, 64, 68, 204, 236, 237, 245, 249, 255, 283, 289, 296, 326, 350
– catabolism   245, 276, 350
– pyrrolase   330
Tubercle bacillus   258
Tubifex   218–220, 272
Tubularia   95
Tunicata   75, 88, 91, 205
Turacin   57, 99, 159, 201, 245
Turbellaria   61–63, 74, 87, 96, 97, 115
Turnover of biochromes   243 ff., 305
Turtle   309
Tylopeis   132
Tyrian purple   59, 80, 202
Tyramine   153
Tyrosinase (dihydroxyphenoloxidase)   49, 50, 61, 122, 195, 256, 313, 325
Tyrosine   60, 63, 232–234, 236, 250, 254, 255, 283–285, 288, 296, 325

Ubiquinone   37, 40, 42, 43, 81, 85, 86, 91, 183, 185–187, 199, 200, 251, 265, 271, 284, 304
Uca   139–141, 145–147, 154, 156–158, 160, 264
Ultraviolet absorption   7, 35, 56, 57, 69, 76, 169
Unconjugated pterin   204
Unidentified biochromes   36, 73 ff., 79, 90, 93, 97, 118, 184, 205, 206, 266, 269, 274, 338, 354
Unique properties   147, 239, 326, 342, 345
– system   184, 228
Universally distributed biochromes   90
Unpaired electron spins   17, 234, 236
Unsaturated compound   11, 12, 60, 202
Unstable biochrome   291
Uptake of biochromes from gut   241, 253, 302, 305, 306, 314
Uranidine   19, 37

Urechis   65, 89, 91, 203, 212, 216, 219, 266, 270, 273
Uric acid   67, 243, 246, 290
Uricase   49, 50
Uricotelism   143, 144
Urinary chromes   88, 89, 106, 110, 245
Urn   105
Urobilin IX α   52, 56, 78
Urocercus   93
Urochrome   37, 75, 106, 205
Urodela   92, 118, 165, 275, 313, 336
Urohaem   88, 243
Urophysis   217
Uroporphyrin   53, 57, 88, 97, 106, 159, 248, 286
Uroporphyrin I   53, 87, 243, 286
– III   53, 286
Urorosein   249
Useless biochrome VII,   132, 242 ff., 269, 354
Ustilago   61

Vacuolar chrome   113 ff.
Valency range of transitional metals   228 ff.
Vanadium compounds   33, 73, 205, 238, 239
– porphyran   336
Vanadochrome   29, 73, 75, 90, 184, 205, 213, 218, 266, 352
Vanadocyte   118
Vanadularia   95
Van den Bergh's diazo reaction   253
Vanessa   246, 313
Vascular system   95
Vasoperitoneal tissue   106
Vector properties   16, 175, 186, 203, 315, 326, 328
Velella   31, 79, 157
Verdoflavin   72
Verdoglobin   54
Verdohaem   54, 58, 88, 107, 306
Vermifuge chrome   159
Versatility of quinones   46, 79, 199
– other biochromes   195, 202, 235, 340, 351
Vertebrates   52, 60, 82, 86, 87, 89, 90, 100–103, 106–108, 138, 140, 142, 143, 154, 163 ff., 167, 169, 170, 199, 206, 213, 241, 245, 250, 252, 265, 270, 272, 303, 306, 309, 330
Violacein   59, 60, 64
Viscosity and molecular size   211, 237
Visible light range   8, 12, 19, 57
– light and indirect response   136, 139, 141, 142, 154, 155, 156, 157, 163 ff., 312, 313
– spectrum   5, 21, 57, 61, 67
Visual effect functions VIII,   3, 81, 90, 97, 131 ff., 206, 209, 242, 245, 341–343
Visual-pituitary-gonad axis   268
Vitamin A   35, 82, 108, 163, 167, 176, 177, 209, 265, 302, 349
– B_1 *(thiamine)*   67, 209
– B_2 (see ribitylflavin)
– B_3 (see pantothenic acid)

Vitamin B$_5$ (see nicotinamide)
- B$_6$ 209   (see pyridoxal)
- B$_{10}$ (see pteroylglutamic acid)
- B$_{12}$ (see cobamide)
- coenzymes   209
- D   9, 157, 209
- E (see α-tocopherol)
- K   37, 42, 44, 81, 181, 186, 188, 199, 200, 209, 251
Vitamins that are biochromes   209, 210, 250–252, 290
Vitiligo   254, 349
Volsella   31

W-substance   143, 146
Warning coloration   131, 138
Wasp   93, 131
Water, absorption of radiation by   7, 9, 343
- as ligand   229
- as stabiliser   229
- solubility   28, 44, 57, 60, 76, 88, 186, 191, 192
Wattles   141
Wave-length discriminator   6, 168, 169
Weber-Fechner relation   137, 176
Whales   82, 108
White mutant of *Drosophila*   326, 329
Wilson's disease   195, 196
Wing of Lepidoptera   32, 37, 68, 74, 79, 82, 89, 92, 115, 133, 160, 276, 318, 350
Wittenberg effect   222, 223, 227
Woad   59
Woodlice   107
Wool   88

Wound-healing   319

X-organ   146,
X-rays   8, 51
Xanthaemia   112, 251
Xanthia   267
Xanthine   67, 246
- oxidase   51, 188–190, 193, 292
Xanthoaphin   18, 44, 45, 47, 78
Xanthocyte   122
Xanthommatin   63 ff., 91, 172, 183, 203, 259, 270
Xanthophore   122, 145, 146, 147
Xanthophyll   18, 31, 32, 33, 176, 323, 337
Xanthopterin   18, 68–70, 93, 273, 291, 292
Xanthosome   143
Xanthurenic acid   64, 67
Xenopus   141, 165
Xeroderma   251
- pigmentosum   249, 326

Yeast   187, 189
Yellow fat mutant   323
Yellowing of lens   170
Yolk platelet   114

Zeaxanthin   31, 265, 304
Zebra   135, 316
- finch   160
Zinc   62, 173, 238, 239
Zoochrome VII.   4
Zoopurpurin   18, 41, 43–45, 47, 78, 183, 201, 247, 248

New Series Zoophysiology and Ecology

Managing Editor: Professor Donald S. Farner, Dept. of Zoology,
University of Washington, Seattle, WA
Editors: W. S. Hoar, J. Jacobs, H. Langer, M. Lindauer
Co-Editors: G. A. Bartholomew, D. Burkhardt, W. R. Dawson, D. Kennedy,
P. Klopfer, P. Marler, C. L. Prosser, L. B. Slobodkin, H. Waring, K. E. F. Watt,
J. Wersäll, W. Wickler

Vol. 1 P. J. Bentley Endocrines and Osmoregulation

A Comparative Account of the Regulation of Water and Salt in Vertebrates

By Peter John Bentley, Professor of Pharmacology, Mount Sinai School of Medicine of the City University of New York

With 29 figures
XVI, 300 pages. 1971
Cloth DM 58,—; US $26.10
ISBN 3-540-05273-9

Prices are subject to change without notice

■ Prospectus on request

In the first two chapters the two major topics as denominated in the title are introduced and summarized at a general level, while, simultaneously, the diversity of such processes in the vertebrates is emphasized. In the second chapter, the author includes a section on the "criteria, methods and difficulties" of experimental comparative endocrinology, which may be of particular use to the younger worker.

Those interested in evolutionary aspects will appreciate the presentation of the vertebrate groups within the framework of their customary classification. Non-endocrine functions involved in the regulation of water and salt balance are brought in, inasmuch as they help to make understandable the role of the endocrine system. Throughout, the physiology of the various animals is related both to their natural environment and the underlying molecular events.

Springer-Verlag
Berlin · Heidelberg · New York

München Johannesburg London New Delhi
Paris Rio de Janeiro Sydney Tokyo Wien

Vol. 2

L. Irving
Arctic Life of Birds and Mammals
Including Man

By **Laurence Irving**, Professor of Zoophysiology, Institute of Arctic Biology, University of Alaska, Fairbanks, Alaska, USA

With 59 figs. XI, 192 pp. 1972
Cloth DM 44,—; US $19.80
ISBN 3-540-05801-X

Prices are subject
to change without notice

■ Prospectus on request

In the Arctic Zone three continents and large islands surround ice-covered Arctic seas. The region is climatically distinguished by large seasonal changes in light and temperature, and transitions of water between its fluid and icy states. A few species of birds and mammals are principally Artic residents and these tend toward world-wide Arctic distribution. Other residents represent northward extensions from temperate lands by a few species that are versatile in their adaption to climate and food. In summer many birds migrate from wintering on southern lands and seas to nest in the Arctic. Major populations of marine mammals resort to summer Arctic seas for their principal nourishment. These migrating populations pursue programs in time and space that are critically timed to reach specific areas at suitable seasons.

Maintenance of warmth is a requirement that must be met by warm-blooded metabolic economy. This constraint has evoked surprising adaptability in habits, behavior, and physiology that enables them to pursue their mammalian and avian lives on Arctic lands and waters.

Observations from which the information was derived were made in the company of Arctic residents and colleagues carrying out modern scientific exploration in the Arctic.

Contents

Introduction. — Environment of Arctic Life. — Mammals of the Arctic. — Arctic Land Birds and Their Migrations. — Maintenance of Arctic Populations: Birds. — Maintenance of Arctic Populations of Mammals. — Warm Temperature of Birds and Mammals. — Maintenance of Warmth by Variable Insulation. — Metabolic Supply of Heat. — Heterothermic Operation of Homeotherms. — Size and Seasonal Change in Dimensions. — Insulation of Man. — Subject Index.

Springer-Verlag
Berlin · Heidelberg · New York
München Johannesburg London New Delhi
Paris Rio de Janeiro Sydney Tokyo Wien